Benchmark Papers in Systematic and Evolutionary Biology Series

Editor: Carl Jay Bajema — Grand Valley State College

T.M

(CLADISTIC THEORY AND METHODOLOGY)

Edited by

THOMAS DUNCAN
University of California

and

TOD F. STUESSY
Ohio State University

A Hutchinson Ross Publication

 VAN NOSTRAND REINHOLD COMPANY
New York

Copyright © 1985 by **Van Nostrand Reinhold Company Inc.**
Benchmark Papers in Systematic and Evolutionary Biology, Volume 8
Library of Congress Catalog Card Number: 84-29175
ISBN: 0-442-21845-1

Manufactured in the United States of America.

Published by Van Nostrand Reinhold Company Inc.
135 West 50th Street
New York, New York 10020

Van Nostrand Reinhold Company Limited
Molly Millars Lane
Wokingham, Berkshire RG11 2PY, England

Van Nostrand Reinhold
480 Latrobe Street
Melbourne, Victoria 3000, Australia

Macmillan of Canada
Division of Gage Publishing Limited
164 Commander Boulevard
Agincourt, Ontario MIS 3C7, Canada

15 14 13 12 11 10 9 8 7 6 5 4 3 2 1

Library of Congress Cataloging in Publication Data
Main entry under title:
Cladistic theory and methodology.
    (Benchmark papers in systematic and evolutionary biology; v. 8)
    "A Hutchinson Ross Benchmark book."
    Includes index.
    1. Cladistic analysis—Addresses, essays, lectures.
I. Duncan, Thomas, 1948-   . II. Stuessy, Tod F. III. Series: Benchmark
papers in systematic and evolutionary biology; 8.
QH83.C48 1985   575   84-29175
ISBN 0-442-21845-1

CONTENTS

Contents

PART VII: CRITIQUES OF CLADISTIC THEORY AND METHODOLOGY

SERIES EDITOR'S FOREWORD

Our classifications will come to be, as far as they can be so made, genealogies.

Charles Darwin (1859, 486)

The volumes of the Benchmark Papers in Systematic and Evolutionary Biology series reprint classic scientific papers on the evolution and systematics of organisms. These Benchmark Papers volumes do more than just provide scholars with facsimile reproductions or English translations of classic papers on a particular topic in a single volume. The interpretative commentaries and extensive bibliographies prepared by each Benchmark Papers volume editor provide busy scholars with a review of the primary and secondary literature of the field from a historical perspective and a summary of the current state of the art.

Charles Darwin was the first scientist to champion the theory that all organisms have descended from one or at most a few common ancestors by a process that involved the splitting of parental species into one or more daughter species (Mayr, 1982, 307)). Darwin (1859, 420) contended that classification should contain information about the amount of evolutionary divergence species have undergone as well as information about the genealogies of species. The systematists whose classic papers are reprinted in this volume have attempted to develop and apply scientific methodologies to reconstruct patterns of branching in phylogeny. This approach to systematics has become known as cladistics. I wish to thank Drs. Thomas Duncan and Tod Stuessy for editing this volume of benchmark papers on one of the most important topics in systematics.

Readers who wish to understand better the philosophical and scientific dimensions of the controversies over the role that cladistic taxonomy plays in systematics are referred to the writings of Beatty (1982), Janvier (1984), Mayr (1981, 1982), and Sober (1983) as well as to the pages of Systematic Zoology.

CARL JAY BAJEMA

REFERENCES

Beatty, J., 1982, Clades and Cladists, *Syst. Zool.* **31**:25–34.

Darwin, C., 1859, *On the Origin of Species by Means of Natural Selection, or the Preservation of Favoured Races in the Struggle for Life,* 1st ed., Murray, London, 502p. (facsimile reprint, Harvard University Press, 1964)

Janvier, P., 1984, Cladistics: Theory, Purpose and Evolutionary Implications, in *Evolutionary Theory: Paths Into the Future,* J. Pollard, ed., Wiley, New York, pp. 39–75.

Mayr, E., 1981, Biological Classification: Toward a Synthesis of Opposing Methodologies, *Science* **214**:510–516.

Mayr, E., 1982, *The Growth of Biological Thought,* Harvard University Press, Cambridge, 974p.

Sober, E., 1983, Parsimony in Systematics: Philosophical Issues, *Annu. Rev. Ecol. Syst.* **14**:335–357.

PREFACE

The past decade has seen a tremendous increase of interest in cladistics: the reconstruction of evolutionary history by more precise and explicit means, and the development of evolutionarily based classifications from these estimates. With increased interest, the literature has grown proportionately. Many papers have appeared in the journal *Systematic Zoology,* but most of the earlier literature is scattered in other journals and books. It seems useful to bring together some significant early papers to introduce the subject of cladistics to those in the systematic biology community who wish to gain an overview of the basic ideas and principles as well as the controversies in the field.

An understanding of the current debates about cladistic theory and methodology requires an acquaintance with the early literature. We have selected what we consider seminal papers that have led to the development of the various forms of cladistics and that have provided focal points for discussion within the field. This anthology not only makes these historical papers more accessible, but illustrates the diverse sources from which cladistics began. It did not develop through a single person or from a single paper, although Willi Hennig's revised English version of his book has certainly been most influential. Rather, cladistics began independently in many different ways, in part with pheneticists who were trying to be more objective and repeatable in their classifications, and in part with practicing taxonomists who were struggling to construct phylogenies in a clearer and more reasoned fashion.

The selection of papers for this volume has been difficult because many worthy and important publications are available and because cladistics is a very diverse field. Cladistics is *not a unified* approach to the reconstruction of evolutionary history, but a *general* approach to understanding historical patterns of diversification of organisms through the use of extant and fossil organisms to reconstruct the speciation events that led to these patterns of variation. The papers included here represent different viewpoints and approaches, to allow the interested reader to attain a broader understanding.

We hope the reader finds this volume not only interesting from the historical standpoint but also useful for generating ideas on how some aspects of cladistics might be applied to understanding relationships in particular taxa. Cladistics offers broader opportunities for data analysis and for formulating explicit statements concerning evolutionary relationships. We hope this volume will provide a sound basis for the careful use and successful application of cladistic methods, and a framework for analyzing the more recent literature.

Acknowledgement is made to Mr. Robert Graham for his invaluable assistance in preparation of the subject index.

THOMAS DUNCAN
TOD. F. STUESSY

CONTENTS BY AUTHOR

CLADISTIC THEORY
AND METHODOLOGY

INTRODUCTION

Cladistics has two major aspects: (1) the concepts and methods for the estimation of evolutionary history expressed as branching diagrams (cladograms), and (2) the use of these estimates in the formulation of a classification. Its use derives from Rensch (1954, 1959), who contrasted two principal modes of evolution: *cladogenesis* (as "kladogenesis"), or the branching events of phylogeny; and *anagenesis,* the progressive change within the same evolutionary line through time (sometimes referred to as *phyletic evolution;* e.g., Simpson, G., 1953, 1961; Simpson, B., 1973). A related term, *stasigenesis* (Huxley, 1957), describes lineages that persist in time without splitting or changing. Cain and Harrison (1960) used cladistics to refer to a relationship expressing relative recency of common ancestry. According to Sokal and Sneath (1963, 220), "cladistic relationship refers to the paths of the ancestral lineages and therefore describes the sequence of branching of the ancestral lines." As applied to a method of classification, Mayr (1969, 70) called it *cladism.* Most workers now refer to this approach as cladistics (e.g., Eldredge and Cracraft, 1980; Funk and Brooks, 1981; Nelson and Platnick, 1981; Duncan and Stuessy, 1984). Some workers, however, have preferred the term *phylogenetic systematics* or *phylogenetics* instead of cladistics (e.g., Hennig, 1950, Paper 1, 1966; Bremer and Wanntorp, 1978, 1982; Farris, 1979; Wiley, 1981) to emphasize the reliance on phylogeny for classification.

Given the variety of definitions employed and the differences in emphasis placed on the two components of cladistics, it is clear that disagreement remains about the definition, circumscription, and importance of the field. We do not attempt to resolve these issues in this volume. The papers we have chosen will provide the background to this ongoing debate and give the reader some perspective for understanding the current literature. A brief history of cladistics will illustrate the development of the different views that are currently being debated.

Voss (1952) traces the idea of using branching diagrams (or phylogenetic trees, or cladograms) to express evolutionary relation-

1

ship to Lamarck's *Philosophie Zoologique* in 1809. As noted by Voss, earlier authors who used tree diagrams were not expressing evolutionary relationship. Throughout the nineteenth century and until the mid-twentieth century, many authors drew branching diagrams to represent evolutionary relationship. These diagrams were not based on explicit methods, and as a result it was very difficult to interpet, analyze, or compare them. Not until Hennig (1950) and Wagner (Paper 2) were explicit methods developed for producing such reconstructions.

Following the initial work of Hennig and Wagner, the 1960s were a period of rapid development of methods and discussion of philosophy. Edwards and Cavalli-Sforza (Paper 17) offered the first "method of minimum evolution" using continuous human blood-group data. In the following year, Camin and Sokal (Paper 10) published the first numerical cladistic technique for discrete data. Wilson (Paper 14) emphasized the selection of unique and unreversed character states in his "consistency test" for constructing cladograms. Throckmorton (Paper 6) provided the first clear contrast between phenetic and cladistic approaches.

In 1965, Hennig (Paper 1) published the first English summary of his argumentation method, based on his 1950 book, *Phylogenetic Systematics*. The appearance in 1966 of the English translation (and revision) of that work marked the first fully documented statement on the philosophy of a cladistic method for evaluating characters for use in defining monophyletic groups using shared uniquely derived character states (synapomorphies). From these nested patterns of synapomorphy a cladogram is constructed. This work has had a major impact because it was a fully reasoned exposition of Hennig's philosophy and methods (see Kavanaugh, 1972, 1978, for summaries). Nelson (e.g., Paper 11; Nelson, 1971, 1972, 1973) deserves principal credit for bringing attention to Hennig's ideas.

While these events were occurring, Wagner's manual cladistic technique was used for teaching evolution and was developed in a work on the fern genus *Diellia* (Wagner, 1952). Paper 9 provides a history and detailed explanation of Wagner's method and its applications. Wagner's Groundplan-divergence method represented the major use of cladistics among botanists at that time.

In 1967, Fitch and Margoliash (Paper 18) published a method of tree construction using molecular sequence data (amino acid sequences of cytochrome c). This paper describes a statistical approach to cladogram construction and combines elements of cladistics and phenetics. Estabrook (1968) refined the Camin-Sokal approach by offering a mathematical solution to the problem of selecting the most efficient (or parsimonious) trees (further improved by Nastansky, Selkow, and Stewart, 1974).

In the early 1970s, an important paper for botanical systematics was published by Solbrig (Paper 13), in which Prim-Kruskal (Sokal and Sneath, 1963) and Wagner Groundplan-divergence methods were used to reconstruct the phylogeny of *Gutierrezia* (Compositae). This early effort is now considered the "eclectic," "evolutionary," or "syncretistic" approach to cladistics, which encourages use of different methods to gain the maximum insights on the phylogeny of a group (e.g., Funk and Stuessy, 1978; Duncan, 1980; Fitch, 1984).

During the 1970s two methodological approaches—parsimony and compatibility analysis—were developed and discussed. The first generalized parsimony method for numerical cladistics was presented by Kluge and Farris (Paper 11) under the rubric of *quantitative phyletics*. Farris elaborated on the Wagner parsimony methods, as a generalization of Wagner's Groundplan-divergence method, in the following year (Paper 12). Hennig's original method was modified to include a criterion of parsimony in Farris, Kluge, and Eckardt (1970). One issue currently under debate concerns the relationship of parsimony to Hennig's method. Churchill, Wiley, and Hauser (1984) suggest that the Wagner parsimony techniques are equivalent to Hennig's method. Duncan (1984) discusses the nature and implications of the changes to Hennig's method required to accommodate parsimony as a criterion to choose among cladograms for a particular group. Whiffin and Bierner (1972) and Nelson and Van Horn (1975) both offered manual adaptations of the Wagner algorithms developed by Farris. More recently Jensen (1981) has summarized many of these methods.

LeQuesne (Paper 15) stressed selection of uniquely derived characters for phylogeny reconstruction. This idea was pursued by Estabrook (1972; Estabrook, Johnson, and McMorris, 1975, 1976a, 1976b), who developed the idea of using a suite of compatible characters that have uniquely derived states. This has led to the character compatibility approach to phylogeny reconstruction. Paper 16 offers both an overview of the method and its application. Meacham (1981) has recently summarized how character compatibility can be done manually and Meacham (1984) discusses compatibility analyses using undirected character state trees. Discussions, pro and con, of this method have been offered by Kluge and Farris (1979), Estabrook (1978), Wiley (1981), Churchill, Wiley, and Hauser (1984), and Duncan (1984).

In 1973, Felsenstein presented a maximum-likelihood statistical model of parsimony and compatibility methods. More recent papers have compared the various methods (especially parsimony and character compatibility) and stressed the value of the maximum-likelihood algorithm (Felsenstein, 1978, 1981, 1983, 1984).

From the late 1970s to the present, a polarization of viewpoints has prevailed. These various viewpoints are expressed in the many

3

symposium volumes and books published during this time (Stuessy and Estabrook, 1978; Cracraft and Eldredge, 1979; Stuessy, 1980; Eldredge and Cracraft, 1980; Funk and Brooks, 1981; Wiley, 1981; Nelson and Platnick, 1981; Hennig, 1981; Joysey and Friday, 1982; Platnick and Funk, 1983; Duncan and Stuessy, 1984). A number of critiques were also published during the 1970s. Three of the best are included here (Papers 20, 21, and 22).

In the 1980s, cladistics continues to offer systematic biologists a significant alternative to the traditional process of biological classification. Workers in this field hold many different viewpoints on nearly every aspect of method and on the concepts and philosophy behind the methods. These views range from the dogmatic to the eclectic (e.g., Van Valen, 1978; Cartmill, 1981; Duncan, 1980; Fitch, 1984). Some workers have even come to view cladistics as a system of organizing information that does not necessarily have to deal with evolution ("transformed cladists," e.g., Patterson, 1980, Platnick, 1980; also called "natural order systematists," Charig, 1982). These diverse viewpoints have yet to be reconciled. Cladistics is currently in a period of active development and vigorous debate.

REFERENCES

Bremer, K., and H. E. Wanntorp, 1978, Phylogenetic Systematics in Botany, *Taxon* **27:**317–329.

Bremer, K., and H. E. Wanntorp, 1982, Phylogenetic Systematics, *Sven. Bot. Tidskr.* **76:**177–183.

Cain, A. J., and G. A. Harrison, 1960, Phyletic Weighting, *Zool. Soc. London Proc.* **135:**1–31.

Cartmill, M., 1981, Hypothesis Testing and Phylogenetic Reconstruction, *Z. zool. Syst. Evolut.-forsch.* **19:**73–96.

Charig, A. J., 1982, Systematics in Biology: A Fundamental Comparison of Some Major Schools of Thought, in *Problems of Phylogenetic Reconstruction,* K. A. Joysey and A. E. Friday, eds., Academic Press, London, pp. 363–440.

Churchill, S. P., E. O. Wiley, and L. A. Hauser, 1984, A Critique of Wagner Groundplan-Divergence Studies and a Comparison with Other Methods of Phylogenetic Analysis, *Taxon* **33:**212–232.

Cracraft, J., and N. Eldredge, eds., 1979, *Phylogenetic Analysis and Paleontology,* Columbia University Press, New York.

Duncan, T., 1980, Cladistics for the Practicing Taxonomist—An Eclectic View, *Syst. Bot.* **5:**136–148.

Duncan, T., 1984, Willi Hennig, Character Compatibility, Wagner Parsimony, and the "Dendrogrammaceae" Revisited, *Taxon* **33:**785–791.

Duncan, T., and T. F. Stuessy, eds., 1984, *Cladistics: Perspectives on the Reconstruction of Evolutionary History,* Columbia University Press, New York.

Eldredge, N., and J. Cracraft, 1980, *Phylogenetic Patterns and the Evolu-*

tionary Process: Method and Theory in Comparative Biology, Columbia University Press, New York.

Estabrook, G. F., 1968, A General Solution in Partial Orders for the Camin-Sokal Model in Phylogeny, *J. Theor. Biol.* **21:**421–438.

Estabrook, G. F., 1972, Cladistic Methodology: A Discussion of the Theoretical Basis for the Induction of Evolutionary History, *Annu. Rev. Ecol. Syst.* **3:**427–456.

Estabrook, G. F., 1978, Some Concepts for the Estimation of Evolutionary Relationships in Systematic Botany, *Syst. Bot.* **3:**146–158.

Estabrook, G. F., 1976a, An Algebraic Analysis of Cladistic Characters, *Discrete Math.* **16:**141–147.

Estabrook, G. F., 1976b, A Mathematical Foundation for the Analysis of Cladistic Character Compatibility, *Math. Biosci.* **29:**181–187.

Estabrook, G. F., C. S. Johnson, Jr., and F. R. McMorris, 1975, An Idealized Concept of the True Cladistic Character, *Math. Biosci.* **23:**263–272.

Farris, J. S., 1979, On the Naturalness of Phylogenetic Classification, *Syst. Zool.* **28:**200–214.

Farris, J. S., A. G. Kluge, and M. J. Eckhardt, 1970, A Numerical Approach to Phylogenetic Systematics, *Syst. Zool.* **19:**172–191.

Felsenstein, J., 1984, The Statistical Approach to Inferring Evolutionary Trees and What It Tells Us about Parsimony and Compatibility, in *Cladistics: Perspectives on the Reconstruction of Evolutionary History,* T. Duncan and T. F. Stuessy, eds., Columbia University Press, New York, pp. 169–191.

Felsenstein, J., 1978, Cases in Which Parsimony or Compatibility Methods Will Be Positively Misleading, *Syst. Zool.* **27:**401–410.

Felsenstein, J., 1981., Evolutionary Trees from DNA Sequences: A Maximum-Likelihood Approach, *J. Mol. Evol.* **17:**368–376.

Felsenstein, J., 1983, Numerical Methods for Inferring Evolutionary Trees, *Q. Rev. Biol.* **57:**379–404.

Felsenstein, J., 1984, The Statistical Approach to Inferring Evolutionary Trees and What It Tells Us about Parsimony and Compatibility, in *Cladistics: Perspectives on the Reconstruction of Evolutionary History,* T. Duncan and T. F. Stuessy, eds., Columbia University Press, New York, pp. 169–191.

Fitch, W. M., 1984, Matrix Methods in Classification, in *Cladistics: Perspectives on the Reconstruction of Evolutionary History,* T. Duncan and T. F. Stuessy, eds., Columbia University Press, New York, pp. 221–252.

Funk, V. A., and D. R. Brooks, eds., 1981, *Advances in Cladistics: Proceedings of the First Meeting of the Willi Hennig Society,* New York Botanical Garden, New York.

Funk, V. A., and T. F. Stuessy, 1978, Cladistics for the Practicing Plant Taxonomist, *Syst. Bot.* **3:**159–178.

Hennig, W., 1950, *Grundzüge einer Theorie der phylogenetischen Systematik,* Deutscher Zentralverlag, Berlin.

Hennig, W., transl. D. Davis and R. Zangerl, 1966, *Phylogenetic Systematics,* University of Illinois Press, Urbana (reprinted 1979).

Hennig, W., transl. A. C. Pont, 1981, *Insect Phylogeny,* John Wiley and Sons, New York.

Huxley, J. S., 1957, The Three Types of Evolutionary Process, *Nature* **180:**454–455.

Jensen, R. J., 1981, Wagner Networks and Wagner Trees: A Presentation of Methods for Estimating Most Parsimonious Solutions, *Taxon* **30:**576–590.

Joysey, K. A., and A. E. Friday, eds., 1982, *Problems of Phylogenetic Reconstruction,* Academic Press, London.

Kavanaugh, D. H., 1972, Hennig's Principles and Methods of Phylogenetic Systematics, *Biologist* **54:**115–127.

Kavanaugh, D. H., 1978, Hennigian Phylogenetics in Contemporary Systematics: Principles, Methods, and Uses, in *Beltsville Symposia in Agricultural Research. 2. Biosystematics in Agriculture,* Allenheld, Osmun, and Company, Montclair, N.J., pp. 139–150.

Kluge, A., and J. S. Farris, 1979, A Botanical Clique, *Syst. Zool.* **28:**400–411.

Mayr, E., 1969, *Principles of Systematic Zoology,* McGraw-Hill, New York.

Meacham, C. A., 1981, A Manual Method for Character Compatibility Analysis, *Taxon* **30:**591–600.

Meacham, C. A., 1984, The Role of Hypothesized Direction of Characters in the Estimation of Evolutionary History, *Taxon* **33:**26–38.

Nastansky, L., S. M. Selkow, and N. F. Stewart, 1974, An Improved Solution to the Generalized Camin-Sokal Model for Numerical Cladistics, *J. Theor. Biol.* **48:**413–424.

Nelson, C. H., and G. S. Van Horn, 1975, A New Simplified Method for Constructing Wagner Networks and the Cladistics of *Pentachaeta* (Compositae, Astereae), *Brittonia* **27:**362–372.

Nelson, G. J., 1971, "Cladism" as a Philosophy of Classification, *Syst. Zool.* **20:**373–376.

Nelson, G. J., 1972, Phylogenetic Relationship and Classification, *Syst. Zool.* **21:**227–230.

Nelson, G. J., 1973, Classification as an Expression of Phylogenetic Relationships, *Syst. Zool.* **22:**344–359.

Nelson, G. J., and N. Platnick, 1981, *Systematics and Biogeography: Cladistics and Vicariance,* Columbia University Press, New York.

Patterson, C., 1980, Cladistics, *Biologist* **27:**234–240.

Platnick, N. I., 1980, Philosophy and the Transformation of Cladistics, *Syst. Zool.* **28:**537–546.

Platnick, N. I., and V. A. Funk, 1983, *Advances in Cladistics: Proceedings of the Second Meeting of the Willi Hennig Society,* Columbia University Press, New York.

Rensch, B., 1954, *Neuere Probleme der Abstammungslehre,* 2nd ed., Enke, Stuttgart.

Rensch, B., 1959, *Evolution above the Species Level,* Columbia University Press, New York.

Simpson, B. B., 1973, Contrasting Modes of Evolution in Two Groups of *Perezia* (Mutisieae; Compositae) of Southern South America, *Taxon* **22:**525–536.

Simpson, G. G., 1953, *The Major Features of Evolution,* Columbia University Press, New York.

Simpson, G. G., 1961, *Principles of Animal Taxonomy,* Columbia University Press, New York.

Sokal, R. R., and P. H. A. Sneath, 1963, *Principles of Numerical Taxonomy,* W. H. Freeman, San Francisco.

Stuessy, T. F., 1980, Cladistics and Plant Systematics: Problems and Prospects: Introduction, *Syst. Bot.* **5:**109–111.

Stuessy, T. F., and G. F. Estabrook, 1978, Cladistics and Plant Systematics: Introduction, *Syst. Bot.* **3:**145.

Van Valen, L., 1978, Why Not to Be a Cladist, *Evol. Theory* **3:**285–299.

Voss, E., 1952, The History of Keys and Phylogenetic Trees in Systematic Biology, *J. Sci. Lab* (Dennison Univ.) **43:**1–25.

Wagner, W. H., Jr., 1952, The Fern Genus *Diellia:* Structure, Affinities, and Taxonomy, *Univ. Calif. Publ. Bot.* **26:**1–212.

Whiffin, T., and M. W. Bierner, 1972, A Quick Method for Computing Wagner Trees, *Taxon* **21:**83–90.

Wiley, E. O., 1981, *Phylogenetics: The Theory and Practice of Phylogenetic Systematics,* John Wiley and Sons, New York.

Part I

THEORETICAL ISSUES

Editors' Comments
on Papers 1, 2, and 3

1 HENNIG
Phylogenetic Systematics

2 WAGNER
Problems in the Classification of Ferns

3 WILEY
Karl R. Popper, Systematics, and Classification: A Reply to Walter Bock and Other Evolutionary Taxonomists

The extensive discussion of theoretical issues in cladistics makes it difficult to select three papers to illustrate early development in this area. Nonetheless, it seems instructive to include the earliest English contribution by Willi Hennig and the earliest paper by W. H. Wagner, both of whom stimulated numerous studies in cladistics over the past two decades. The three papers we have chosen represent early statements on the goals and methods of cladistics by Hennig and by Wagner and the first paper to tie systematic methodology to the hypothetico-deductive scientific method of Popper.

Hennig's paper (Paper 1) summarizes ideas contained in his 1950 book in German, as well as the English revision of his book (Hennig, 1966), which has been very influential in encouraging further theoretical studies. Hennig's principles of grouping by synapomorphy for cladogram construction and the use of synapomorphies to define monophyletic groups provided the principal stimulus for the development of cladistics (he called it phylogenetic systematics) as a major subdiscipline of systematic biology. These ideas are the basis of the Hennigian argumentation method. The influence of this approach on systematic biology has been enhanced greatly by the attention given to it by Gareth Nelson (see Paper 8), Norman Platnick, and Donn Rosen, all of the American Museum of Natural History. For classification, Hennig's basic contentions were that classification should be based on degrees, of evolutionary relationship because evolution has produced the organic diversity summarized by these classifications. Because evolution seems to have been largely dichotomous (at least in most animals), branching diagrams should reflect reasonably well the course of evolution of most groups. It follows that classifications should reflect these branching patterns closely.

At about the same time, Wagner (Paper 2) developed a method for reconstructing phylogeny called the Groundplan Divergence method. Wagner's method was developed initially as a teaching device, and many students at the University of Michigan and elsewhere used it in their research. Although this approach has only recently been documented fully (Paper 9), the short paper reproduced here gives the basic ideas.

One of the most debated issues in cladistics has centered on the coupling of Hennig's principles and Karl Popper's hypothetico-deductive falsificationist view of science. To place both cladistics and systematics in a position of greater scientific credibility, repeatability, and testability, Popper's ideas offered a context in which cladistic hypotheses might be discussed and evaluated. The first statement of cladistics from a Popperian perspective was by Wiley (Paper 3). Wiley's view is that Hennigian cladistics offers the only truly scientific approach to phylogeny reconstruction and the development of classifications, with phenetics and evolutionary systematics being deficient in various ways.

REFERENCE

Hennig, W., transl. D. Davis and R. Zangerl, 1966, *Phylogenetic Systematics,* University of Illinois Press, Urbana (reprinted 1979).

1

PHYLOGENETIC SYSTEMATICS[1]

By Willi Hennig

Staatliches Museum für Naturkunde in Stuttgart, Germany

Since the advent of the theory of evolution, one of the tasks of biology has been to investigate the phylogenetic relationship between species. This task is especially important because all of the differences which exist between species, whether in morphology, physiology, or ecology, in ways of behavior, or even in geographical distribution, have evolved, like the species themselves, in the course of phylogenesis. The present-day multiplicity of species and the structure of the differences between them, first becomes intelligible when it is recognized that the differences have evolved in the course of phylogenesis; in other words, when the phylogenetic relationship of the species is understood.

Investigation of the phylogenetic relationship between all existing species and the expression of the results of this research, in a form which cannot be misunderstood, is the task of phylogenetic systematics.

The problems and methods of this important province of biology can be understood only if three fundamental questions are posed and answered: what is phylogenetic relationship, how is it established, and how is knowledge of it expressed so that misunderstandings are excluded?

The definition of the concept "phylogenetic relationship" is based on the fact that reproduction is bisexual in the majority of organisms, and that it usually takes place only within the framework of confined reproductive communities which are genetically isolated from each other. This is especially true for the insects, with which this paper is mainly concerned. The reproductive communities which occur in nature we call species. New species originate exclusively because parts of existing reproductive communities have first become externally isolated from one another for such extended periods that genetic isolation mechanisms have developed which make reproductive relationships between these parts impossible when the external barriers which have led to their isolation are removed. Thus, all species (= reproductive communities) which exist together at a given time, e.g., the present, have originated by the splitting of older homogeneous reproductive communities. On this fact is based the definition of the concept, "phylogenetic relationship": under such concept, species, B, is more nearly related to species, C, than to another species, A, when B has at least one ancestral species source in common with species C which is not the ancestral source of species A [Hennig (8)].

"Phylogenetic relationship" is thus a relative concept. It is pointless (since it is self-evident) to say, as is often said, that a species or species-group is "phylogenetically related" to another. The question is rather one of

[1] The survey of the literature pertaining to this review was concluded in 1963.

knowing whether a species or species-group is more or less closely related to another than to a third. The measurement of the degree of phylogenetic relationship is, as the definition of the concept shows, "recency of common ancestry" [Bigelow (1)]. A phylogenetic relationship of varying degree exists between all living species, irrespective of whether we know of it or not. The aim of research on phylogenetic systematics is to discover the appropriate degrees of phylogenetic relationship within a given group of organisms.

The degree of phylogenetic relationship which exists between different species, and thus also the results of research on phylogenetic systematics, can be represented in a visual form which is not open to misinterpretation as is a so-called phylogeny tree (dendrogram). To be able to discuss this, not only the species but also all of the monophyletic groups included in the diagram, must be given names. "Monophyletic groups" are small or large species-groups whose member species can be considered to be more closely related to one another than to species which stand outside these groups [Hennig (8)]. When a phylogeny diagram, conforming to this postulate, has been rendered suitable for discussion by the naming of all of the monophyletic groups, then the diagram can be discarded and its information may be expressed solely by ranking the names of the groups:

A. Myriopoda
B. Insecta
    B.1 Entognatha
        B.1a Diplura
        B.1b Ellipura
            B.1ba Protura
            B.1bb Collembola
    B.2 Ectognatha

Such arrangement of monophyletic groups of animals according to their degree of phylogenetic relationship is called, in the narrower sense, a phylogenetic system of the group in question. Such a system belongs to the type called a "hierarchical" system. Since "system" in the wider sense means every arrangement of elements according to a given principle, the phylogeny tree, too, can be termed a phylogenetic system. Phylogeny diagrams and arrangement of the names of monophyletic groups in a hierarchical sequence are merely different but closely comparable forms of presentation whose content is the same. Therefore, everything which can be said about the methods of phylogenetic systematics (see below) applies irrespective of whether the results sought by the use of these methods are expressed only as a phylogeny tree or, as a phylogenetic system in the narrower sense, in a hierarchically arranged list of the names of monophyletic groups.

In some cases, a hierarchical arrangement of group names, that is, a phylogenetic system in the narrower sense, is to be preferred to a phylogeny tree. One can, for instance, in a catalogue or check-list of Nearctic Diptera, give expression to all that one thinks is known about the phylogenetic relationship of all Nearctic species of Diptera in a form which can in no way be misinterpreted, without using a single phylogeny tree.

However, considerable difficulties arise because systems of the hierarchical type have also been used in biology with intentions other than of expressing the phylogenetic relationship of species. Long before the advent of the theory of evolution, "systematics" existed as the branch of biological science which had adopted as its aim an orderly survey of the plurality of organisms. Naturally, the principle of classification in systematics could not then be the phylogenetic relationship of species, which was still unrecognized, but only a morphological resemblance between organisms. This morphological systematics also used the hierarchical type of system to express its results although Linnaeus already held the view that morphological resemblances between organisms corresponded to a multidimensional net. Numerous attempts have also been made to introduce other types of system, which differ from the hierarchical, into biological systematics [see Wilson & Doner (21)]. But they have not been successful.

Today, there are still many authors who consider that the purpose of biological "systematics" is to classify organisms according to their morphological resemblance, and who use a system of the hierarchical type to this end. It is hardly surprising that misunderstandings and serious errors can be produced by this formal identity between morphological and phylogenetic systems.

The source of danger in the formal identity between systems based on such different principles of classification is that, in a hierarchical system, each group formation relates to a "beginner," which is linked in "one-many relations" with all of the members of that group and only those [Gregg (3)]. In morphological systems, the "beginner" which belongs to each group is a formal idealistic standard ("Archetype") whose connections with the other members of the group are likewise purely formal and idealistic. But, in a phylogenetic system, the "beginner" to which each group formation relates is a real reproductive community which has at some time in the past really existed as the ancestral species of the group in question, independently of the mind which conceives it, and which is linked by genealogical connections with the other members of the group and only with these. One could, without difficulty, adduce many examples from the literature in which the formal beginner ("Archetype") of a group, conceived according to the principles of morphological systematics, has been erroneously taken, with all of the consequences of such an error of logic, as the real beginner (ancestral species) of a monophyletic group.

This dangerous difference between a formal morphological (typological) hierarchical system and the equally hierarchical system of phylogenetic systematics, would not arise if the degree of morphological resemblance were an exact measurement of the degree of phylogenetic relationship. But this is not the case. Furthermore, there is yet no definition of the concept of morphological resemblance which is not open to theoretical objection, nor any method which can be accepted as the one and only method which achieves a satisfactory determination of more than the threshold of morphological resemblance, that is, the degree of resemblance between relatively similar species which agree in very many characters.

In these circumstances, the dangers which arise from the formal identity of phylogenetic and morphological systems will be avoided if agreement can be reached on whether or not the branch of biological science known simply as systematics will, in future always try to express the morphological resemblance of organisms or their phylogenetic relationship in the system in which it works.

It has often been stated, in defense of a system of morphological resemblance, that this has historical primacy over endeavors to express phylogenetic relationship in a system, because the morphological system had already existed as the aim of "systematics" before the advent of the theory of evolution. Even today, this reasoning is often augmented with the argument that the theory of evolution was established with the help, among other things, of the system of graduated morphological resemblances between organisms, and that therefore one is prescribing a circle if, in reverse, one wishes to take the theory of evolution and the notion of the phylogenetic relationship of organisms which follows from it as the theoretical starting-point of their classification in a system [Sokal (17); Blackwelder, Alexander & Blair (2)]. This "ebenso halt-wie heillose Einwand" [Günther, in discussing the work of Sokal (12)], has already been so often refuted that one can only attribute, to authors who persist in asserting it today, a lack of information.

It is certainly correct that the classification of organisms according to their morphological resemblance has led to the theory of evolution. This was possible only because the morphological differences between organisms are the result of a historical (phylogenetic) development and because, at least in rough terms, very similar organisms are, in fact, generally more closely related than are very different ones. It was therefore inevitable that the classification of organisms according to their morphological resemblance, in association with certain features of their ontogenetic development and their geographical distribution, would sooner or later lead to the discovery of their successive degrees of phylogenetic relationship and thus to the theory of evolution.

However, there are historical origins not only of the morphological differences between organisms in the narrower sense, but also differences in their physiological functions, their ways of behavior and, in addition to these physical ("holomorphological") attributes, differences in their distribution in geographical and ecological space. Since it has been recognized and, more-over, become widely known, that there are not the same degrees of agreement and difference in the various holomorphological and chorological resemblances which connect organisms, the way is open for establishing the phylogenetic relationship itself of organisms as the principle of classification, instead of successive degrees of resemblance in a single category of characters: for, only from the phylogenetic relationship is it possible to establish direct connections with all other thinkable kinds of agreement and difference between organisms. The demand for a phylogenetic system is thus not so much a renunciation of pre-phylogenetic resemblance, systematics, but its consequential further development.

The claim of the phylogenetic system to elevation into the universal reference system of biology has a logical, even if not historical, foundation, and arises because few areas of research can be conceived which do not bear fruit and lead to more profound conclusions through a knowledge of the phylogenetic relationship of its objects, and which cannot, in turn, lead to the discovery of hitherto unknown relationships in the course of mutual exchange of information. This is not true to the same extent for any other system built on any other principle of classification. Other systems may also have their value as knowledge; but this value is, in each case, restricted to answering particular questions.

The logical primacy of the phylogenetic system also arises because it alone provides all parts of the field studied by biological systematics with a common theoretical foundation [Kiriakoff (14)]. It is true that phylogenetic relationship exists only between different species, and species are not the simplest elements of biological systematics. These are not even the "individuals," but the individuals in given short periods of their lifetime ("semaphoronts"). The first and basic task of systematics is to establish that different individuals, or rather "semaphoronts," belong to particular species. The difficulty within this task rests in the fact that the species, which exist in nature as real phenomena independent of the men who perceive them, are units which are not morphologically but genetically defined. They are communities of reproduction, not resemblance. Of course, the morphological resemblance between members of a species is not unimportant for the practical establishment of specific limits. But it has only the significance of an auxiliary criterion whose capabilities of use are limited. This is because the definition of the phylogenetic relationship between species, as well as the definition of the species-concept, is deduced from the fact that the reproduction of species generally takes place only within the framework of defined communities which cannot be unqualified communities of resemblance if, in the demand for a phylogenetic system, biological systematics has acquired for all its spheres of activity a common aim, that is, the discovery and recording of the "hologenetical" connections which exist between all organisms. In contrast with this, morphological resemblance-systematics, though not denying the modern genetic species-concept, employs different principles of classification above and below the specific level.

It would, of course, be meaningless to extol the need for a phylogenetic system, however well founded it might be theoretically, if this demand could not be put into practice. There is, in fact, a widespread notion that phylogenetic systematics, at least in those groups of animals for which no fossil finds are available, possesses no method of its own, but can only interpret the

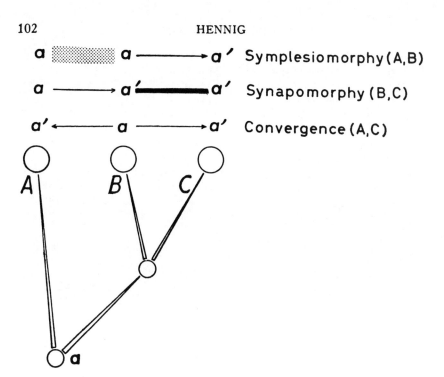

FIG. 1. The three different categories of morphological resemblance. *a* plesio-morph; *a'* apomorph expression of the morphological character *a*. Agreement may rest on sympleisiomorphy (*a-a*), synapomorphy (*a'-a'*) or convergence (*a'-a'*).

results of morphological systematics according to the principle that the degree of morphological resemblance equals the degree of phylogenetic rela-tionship. This notion is false. The fundamental difference between the method of morphological and phylogenetic systematics is that the latter breaks up the simple concept of "resemblance." (Fig. 1).

It is a consequence of the theory of evolution that the differences between various organisms must have arisen through changes of characters in the course of a historical process. Therefore it is not the extent of resemblance or difference between various organisms that is of significance for research into phylogenetic relationship, but the connection of the agreeing or divergent characters with earlier conditions. It is valid to distinguish different cate-gories of resemblance according to the nature of these connections.

The division of the concept of resemblance into various categories of resemblances probably began, in the history of systematics, with the intro-duction of the concept of convergence. Often this concept was linked with the distinction between analogous and homologous organs. Convergence is, in fact, commonly manifested by similar organs having arisen in adaptation to the same functions from different morphological foundations in different

organisms. But there are also cases where virtually complete agreement in
the form of homologous organs rests on convergence. "Convergence" means
resemblance between the characters of different species which has evolved
through the independent change of divergent earlier conditions of these characters. It shows how species which differed from one another are ancestors of
species which have become similar to one another. If one associates in a
group the species whose resemblance rests on convergence, then this is not a
monophyletic but a polyphyletic group. There are few authors today who
would specifically support the inclusion of demonstrably polyphyletic groups
in a system. "Convergence" and "polyphyletic groups" are concepts which
presuppose acceptance of the theory of evolution. Therefore, some systematists think they are already working with a "phylogenetic system" when, in

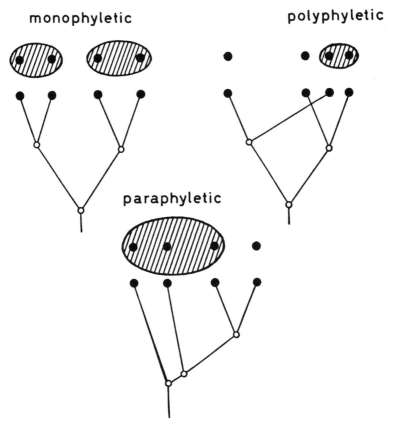

FIG. 2. The three different categories of systematic group formations corresponding to the resemblance of their constituents resting on synapomorphy (monophyletic groups), convergence (polyphyletic groups), or symplesiomorphy (paraphyletic groups). For comparison with Figure 1.

their evaluation of morphological resemblance, they exclude convergence and thus polyphyletic groups from their system.

But even when purged of convergence, morphological resemblance is still not a satisfactory criterion for the degree of phylogenetic relationship between species. It still does not provide one with exclusively monophyletic groups, such as a phylogenetic system demands. This arises from the fact that characters can remain unchanged during a number of speciation processes. Therefore, it follows that the common possession of primitive ("plesiomorph") characters which have remained unchanged cannot be evidence of the close relationship of their possessors.

Often, a given species can be phylogenetically more closely related to a species which possesses a particular character in a derivative ("apomorph") stage of expression than to species with which it agrees in the possession of the primitive ("plesiomorph") stage in the expression of this character. Therefore, a resemblance which rests on symplesiomorphy is of no more value in justifying a supposition of closer phylogenetic relationship than is a resemblance which has occurred through convergence. If, in a system, one associates in a group species whose agreement rests on convergence, a polyphyletic group is thereby formed, as has been established above and is generally recognized. If one associates species whose agreement rests on symplesiomorphy, then a paraphyletic group is formed (Fig. 2). Paraphyletic groups among insects are the "Apterygota" and Palaeoptilota (= Palaeoptera), if one considers the closer relationship of the Odonata with the Neoptera as established. Paraphyletic vertebrate groups are the "Pisces" and the "Reptilia."

The supposition that two or more species are more closely related to one another than to any other species, and that, together they form a monophyletic group, can only be confirmed by demonstrating their common possession of derivative characters ("synapomorphy"). When such characters have been demonstrated, then the supposition has been confirmed that they have been inherited from an ancestral species common only to the species showing these characters.

It must be recognized as a principle of inquiry for the practice of systematics that agreement in characters must be interpreted as synapomorphy as long as there are no grounds for suspecting its origin to be symplesiomorphy or convergence.

The method of phylogenetic systematics, as that part of biological science whose aim is to investigate the degree of phylogenetic relationship between species and to express this in the system which it has designed, thus has the following basis: that morphological resemblance between species cannot be considered simply as a criterion of phylogenetic relationship, but that this concept should be divided into the concepts of symplesiomorphy, convergence, and synapomorphy, and that only the last-named category of resemblance can be used to establish states of relationship.

The differences between the phylogenetic system and all other systems

19

which likewise classify species on the basis of their morphological resemblance, are as follows: (A) Systems which employ the simple criterion of morphological resemblance. Such systems include polyphyletic, paraphyletic, and monophyletic groups. (B) Systems which employ the criterion of morphological resemblance, but fail to consider characters whose agreement rests on convergence. In such systems, polyphyletic groups are excluded but paraphyletic as well as monophyletic groups are admitted. (C) Phylogenetic system. Characters whose agreement rests on convergence or symplesiomorphy are not considered. Therefore, polyphyletic and paraphyletic groups are excluded and only monophyletic groups admitted.

The systems named under (B) have also often been termed phylogenetic systems in the literature [e.g., Stammer (18); Verheyen (20)]. But it is thereby overlooked that the paraphyletic groups admitted in these "pseudophylogenetic" or "cryptotypological" systems [Kiriakoff (14)] are similar in many respects to polyphyletic groups. No one would think of considering polyphyletic groups in studies concerned with the course and eventual rules of phylogenesis (zoogeographical studies, for instance, belong here), since they have no ancestors solely of their own and therefore no individual history. Exactly the same holds true, however, for paraphyletic groups. The sole common ancestors of all of the socalled "Apterygota," for instance, were also the ancestors of the Pterygota, and the beginning of the history of the Apterygota was not the beginning of an individual history of this group, but the beginning of the individual history of the Insecta, which were at first Apterygota in the morphological-typological sense. Also, the concept of "extinction" is different in paraphyletic and monophyletic groups. Only monophyletic groups can become "extinct" in the sense that from a particular point in time no physical progeny of any member of the group have existed. But if, however, one says that a paraphyletic group has become "extinct," this can only mean that after a particular point of time no bearers of the morphological characters of this group have existed. But physical progeny of many of its members may, with changed characters, continue to live. Monophyletic and paraphyletic groups thus cannot be compared with each other in any question concerning their history. Failure to take account of this fact and invalid uncritical comparison of paraphyletic and monophyletic groups has led to some false conclusions in studies about the "Grossablauf der phylogenetischen Entwicklung" [Müller (15)], and the history of the distribution of animals.

From the premise that morphological agreement only confirms a supposition that the species concerned belong to a monophyletic group when it can be interpreted as synapomorphy. is derived for the practical work of the systematist, the "Argumentation plan of phylogenetic systematics" (Fig. 3). This plan shows that in a phylogenetic system which must contain only monophyletic groups, every group formation, irrespective of the rank to which it belongs, must be established by demonstration of derivative ("apomorph") characters in its ground plan. But it also shows clearly that in two

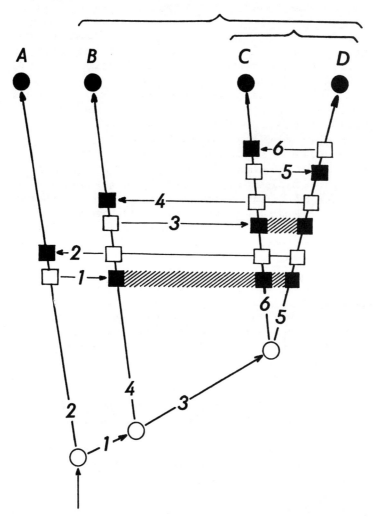

Fig. 3. Argumentation plan of phylogenetic systematics. □ plesiomorph, ■ apo-morph expression of characters. Equal numbers indicate how sister-group relations are established by the distribution of relatively plesiomorph (white) and relatively apomorph (black) characters ("heterobathmy of characters"). Adapted from Hennig (11).

monophyletic groups which together form a monophyletic group of higher rank and are therefore to be termed "sister-groups," one particular character must always occur in a more primitive (relatively plesiomorph) condition in one group than in its sister-group. For the latter, the same is true in respect to other characters. This mosaiclike distribution of relatively primitive and relatively derivative characters in related species and species-groups [Spezial-isationskreuzungen, Heterobathmie der Merkmale: Takhtajan (19)] is a fact which has long been known. But one still finds it occasionally mentioned in the literature as a special peculiarity of some groups of animals that the classification of their constituent groups cannot be achieved in a definite sequence, because there are no solely primitive and no solely derivative species or species-groups. In a phylogenetic system there can indeed be no solely primitive and no solely derivative groups. The possession of at least one derivative (relatively apomorph) ground-plan character is a precondition for a group to be recognized at all as a monophyletic group. But it also follows from this that this same character in the nearest related group must be present in a more primitive (relatively plesiomorph) stage of expression. The exclusive presence of relatively plesiomorph characters is indicative of paraphyletic groupings: these are to be found only in pseudophyletic (see above under B) and purely morphological systems (see above under A), but not in phylogenetic systems. Heterobathmy of characters is therefore a precondition for the establishment of the phylogenetic relationship of species and hence a phylogenetic system.

It is sometimes said that the aims of phylogenetic systematics are not only practically but also theoretically unattainable, because the comparison of species living in a given time-horizon, such as the present, cannot in any way reveal their phylogenetic relationship which refers to a completely different dimension. This view is false. Just as two stereoscopic views of a landscape, which themselves assume only a two-dimensional form, together contain exact information about the third spatial dimension, so the mosaic of heterobathmic characters in its distribution over a number of simultaneously living species contains reliable information about the sequence in which the species have evolved from common ancestors at different times. The study and use of the methods which serve to reveal this information needs, it is true, a far greater amount of knowledge and experience than some systematists are willing to employ. The theoretical foundation and refinement of these methods forms a special chapter in the theory of phylogenetic systematics which can only just be touched upon in the present brief paper.

It is sometimes alleged that consideration of as many characters as possible which have so far not been studied is a prerequisite for the progress of phylogenetic systematics. In particular, the restriction of entomological systematics to comparatively easily recognizable characters of the external skeleton which lie open to view is often not highly regarded. This has some justification. The phenomena of convergence (particularly in its variant known under the name "parallel development"), reversed development of characters and paedomorphosis, which leads to pseudoplesiomorph conditions, make the establishment of true synapomorphy difficult. The more complex is the mosaic of heterobathmic characters which we have at our disposal in a chosen group of species, the more surely can their phylogenetic relationship be deduced from it.

Consideration of new and hitherto unobserved characters can, however, represent progress only if these are analyzed with the special methods of phylogenetic systematics. Thus, it is also necessary to distinguish between plesiomorph and apomorph expressions of characters in the internal anatomy and chemical structure, physiology, and serology and when considering different ways of behavior. Symplesiomorphy must be excluded just as much as convergence. If this is not observed, then consideration of however many characters leads, at best, only to a more precise determination of the overall similarity of the bearers of all of these characters, but not to a more precise establishment of their degree of phylogenetic relationship.

This becomes particularly obvious in animal groups such as the insects in which the life of the individual is subject to the phenomenon of metamorphosis. This is the cause of the incongruences which are so often discussed between larval, pupal, and imaginal classification in morphological and pseudophylogenetic systematics. A theoretically acceptable solution of such "incongruences" is possible only in phylogenetic systematics. It can indeed be the case that particular instances of synapomorphy, and therefore of monophyletic groups, can be recognized only in the larval or pupal stages and others only in the imaginal stage. But this is not a true incongruence, for the phylogenetic system does not try to classify organisms according to their degree of resemblances, but species according to their degree of phylogenetic relationship. It does not matter therefore which stage of development is used to establish relationship on the ground of synapomorphy. A monophyletic group remains such even if it can be established only with the characters of a single stage of development [for more detailed exposition see Hennig (11)].

The fact that not resemblance as such, but only agreement in a particular category of characters is significant for the study of phylogenetic relationship, also makes it possible for phylogenetic systematics to adduce for its purposes features other than physical (holomorphological) characters. Such nonholomorphological characters are the life history and geographical distribution of species. Phylogenetic systematics can, for instance, proceed from the plausible hypothesis that species which show a clearly derivative ("apooec") life history, and for which a certain relationship is probable on other grounds, form a monophyletic group. This is, for instance, often true with parasites. However, hypotheses of this kind must always be verified by close morphological studies, for it is particularly with similar life histories that adaptive convergence is common.

A particularly great importance for phylogenetic systematics is presently often ascribed to parasites and to monophagous and oligophagous plant-feeders which are to be equated with them from the standpoint of phylogenetic method. The theoretical justification for this is supplied by the so-called parasitophyletic rules. Particularly important among these is the so-called "Fahrenholz rule," which supposes a marked parallelism between the phylogenetic development of parasitic groups of animals and their hosts in the majority of cases. If this is correct, then it might be concluded from the restriction of a monophyletic group of parasites to a particular group of host species that the latter, too, form a monophyletic group. But it can easily be shown that this conclusion would be correct only if one could assume that the ancestral species of the host group was attacked by one parasite species and that thereafter each process of speciation in the host group has been accompanied by one speciation process in the parasites. Clearly, this precondition is only rarely fulfilled, since the evolution of parasites often seems to be retarded in comparison with that of their hosts, both in respect to character changes and speciation. The result of this is that paraphyletic host groups can also be attacked by monophyletic groups of parasites. Moreover, it happens that parasites can transfer secondarily (without being passed from ancestors to progeny in the course of speciation) to host species which offer them similar conditions of life. This, too, is often seen as an indication of close phylogenetic relationship between host species which are exclusively attacked by particular parasite species or a monophyletic group of parasites. But this assumption would be valid only if one could assume that the "degree of resemblance" of different species and the "degree of their phylogenetic relationship" corresponded closely with each other. As has been shown, this is not the case. Resemblance can also be based, for instance, on symplesiomorphy, and this cannot be assumed to establish phylogenetic relationship. Since one cannot assume that parasites distinguish, in their choice of host range, the categories of resemblance connections (symplesiomorphy, synapomorphy and convergence) whose differences are important for phylogenetic systematics the greatest care is necessary in attempting to draw conclusions about the phylogenetic relationship of their hosts from the occurrence of monophyletic groups of parasites. The importance of parasitology for phylogenetic systematics is considerable. But on the grounds given it is not so great as is sometimes supposed. In particular there is still no really satisfactory clarification of this whole complex of questions.

The geographical distribution of organisms is also of restricted though not to be underestimated importance for phylogenetic systematics. This can often proceed from the hypothesis that parts of a group which are restricted to a defined, more or less separated, part of the total range, whose ancestors may be assumed to have arrived from other regions, form a monophyletic group. This is particularly valid for the fauna of the marginal continents (Australia and South America), whose ease of accessibility has been different at different periods of the earth's history, and for some islands (e.g., Madagascar, New Zealand). One can, for instance, proceed on the working hypothesis that the Marsupialia of Australia form a monophyletic group, and then seek either to sustain or refute this hypothesis with the morphological

methods of phylogenetic systematics. With groups of animals with disjunctive distribution, one may proceed on the hypothesis that both parts of the
range (Australia and South America in the case of pouched mammals) have
been settled by monophyletic subgroups and that between these a sister-
group relationship exists. Extensive investigations of the phylogenetic development of animal groups (e.g., Hofer on the Marsupialia) often in themselves remain fruitless, since they do not proceed from a working hypothesis
of this kind and as a result contain no statements which serve to answer the
questions which first come clearly to light in such an hypothesis. This is often
of even greater importance in studies of the history of the settlement of
geographical space. Discussions about the earlier existence of direct land
connections between now separate regions [Madagascar and the Oriental
Region, Günther (5); New Zealand and South America, Hennig (12)] have
somewhat the same significance as have attempts to sustain or refute hypotheses about the monophyletic, paraphyletic, or polyphyletic character of
particular groups of animals. The inadequacy of morphological or pseudophyletic systems is shown here with particular clarity.

A special chapter in the theory of phylogenetic systematics which can only
be touched upon here, is the position of fossils in the system [Hennig (9)].
Despite a widely held opinion, establishing the phylogenetic relationships of
fossil animal forms is usually more difficult than that of recent species. The
cause of this is that in fossil finds, usually only a small, often extremely small,
section is available from the character structure of the whole organism. But,
since the methods of phylogenetic systematics have a numerical character
insofar as the certainty of their conclusions grows as the number of characters at their disposal increases (see above), it follows necessarily that the
reliability with which relationships can be established cannot usually be as
great with fossils as with recent species. In the sphere of the lower categories
of the system, the species and their subunits, palaeosystematics is, in addition, at a decisive disadvantage because it can never observe its objects
alive, and can therefore only solve its problems with the help of relatively
unreliable morphological criteria. It is true that the systematics of recent
organisms also satisfies itself mainly with morphological criteria to help it
establish the limits of species. However, there is always the possibility, in
principle, of testing in important cases, that individuals of similar or different
appearance actually belong to one or to different reproductive communities
by observation of their life in nature or by breeding and crossing experiments.
In species with seasonal and sexual dimorphism and those in which the life of
the individual contains a metamorphosis, systematics depends upon such
methods. But, in palaeontology, they cannot be employed. Here systematics
can establish the specific limits only with a much lower degree of accuracy
than with recently known organisms. It would, however, be completely false
to deduce from this, as is sometimes done, that palaeontological systematics
operates with other concepts (e.g., a different species-concept) and other
methods. It differs from the systematics applicable to recent animal forms

only in the lesser degree of certainty and accuracy with which it is able to apply itself.

This applies to inquiry into specific limits just as it does to establishing the degree of the phylogenetic relationship between species. If the purpose of systematics does not consist exclusively of conducting a survey of the animal forms which have existed on the earth at any time, then palaeontology must also try to relate its objects to the phylogenetic system of recent organisms, that is to include them in this system. But this can be meaningful and fruitful only if the limits of the knowledge it can supply are known very precisely and are clearly expressed in each particular case.

Subject to these conditions, the value of fossil finds lies in enabling one to interpret character agreements in recent species when this cannot be done solely from a knowledge of these recent forms. There are, in the recent fauna, monophyletic groups which agree in certainly derivative (apomorph) characters with other diverse groups which are just as surely monophyletic. Some of these agreements must therefore rest on convergence. But it is often impossible to decide with certainty which of these agreements are based on convergence and which are to be considered as true synapomorphy. The possibility of decision in such cases depends on a knowledge of the sequence in which the characters in question evolved. This is sometimes clarified by fossils. An example of this kind is supplied by the sea urchins (Echinoidea).

The Cidaroidea, which are shown to be a monophyletic group by their peculiar spine formation, agree completely with most other recent sea urchins in their possession of a rigid corona. The more primitive expression of this character, a flexible corona, is present only in the Echinothuridae. On the other hand, the Echinothuridae agree completely, in their possession of external gills, with the sea urchins which do not belong to the Cidaroidea. This is likewise a derivative character. This character distribution allows no decision on the question of whether the Cidaroidea or the Echinothuridae are more closely related to the bulk of recent sea urchins. One of the two derivative characters, the external gills or the rigid corona, must thus have evolved through convergence at least twice independently. The oldest fossil Cidaroidea, which are shown to belong to this group by their spine formation, possess a flexible corona. This is decisive evidence that the rigidity of the corona in recent Cidaroidea and in the remaining recent sea urchins (except the Echinothuridae) has evolved through convergence. Concerning the external gills, there are no reasons to suggest convergent evolution. Their presence in recent sea urchins which do not belong to the Cidaroidea may therefore be regarded as synapomorphy. However, it must also be said that they have often been lost secondarily. In other cases, only fossil finds make it possible to establish which expression of a character should be regarded as plesiomorph in a group and which as apomorph.

The importance of fossils thus lies, not so much in the fact that they reduce the morphological gap between different monophyletic groups of the recent fauna, but in that they help to make it possible to decide the categories of resemblance (symplesiomorphy, synapomorphy, or convergence) to which particular agreements of character belong.

Still greater is the value of fossils for determining the age of animal groups. But in this context it should be realized that age determinations have a meaning only in monophyletic groups, since only they have a history of their own (see above). It can be difficult, however, to demonstrate the relationship of a fossil to a given monophyletic group of animals. As has been shown above, heterobathmy of characters is characteristic for nearly related monophyletic groups. Therefore, it often happens that one of two sistergroups can be established as a monophyletic group only by a few apomorph characters which are difficult to verify or only present at a particular stage of metamorphosis. For the distinction of the two groups and the identification of the species belonging to them, this has no significance, because plesiomorph characters can also be employed for diagnosis, though they must be left out of consideration in establishing the monophyly of a group. One can, for instance, recognize at once that a recent arthropod species belongs to the Myriopoda from its possession of homonomous body segmentation with jointed appendages on more than three of its trunk segments, although both are plesiomorph characters and cannot be used to justify the supposition that the Myriopoda are monophyletic. But this is not the case with fossils. One cannot assume without qualification that fossils, especially from the early Palaeozoic, belong to the Myriopoda if they possess a homonomous segmentation and jointed appendages on more than three trunk segments. Both are plesiomorph characters which must also have been present in the common ancestors of the Insecta and Myriopoda. To demonstrate that fossils in fact belong to the Myriopoda, one must demonstrate in them those apomorph characters in the ground plan of the group which suggest its monophyly, i.e., the absence of ocelli and compound eyes. Such demonstration is often very difficult, since these characters are not preserved for us in the fossils. If, in this case, one proceeds uncritically, and classifies fossils on the basis of plesiomorph characters which suffice as diagnostic characters for the certain recognition of all recent species of a monophyletic group, then it can happen that the group will become a paraphyletic group solely through its acquisition of fossils. This can then become the source of all the errors which necessarily arise if one compares monophyletic and paraphyletic groups with one another in phylogenetic studies (see above).

When, however, it has been firmly established that a fossil belongs to a given monophyletic group, that fossil can then be of importance not only for determining the minimum age of the group to which it belongs, but also for determining the minimum age of related groups, of which no fossil finds are available. The existence of *Rhyniella praecursor* in the Devonian not only proves that Collembola, the group to which *Rhyniella* belongs, already occurred then, but from our relatively certain knowledge of the phylogenetic relationships of the principal monophyletic groups of insects it follows that at

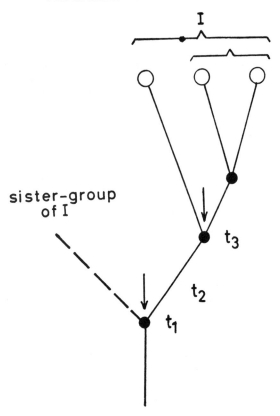

Fig. 4. The three different meanings of questions about the "age" of an animal group.

t_1 age of origin (separation of group I from its sister-group),

t_2 first appearance of the "typical" characters of group I,

t_3 age of division (last common ancestor of all recent species of group I)

the same period the Protura, Diplura, and Ectognatha must also have existed, although, of course, not in the form of their present-day progeny.

In determining the age of animal groups, another factor should be considered as well. In the history of a monophyletic group of animals, there are two points of time which are especially important (see Fig. 4): one is the time at which the group in question was separated from its sister-group by the splitting of their common ancestor (age of origin), and the other the time at which the last common ancestral species of all recent species of the group ceased to exist as a homogeneous reproductive community (age of division). The distinction between these two points of time is especially important in those groups whose recent species are distinguished from species of other groups by their agreement in a large number of derivative characters. One must assume that these characters were already present in the last common ancestral species to whose progeny they have been transmitted unchanged or, in part, further developed. These characters must have evolved in the period between the two named points of time.

Speculation upon the age of a particular group of animals can have three appropriate but different meanings. The following may be intended: (*a*) When did the last common ancestral species of all the recent species of this group which have inherited their derivative characters from it, live? (question about the group's age of division). (*b*) When was the group separated from its sister-group? (question about the group's age of origin). (c) When, in the period between these two points of time, did species for the first time occur with the characters which justify their ascription to the "type" represented by the recent species?

It is seldom clear which of these three essentially different questions is intended when questions are asked about the age of fleas, lice, or other animal groups. This fact, in conjunction with the custom of seeing in phylogenesis mainly the emergence of particular "types" or "Baupläne" whose delimitation is dependent on subjective criteria, is the cause of endless and fruitless debate on the question of whether or not certain fossils should be considered "reptiles," "birds," "mammals," or "men," and when these groups evolved.

It might seem that questions about the age of animal groups lie outside the field of systematics. But this is not the case. The examples quoted should have shown that answering these questions has the same significance as systematically classifying fossils in particular groups, and that the meaning of an answer depends on the classificatory principle used in forming them.

The age of animal groups also has yet another significance for phylogenetic systematics, under some circumstances. It has been said above that the phylogeny diagram and the hierarchical system are closely corresponding kinds of presentation whose content is one and the same. The phylogeny tree presents, as the most important factor, the time dimension in which the degree of phylogenetic relationship between species or monophyletic groups of species is expressed by the sequence in which they have evolved from each common ancestral species (i.e., recency of common ancestry); in a hierarchical system this is shown by the sequence of subordination in the group categories. It is a justifiable aim to perfect the phylogeny diagram by giving, not only the relative sequence of origin of the monophyletic groups, but also the actual time of their origin. This detail of a perfected phylogeny diagram can also be reproduced in a hierarchical system by means of the absolute rank of its group categories. In a hierarchical system, not only are the names of the monophyletic groups quoted but they are also given a specific absolute rank (class, order, family, etc.). Some clearsighted authors [e.g., Simpson (16)] have quite correctly realized that the absolute rank which is attributed to a given group (e.g., family) does not generally mean that this group can be compared with any other of the same rank in any particular respect. Only within one and the same sequence of subordination is it true that the lower ranks show a higher degree of phylogenetic relationship than the higher. This situation can be accepted without injury to the basic principles of phylogenetic systematics. It could be changed, without injury to these principles, only if the absolute rank of categories was linked to their time of origin, just as in geology the sequence of strata in different continents is made comparable by its correlation with specific periods of the earth's history (e.g., Triassic, Jurassic, Cretaceous). Some authors [e.g., Stammer (18)] think that one must take into account, when according absolute rank to systematic groups, their different rates of evolution which have led to greater or lesser morphological "differentiation." But it needs little reflection to see that this is incompatible with the theoretical foundations of phylogenetic systematics and necessarily leads to pseudophylogenetic systems. This should already have been shown by the fact that sister-groups must have the same rank in a phylogenetic system, entirely without regard for the way in which this rank is established; for sister-groups can, of course, have morphologically unfolded (i.e., diverged from the form of their common ancestors) with completely different rates of evolution.

Biological systematics can no more do without a theoretical foundation for its work than can any other science. The theory of phylogenetic systematics is a comprehensive and complex edifice of thought, which here can only be touched upon lightly, even in its most important aspects. In this edifice there is, as always, a logical arrangement of individual problems. In critical expositions, this logical order must be observed. It is not permissible, as sometimes happens, to confuse the critique for answering logically subordinate questions with the critique concerning the principles of the phylogenetic system. From a thoroughgoing theory of phylogenetic systematics, there arise necessarily some unexpected demands on the practical work of the systematist. If the theory as such is accepted in principle it is not permissible to refuse these demands or leave them unconsidered merely because they conflict with certain customary methods obtaining at the time when systematics had no theory. There are many problems in biology whose solution presupposes knowledge of the phylogenetic relationship of one or many species; that is a phylogenetic system of one or more groups of animals. To avoid false conclusions it is therefore especially important that every author of a system should make it easy to recognize whether, or rather to what extent, his system ought to meet the demands imposed by the theory of phylogenetic systematics. But even when these demands should be met in a system, according to the expressed wish of its author, there will always be differences of opinion over the actual relationships of some species or species groups. The person who requires a phylogenetic system as a premise for his own work, will then have to decide on which side lie the better arguments; the criteria for this must again emerge from the theory of phylogenetic systematics. Differences of opinion on matters of fact are not, however, a special defect of phylogenetic systematics but the universal mark of every science.

It is impossible in a short paper to treat even sketchily the extensive field of phylogenetic systematics with all of the questions of detail which are important for the practical work of the systematist. A more comprehensive account in Spanish and another in English, with detailed bibliography, are in the course of preparation. Excellent introductions on its theoretical and methodological foundations with many critical comments on recent systematic works are given in the writings of Günther (4, 6). A valuable study on the philosophical foundations of biological systematics has very recently been published by Kiriakoff (14).

LITERATURE CITED

1. Bigelow, R. S. Monophyletic classification and evolution. *System. Zool.*, 5, 145–46 (1956)
2. Blackwelder, R. A., Alexander, R. D., and Blair, W. F. The data of classification, a symposium. *System. Zool.*, 11, 49–84 (1962); [Critical review by Günther, K., in *Ber. Wiss. Biol.*, 1963]
3. Gregg, J. R. *The language of taxonomy. An application of symbolic logics to the study of classificatory systems.* (Columbia Univ. Press, New York, 70 pp., 1954)
4. Günther, K. Systematik und Stammesgeschichte der Tiere 1939–1953. *Fortschr. Zool.* (N.F.), 10, 33–278 (1956)
5. Günther, K. Die Tetrigidae von Madagaskar, mit einer Erörterung ihrer zoogeographischen Beziehungen und ihrer phylogenetischen Verwandtschaften. *Abhandl. Ber. Staatl. Mus.Tierkde Dresden*, 24, 3–56 (1959)
6. Günther, K. Systematik und Stammesgeschichte der Tiere 1954–1959. *Fortschr. Zool.* (N.F.), 14, 268–547 (1962)
7. Hennig, W. *Grundzüge einer Theorie der phylogenetischen Systematik.* (Deutscher Zentralverlag, Berlin, 370 pp., 1950)
8. Hennig, W. Kritische Bemerkungen zum phylogenetischen System der Insekten. *Beitr. Entomol.*, 3, 1–85 (1953)
9. Hennig, W. Flügelgeäder und System der Dipteren, unter Berückschtigung der aus dem Mesozoikum beschriebenen Fossilien. *Beitr. Entomol.*, 4, 245–388 (1954)
10. Hennig, W. Meinungsverschiedenheiten über das System der niederen Insketen. *Zool Anz.*, 155, 21–30 (1955)
11. Henig. W. Systematik und Phylo-genese. *Ber. Hundertjahrf. Deut. Entomol. Ges. (Berlin, 1956)*, 50–71 (1957)
12. Hennig, W. Die Dipterenfauna von Neuseeland als systematisches und tiergeographisches Problem. *Beitr. Entomol.*, 10, 221–329 (1960)
13. Hennig, W. Veränderungen am phylogenetischen System der Insketen seit 1953. *Ber. Wandervers. Deut. Entomol. (Berlin, 1951)*, 9, 29–42 (1962)
14. Kiriakoff, S. G. Les fondaments philosophiques de la systématique biologique. *Natuurw. Tijdschr. (Ghent)*, 42, 35–57 (1960)
15. Müller, A. H. *Der Großablauf der stammesgeschichtlichen Entwicklung.* (Gustav Fischer, Jena, 50 pp., 1955)
16. Simpson, G. G. Supra-specific variation in nature and in classification from the view-point of paleontology. *Am. Naturalist*, 71, 236–67 (1937)
17. Sokal, R. R. Typology and empiricism in taxonomy. *J. Theoret. Biol.*, 3, 230–67 (1963); (Critical review by Günther, K., in *Ber. Wiss. Biol.*, A, 191, 70, 1963)
18. Stammer, H. J. Neue Wege der Insektensystematik, *Verhandl. Intern. Kongr. Entomol., Wien, 1950*, 1, 1–7 (1961)
19. Takhtajan, A. *Die Evolution der Angiospermen* (Gustav Fischer, Jena, 1959)
20. Verheyen, R. A new classification for the non-passerine birds of the world. *Bull. Inst. Roy. Sci. Nat. Belg.*, 37(27), 36 pp. (1961)
21. Wilson, H. F., and Doner, M. H. The historical development of insect classification (Planographed by John S. Swift Co., Inc., St. Louis, Chicago, New York, Indianapolis, 1937)

2

Reprinted from *Recent Advances in Botany, Int. Bot. Cong., 9th, Symp, Montreal, 1959*, vol. 1, University of Toronto Press, 1961, pp. 841–844

PROBLEMS IN THE CLASSIFICATION OF FERNS

WARREN H. WAGNER, JR.

University of Michigan

MANY OF THE same problems of phylogenetic deduction and classification are found in all plant groups, but the *Filicineae* furnish an unusually good illustrative group. Fern classification is in a state of upheaval and has been since the time of F. O. Bower, but the disagreements—however inconvenient they may be to floristicians—do not make an entirely unhealthy state. They in fact provide valuable hypotheses that may be subjected to test. Christensen (1938), Ching (1940), Holttum (1946), Copeland (1947), Alston (1956), and Pichi-Sermolli (1958) have all contributed different systems. Christensen had one family with fifteen subfamilies where Holttum had five families with thirteen subfamilies; Copeland ten families, no subfamilies; and Ching holds the record with thirty-three families (no subfamilies). The old *"Polypodiaceae"* is of course the largest area of disagreement. Bower believed that parallel evolution was rampant among these plants and that similar structures could arise in different lines; this has since been repeatedly confirmed, and is one of the important reasons why many characters should be used in working out taxonomy, including "difficult" or technical ones. At the generic level there are many disagreements as well; for example, Copeland in the *Hymenophylaceae* had 34 genera where Christensen had 2. At the species level the main questions at present deal with the so-called "biological species," i.e., plants which cannot form fertile hybrids with each other because of polyploid changes. The problems may be broken down as follows: (*a*) the data of comparison; (*b*) the methods of synthesizing the data; and (*c*) the application of categories. The latter two especially concern other plants as much as ferns.

For comparative purposes we have large gaps in our knowledge that will require much effort to fill in. New and simple techniques, like those of clearing and staining leaves and sori, of lactic acid or diaphane for spore studies, and the squash technique for chromosomes, are yielding information of major importance. The more laborious studies of gametophytes and young sporophytes, and the detailed ontogeny of organs are no less important, however. Taxonomists generally prefer to think in terms of key characters, the single or few characters that make for easy identification. The position and shape of the sorus is the time-honored key character in leptosporangiate ferns. It should be pointed out, however, that we really do not know what a single character is. How many genes are involved? Is a character in one part of the plant determined by another in another part? Assuming, though, that we *do* know what a "single character" is, then what is its value? The value is supposed to be measured by its constancy, i.e., reliability; what this means, I believe, is that it is reliable for identification or for keying purposes. Yet, as I have pointed out earlier, in *Pteris* with roughly 280 species defined by a coenosorus, *P. lidgatii* has a dissected sorus; in *Athyrium* with 600 species with dorsal linear sori, *A. proliferum* has dennstaedtioid sori; and in *Elaphoglossum* with 400 paddle-leaved species, *E. cardenasii* has complex pedate fronds. In fact, between ferns with radically different key characters we may sometimes get hybrids; and the key characters are so different that they do not even combine or blend except by producing highly irregular morphology, as shown by the sori of the intermediates between *Aspidotis* and *Onychium*, the bulblets of the intermediates between *Cystopteris bulbifera* and *C. fragilis*, the leaves of *Tectaria* × *Dictyoxiphium* or *Asplenium* × *Camptosorus*, and many other examples.

The point is that single characters may be reliable and convenient for keys, but for phylogenetic research they must be used only in coordination with as many as possible other features. Reniform leaves have appeared by parallel evolution in wholly unrelated ferns. One form of *Cystopteris fragilis* has spores like the genus *Woodsia*. Even chromosome number is constant in some groups (for example, the "X" numbers of *Dryopteris*, *Asplenium*, and *Botrychium*) and variable in others (*Lindsaea*, *Hymenophyllum*, *Woodsia*, *Thelypteris*, and *Blechnum*). Much of the phylogenetic application of the "Telome Theory" has been without regard to other characters, a "disembodied phylogeny." The recent work of K. A. Wilson on the morphology of the leptosporangium exemplifies good coordination with other data; he did not evaluate relationships solely on the single feature he studied, but considered all the other facts together. Whether or not floristicians may prefer to think in only a few key characters, the problems of relationship can be solved only by working with large ensembles of data from all aspects of the plant.

Objective methods to synthesize the comparative facts may merely confirm the intuition of the good taxonomist, of course; but they may also greatly improve the reliability of our correlations. Basic ground plans of similarities underlie all phylogenetic groupings and this, to me, is the essence of systematic research. Various devices have been proposed recently for assessing relationships and evolutionary lines, but the visual ground plan correlation method I developed some time ago for teaching purposes and as an aid in research seems to be as simple as any of them, and based upon sound principles. A number of researchers (e.g., D. F. Brown on *Woodsia*; J. W. Hardin on *Aesculus*; R. L. Hauke on *Equisetum* Subg. *Hippochaete*; and R. F. Blasdell on *Cystopteris*) have adopted it as a useful tool. It involves three assumptions that would seem to apply to any natural and diverse group of plants: common ancestry—plants which have in common a majority of similar characteristics have the same common ancestry; evolutionary divergence—evolution proceeds normally in various directions, and different lines therefore change in different characters and different character-complexes; and inequality of evolutionary rates—evolution occurs at different rates at various times and in different lines. Some forms remain stereotyped and resemble the common ancestor, while others may change radically during the same time period. The ever-present pitfalls of reticulate and parallel evolution may be revealed only by general correlations of many characters. The more characters that are used, therefore, the more accurate are the conclusions.

To work out a phylogenetic problem three broad phases are involved: (*a*) systematic or comparative analysis of the plants in question to find and understand their contrasting characters; (*b*) determination of ground plans to

33

find the character states common to all or most of the plants in order to deduce the most probable ancestral or primitive states; and (*c*) phylogenetic synthesis to assemble the taxa according to their respective deviations from the basic ground plan and from each other. The detailed steps are as follows: (1) to compare and study all the variable characters among the taxa; (2) to determine the generalized or primitive conditions on the principle that characters found in most or all of a number of related taxa are inherited essentially unchanged from the common ancestor, using data also from related taxonomic groups of the same level. (If no obvious trend can be determined in a given character that character may be used only for grouping purposes.) (3) to assign for each character the value 0 for the generalized or primitive condition, and 1 for the specialized or secondary condition (the intermediate states being assigned the value 0.5); (4) to list in tabular form the taxa and for each give the divergence values from the ground plan, both for individual characters and in total; and (5) to determine the mutual character groupings between taxa and then arrange them in sequence according to these groupings on a concentric chart or graph, the radii and branchings to be determined by the mutual character complexes, and the distances by the divergence indices. So that the facts may be made readily visual, the secondary or advanced states of each character should be expressed by letters (intermediate conditions, lower case; fully developed changes, upper case). Taxa are connected to each other by their ensembles of common features, which are plotted as the points of separation, i.e., as the most probable common ancestors. Such a method as this (though certainly subject to improvement and refinement) helps to solve problems. We can find correlations that had been overlooked. We are forced to use all the available data and other workers can repeat our results with the same information. My method also shows at a glance the character groupings of the most probable common ancestors and thus outlines the pathways of phylogeny.

Authors have stated that one must know all the species of a genus in order to work out phylogeny, but how can this be so? The majority of species in any genus have probably disappeared from the earth anyway. If our methods have validity at all, it should be possible with considerable probability to assess relationships where large gaps exist—the fewer the lacunae, of course, the more valid our conclusions, but this does not mean that where they do exist our objective efforts are worthless. The idea that the paleobotanists alone hold the keys that can reveal the course of evolution seems to me to be a negativistic one, and essentially denies the worth of our methods of determining relationships. Phylogenetic relationships exist, of course, between times as well as at the same time. They may be considered, in fact, entirely independently of time, if by phylogeny we mean evolutionary changes and pathways. The primary aim of the phylogenist should be to determine the pathways of relationship, and it is immaterial whether the data come from plants which are living or fossil—both should be used. Ideally the phylogenist should embody all the data, from past and present, in his conclusions. The dating and correlation of phylogenetic pathways with geological horizons is secondary and must rest first on valid conclusions concerning relationships. All efforts should be bent, therefore, towards improving the objectivity of our determination of phylogenetic trends and relationships with the highest degree of probability; all other deductions are subsidiary to this.

Our biggest problem in fern taxonomy is a purely mechanical one which applies as well to other plant groups, and which I shall refer to as "hierarchical

34

inflation." It is my opinion that the "sub" categories are passing into disuse at all levels, a situation that not only tends to blur the subtleties of expression of relationship for which the categories were designed, but also leads to inflation. Harold Bold now has 25 phyla (divisions) for the plant kingdom. In monocots Bessey had 8 orders, Engler 10, but Hutchinson has 29. We all know the situation in the ferns. Two circumstances especially tend toward inflation: the group is extremely well studied, and the group has numerous members.

At the familial level, recent research and correlation of evidence has yielded numerous intriguing suggestions of relationships, such as the following: The gymnogrammeoid ferns may actually be much more separate from the *Dennstaedtia* group than we formerly assumed. A strong separation of the thelypteroid ferns from the aspidioid group may be questioned. The *Davallia-Oleandra* assemblage may be an epiphytic offshoot from the aspidioid groups; the Blechnum and Elaphoglossum groups are evidently also aspidioid. Whether *Dipteris* and *Cheiropleuria* are truly polypodioid ferns is dubious. The degree of taxonomic separation of the *Grammitis* group demands further evidence. However, as new differences are found, these must be balanced against the broad, traditional taxonomy of ferns. Rather than setting up whole new families without reference to the level of existing families, the following, possibly more accurate format is illustrated by hypothetical examples:

Fam. Aspidiaceae
  Subf. Aspidioideae
  Subf. Thelypteridoideae
  Subf. Elaphoglossoideae
  Subf. Davallioideae

Fam. Polypodiaceae s.s.
  Subf. Polypodioideae
  Subf. Grammitidoideae
  Subf. Loxogrammeoideae

At the level of species, the problem of hierarchical inflation also threatens: very minor differences have been held to warrant species recognition. The separation of *Asplenium cryptolepis* from *A. ruta-muraria* is an example. This is a case where the concept of allopatric subspecies is ideally suited to the expression of relationships. The separation of the American *Phyllitis fernaldiana* from the Old World *P. scolopendria* as a distinct species is complicated by differences in polyploidy and physiology, but again the use of subspecies for their designation seems more reasonable and closer to the facts. The "aggregate species" comprising different levels of polyploidy (e.g., *Cystopteris fragilis*, *Asplenium trichomanes*, *Polypodium vulgare*) constitute a special theoretical problem. Different polyploid levels exist in plants which are otherwise nearly indistinguishable; is it possible that the pairing of chromosomes in these plants is determined by simple genetic factors rather than a multitude of differences in homology? Before we set up different "chromosome races" as separate species and upset our taxonomic traditions, more detailed knowledge of the factors that control chromosome pairing should be obtained.

In summary, many of our problems in fern classification may be brought nearer to solution by (*a*) obtaining new data, in many cases through the application of a variety of techniques, and avoiding, thereby, the snares of single- or few-character taxonomy; (*b*) using more objective methods for correlating phylogenetic data, such as the visual ground plan technique described above; and finally (*c*) working toward a reasonable application of the taxonomic categories that invoke the subtleties of the subcategories and conform as much as possible to the traditional standards.

REFERENCES

Alston, A. H. G., 1956, The subdivision of the polypodiaceae, *Taxon* **5:**23-25.

Christiansen, C., 1938, Filicineae, in F. Verdoorn, ed., *Manual of Pteridology,* pp. 522-550, The Hague, Nijhoff.

Ching, R. C., 1940, On natural classification of the family "polypodiaceae," *Sunyatsenia* **5:**201-268.

Copeland, E. B., 1947, *Genera Filicum, the Genera of Ferns,* Cronica Botanica, Waltham, Mass.

Holtum, R. E., 1947, A revised classification of leptosporangiate ferns, *J. Linn. Soc. (Bot.)* **53:**123-158.

Sermoli, P. T., 1958, The higher taxa of the pteridophyta and their classification, in O. Hedberg, ed., *Systematics of Today,* Proceedings of a symposium held at University of Uppsala in commemoration of the 250th anniversary of the birth of Karolus Linnaeus, Uppsala University, Arsskrift 1958, no. 6, pp. 70-90.

3

KARL R. POPPER, SYSTEMATICS, AND CLASSIFICATION: A REPLY TO WALTER BOCK AND OTHER EVOLUTIONARY TAXONOMISTS

E. O. Wiley

Popper (1968a, p. 37) has stated that *"the criterion of the status of a* (scientific as opposed to a metaphysical) *theory is its falsifiability, or refutability, or its testability."* This basic criterion of testability is held by the majority of scientific philosophers, Popper differing only in the specific way in which testing is carried out. Bock (1973) has argued that Popper's philosophy should be adopted for biological classification. I will argue that a theory of phylogeny via genealogical descent conforms to the general criterion of testability and the specific methods of testing outlined by Popper (1968a, b) and thus can serve as a basis for classification. I will also argue that the theory and methodology of evolutionary taxonomy as outlined by Mayr (1969, 1974), Bock (1973), and Ashlock (1974) fits neither the basic criterion of testability nor Popper's specific philosophy and thus cannot serve as a basis for deriving classifications which purport to reflect or communicate scientific inferences. I do not claim that a genealogical theory is the only one which conforms to Popper's philosophy, but I maintain that Hennig's (1966) methodology is testable within Popper's philosophy. The basis for this conclusion rests on the relationship between homology and phylogeny. I will explore this relationship within inductive and deductive systems of hypothesis testing. Certain axioms will be suggested for a phylogenetic system. A restricted definition of homology is defended as best suited for application to estimates of phylogeny. The relationship between homology and phylogeny is then defended as being non-circular. After demonstrating the usefulness of Hennig's (1966) methods within Popper's philosophy, some logical aspects of "classical evolutionary taxonomy" (Bock, 1973) are discussed. Finally, the types of phylogenetic classifications which can exist under Popper's philosophy are discussed.

INDUCTIVE AND DEDUCTIVE HYPOTHESIS TESTING

Induction and deduction play central roles in any science. Induction is the fact or observation gathering process which enables an investigator to pose meaningful hypotheses about the world of nature. Observations comprise our world of experience. Deduction also plays a central role in science for it is through this process that tests are made which corroborate or refute the hypotheses formulated from our world of experience.

Inductive and deductive methods of hypothesis testing are empirical methods of varifying, or corroborating, or refuting hypotheses. Popper (1968b, pp. 27–48) states that inductive methods of hypothesis testing should be rejected because they require adoption of a doctrine of apriorism. In a system of inductive hypothesis testing, attempts to evade apriorism, that is, attempts to evade justification of certain statements as *a priori* true, lead to infinite regression (Popper, 1968b, p. 30). Thus, if we do not regard a universal statement as true based on our world of experience, we would have to justify it by making some inductive inferences. Yet, to make these inductive inferences we must assume some inductive principle of a higher order, and to justify this higher principle we must assume yet another inductive principle of yet an even higher order, and so on to infinity. For example, if we do not take as true a statement that the feet of two species are homologous, we would have to justify the statement by saying that all feet which are attached to the hind leg are homologous. Unwilling to accept this statement as true, we might further qualify our statement by saying that all feet which are attached to the hind leg and are composed of three bones are homologous. To justify this assertion we might add that the muscles of the foot must be arranged in a certain way. Unless we stop and say "I now accept this homology" we will have to go on investigating the structure of the two feet on a finer and finer level, eventually reaching the atomic and subatomic levels, and this eventually leads to the infinite. Appeal to probability (logical, not statistical) does not solve the problem of inifinite regression because logical probability statements are themselves based on higher order inductive principles and also lead to infinite regression. Popper (1968a, b) suggests that deductive methods of hypothesis testing do not lead to infinite regression and do not depend on apriorism and thus can provide a suitable separation between science and metaphysics. Such methods systematically exclude all attempts to *avoid falsification*. The system consists of posing and testing a series of statements or hypotheses. The result of the test can take one of three forms: (1) consistency with the hypothesis, (2) inconsistency, and (3) irrelevancy of the test.

AXIOMS OF PHYLOGENETIC SYSTEMATICS UNDER POPPER'S SYSTEM

Axioms may take two forms, conventions and hypotheses. Conventions are taken as true (not testable) and are to be avoided. Hypotheses which occupy the highest levels of universality within a particular theoretical system are termed the axioms of that system. Hypotheses at a lower level of universality are simply termed hypotheses. It is important to note that Popper conceives of observations as low level hypotheses, not as facts, because he maintains that observations exist only as interpretations of the facts of nature in light of present theories, not as the facts of nature themselves (Popper, 1968b, p. 107).

Three nested axioms occupy the highest levels of universality in the phylogenetic system: (1) evolution occurs; (2) only one phylogeny of all living and extinct organisms exists, and this phylogeny is the result of genealogical descent; (3) characters may be passed from one generation to the next generation, modified or unmodified, through genealogical descent.

HOMOLOGY IN INDUCTIVE AND DEDUCTIVE SYSTEMS OF HYPOTHESIS TESTING

Under the axioms stated above, phylogenetic definitions of homology will suffer from circularity in an inductive system of hypothesis testing unless the homologies are recognized as empirical facts. Because homologies are not empirical facts but hypotheses, the relationship between a homology and a phylogeny would become circular. It is also possible to establish a circular relationship between homology and phylogeny in a deductive system. I maintain that a phylogenetic definition of homology can be produced in the deductive system which can be applied to phylogenetic hypotheses and classifications in a non-circular way.

There is a problem in selecting a definition of homology to be used within the deductive system because many definitions exist, e.g., phenetic (Sneath and Sokal, 1973), cladistic (Hennig, 1966), evolutionary (Simpson, 1961; Mayr, 1969; Bock, 1969), classical (Owen, 1848), etc. Given the choice among two or more definitions I prefer one which provides the best vehicle or tool for testing the hypothesis that two or more characters are homologous. Popper (1968b, p. 115) has stated:

> "a statement x is said to be 'falsifiable to a higher degree' or 'better testable' than a statement y, . . . if and only if the classes of potential falsifiers of x includes the classes of potential falsifiers of y as a *proper subset*."

For example, a phylogenetic definition of homology may be considered more falsifiable than a phenetic definition and therefore preferable if it leads to a hypothesis of homology which includes all the potential falsifiers provided by phenetic comparisons as well as the potential falsifiers provided by phylogeny.

Within this system two (or more) characters are said to be homologous if they are transformation stages of the same original character present in the ancestor of the taxa which display the characters (modified from Hennig, 1966). There are two types of homologous characters: (1) those which are derived from the immediate ancestor of two taxa, and (2) those derived from an ancestor more genealogically distant than the immediate common ancestor. Apomorphic character states are those which are hypothesized to be derived from the immediate ancestral species and absent from earlier common ancestors (two taxa which share them are said to display a synapomorphy). Plesiomorphic character states are those which are hypothesized to be derived from an earlier ancestor and retained in all later ancestors (two taxa

which share a plesiomorphic state are said to display a symplesiomorphy). Finally, those character states that are not derived from the same original state in a common ancestor are said to be non-homologous.

LOGICAL IMPLICATIONS OF A PHYLOGENETIC DEFINITION OF HOMOLOGY IN A METHODOLOGY OF PHYLOGENETIC ANALYSIS

I shall explore the relationship between homology, as defined above, and phylogeny within the methodology of phylogenetic hypothesis testing developed by Hennig (1966).

A hypothesis of homology at a given level of universality contains a *minimum* of two character states, apomorphic and plesiomorphic. Whether a character state is plesiomorphic or apomorphic is relative depending on the level of universality (Fig. 1) of the phylogenetic hypothesis to which it belongs *as a proper subset*. Homologies can be tested only at the level of universality at which they are hypothesized to exist as synapomorphies because the best test of homology is common ancestry. Supposed homologies, therefore, necessarily take on the characteristics of axiomatic conventions at the level of universality at which they exist as symplesiomorphies. Hence, the only valid test of homology under this system is to hypothesize that the supposed homology is a synapomorphy.

Homologies themselves exist at different levels of universality depending on the number of potential falsifiers they contain. If two hypotheses of homology have classes of potential falsifiers which do not intersect they may be thought of as independent of each other. The character states of a transformation series have classes of potential falsifiers which do intersect and comprise no more than a single potential falsifier of a phylogenetic hypothesis. Different transformation series have non-intersecting classes of potential falsifiers. Although they may not be compared as to their suitability as potential falsifiers of a phylogenetic hypothesis, they can be compared as to their

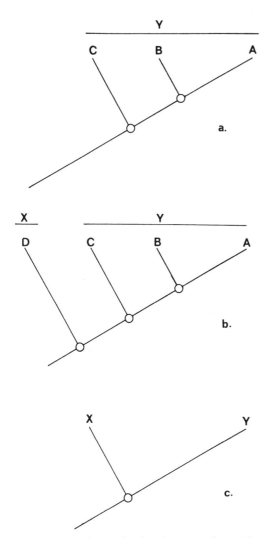

FIG. 1.—Relative levels of universality. Phylogeny "b" exists at a higher level of universality than phylogeny "a" because it has the potential falsifiers of "a" plus an extra and independent set of falsifiers associated with taxon D. Phylogeny "c" exists at the same level of universality as phylogeny "b."

relative strength for corroborating or rejecting that hypothesis. An hypothesis of synapomorphy which has a greater number of potential falsifiers may be said to exist at a higher level of universality than another hypothesis of synapomorphy which has fewer potential falsifiers and the former

would have more weight than the latter *when they conflict.*

Under the theoretical system developed by Popper the mechanism by which an original hypothesis is formulated is not important. But in the practical world of systematics most hypotheses are generated from our world of experience, inductively. An investigator usually avoids trivial hypotheses by experience and a clear formulation of the objectives he has in mind. This may be thought of as the "pre-hypothesis" stage of systematics and is not scientific, strictly speaking, under Popper's philosophy.

Once a hypothesis of homology is formulated from the world of experience it is tested in two phases: by its own set of potential falsifiers and by a set of potential falsifiers of the phylogenetic hypothesis to which it belongs as a proper subset (i.e., it is tested by other hypotheses of synapomorphy through the testing of the phylogenetic hypotheses which they corroborate). Both phases of testing must be done under the rules of parsimony, not because nature is parsimonious, but because only parsimonous hypotheses can be defended by the investigator without resorting to authoritarianism or apriorism.

In the first phase of testing (= attempting to falsify), any potential falsifiers thought to form proper subsets of the hypothesis of homology may be used (without reference to a phylogeny of which it may be a proper subset). Most of the potential falsifiers are morphological similarities and dissimilarities between the characters compared. This does not imply acceptance of a phenetic definition of homology such as that applied by the numerical school of taxonomy. Nor does it imply that only morphological attributes may be used. Rather, any observable differences and similarities between characters, including phenetic criteria, may be used to test an initial proposition that the characters compared are worthy of consideration as possible homologies and thus worthy of consideration as possible falsifiers of phylogenetic hypotheses. Thus, this phase may be thought of as a tentative initial test which may be useful in sorting out those hypotheses which are not worthy of consideration.

Various criteria which may be applied during this phase of testing have been discussed by Simpson (1961), Hennig (1966), Mayr (1969), Sneath and Sokal (1973) and others. Any applicable criterion that was not used to formulate the hypothesis in the first place may be used during this phase in an attempt to falsify the hypothesis. This avoids circularity. For example, if similarity of topographic position is used to hypothesize homology, corroboration must be achieved by testing the hypothesis with some other criterion. Popper's philosophy does not provide for absolute falsification or absolute corroboration. Every species, and indeed every individual of a species, differs in some morphological respect from every other species. Any investigator can demonstrate differences and similarities between structures. The question of "how different" non-homologous characters are, and "how similar" homologous characters are becomes the *opinion* of the investigator. Hypotheses of homology based on overall similarity can always be rejected because no two structures are exactly similar. But, that hypothesis of homology based on morphological comparisons which has been rejected the least number of times relative to other possible hypotheses is to be preferred over these other hypotheses. Another way of looking at this process may be summarized thus: a hypothesis is proposed that two characters are homologous, and that hypothesis of homology inherently carries certain predictions. For example, it might predict that the structure will continue to be similar at finer and finer levels of morphological comparisons, or perhaps two rather dissimilar structures can be traced back to the same embryological structure. The greater the number of these predictions fulfilled by the hypothesis, the stronger will be the hypothesis of homology.

I think it is important to precision of methodology that some form of testing be done at this lower level of the problem rather than simply applying characters that look similar to a phylogenetic hypothesis. A good heuristic rule is that any chance to test should be taken. I do not suggest that morphological testing proves homology. Instead, testing may eliminate certain ill-founded ideas of homology at an early stage in the investigation, thus strengthening phylogenetic statements with which other homologies are associated. Finally, establishment of testing on this level may prove useful in further testing of incongruent synapomorphies as discussed later.

The highest level of corrobration during this phase of testing is failure to refute the characters as synapomorphies, that is, failure to observe any differences between the states of the characters *as formulated*. Another level might be reached in which two of three character states are not exactly similar as stated, but they are more similar to each other than either is to the third. Thus, we might like to investigate the possibility that these two characters are part of a transformation series (with one state plesiomorphous and the other apomorphous, although no initial judgment need be made as to which of the states is apomorphous). Lack of general and specific similarity of the characters of the organisms might be reason to reject the hypothesis of homology. If the hypothesis survives this round of testing it may be thought of as an unrejected morphological hypothesis of homology.

Several examples might be discussed. I might, for example, formulate a character state "spines present in the dorsal fin" and find that two of three taxa I am considering have spines while the other does not. It makes no difference that one species has three spines while the other has twelve because the character state specifies only the presence of spines and not their number. I would, based on this observation, be unable to reject a hypothesis of synapomorphy and might like to corroborate this hypothesis by checking the fine structure of the spines and their ontogeny. If the hypothesis is corroborated, then the alternate state, without spines on the dorsal fin, would assume the alternate plesiomorphous state in the transition series. This is because it is assumed (as a convention) that the dorsal fin itself is homologous in all three species at a level of universality that the investigator does not wish to consider. In comparing the foot of *Australopithecus* and *Homo* I might reject the structure of the foot as being synapomorphous because it is not exactly similar in the two genera. But, I might also observe that while there is not exact correspondence, these two structures are much more similar to each other than either is to the foot of *Pan*. Thus, we might suspect that they are parts of a transformation series. Failure to find synapomorphy in this example might be a result of our inability to formulate the correct hypothesis at the correct level of universality, but it is valuable nevertheless in pointing out characters and structures which may be potential synapomorphies or autapomorphies. As a last example, we might reject a hypothesis of homology between the anal fin modifications of osmerid and poeciliid fishes because of a lack of similarity in anything except their being anal fins (that is, although both are modified anal fins, they are modified in different ways).

TESTING PHYLOGENETIC HYPOTHESES AND
THEIR PROPER SUBSETS

From an unrejected hypothesis of synapomorphy the investigator can proceed in two ways. (1) If there is no previous hypothesis of relationship the investigator may generate a hypothesis via the inductive process, and the synapomorphy logically becomes a proper subset of the phylogenetic hypothesis. *This does not provide a test of either the homology or the phylogeny.* It simply provides the investigator with a hypothesis of phylogenetic relationships with which he can proceed to the second step. (2) If a hypothesis of phylogenetic relationship already exists, then the hypothesis of synapomorphy can be applied to this phylogeny as a test of that phylogenetic hypothesis and its proper subsets (that is, the synapomorphies which corroborate the phylogeny). It is a valid test only if its potential falsifiers do not intersect the potential falsifiers already present in the phylogenetic statement, i.e., it must be independent. Two outcomes are possible: (a) it fails to refute the hypothesis of phylogeny as stated and thus becomes another proper subset of the hypothesis; or (b) it refutes the hypothesis and its proper subsets.

If (a) is the result, the hypothesis has "proved its mettle" (Popper, 1968b, pp. 265–268), or it has been corroborated. The greater number of hypotheses of synapomorphy that are applied to a hypothesis of phylogeny without falsifying it, the more strongly corroborated is the phylogenetic hypothesis. The best corroborated hypothesis is perferred over its alternates.

If (b) is the result, the phylogenetic hypothesis has been falsified unless, of course, the test can be shown to be invalid. If the "incongruent synapomorphy" is shown to be either (1) a symplesiomorphy or (2) a non-homology, then it can be rejected as a valid test of the phylogenetic hypothesis, i.e., it would not be a valid refutation of the phylogenetic hypothesis.

In Figure 2 a hypothetical example of rejection of a character state as a synapomorphy is shown. In Figure 2a and 2b, phylogenies "a" and "b" were generated by induction from hypotheses that characters "1" and "2" are synapomorphies of their respective phylogenies. Both phylogenies have therefore been rejected. If the axiom of only one true phylogeny in nature is to be upheld then at least one of these phylogenies and its associated synapomorphy must be false. It is possible, of course that both are false. In attempting to refute both hypotheses (Fig. 2c) we find three

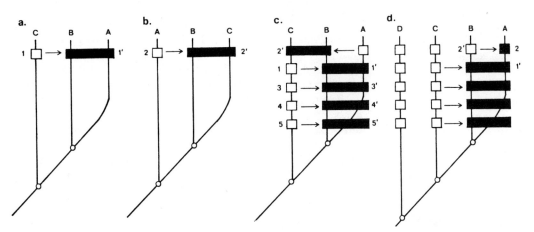

FIG. 2.—Rejection of an "incongruent synapomorphy" using the holomorphological method of Hennig (1966). Hypothesized synapomorphies and autapomorphies are shaded black while the alternate symplesiomorphies are open squares. See text for explanation.

additional proper subsets for phylogeny "a" but no additional subsets for phylogeny "b." We may begin to suspect that synapomorphy "2" is not a valid synapomorphy and therefore not a valid test with which we can reject phylogeny "a." Casting about in our world of experience we come up with another taxon (D, Fig. 2d) which has character state "2" but none of the other synapomorphous character states shared by taxa C and B. By adding taxon D to our analysis we have *raised the level of universality* of the phylogenetic hypothesis and we have demonstrated that it is more parsimonous to consider "2" as a symplesiomorphy at our original level of universality than to consider it a synapomorphy. It is interesting to note that the alternate character state "2'" now becomes an autapomorphy of taxon A (or a synapomorphy if, from our world of experience, we know that taxon A is composed of 2 or more species) and that "2" becomes a *possible* synapomorphy at a still higher level of universality.

We might also reject "2" as a synapomorphy without raising the level of universality of the phylogenetic hypothesis by raising the level of universality of the morphological testing. For example, further ontogenetic investigation may show that the structures expressed as "2" in the adult develop from different germ layers in the two taxa and thus cannot be considered homologous (although the germ layers might be).

Hennig (1966) presents several criteria which help the investigator to decide which of the alternate character states in a transformation series is plesiomorphous and which is apomorphous. It is my contention that application of such criteria will automatically raise the level of universality of the phylogenetic hypothesis. Thus, "2" might be determined to be the plesiomorphous expression because it appears earlier in development than its alternate "2'." Application of this criterion automatically raises the phylogenetic level of universality be calling on our experiences with animals and/or plants outside the group of immediate interest. Calling on such experiences is perfectly valid as long as these experiences can be tested. It is important that such statements be made clearly so that authority is called upon only as a reservior of specialized knowledge and not as apriorism.

BOCK, HOMOLOGY, AND CLASSIFICATION

Bock (1973, pp. 386–389) discusses various aspects of homology, namely, the criteria by which homologies are corroborated or falsified, and the relationship between homology and classification.

Bock (1973, p. 387) states: "similarity between features is the only criterion by which homologues can be recognized." This concept, says Bock, leads to difficulties in distinguishing similar but nonhomologous characters from similar and homologous characters: "Unfortunately, the methods by which homologues are recognized and distinguished from nonhomologous features have low resolving powers." Now, much depends on what Bock means by the word "recognize." If recognize means to be aware of the similarity between two characters and thus to be aware that a potential homology exists, then I would agree that similarity is the sole recognizing criterion of homology. In other words, how can one be aware of a potential homology without perceiving similarities? But, Bock gives no means of testing our perceptions. On the other hand, if Bock uses recognize as a synonym for corroborate, then I must disagree. In this case, Bock would be saying that similarity is the only criterion for corroborating or falsifying hypotheses of homology. Similarities are important, but similarities are neither the only, nor the most severe test for homology. Hypotheses of homology may be tested with other hypotheses of homology. The process is reflected in building and testing hypotheses of phylogeny and their associated synapomorphies. Thus, homologies are *potential falsifying hypotheses* (Popper, 1968b, p. 87) of other independent homologies and the phylogenies with which these homologies are associated. Similarity is important in providing corroboration for the hypothesis of homology, thus qualifying the hypothesis as a falsifying hypothesis. Such a process is not circular, but is the process of reciprocal illumination (see Hull, 1967, for the difference). It provides a deductive test for a phylogenetic hypothesis of homology.

I have tried to show that my concept of the relationship between homology and phylogeny/classification is non-circular. In contrast, I find Bock's (1973, p. 389) concept of the relationship between homology and classification distinctly circular: "Classifications are deduced on the basis of previously established homologies, . . ., which then serve as falsification tests for the classification." If one is to follow Popper, then one cannot say that evidence used to formulate a hypothesis can also be used to corroborate the hypothesis. That is circular inductive testing. Bock approaches a deductive test when he states (1973, p. 389): "more severe falsification tests may be undertaken by careful study of certain features." But, Bock has missed an important point: *no test existed in the first place.*

EVOLUTIONARY AND CLADISTIC CLASSIFICATIONS

It now appears, based on the recent papers of Mayr (1974) and Ashlock (1974), that evolutionary taxonomists accept Hennig's (1966) methodology as best suited for testing cladistic hypotheses. They contend, however, that evolutionary classifications convey more information about the classified organisms than do strictly cladistic classifications. On the surface, we might conclude that since evolutionary classifications supposedly contain more information, they would be preferred over cladistic classifications. This might especially be concluded under Popper's philosophy because the evolutionary classification would be the "bolder" hypothesis. But, as Popper has stated (1968b, p. 267): ". . . it is not so much the number of corroborating instances which determines the degree of corroboration as the severity of the various tests to which the hypothesis in question can be, and has been, subjected. But the severity of the tests, in turn, depends on the degree of testability, and thus on the simplicity of the hypothesis." I hasten to add that being simple does not imply being simplistic. The real question is: are evolutionary classifications testable?

Mayr (1969, 1974), Bock (1973), and Ashlock (1974) have stated that some form of genetic similiarity, or inferred genetic similarity, should be incorporated into the phylogeny and thus into the classification of the taxa studied. The incorporation of these data is the fundamental difference between evolutionary and cladistic classifications and thus is central to the discussion. Problems associated with the application of these data to the modification of cladograms must lead to problems in testing the resultant classifications. One major problem is the lack of a precise and specified methodology in applying a measure of genetic similarity to the actual modification of a cladogram (Rosen, 1974). Those methods actually used seem very vague, even "arty" (Hull, 1970). Another major problem is the loss of information when phenetic and cladistic systems are combined (Hennig, 1966; Griffiths, 1972; Sneath and Sokal, 1973; Cracraft, 1974). Both Bock (1973) and Mayr (1974) concede this when they talk of maximizing the two "semi-independent" variables of phenetic (or genetic) and cladistic information. Can one maximize two semi-independent variables without losing some information from both? A third difficulty lies in the assumption that phenetic divergence may be equated with some determinable amount of genetic divergence. That this is not necessarily the case has been pointed out by Crowson (1970), and Rosen (1974).

Ashlock (1974) has outlined a general methodology for producing evolutionary taxonomic classifications. While he avoids specific methods of modifying cladograms, he does suggest that evolutionary phylogenies and classifications are produced in a two step process. First, a cladistic analysis is performed. This is followed by an anagenetic analysis. The result supposedly gives a measure of relative evolutionary change. Ashlock (1974, p. 96) states: "as the cladogram was established on the basis of these relatively few characters (i.e., synapomorphies or autapomorphies), anagenetic anal-

ysis would attempt to place as many other characters on the dendogram as possible." He makes it clear that these are plesiomorphous or ambiguous character states left over from the cladistic analysis. The problem is that such character states are *untestable hypotheses*. They can only be made testable by raising the level of universality of the phylogeny to the level at which each exists as a synapomorphy or autapomorphy. Thus, Ashlock suggests that including untestable hypotheses of homology into a testable hypothesis of cladistic relationships results in a better classification. Can such a procedure result in a better hypothesis of classification? Is a hypothesis of classification derived from such a methodology testable? No, and I submit that such an analysis would obscure the only testable elements included in it, the cladistic relationships.

Would we not be better off if we raised the level of universality of the problem? And if we raised the level of universality to the ultimate degree, by inclusion of every known taxon, then where would these "anagenetic" character states exist? They would not exist at all because every character state of every character would already be incorporated into the analysis at the level at which it exists as an apomorphous character state, or, it would be so ambiguous as to defy analysis. Why should we incorporate, or try to incorporate, such anagenetic elements into our analyses? Cannot the description of taxa include all of the character states, apomorphous and plesiomorphous, that the investigator feels are important to the full understanding of the taxa he studies? Such a procedure would save the testability of the hypothesis of classification and permit the investigator to list those character states he or she feels are biologically meaningful.

Besides the difficulties outlined above, evolutionary taxonomy has an additional characteristic which seems to nullify Bock's contention that the theory of evolutionary taxonomy is consistent with "Popper's basic ideas" (Bock, 1973, p. 382). The nullification of Bock's contention stems from the lack of the additional potential falsifier which is necessary to test evolutionary classifications as they are put forth today.

A classification purporting to show more than cladistic relationships would be preferred under Popper's philosophy if that classification had all the potential falsifiers of genealogical descent plus at least one additional falsifier independent of genealogical descent. If there is no independent falsifier then the falsifiability of a statement of classification, $f(cl)$, would equal the falsifiability of a statement of phylogeny, $f(phy)$. Or, $f(cl) = f(phy)$. A logical implication of Bock's line of thought is that $f(cl) > f(phy)$ and that overall genetic similarity provides the additional falsifier. But, is this an independent falsifier? Bock (1973) and Mayr (1974) have termed genetic similarity and phylogeny semi-independent variables. Is genetic similarity even a semi-independent variable? I suggest that genetic similarity is either the product of descent from a common ancestor or that it results from convergence at the genome level. That is, it is either a proper subset of genealogical descent or has nothing to do with phylogenetic relationships at all (except to produce error). So, genetic similarity is like any other kind of similarity, it is either apomorphous, plesiomorphous, or nonhomologous. Thus, it cannot provide the additional falsifier. I conclude that $f(cl) = f(phy)$. A logical consequence of this conclusion is that classifications and phylogenies must mirror each other and that the method of falsifying a classification is to refute the phylogeny with which it is associated.

The difficulties embodied in that part of evolutionary taxonomy which differs from phylogenetic systematics lead to a general lack of testability of evolutionary classifications. Claims that evolutionary taxonomy can exist within Popper's philosophy are, in my opinion, invalid. Until such time as evolutionary taxonomists demon-

strate the testability of their classification, we may take the advice of Karl Popper when he states (1968b, p. 277): "those theories which are at too high a level of universality, as it were (that is, too far removed from the level reached by the testable science of the day) give rise, perhaps, to a 'metaphysical system.' In this case, even if from this system statements should be deducible . . . which belong to the prevailing scientific systems, there will be no *new* testable statements among them, which means that no crucial experiment can be designed to test the system in question."

SUMMARY WITH A RECOMMENDATION

(1) Although the relationship between phylogenetic homology and phylogeny may be circular within an inductive hypothesis testing philosophy of science, it is not circular within the deductive hypothesis testing philosophy advocated by Popper (1968a, b).

(2) The terms apomorphous and plesiomorphous (and their derivatives) convey precise concepts which are logical derivations of a phylogenetic definition of homology. As such, they should be substituted for the word homology in systematic studies. This distinguishes the concept of homology used in the study from all other concepts of homology. It also makes the conditional phrase of Bock (1969, 1973) unnecessary.

(3) Only synapomorphies can be used to test hypotheses of phylogeny, and a synapomorphy which corroborates a phylogeny becomes a proper subset of that phylogeny.

(4) Tests of phylogenetic hypotheses are valid if the potential classes of falsifiers of the synapomorphy used to test the phylogeny do not intersect the potential classes of falsifiers which are already proper subsets of the phylogenetic hypothesis.

(5) Production of a phylogenetic hypothesis via induction does not constitute a valid test of the phylogeny or a test of the synapomorphy used to generate the phylogeny.

(6) Hypotheses of synapomorphy which refute a phylogeny also refute all hypotheses of synapomorphy which form proper subsets of the rejected phylogeny. Such a test is valid unless the supposed synapomorphy is demonstrated to be either a plesiomorphy or a nonhomology at the original level of universality of the phylogenetic hypothesis. This demonstration can only be accomplished by raising the level of universality of the problem.

(7) The phylogenetic hypothesis which has been rejected the least number of times is preferred over its alternates.

(8) The classical evolutionary classification system advocated by Bock (1973), Mayr (1974), and Ashlock (1974) is invalid under Popper's philosophy and will remain invalid because concepts such as genetic similarity, phenetic similiarity, adaptive breakthrough, and evolutionary divergence are not independent of genealogical descent and cladistic relationships. Thus, they cannot independently alter a classification based on genealogical relationships. Attempts to alter these relationships will lead to apriorism because of the vagueness of the methodology employed. Thus, this system is rejected in favor of a system wherein classification mirrors phylogeny.

ACKNOWLEDGMENTS

I accept full responsibility for the conclusions presented but I do not claim originality for the ideas. This paper has grown from dialogues with my colleagues at the American Museum of Natural History and their views have significantly influenced my own. Drs. Donn E. Rosen and Gareth Nelson were particularly helpful in critically evaluating my ideas. Both they and Michael K. Oliver read and improved earlier versions of the manuscript. Financial support from the American Museum of Natural History and City University of New York is gratefully appreciated.

REFERENCES

ASHLOCK, P. D. 1974. The uses of caldistics. Ann. Rev. Ecol. Syst. 5:81–99.

BOCK, W. J. 1969. The concept of homology. Ann. N.Y. Acad. Sci. 167:71–73.

BOCK, W. J. 1973. Philosophical foundations of classical evolutionary classification. Syst. Zool. 22:375–392.

CRACRAFT, J. 1974. Phylogenetic models and classification. Syst. Zool. 23:71–90.

CROWSON, R. A. 1970. Classification and biology. Heinmann Educational Books, Ltd., London.

HENNIG, W. 1966. Phylogenetic systematics. Univ. Illinois Press, Urbana.

HULL, D. L. 1967. Certainty and circularity in evolutionary taxonomy. Evolution 21:174–189.

HULL, D. L. 1970. Contemporary systematic philosophies. Ann. Rev. Ecol. Syst. 1:19–54.

MAYR, E. 1969. Principles of systematic zoology. McGraw-Hill, New York.

MAYR, E. 1974. Cladistic analysis or cladistic classification? Z. Zool. Syst. Evolut.-forsch. 12: 94–128.

OWEN, R. 1848. Report on the archtype and homologies of the vertebrate skeleton. Rep. 16th Meeting British Assoc. Adv. Sci.:169–340.

POPPER, K. R. 1968a. Conjectures and refutations: the growth of scientific knowledge. Harper Torchbooks, New York.

POPPER, K. R. 1968b. The logic of scientific discovery. Harper Torchbooks, New York.

ROSEN, D. E. 1974. Cladism or gradism?: a reply to Ernst Mayr. Syst. Zool. 23:446–451.

SIMPSON, G. G. 1961. Principles of animal taxonomy. Columbia Univ. Press, New York.

SNEATH, P. H. A., AND R. R. SOKAL. 1973. Principles of numerical taxonomy. W. H. Freeman and Co., San Francisco.

E. O. WILEY

Department of Ichthyology
The American Museum of Natural History
New York, New York 10024

Part II

CHARACTER ANALYSIS

Editors' Comments
on Papers 4, 5, and 6

Selection of characters and character states is the most important part of a cladistic analysis. Before branching sequences can be constructed, characters must be chosen that are presumed to have evolutionary import, states of these characters must be delimited carefully, the character states must be judged to be homologous, character state trees must be constructed, and polarity (or evolutionary directionality from primitive to derived) must be estimated for these character state trees. In some methods it is possible to delay determination of polarity until after cladograms have been generated; a taxon assessed as primitive must then be selected by application of these same types of ideas for the individual characters and states. Reconstruction of the branching sequences of phylogeny and whatever classifications may be derived from it are entirely dependent on the careful and skillful treatment of the characters and states.

Inglis (Paper 4) focuses on the difficult problem of determining homologies of character states, and particularly on problems with phenetic classification. These same difficulties arise in both cladistic and evolutionary classification. One of the nagging problems with homology has been the circularity that comes with attempting to determine structures that (1) have come by descent from a common ancestor and (2) are also used to determine descent of taxa from a common ancestor. One of Inglis's helpful suggestions is to treat characters as inseparable parts of the whole organisms that have had ancestral-descendent histories. Further, he offers useful definitions and perspectives on the terms *homology, analogy,* and *homoplasy.*

Maslin's work (Paper 5) is an outstanding early effort to determine polarities of character states. His interest was in determining evolutionary directionality of character states for use in evolutionary classifications, and in providing a more explicit basis for these decisions. Although this paper was often cited, its complexity precluded easy interpretations of the arguments presented. Moreover, some of Maslin's "principles" deal with assumptions about the process of evolution and do not deal with polarity (for more complete referral of his principles to more recently accepted criteria of polarity, see Crisci and Stuessy, 1980).

Throckmorton (Paper 6) attempts an analysis of characters and states on an individual basis to learn their in-group distributions for constructing cladistic branching patterns. Like that of Inglis, this paper was written to contrast numerical phenetics with traditional evolutionary classifications. This paper also outlines a manual parsimony method, somewhat intuitive with regard to construction of the branching points, but advocating a careful examination of characters and states before developing cladistic relationships.

These three historical papers do not provide insights on current ideas regarding character analysis, but they do give a perspective on early attempts to organize thoughts on exactly what the traditional worker was doing intuitively with characters and states in the development of phyletic insights. For recent ideas on character analysis, see Crisci and Stuessy (1980), Jong (1980), and Stevens (1980).

REFERENCES

Crisci, J. V., and T. F. Stuessy, 1980, Determining Primitive Character States for Phylogenetic Reconstruction, *Syst. Bot.* **5:**112–135.

Jong, R. de, 1980, Some Tools for Evolutionary and Phylogenetic Studies, *Z. zool. Syst. Evolut.-forsch.* **18:**1–23.

Stevens, P., 1980, Evolutionary Polarity of Character States, *Annu. Rev. Ecol. Syst.* **11:**333–358.

4

Reprinted from *Syst. Zool.* **15**:219-228 (1966)

The Observational Basis of Homology

WILLIAM G. INGLIS

Introduction

The long-standing argument about the meaning of the term *homology* has acquired added impetus with the recent use of electronic computors in the preparation of biological classifications and the associated attempts to criticise the methods of so-called "classical taxonomy." Much of the argument is semantic and appears to be largely attributable to misunderstandings about what has been done in the past, is being done at present, and must inevitably continue to be done in future. Briefly, when two structures are described as homologous (or homologues), an absolute statement is made about their identity, after which different states of such homologues may be used as wholly equivalent in constructing a classification.

This statement of homology is the result of a decision about the degree of resemblance between the structures concerned, a decision taken only after a complex comparison has been carried out. In other words, when a comparison of resemblances shows that a certain level of resemblance exists this resemblance ceases to be a variable and becomes an absolute. Thus it is possible to make a wholly meaningful statement such as: something resembles something else to a greater or lesser degree, relative to a third. But it is only possible to state that two (or more) things are homologous, never more or less homologous than something else.

Unfortunately I know from discussions that what I have just said is liable to be misunderstood because of a tendency automatically to equate the term homology with some phylogenetic concept or as implying some evolutionary background. Such evolutionary implications are not intended, and I intend it to refer only to morphological equivalence with no implications of phylogenetic or evolutionary origins (but see final definitions).

A further semantic difficulty must be cleared up before I continue. This is the meaning of the word similarity. The meaning of the word as used to describe the degree of phenetic relationship reflected by a classification differs from the meaning it has in a sentence such as "Homology is similarity attributable to common ancestry." The first usage involves a relative assessment of a variable or variables on the basis of which more than two things are considered and ranked, while the second use of the word indicates an assessment of the degree of likeness between two things only. I therefore propose to use the word resemblance to describe what is assessed in this second procedure.

The problem of establishing what is meant by the noun *homology* is two-fold: (1) How do we compare two things to establish the resemblance(s) between them? (2) What degree of resemblance is considered sufficient to justify the absolute relationship called by the noun *homology*? That both things can be done and regularly are done is obvious. The difficulties are introduced when an attempt is made to describe what is done and to define the various levels at which conclusions are reached. Owen's original argument (see Appendix 2) that homologous structures are structures in different organisms that resemble each other so much that they warrant the use of the *same name* in their discussion was, and still is, a reasonable description of what happens in practice and is comparable to the definition of a species given by Regan (1926). Two structures are homologous when in the opinion of a competent comparative anatomist they are homologues.

That this definition is philosophically and operationally unsatisfying is obvious, and Owen himself introduced his archetype concept as one way of overcoming its weakness. Similar arguments have continued to be advanced much more recently in which homologous structures are defined by reference to some basic plan (morphotype, bauplan, archetype, etc.). Later definitions, which have caused much of the current controversy, incorporate some concept of common ancestry. This improves the apparent logical satisfaction content of the definition but not the operational value. The truth of this contention is exemplified by most discussions (descriptions) of how to recognize homologies (in an evolutionary sense), which are simply cook-book instructions on how to establish sufficient resemblances to warrant the absolute decision, based apparently on an unexplained personal judgement of empirically derived resemblances, that two structures are homologous.

The Process of Comparison

When a feature shown by one organism is compared with a feature of a different organism and the two are considered to be absolutely equivalent, it is because they are the same shape (approximately, but not always) in both, and occur in the same position relative to other features in both organisms and to the total outline of both organisms. But this implies that the further features being used to support the conclusion about the first pair of features can themselves be reliably considered absolutely equivalent. The latter features are generally used in this way because they themselves lie in a definite, corresponding spatial relationship (and frequently show a likeness of shape and structure) to yet other features occurring in both organisms, and of course have a corresponding relationship in both organisms with the first features compared. In other words, any

organ or feature of an organism is compared with some organ or feature of another organism in terms of the distribution, patterns, and spatial relationships they show towards other as yet uncompared organs or features. Thus a statement of absolute resemblance involves a system of comparison, using as reference points as yet uncompared organs or features, the validity of whose use in this comparison is only justified by later reference back to the first organs or features compared (see detailed discussion in Appendix 1). Comparison leading to the recognition of absolute resemblances is therefore circular and involves whole organisms.

Such circularity appears to be inherent in any discussion of universals, and an attempt can be made to explain it in two ways: (1) by reference to some generalized or basic plan, or (2) by conversion to an infinite regress. Let us consider both these explanations in order, because both have been advanced to justify and explain what are called homologies.

Basic plan : Bauplan : Morphotype

The idea of explaining what has been recognized in terms of reference to some basic plan is very old and has been advanced more recently as an explanation of homology by several authors, e.g., Zangerl (1948), Woodger (1945), and Meyer-Abich (1964). The value of a basic plan is usually considered to be that it allows the existence of those absolute relationships called homologies to be explained logically. The problem is that, in spite of the symbolic presentation of the problem given by Woodger (1945), it adds nothing to the logical argument and removes nothing from the logical difficulties. First one establishes certain absolute relationships (homologies), then abstracts a generalized basic plan, and then defines or explains the homologies used in the erection of this plan by reference back to it. This is clearly unsatisfactory, and it is interesting to ask the rhetorical question: Are the homologies derived by relating them to the basic plan indistinguishable from those used in the invention of the basic plan?

That such basic plans can be erected is not in doubt. But that they can be used to define homologies and can be in any way an operational tool in the initial determination of homologies from empirically obtained data is certainly not true. Their values lies in being a useful and expedient summary of the position of current knowledge about the morphology of a group, to which reference can be made in describing and naming the parts of some previously undescribed organism.

Infinite Regress

The apparent value of transforming the circularity to an infinite regress is that instead of coming round in a circle in explaining the establishment of homologies, comparison is made with a third, fourth . . . *n*th individual organism. The problem here is the obvious one that there must be some final point at which the regress begins, or began. Thus, in the limit whether one argues that a basic plan is given by a Deity (or some equally undefined source) or that an infinite regress began with some action by a Deity appears to me to make little logical difference within the context of the present argument. Nevertheless, in practice, the regress is the more satisfactory concept in being referable to (or equatable with) a Theory of Evolution, which helps to explain and unify so many aspects of biological thought. (But I pass over the question of an Unmoved Mover and a First Cause.)

There is no *observational* method by which the circularity can be broken, and certainly no basic-plan concept is satisfactory. The Theory of Evolution does, however, explain the existence of the unity of organization implied by homology, and its use as a non-observational explanatory appendage is inevitable.

Homologous Structures

An organ can never be in itself homologous or a homologue; it can only be so described relative to some organ in another organism, even if the other organism(s) is (are) summarized in a basic plan. This rigid dependence upon a one-to-one comparison is usually hidden by an explanatory appendage about evolutionary origins or by reference to some basic plan (or one of the many synonyms of this concept). The fact that an evolutionary appendage to any definition of homology is operationally unnecessary, although logically satisfying, has been, and still is, stressed by the proponents of the basic-plan concept. This I accept. But the concept of a basic plan is an equally unnecessary addition to any definition of homologous structures. Both arguments are the result of approaching the problem at too many removes from the basic operational level since all the discussions depend upon the prior establishment of absolute relationships. An attempt is then made to explain the intrinsic metaphysical meaning of such absolutes by phylogenetic sequences on the one hand and basic plans on the other.

Simpson (1961: 81–82) very reasonably points out "If, as indeed seems to be the case in some polemics, the argument were only about which concept is to bear the name 'homology' it would be quite senseless." I would insist that because the relationships being discussed by both schools are arrived at in exactly the same way and that the differences are purely those of *a posteriori* explanation, there is no difference in concept but only one of semantics. I therefore argue that the terms homology, paralogy (sensu Hunter, 1964), operational homology (Sokal and Sneath, 1963) and morphological correspondence (Woodger, 1945) are absolutely equivalent. That is, they are "homologous." They all refer to features established in the same way, and the only differences between them are in the presence or absence of an explanatory appendage about common ancestry, which appendage appears to be generally accepted by all the contenders anyway.

To return to the question upon which all the rest depends, how are the absolutes called homologies recognized? The establishment of such absolutes is circular, in the sense demonstrated above, and most discussions of their meaning have concentrated upon explaining the origins of such absolutes, the absolutes themselves being largely taken for granted. The *origin* of the absolutes is satisfactorily *explained* by the Theory of Evolution, but this does not give any method of recognizing them in previously undescribed pairs of organisms. They can only be explained after it has been established that they exist. A basic plan is certainly an expedient reference against which to recognize homologies when later organisms are being examined. But such a basic plan depends upon the prior establishment of homologies without the assistance of any *a priori* basic plan and cannot *explain* the homologies after they have been recognized. Thus, the use of a basic plan as a non-observational explanation is logically unacceptable.

Further, in the process of comparison that leads to the development of a circular sequence, the organs involved in the development of this sequence are *all* homologous. Because so many homologies have been established for so long, this is generally overlooked, and most discussion of homology implies the existence of *known* homologies or basic plans. Even when such plans are used, the process is still that of a one-to-one comparison.

In other words, if we consider those publications in which the concepts of basic plans are discussed in detail (e.g., Zangerl, 1948; Woodger, 1937, 1945; Meyer-Abich, 1964), ignore the metaphysical content of some of them, and substitute the words "organism with which the other organism is being compared" for the term "basic plan" (or one of its synonyms), we have a reasonably accurate description of how homologies are established and what

they are. The problem which lies behind all the arguments and discussion on how homologies are recognized has been concealed by theories which have attempted to answer quite a different question, namely, why do homologies exist?

It follows that it is impossible to establish that two structures are homologous *in vacuo*. A comparison of the whole organism is involved, or more frequently is implied. Further, the comparison is always one-to-one, although the impression is frequently given that this is not so and that several organisms, usually grouped as taxa, are involved (see Woodger, 1945: 106; Haas and Simpson, 1946: 321). This applies equally to the sequences of changing forms discussed below since such cases depend upon a one-to-one comparison in terms of the changing feature(s), while the homology of the other parts of the organisms is accepted without further discussion.

Homoplasious Structures

As a corollary of what is argued above, the definition of homoplasious structures (Lankester, 1870; and see Haas and Simpson, 1946) becomes simple. They are defined by Haas and Simpson (1946) as "organs showing similarities due not to common ancestry but to independent acquisition of the similar characters." The reference to "not due to common ancestry" is only a consequence of the usual evolutionary appendage to definitions of homology and is of no use in the actual recognition of homoplasious resemblance.

I would contend that certain kinds of resemblance are considered to be homoplasious because, although the organs being compared resemble each other in some way or another, it is impossible to establish resemblances between other features of the organisms, so that no circularity can be established (see Appendix 1). In other words, homoplasy simply implies an unconfirmed hypothesis of homology and is a consequence of a definite negative decision on homology and not a positive decision on homoplasy (but see later, paralogy).

Homoplasious resemblance can be qualified by three terms: analogy, parallelism, and convergence. Simpson (in Haas and Simpson, 1946) stresses the difference between the first of these terms and the other two, but all three refer to some explanatory appendage added to an earlier decision of not-homologous. Because it depends on interpretation and explanation, I cannot agree that analogy is a term equivalent in rank to homology and homoplasy, both of which result from direct observation. On the other hand, there is a marked practical difference between the terms analogy and parallelism and convergence, in that only analogy can be suggested from a study of only two organisms. The other terms imply that some phylogenetic sequences, involving more than two organisms, have been established. The only one of these terms of direct practical consequence to comparative anatomy is, therefore, analogy, while the other two form part of the descriptive vocabulary used in discussing the interrelationships suggested by phylogeny (Fig. 1).

In one respect, when it is decided that some resemblance is not circular, and so not homologous, a definite, although negative, decision has been taken. A condition can, however, occur in which it is not possible to reach any definite conclusion, e.g., due to lack of information or gross distortion or simplification of the body, with insufficient linking forms. This is fairly frequent in poorly known invertebrate groups and in many parasites. In such cases, although it is impossible to establish either homology or homoplasy, it is nevertheless useful, if only as an expedient simplification in morphological naming, to treat structures as equivalent. In such cases the term paralogy (suggested by Hunter, 1964) may have a part to play, but only as the description of an agnostic condition (provisional homology) which will disappear when sufficient information is available to enable a definite decision to be taken.

Homology and Classification

The formation, or invention, of biological classifications (as with any other) involves the prior establishment or recognition of those features of the organisms to be classified which can be considered absolutely equivalent. In other words, the classification of absolute resemblance must precede the classification of relative similarity recognized formally (in the present context) as a biological classification. Sokal and Sneath (1963: 72) approach this conclusion with reference to morphological evidence when they say, "It is somewhat embarrassing to find that within the concept of natural groups there are similar concepts of natural organ groups . . . ," but they do not consider it further.

The two ways of explaining the absolutes called homologies are still currently in use since they are expedient and useful. Thus the basic plan is a very useful way of summarizing a mass of morphological information and enables short descriptions to be given by reference to it. This also explains part of the Sokal and Sneath dilemma since many groups found in biological classifications are simply descriptions of basic plans written as diagnoses of taxa (see particularly Danser, 1950).

The second explanation of the origin of homologies, the use of a regress, has been stressed by many authors, particularly when pointing out that homologies are not simply based on the grouping of organs or features that look alike (particularly Remane, 1956; see also Simpson, 1961). This is overcome by the reference of one organism to another to produce a chain of forms the contiguous members of which frequently show a high level of resemblance, with a concomitant ease in the recognition of homologies (or at least of decision making), although the terminal members of the chain may be very different in appearance. Such sequences have been discussed in detail and their apparent practical importance in the establishment of evolutionary sequences assessed by, among many others, Maslin

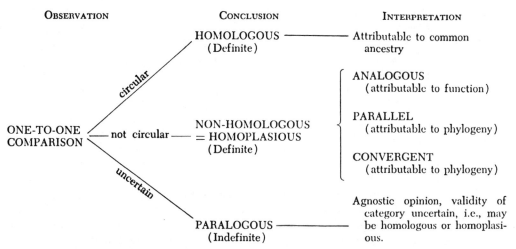

FIG. 1. Diagrammatic presentation of the stages involved in the recognition and interpretation of morphological resemblance.

(1952) who proposes the name "morphoclines" for them. This is, of course, simply a practical application of a regress, but it has certain consequences.

The use of such chains of forms allows the recognition of homologies between apparently diverse structures in sequences of adult organisms. But the same principle is also the basis of using embryological data in suggesting or supporting homologies (Woodger, 1945; Lam, 1959). In the case of chains of adult forms, a claim to recognise homologies is clearly based on a hypothesis that interconnecting forms exist, or can be imagined as existing or having existed at some time. The hypothesis may then receive further support by the discovery of more organisms of the same date, by the analysis of the embryology of sufficient organisms, or by the discovery of connecting fossils (Simpson, 1961:97, footnote). The practical value of extrapolating sequences of changing forms and of inserting hypothetical forms into a chain is certainly great, but it must be noted that all this can be done only if there is a background of prior homologies established for other organs of the organism.

However, such sequences are frequently equated (with varying amounts of reserva-

tion) with, or used as a basis for the invention of, phylogenies, and it must be stressed that they are, or at least can be, established during the assessment of the resemblances which leads to the recognition of homologies. But this assessment precedes the process of classification, so that in such cases a decision has been taken on the morphological sequences, later equated with phylogenies, prior to the erection of any classification. This implies that phylogenies (or at least paraphylogenies) can be erected prior to classification, in contradistinction to the arguments advanced by many authors of the Neoadansonian school.

The real problem currently causing so much dialectic is to what extent such sequences, or even homologies, can be equated with the phylogenetic history of the organisms being studied if there is no fossil record. I do not wish to enter into it at this time, but I would prefer that the highly speculative nature of such sequential classifications be recognised. They are certainly not phyletic, neither are they phenetic. I suggest that they be called *akoluthic*.

I am not prepared to go so far as to suggest that the sequences so often produced can give no indication of the phylogenetic history of the organisms under consideration since they demonstrably can. Even when such sequential classifications have not been sufficiently analyzed to allow any phylogenetic interpretations of value to be reached, they do supply an expedient framework around which to arrange a classification. The term akoluthic is proposed to indicate that a classification is based on sequences of contemporaneous forms which need not be causally related (i.e., show ancestor : descendent relationships) but which only reflect sequences and nothing else. The analysis and interpretation of such sequences is complex and time-consuming, and an examination of the published literature shows that in many cases so-called phylogenetic classifications are only akoluthic.

Concluding Remarks

This paper is the result of listening to and taking part in arguments on phenetic and phylogenetic classifications, during which it has been clear that little difference exists in what is done but that there is much argument about what is claimed to be done. Whether an evolutionary appendage is added to the definition of homology or not makes no difference to what is done or to the way in which homologies are established. Likewise, no amount of argument about the objectivity of the methods used to produce a classification will change the fact that the validity of the data on which the classification is based is itself based on a prior subjective assessment of homology or non-homology.

The great strength and the most remarkable feature of most biological classification is its predictive content, which is a bye-product of the intra-individual correlations which have been analyzed and assessed in the establishment of homologies. The value of all biological classification depends upon the validity of the homologies upon which it is built and great disservice may be done to biological studies as a whole if this dependence is overlooked or deprecated. The recognition of homologies is not simply by saying that two structures look alike; it is the result of a long, and often wearisome, process of study and comparison. It is as a consequence of this previous labour that computors, whether human or electronic, can produce the groupings, with their high predictive and informative content, upon which so much further work and study depend.

One final point is worth making, very briefly, about the use of chemical characters in biological classification. In such cases chemical absolutes, or universals, are being used instead of, or alongside, the morphological universals of homology. Both kinds of universal are then considered to be equivalent forms of evidence, which they certainly are not. The recognition or identification of chemical compounds has depended upon the prior establishment by chemists of the chemical universals (e.g., elements) which can be recognized by tests established by chemists. Carrying out a series of tests that gives results of value to chemists, however, does not allow us to say, without further evidence, what the results mean in a living organism. More information is required. For example, we should know the processes by which the identified chemical compounds have been produced by the living organisms since the same compound could be the end product, an intermediate product, or a bye-product of a series of chemical changes. Without further evidence chemical absolutes must be treated as biological paralogies, i.e., their status and value are unknown.

Conclusions

1. A statement that two organs are homologous indicates that a process of comparison has been carried out between two organisms, as a result of which a circularity of homologous structures has been established. Thus a statement of homology indicates a conclusion referring to whole organisms and not simply to one pair of corresponding organs. The absolute resemblance (a morphological universal) so rec-

ognized may best be explained by an evolutionary interpretation.

2. Where some resemblance exists which does not form part of a circular sequence when the whole organisms are compared, such a resemblance is homoplasious. Homoplasy can then be explained in one, or a combination of two, of three ways: analogy, which is homoplasy attributable to similarity of function; parallelism or convergence, which are homoplasy interpreted in terms of some previously established phylogenetic sequences.

3. Where some resemblance exists about which no definite conclusion can be reached and it is expedient to use the same name for the structures involved, or to use them in producing a classification, the term paralogy is available to describe the condition.

4. Classifications built around sequences of changing morphological forms (i.e., morpho-clines) for which no time scale is available should be called akoluthic, and not phylogenetic.

Acknowledgments

Acknowledgments are due several colleagues for many interesting and valuable discussions on the subject matter of this paper, particularly to Dr. J. G. Sheals for his many apt and cryptic remarks.

Appendix 1: Comparison Leading to Homology

The circularity established in reaching a conclusion of homology is here demonstrated in detail, together with a concomitant demonstation of the non-circularity which leads to a conclusion of homoplasy. It should be noted that for simplicity the first two hypothetical organisms are identical while the upper regions of all three organisms are identical (Fig. 2). This does not affect the argument but simplifies the discussion.

It is clear that the first step is to compare two things and reach a preliminary, provisional decision that they are sufficiently alike in appearance to warrant further de-

tailed study. This step hardly need be regarded as a provisional hypothesis; it is simply a method of saving time. The next stage is to select definite corresponding features in the two organisms under consideration and, if warranted, erect a provisional hypothesis of homology with respect to them both. Thus, let us consider first of all structure a (any other would serve, as will become clear later) of organism 1, which can be considered provisionally homologous with structure a' of organism 2 (Fig. 2). *Why* are these structures homologous, or *why* can they be considered equivalent features of the two organisms? *Because* they occur in the same position on (or relative to) plates A and A'. But this supposes that the plates are comparable or absolutely equivalent. *Why* are the plates equivalent (i.e., homologous)? *Because* they occur on similarly shaped anterior regions of the body of both organisms, which in turn lie in the same positions relative to the rest of the body in both organisms; and, of course, they both carry a structure a (or a'). But this supposes that the anterior part of the body of each organism is comparable. *Why* are they equivalent? *Because* each contains two plates with the same relationship to each other (i.e., one in front of the other); and the more forward plate of each carries two structures a and a', which occur on plates A and A', which occur in the same relationship to the other plate in both organisms and, of course, in the same relation to the total outline of the anterior part of each organism. This can be extended indefinitely, justifying one conclusion on absolute relationship by reference to another conclusion on absolute relationship, which can only be justified by reference to a further set or by reference back to an already used pair.

To summarize, the homology of structures a and a' is shown by the use of a chain of features running from a:a'–A:A'–Ant: Ant'–B:B'–b:b'–c:c'–B:B'–etc. Thus we are justified in referring to the resemblances between the anterior regions of organisms 1 and 2 as homologous. By a similar se-

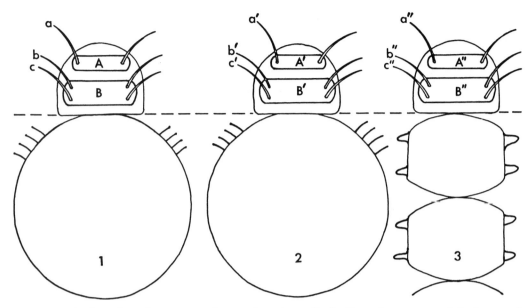

FIG. 2. The hypothetical organisms discussed in Appendix 1. Note that organism 3 is shown only in part. The term anterior as used in the text refers to all structures drawn above the dashed line. The lettering is explained in the text.

quence the same conclusion can be reached between the anterior regions of organisms 1 and 3 or 2 and 3. But extension of the comparison shows that the anterior region in organism 3 is not homologous with that in organisms 1 and 2 since the comparison cannot continue between the remainder of the two organisms of any pair involving organism 3. Therefore, the similarity between the anterior region of organism 3 and that of either of the others is homoplasy.

Appendix 2

Much confusion is apparent about the original concept of homology as used by Owen. Reference is usually made to Owen's definition of the term, and his other comments on the concept have been overlooked. As a discussion of this would have interrupted the flow of the arguments presented above it is presented here.

When Owen (1843) first defined the term "homologue" he simply referred to it as meaning "the same organ in different animals under every variety of form and function." This has been quoted almost solely ever since. As a definition it is clearly unsatisfactory, as Simpson (1961) and Haas and Simpson (1946), among others, have pointed out. However, what has been overlooked is that even in this definition Owen gave a clear indication of the ideas lying behind his concept of homologues since he gave the derivation of the word as "Gr. homos; logos, speech." This concept was expressed more clearly in 1848, when he wrote (p. 5) "The corresponding parts in different animals being thus made namesakes, are called technically 'homologues.' The term is used by logicians as synonymous with 'homonyms' . . . ." The clearest expression of the ideas lying behind the term does not appear until 1866, when he says (p. vii) " 'Homological Anatomy' seeks in the characters of an organ and part those, chiefly of relative position and connections, that guide to a conclusion manifested by applying the *same name* to such parts or organs, . . ." (italics Owen's), and later (p. xii) "A 'homologue' is a part or organ in one organism so answering to that in another as to require the same name."

REFERENCES

DANSER, B. H. 1950. A theory of systematics. Bibliotheca Biotheor. 4:113–180.

HAAS, O., AND SIMPSON, G. G. 1946. Analysis of some phylogenetic terms, with attempts at redefinition. Proc. Amer. Philos. Soc. 90:319–349.

HUNTER, I. J. 1964. Paralogy, a concept complementary to homology and analogy. Nature 204:604.

LAM, H. J. 1959. Taxonomy, general principles and angiosperms. In W. B. Turrill [ed.], International series of monographs in pure and applied biology. Division: Botany. Vol. 2, pp. 3–75. Pergamon Press, London.

LANKESTER, E. R. 1870. On the use of the term homology in modern zoology, and the distinction between homogenetic and homoplastic agreements. Ann. Mag. Nat. Hist. (4)6:34–43.

MASLIN, T. P. 1952. Morphological criteria of phyletic relationships. Systematic Zool. 1:49–70.

MEYER-ABICH, A. 1964. The historico-philosophical background of the modern evolution-biology. Acta Biotheor, (Suppl.2):1–170.

OWEN, R. 1843. Lectures on the comparative anatomy and physiology of the invertebrate animals delivered at the Royal College of Surgeons in 1843. Longman, Brown, Green and Longmans, London.

1848. On the archetype and homologies of the vertebrate skeleton. John Van Voorst, London.
1866. The anatomy of vertebrates. I. Fishes and reptiles. Longman, Green & Co., London.

REGAN, C. T. 1926. Organic evolution. Presidential address, Section D. Rep. British Assoc. Adv. Sci. 1925:75–86.

REMANE, A. 1956. Die Grundlagen des naturlichen Systems, der vergleichenden Anatomie und der Phylogenetik. Theoretische Morphologie und systematik. (Second Ed.) Geest und Portig, Leipzig.

SIMPSON, G. G. 1961. Principles of animal taxonomy. Columbia Biological Series. Columbia University Press, New York.

SOKAL, R. R., AND P. H. A. SNEATH. 1963. Principles of numerical taxonomy. W. H. Freeman and Company, San Francisco and London.

WOODGER, J. H. 1937. The axiomatic method in biology, with appendices by Alfred Tarski and W. F. Floyd. University Press, Cambridge.
1945. On biological transformations, p. 95–120. In Essays on growth and form presented to D'Arcy Wentworth Thompson. Oxford University Press, Oxford.

ZANGERL, R. 1948. The methods of comparative anatomy and its contribution to the study of evolution. Evolution 2:351–374.

WILLIAM G. INGLIS is Head of the Aschelminthes Section of the British Museum (Natural History), Cromwell Road, London S.W.7, England. He is interested in the comparative and functional anatomy of nematodes, their taxonomy and their adaptations to parasitism.

5

Copyright © 1952 by the Society of Systematic Zoology
Reprinted from *Syst. Zool.* **1:**49-70 (1952)

Morphological Criteria of Phyletic Relationships

T. PAUL MASLIN

SEVERAL years ago the author made a taxonomic study of a relatively large genus of snakes. The number of available specimens was small, data on the natural history of the various forms were lacking, and no recourse could be had to paleontological evidence. The attempt to establish the phyletic relationships of the known species was made from morphological and incomplete geographic data alone. This situation was far from unique; most taxonomists at one time or another are faced with the same problem and they formulate certain principles to aid them in their work. But these principles, seldom clearly voiced, are frequently tenuous or are inconsistently applied. This paper is an attempt to enunciate some of these principles and to formulate criteria which may be of value in phyletic studies.

The basic assumption upon which all taxonomic practices rest is that similar organisms are related. This may be called the principle of similarity. It has been stated repeatedly in various ways, each statement qualifying or buttressing certain phases of the principle but in no way altering its basic meaning. The assumption as stated is open to criticism; what is meant by the phrase, "similar organisms"? The phrase ordinarily implies that similar organisms are made of the same substances, have developed under the control of the same genetic factors, and have developed in the same way. But if all of these processes were truly the same the organisms would, in fact, be identical. The very use of the term "similar" carries with it a connotation of degree. Organisms are similar if they are formed from similar substances under the control of similar modifiers in a similar way. This may seem a cyclic argument—organisms are similar if they are similar—but it is more than that. The intended implication is that organisms are similar in direct proportion to the degree of similarity of all phases of their existence. As there are degrees of similarity, so must there be degrees of relationship. This should be expressed as a percentage of developmental and metabolic factors shared by two organisms; the greater the percentage of common factors, the closer the relationship. The taxonomist is, however, seldom in a position to determine the physiological factors shared by two organisms. He assumes that if two structures or organisms are similar, then the factors controlling the development of each structure or organism must in part be identical, and that the organisms are related.

However, it is not always easy to apply this principle of similarity. Often superficially similar organisms are not as closely related as dissimilar ones. Characters such as color, size, and shape, are usually subject to rigorous selection by the fluctuating environment, with the result that only a limited variety of forms is tolerated in any particular complex of environmental factors. Thus similar organisms, when exposed to the same environment, may tend to approach each other morphologically. These superficial similarities obscure their true relationship.

Internal characters, which are often less variable, are of much greater value than most external ones in establishing relationship, but frequently are less accessible to the taxonomist. Camp (1916) discovered a new salamander in the Yosemite of California. He placed this species, *platycephalus*, in a Mexican genus,

Spelerpes. Later Dunn (1923) transferred the species to the Southern European genus, *Hydromantes,* hitherto represented by a single species, *H. genei* (Schlegel). The change was made on the basis of habits, coloration, dentition, some skull characters, lack of a basal constriction of the tail, webbing of toes, and the expansion of the terminal phalanges. The allocation to the same genus of two species, each highly restricted in its range on its own continent, was remarkable, especially in view of the characters employed. But a careful myological study of the head and shoulder region (Adams, 1942) has corroborated Dunn's decision. Here the relationship between two geographically remote species was clearly established on the basis of the remarkable similarity of internal characters.

Although the primary concern of taxonomists is the study of relationships this paper deals with divergence, their second major concern. It deals especially with the patterns by which species or groups diverge from each other and the degrees of divergence involved. Divergence is the result of differential modification of the various components of a taxonomic group. These modifications can better be understood not as an adaptive feature of organisms but as the toleration of an environment for particular forms. Individuals of a population vary. Basically this variation is dependent upon mutations controlling the development of the organism. These mutations and the rate at which they occur ordinarily seem to be under the influence of internal and environmental factors different from those which more directly affect organisms. But the effect of the mutation rate, regardless of its origin, on the rate of morphological change, seems to be of degree, not kind. Through various climatological phenomena the environment is also subject to change, and at various rates. Each step in phylogeny consists of a shift of an organism toward an equilibrium with a changing environment.

This simple concept of an approach to-ward equilibrium may be illustrated diagrammatically (Fig. 1). In this diagram successive environments are represented by grids, the perforations of which progress from truncated triangles to equilateral triangles. Each grid represents a biological barrier through which only certain organisms can successfully pass. Below each grid lies a population of variable organisms. Certain members of the population are tolerated, or selected, by a particular environment, passing through the grid to the level above. Here variation again becomes apparent; the new variants, however, are derived from the mutant forms which were screened by the environment from the earlier populations. This diagram represents the changes which may occur in a single population exposed to a changing environment. Inasmuch as a species or larger group is

Fig. 1

FIG. 1. The evolution in a changing environment of species "square" into species "triangle." The grids represent the selective action of three successive environmental complexes.

recognized by a complex of characters which define its form, the change from one type to another simply represents the collective change of a large number of the individual characters of the complex which defines the species or group. But the phenomena associated with the changes in a complex of characters are the same as those concerned with the modification of a single character. Hereafter, unit characters alone will be considered. The sequence depicted in Figure 1 then can be taken to represent a series of changes involving but a single character instead of a race. This is more simply diagrammed in Figure 2, where a character, square, is modified in time to a different character, triangle, through the sequence A to E.

Such a sequence, known as a chrono-cline, exhibits polarity. That extreme which lies at a higher level in time is derived from forms lying at lower levels and is here called the *derived extreme*. In the particular cline under consideration an arbitrary starting point was selected. The cline might continue back into time, but for the study of any one character, there is usually a fairly well defined starting point previous to which the character was relatively stable for a long period. This starting point in any particular chrono-cline will be designated as the *primitive extreme*. Between the primitive extreme and the derived extreme there will of necessity be a large number of intermediate forms, the number depending upon the temporal duration of the cline, the refinement of the methods of mensuration, and the number of actual mutations involved in the evolution of the cline. These intermediate stages will be called *intervenients*. By the definitions employed in describing this chrono-cline, and by the nature of its inception, it can be seen that the morphological changes involved proceed in one direction only, from the primitive extreme to the derived extreme. Furthermore, if any two of the intervenients or extremes are available, the direction of the cline can be determined

immediately, for each intervenient in a chrono-cline must precede the derived extreme.

Principle of Identity of Chrono- and Morpho-clines

In the above description of a chrono-cline it has been assumed that the race in which the cline has developed cannot escape from a changing environment. Actually, however, escape is often a possible, if not a frequent occurrence. When the environment changes in one geographic area, that of adjacent areas may remain unchanged. Sometimes the environment of a particular area seems to shift in time to a different geographic area without great alteration. Presumably, if an organism remains within this environmental complex it will not change morphologically, thus preserving primitive characters for long periods of time. This concept is also represented in Figure 2. These escapes, which may occur at any level in time, are represented by the phyletic lines A–A', B–B', etc. If a series of temporal intervenients should escape in this fashion, then at any one level in time there would exist a series of populations varying morphologically from one extreme to another. This series forms an

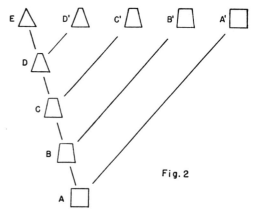

Fig. 2

FIG. 2. Illustration of the principle of identity. The chrono-cline *A–E* is represented at the present time by the morpho-cline *A'–E*, each member of which represents an escape from the orthogenetic chrono-cline.

isochronic cline as represented by the sequence *A'–E*. While escapes ordinarily involve the invasion of different geographical areas, it is possible that only part of the area inhabited by a particular organism might become subject to environmental change. In this event the population remaining in the unaltered area can be considered an "escape" relative to that of the altered. Clines have been designated as eco-clines or geo-clines (Huxley, 1939), but it is quite possible that allopatric forms arising in these ways might later become sympatric. To my knowledge no term has been coined to describe such morphological clines. I propose, therefore, the term morpho-cline to include not only eco-clines and geo-clines as defined by Huxley but also those discontinuous clines remaining should the populations attain complete specific status. As a taxonomist is confronted ordinarily with morpho-clines only, their interpretation is of vital importance. If each intervenient in time should persist unchanged as an intervenient in space at a subsequent time, an ideal situation would exist for determining the phylogeny of a particular population. In analyzing the relationships between races of organisms, the taxonomist assumes that *morpho-clines are partially or entirely identical to the chrono-clines from which they are derived.*

Before proceeding with an analysis of morpho-clines, it should be pointed out that one exception to the above criterion is of frequent occurrence. Sometimes a

morpho-cline represents two chronclines. This possibility is diagrammed in Figure 3. The derived extreme *A* of a chrono-cline *D–A* might fit into a morphocline designated by *W, X', Y', Z'*. The latter cline might represent in part a completely different chrono-cline *W–Z*. This demonstrates *convergence,* and shows that every morpho-cline is not necessarily identical with a single chrono-cline. Methods are developed later in this paper which will assist in recognizing such phenomena.

Principles of Divergence

The principal problem in the analysis of a morpho-cline, the determination of its polarity, scarcely exists in chronoclines. If the geological horizons have been carefully and accurately determined then the forms found in these horizons form a chronological sequence, and the polarity of any cline is apparent. But in morpho-clines of any particular horizon the polarity of such clines cannot so easily be determined. The fact that the derived extreme might be either a degenerate or a specialized condition further complicates the problem, and a worker familiar only with other groups, might erroneously conclude that a derived degenerate condition was primitive.

The examples given below indicate the methods currently employed by taxonomists in the determination of polarity in such morpho-clines.

In his monograph Blanchard (1942) recognized several adaptive characters in the snake genus *Diadophis*. These were: 1, small size; 2, slender bodies; 3, flattened heads; 4, semi-fossorial habits. He also found that large snakes tended to have more rows of body scales than did smaller snakes closely related to them. He concluded from these adaptive features that the most primitive member of the genus should exhibit these characters in the least modified form and should resemble the less specialized members of other

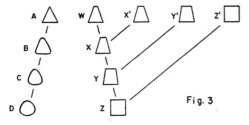

Fig. 3. Diagram to illustrate the origin of a morpho-cline by convergence. The morpho-cline *A–Z'* is derived from two separate chrono-clines, *D–A* and *Z–W*.

genera. The species *dugesii* fulfills these requirements because it has:

1. Large size.
2. The largest number of scale rows in the genus.
3. A robust body.
4. A temporal scale formula of 2-2-1 as opposed to 2-1-1 in the rest of the genus. This reflects the lack of head-flattening in this species.
5. A moderate number of ventral and subcaudal scales.
6. A lack of any extreme modifications.

This lack of divergence from the basal stock from which the genus must have been derived indicates that *D. dugesii* approaches the common condition from which divergence in different genera has occurred. Hence *D. dugesii* is primitive.

Dunn (1917) used similar reasoning in determining the phylogenetic position of *D. quadramaculata* in the American pleth-odont salamander genus *Desmognathus*. He suggests that it is the most primitive member of the genus because it most closely resembles species of other plethodont genera in larval size, retention of vomerine teeth in the male, and a lack of sexual dimorphism in the hooking of the jaws. It is also the largest member of the genus and in external characters closely resembles *Leurognathus marmorata*, a monotypic species in the genus most similar to *Desmognathus*. He concluded from this lack of divergence from a generalized plethodont stock that this species is primitive.

The same logic was employed by Clark (1944) in his discussion of the phylogeny of snakes on the basis of hemipenial structure. The ontogeny of the hemipenes gives no clue to the manner in which they might have evolved. Clark was faced with the problem of determining which hemipenial type constituted the primitive condition. Camp (1923) had suggested, on other anatomical grounds, that snakes probably arose from anguimorph lizards of the platynotid stock. The modern varanid lizards are also members of this stock

and in them the hemipenes are ornamented with flounces and have a bifurcate sulcus spermaticus. Clark therefore suggests that similar hemipenes in snakes represent the primitive condition.

The principle involved in these examples is the same, namely, that members of a particular group have diverged orthogenetically, from some ancestral stock and that the primitive extremes of the morpho-clines concerned are to be sought in the less divergent, or more generalized, representatives of related groups. Figure 2 represents this principle diagrammatically. Assume that E, D', C', and B' are the intervenients and extremes of a morpho-cline found in as many species. The character A' resembles B', but occurs, let us say, in a species which on other anatomical grounds must be included in a different genus. Our problem is the determination of the primitive extreme of the morpho-cline E–B'. If this cline represents a series of escapes in time, as has been suggested above, then the morpho-cline has its homologue in a chrono-cline. This chrono-cline is represented in the diagram by the sequence B–E. Since B represents the character before it was modified, it should resemble that found in contemporary species closely related to it. Each species in turn may give rise to divergent groups. A represents the character as it is found in one of these species. It occurs later in time unchanged as A' in a species of a different genus. This genus, of course, might include other more modified species not shown in the diagram, but the resemblance of the character A' in this genus to the character B' in the first genus strongly suggests that these conditions represent the character in its least modified state. It is highly probable that B' represents the primitive extreme of the morpho-cline B'–E. It can therefore be said that *if one extreme of a morpho-cline resembles a condition found in the less modified members of related groups of the same rank, this extreme is primitive.*

A variation of this principle is employed by some workers in which two or more morpho-clines are considered simultaneously. Dunn's work (1917) on the phylogeny of the American plethodont salamander genus *Desmognathus* again furnishes an example. On the basis of several morpho-clines he has arranged the species and subspecies of the genus in the following manner.

quadramaculata—monticola—fusca fusca

D. quadramaculata is considered primitive for the reasons stated above. Starting with this primitive member, several morpho-clines can be traced through these species and races. But beyond *D. fusca fusca* two sets of divergent morpho-clines become apparent, ending in their extreme conditions in *D. brimleyorum* and *D. o. ochrophoea*, respectively. In the *D. fusca fusca—D. brimleyorum* line:

1. The lateral areas assume importance in the development of color pattern ending in lateral light spots.
2. The parasphenoid teeth patches become progressively shorter.
3. The body becomes elongate and slender.
4. The limbs become weak.

In the *D. fusca fusca—D. ochrophoea ochrophoea* line:

1. The dorsal areas assume importance in the development of color pattern ending in longitudinal stripes.
2. The males develop a hooked mandible.
3. The body becomes more robust and shorter.
4. The legs become stronger and better adapted to terrestrialism.

D. fusca fusca is considered the most primitive of the five species and races concerned because it exhibits the extremes of both series of divergent morpho-clines.

Stull (1940) has also made use of this device in her study of the American colubrid snake genus *Pituophis*. She found a

morpho-cline involving an increase in the number of dorsal spots in the three recognized subspecies of a western form, *Pituophis catenifer catenifer*, *P. c. deserticola* and *P. c. anectens*. On this basis the three forms could be arranged in the sequence: *deserticola—catenifer—anectens*. But another morpho-cline involving decrease in the number of ventral scales, also occurs within this complex. On the basis of this

/ *f. auriculata—brimleyorum*
\ *ochrophoea carolinensis—o. ochrophoea*

morpho-cline the races should be arranged in a different sequence: *deserticola—anectens—catenifer*. As only *anectens* exhibits an extreme of both clines, she assumes that *anectens* is primitive and that the extremes of both clines, as exhibited in *anectens*, are the primitive extremes of the morpho-clines involved.

In these instances two or more morpho-clines appear in different but closely related groups of species. The problem is again the determination of the polarity of these morpho-clines. In each example one species exhibits an extreme of both morpho-clines. This species is considered an escape from an ancestral type which subsequently gave rise to two or more divergent chrono-clines which are represented today by different morpho-clines. The extremes of these morpho-clines which are present in a single species should be considered primitive. Although only relatively small taxonomic categories are involved in these illustrations, the same principle could be applied to larger categories. This principle might be expressed as a criterion: *If two or more morpho-clines appear in different groups of organisms, and an extreme of each morpho-cline occurs in the same taxonomic unit, then these extremes are the primitive extremes.*

A similar method may be employed in determining simultaneously the polarity of two morpho-clines involving the divergent modification of the same character. Blanchard (1942) found four non-integrating groups in the American snake genus *Diadophis*. Two were western, one eastern, and one southern. The southern group, represented by a single species, *dugesii,* is the most isolated morphologically, yet the trends of modification (morpho-clines) in the two western groups approach the characters found in this southern species. Different trends involving the same or different characters in the eastern group also have extremes which approach conditions found in *D. dugesii*. Some trends in the western *D. regalis* and *D. amabilis* groups are:

1. Increase in ventral scale count.
2. Increase in length and relative slenderness.
3. Recession of dark color from lower rows of dorsal scales.
4. Widening of neck ring.

Some trends in the eastern *D. punctatus* group are:

1. Decrease in ventral scale count.
2. Decrease in body length accompanied by greater robustness.
3. Reduction of size of posterior maxillary teeth.
4. Reduction of extension of dark color of head around corner of mouth.

The first two trends of each group involve the same characters; the second two involve different characters which remain unmodified in the other. These latter characters (many others are described by Blanchard) may be used in applying the second principle of divergence. But the first two characters involve morpho-clines which have different extremes in the most modified members of the various groups but tend to approach each other in the less modified members, and finally are identical in the southern form, *D. dugesii*. Blanchard concluded that *D. dugesii* is the most primitive member of the genus and the extremes of the various morpho-clines found in this species are the primitive extremes.

These examples are represented diagrammatically in Figure 4. The primitive extreme in the chrono-clines involved is *A'W'*. Two clines diverge from this common starting point ending in the derived extremes *D* and *Z* which represent dissimilar characters. If a number of intervenients of both clines persists, a double morpho-cline will be formed, represented by the line *D–Z*. In practice, analysis of such a cline will reveal that it can best be described by recognizing three extremes. By using *AW* as one, there will be two clines starting from the same point. One proceeds through a series of intervenients to the extreme *D*, the other through a different series to *Z*. If it is assumed that a morpho-cline is wholly or partially identical to a chrono-cline, the polarity of each of the clines can be determined. It can be assumed that the starting point *AW* is the primitive extreme of two morpho-clines and that *D* and *Z* are the respective derived extremes. A criterion may be formulated to embody this principle of divergence: *If two extremes of two clines are identical and are found in the same taxonomic unit, the identical extremes are primitive and the dissimilar extremes are derived.*

There is an alternative interpretation of such a double cline which is not merely an academic possibility but often presents difficulty in taxonomic studies. If only a single pair of clinous characters is considered, the spatial dispersion of the available intervenients could be explained by

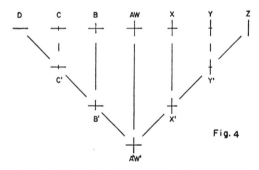

FIG. 4. The third principle of divergence. A character *A'W'* evolves in two directions forming the two chrono-clines *A'W'–Z* and *A'W'–D*. Escapes from these two chrono-clines form the double morpho-cline *D–AW–Z*.

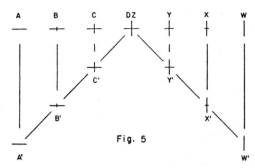

FIG. 5. Diagram to illustrate the possible origin of a double morpho-cline by convergence.

the principle of convergence, as Figure 5 depicts. Two chrono-clines might exist, consisting of two dissimilar primitive extremes, A' and W', which proceed in time through a series of intervenients to a similar endpoint DZ. These clines A'–DZ and W'–DZ might then be represented morphologically by the clines A–DZ and W–DZ. The formation of the double cline A–DZ–W could then be accounted for by two diametrically opposite phenomena, namely, convergence and divergence. Practical experience shows that convergence is relatively rare. Nevertheless, it must always be considered in the analysis of such clines. Criteria which may assist in distinguishing between convergence and divergence are developed later in this paper.

Principle of Precurrence

It has been suggested above that during the development of an orthogenetic series, members may escape at any time, remaining in that environment to which they are adapted. It is these members which, preserving their characters unchanged, constitute the intervenients of morpho-clines. If, however, a temporal intervenient of a phyletic series successfully makes an escape and diverges sufficiently to be specifically isolated, the environment with which it has established an equilibrium may then shift or alter. If the environmental change is not too rapid, selection of mutants within this iso-

lated species may establish a new phyletic series. This series with its galaxy of clines may follow the succession of changes characteristic of the original series or it may take an entirely new direction. The starting point of the new series is then an intervenient of the old one. This process may be repeated, and some part of each old series would be a precursor to the new. Or, to state this differently, a new series with its new pattern of changes arises from a precurrent series.

Noble (1922) in his remarkable studies on the phylogeny of the Salientia has employed this principle. He developed a system of classification which emphasized vertebral structure and the number and arrangement of the thigh muscles. This classification cut across the major groupings of some earlier systems which had placed primary emphasis on the presence or absence of teeth. Noble interpreted the loss of teeth as a secondary adaptation which had occurred independently as a terminal specialization in several different groups. He took exception to Boulenger's opinion (1919) that several toothless genera of the family Pelobatidae are primitive and might have led towards those more advanced families, Cystignathidae and Bufonidae. While the thigh musculature is highly modified in these latter families, the more generalized members have maxillary teeth. In the Pelobatidae the thigh musculature is generalized and all but three genera possess maxillary teeth. In reference to these toothless pelobatid genera, Noble (loc. cit. p. 60) states that "If (these) end states . . . . represent the terminations of a series of orthogenetic changes, they most certainly do not form the ancestral stock from which other groups possessing more primitive features have been derived." Noble has thus shown that the polarity of the morpho-cline involving the presence of teeth is toothed-to-toothless, and that any forms evolving from a toothless progenitor would of necessity also be toothless or if teeth were present they would be pseudo-teeth (a phenomenon that actually occurs in the lower jaw of some Salientia).

Clark (1944) used similar reasoning in his studies of the hemipenes of snakes. He showed that the spinous ornamentation present in many forms is a derived condition. The initial step in the phylogenetic sequence must have been the appearance of spines at the proximal end of the organ, followed by an invasion of the distal portion of the organ which is ornamented with the more primitive calyces or flounces. The extreme of this gradient is the complete replacement of the calyces, as occurs in the Natricinae. Clark points out that without a reversal of the evolutionary process, such a group could not give rise to forms with the calyculate hemipenes of other subfamilies of the Colubridae such as the Colubrinae or Xenodontinae.

Both of these examples use the principle of precurrence in a negative way. The direction of change in a morpho-cline having been established by other methods, it was then concluded that more primitive forms could not evolve from the derived extremes of these morpho-clines. The principle has not been used to determine the polarity of a morpho-cline itself. In Asiatic crotalid snakes of the genus *Trimeresurus* there is a morpho-cline involving the fragmentation of the head scales. As most families and genera of snakes have a crown pattern composed of large, symmetrically arranged plates, any condition in *Trimeresurus* approaching this pattern must be considered primitive on the basis of my first principle of divergence. Among forms in which the platelets are numerous, another morpho-cline is apparent, which involves the conversion of irregular platelets into symmetrical, imbricated scales. And again in those forms in which the fragmented platelets are scale-like, there is a third morpho-cline. Posteriorly the scales become carinated and in extreme instances all of the scale-like platelets are so keeled. It appears to me that those forms which have keeled head-scales are more modified than

those forms which have smooth head-scales only, because keeling could not invade this area unless there were symmetrical scales present which could become keeled. Again, if platelets or scales were present in different species, the former condition is the more primitive because they are essentially normal plates but smaller. In this example one morpho-cline must be in existence before another can appear, and if the polarity of any of these clines can be established, the polarity of all of them is evident. Such a succession has been represented in Figure 6 as a series of chrono-clines rather than as a phyletic series of species. If each temporal intervenient of the three clines depicted was represented by an escape, these escapes would constitute the intervenients and extremes of comparable

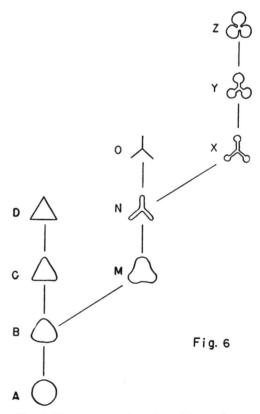

Fig. 6

FIG. 6. Precurrence. A series of three chrono-clines *ABCD*, *BMNO*, and *NXYZ* is shown. The second of these could not appear before the first, nor the third before the second.

morpho-clines, the counterparts of chrono-clines *ABCD, BMNO,* and *NXYZ.* In interlocking morpho-clines it is possible to determine the primitive and derived extremes, because two of these clines are successively dependent on precurrents of the same series for their initiation. This line of reasoning may be summarized as a criterion: *If an intervenient of any cline is identical to an extreme of another, then the first cline is primitive and the intervenient which forms an extreme of the second cline is its primitive extreme.*

Principle of Relicts and Specialists

It frequently happens in the study of groups of organisms that one member is conspicuously different from others of its group in one or more characters. This phenomenon may be due to the lack of intervenients in a morpho-cline, a situation which might arise in various ways. Temporal intervenients leading to the derived extreme might have formed an orthogenetic line without escapes occurring, or escapes might have occurred but become extinct. Either precludes the subsequent formation of a morpho-cline. This situation is depicted in Figure 7. The chrono-cline *A–E'* is incompletely represented by the morpho-cline *A'–E'.* The intervenients *C* and *D* which might have given rise to intervenients in this morpho-cline presumably have failed to do so, leaving a character *E'* isolated from the group of characters *A'* and *B'.* Here the isolated extreme *E'* is the derived extreme and is the most highly modified. For this reason it is said to be *specialized* and the form in which it occurs is a *specialist.* But in a morpho-cline lacking a number of intervenients, the polarity of the cline cannot always be determined. For example, the isolated member might represent an escape which retains its primitive nature as is depicted in Figure 8. Such a character is *primitive* and the form possessing such a character is called a *relict.*

Relicts and specialists are of frequent occurrence in various taxonomic categories, but most taxonomists have little difficulty with them. The problems involved in their differentiation are akin to those of determining the polarity of divergent morpho-clines. For example, Bogert (1947) in naming a geographically remote species of the iguanid lizard genus *Uma* was struck by the number of differences between this form and the three other species of the genus. The question arose as to whether this isolated form, *U. exsul,* was primitive or specialized. Examination of a related iguanid genus, *Callisaurus,* revealed a number of the peculiar characters of *U. exsul,* in particular, the unusually long hind legs, the abnormally long tail, the peculiar color pattern quite alien to the genus *Uma,* and details of plantar scutelation. On the basis of these resemblances *U. exsul* was considered primitive, a relict which is less specialized for the sand habitat than are the other species of *Uma.*

Bogert and Matalas (1945) use this argument in their discussion of the taxonomic position of the elapid snake genus *Ultrocalamus.* They point out that this genus is closely related to three other New Guinea genera and one New Zealand genus and that possibly the five genera should be reduced to one. All show peculiarities in their maxillary dentition which set them distinctly apart from other elapids. The teeth posterior to the fangs are numerous, form a gradient with the fangs, and are not segregated from the fangs by a diastema. As this condition is unique in the Elapidae but approaches the tooth arrangement in the Colubridae, Bogert suggests that this group of genera represents a primitive nucleus within the Elapidae and hence is relict.

Stebbins and Lowe (1949) have also used this principle in their discussion of the phylogenetic position of the monotypic salamander genus *Plethopsis.* On the basis of a number of morphological characters they conclude that *P. wrighti* should be allocated to the genus *Batrachoseps,* which differs from most plethodont genera in having the paired premaxillary bones fused into a single element as well as a high anucleate erythrocyte count (90%). In these two characters *B. wrighti* does not agree and hence resembles the presumably more primitive condition of the closely related genus *Plethodon*. On this ground, and on the basis of other less isolated morphological characters, as well as the geographic position of *B. wrighti*, it is assumed to be primitive and represents a relict species in the genus *Batrachoseps*.

In these examples certain basic assumptions have been made. These may be illustrated by referring again to Figure 8, where *A* is the primitive extreme of a chrono-cline which gives rise to a group of extremes and intervenients *A′*, *D′*, and *E′* of a morpho-cline. These forms are all

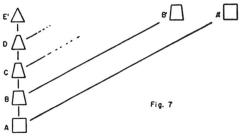

Fig. 7

FIG. 7. Origin of a specialist. The morpho-cline *E′*–*A′* is incomplete; the character *E′* is isolated morphologically from the forms *B′* and *A′*. The character *E′* is specialized and the form bearing it is a specialist.

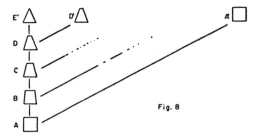

Fig. 8

FIG. 8. Origin of a relict. The chrono-cline *A*–*E′* gives rise by escapes to the imperfect morpho-cline *A′*–*E′* from which certain intervenients are lacking, thus isolating the primitive extreme *A′* from the more specialized characters *D′* and *E′*. The form bearing *A′* is a relict.

derivatives of one primitive character. Prior to the origin of this particular sequence of changes it can be assumed that the character *A* had been relatively static. If this were not so then the cline would be only a partial cline and could be extended back in time to its real or hypothetical primitive extreme. This new point would represent a static condition of character *A*. But, by invoking the principle of similarity, it can be seen that the character *A* at this time level would perforce be shared in an identical, or only slightly modified, form with sister groups of organisms of the same taxonomic rank. Then the species arising from these related forms would also show the same character *A*. Or, if these adjacent forms gave rise to phyletic series in which the character *A* itself was involved, there would be a series of morpho-clines, all of which would have character *A* as a primitive extreme in several different taxonomic groups. In any event, related groups would exhibit the same character or its derivatives at all subsequent time levels, if the *A′* of Figure 8 represents a relict character. This line of reasoning may be formulated into a criterion: *If a morphologically isolated character which occurs in one member of a group of related forms also appears in the primitive members of closely related groups of comparable rank, then the character under consideration is primitive and the form possessing it is a relict.*

A similar line of reasoning has been employed by several authors to determine whether an isolated morphological character constitutes a specialization. Bailey (1928), for example, in his monograph of the American lizard genus *Ctenosaura* found it unnecessary to compare *C. bakeri* with any other members of the genus except *C. palearis* because these two species are morphologically isolated. Bailey points out that Stejneger in naming the species *C. palearis* in 1899 was inclined toward erecting a separate genus for it because of its tremendously developed gular crest or dewlap. He refrained from so doing and two years later commented on his wisdom in the same paper in which he described *C. bakeri,* a species somewhat anectant between *C. palearis* and the rest of the genus. Bailey furthermore points out that while a dewlap does occur sporadically in other genera, it is not a generalized iguanid character, and should be considered a specialization.

Smith (1942) has made a similar argument in his discussion of the Central American snake genus *Adelphicos,* in which a number of morphological gradients exist with their extremes in the same or closely related species. One character concerns the size of the chin-shields and the third infra-orbital scale. In several species the chin-shields are so tremendously enlarged that the third infralabials are strikingly reduced in size. As this condition departs radically from that found in other genera of the same family, he concludes that the character is specialized and that in this genus the smaller the chin-shields, the more primitive the species.

Clark (1944) points out that the hemipenes of snakes of the subspecies *Coluber constrictor constrictor* lack spines which those of all other subspecies of *C. constrictor* possess. In general a lack of spines is considered a primitive character and one might consider *C. c. constrictor* a relict, but Clark argues that it is a specialization representing a secondary loss, because spines normally occur not only in other

races of this species but in the remaining species of the genus as well. He implies that if it were a primitive condition in the species *C. constrictor* one should expect to find a spineless condition in related taxonomic groups of comparable rank. But, as it does not occur in the other species of *Coluber,* the character must have evolved within the species *C. constrictor* alone. It is, therefore, a derived and specialized condition peculiar to this small taxonomic group.

The reasoning which has led these authors to conclude that they were dealing with specializations can be followed with a hypothetical genealogical tree (Fig. 9). In this diagram the Arabic numerals represent species, the capital letters genera, and the Roman numerals families. In genera *A, B,* and *C* of family *I* a morphologically isolated character appears in the three species represented by the figures in bold type. The character appears in family *I* only and not in family *II* or family *III* which are indicated as being related groups. This character, having evolved within the fabric of family *I,* is specialized. If it were primitive, a relict character, the same character or modification of it likely would occur in families *II* and *III* as well. However, it is specialized in one sense only; it is specialized with reference to its occurrence within the family as a unit and with reference to its absence in the two sister units of comparable rank. The sporadic appearance of the character in other groups need not negate this conclusion. Sporadic occurrences may represent parallelisms, a phenomenon discussed in the next section, especially if they occurred in species or groups which, on the basis of other criteria, can be shown to be derived rather than primitive. This principle of specialization can be stated as a criterion: *If a morphologically isolated character is unique or appears only sporadically in the derived forms of related groups of comparable rank, then the character is specialized, and the form possessing the character is a specialist.*

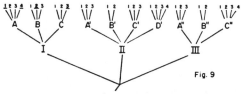

Fig. 9

FIG. 9. The occurrence of specialists. A hypothetical phylogenetic tree consisting of three related families *I, II,* and *III;* a number of genera represented by capital letters, and a series of species represented by Arabic numerals. In family *I* each genus has one or more members showing the same character. This character appears in no related families. Hence the family is specialized in reference to this character.

Principles of Parallelism

The appearance of identical or nearly identical characters in related groups of organisms constitutes the phenomenon of parallelism. Inasmuch as parallelism may be used to determine the polarity of morpho-clines and to establish phyletic relationships, a discussion of its nature seems essential.

It is almost universally accepted that new species arise from pre-existing species. Whether this origin is through macromutation, isolation and subsequent gene drift in extremely small populations, hybridization, the escape of subspecies, or other modes is immaterial to the arguments presented below. The newly derived species will be extremely closely related to the initial species, and, as was pointed out in the discussion of the principle of similarity, this relationship is based upon a high percentage of genetic factors common to both. Sturtevant and Novitski (1941) have compared the chromosomal elements of various species of *Drosophila*. This work and a number of earlier studies is based on the occurrence of selected mutant genes in different species. Homologies between chromosomes have been established through the identity of these mutants. As Sturtevant and Novitski and later Sturtevant (1948) point out, only a relatively few mutant characters were involved, but the implications are that genes for which mutants are not known likewise have homologues in related species. There is a high probability that similar genes in related species mutate in a similar fashion, and shape the development of similar phenotypes. If, now, two species are exposed to nearly identical environments, it is not illogical to assume that identical mutants will be selected over a period of time, and that this selection will result in an identical pattern of phenotypic modification. If this sequence of changes can take place within two closely related species it could occur in a phyletic series of species much less closely related. If inter-venients are selected from two such phyletic lines, it is apparent that two clines affecting the same character could be identified in two different phyletic series. Admittedly, such clines might be modified by adjacent, divergent characters or by the secondary influence of genetic factors not directly associated with the development of a particular character. For this reason two clines probably would never be strictly identical, but would resemble each other sufficiently to be recognized as the phenotypic expressions of similar genetic complexes. In essence, however, the clines would be the same. The importance of this similarity lies in the relative positions of the extremes. Whenever two clines evolve in a similar fashion, the polarity of the clines will be similar. Blanchard (1923) has made use of this principle in establishing the phylogenetic relationship between the two species of the American snake genus *Virginia*. He first demonstrated that the two forms intergrade, constituting a single species, and that the two subspecies *V. valeriae valeriae* and *V. v. elegans* differ from each other in clinous characters.

1. *V. v. valeriae* has fewer subcaudal and ventral scales sex for sex than *V. v. elegans*.
2. *V. v. valeriae* is smaller than *V. v. elegans*.
3. *V. v. valeriae* has 15 rows of scales, *V. v. elegans* has 17.
4. *V. v. valeriae* has hardly a trace of keeling in its dorsal scales while *V. v. elegans* has an occasional keeled scale.

Similar trends and characters are known in other related genera in which larger numbers of species are involved. In these clines the primitive and derived extremes are known. Blanchard concluded that the direction of change in *Virginia* is the same as in these known clines and that while both subspecies are in the process of degenerative change from terrestrial to fossorial types, *V. v. elegans* is the more primitive and *V. v. valeriae* the more degenerate; the latter is, in this case, derived.

Smith and Laufe (1945) employed the same reasoning in their study of relationships within the Mexican snake genera *Toluca* and *Conopsis*. Of the seven species which occur in *Toluca*, one, *T. amphisticha*, lacks a dorso-median row of spots. It was clearly demonstrated on the basis of geographic distribution and other clinous characters that this pattern is a derived condition and that the presence of a well-developed row of spots, as occurs in *T. conica* and *T. megalodon*, is primitive. There is, however, a well-defined morpho-cline with these forms representing the extremes. Smith and Laufe point out that in the closely related, and possibly congeneric, *Conopsis*, one of the two species, *C. nasus,* possesses a large median row of dorsal blotches while the other species, *C. biseriatus,* lacks it. By comparison with the trend in the genus *Toluca*, which was easier to analyse because of the larger number of species, they conclude that the direction of change in *Conopsis* is the same as that in *Toluca*, and that *C. nasus* is primitive and *C. biseriatus* derived.

This principle involving the direction of change in two parallel morpho-clines may be stated as a criterion: *If the same morpho-cline appears in two related groups the direction of change of the clines in both groups is identical.*

The principle of parallelism has been used repeatedly, but a number of taxonomists have been forced to go a step further to account for the appearance of similar characters in distantly related forms. A few of the more recent studies of this nature might be cited. Michener (1949) in his study of parallel evolution in the saturnid moths points out the high frequency of identical derived extremes, many of which appear in subfamilies widely separated geographically. Wood (1947), after many years of work on the paleontology of rodents, has concluded that identical clines appear in the higher ranks of rodents, and that these clines frequently appear at different times or in widely separated geographical areas. Later (1950), he develops this hypothesis in still greater detail to account for the distribution and evolution of modern porcupines. A. Williams indicated in an address delivered before the Society for the Study of Evolution in 1949 that a number of clines leading towards identical derived extremes appeared in strapheodontid brachiopods. The clines were frequently of different duration; the changes were accelerated in some lines and retarded in others; but the intervenients in the various lines occurred at relatively similar intervals. Williams made no attempt to account for this phenomenon but presented strong evidence for its occurrence. Current studies by many workers are disclosing more and more evidence to indicate that identical temporal clines have made their appearance in various groups and that frequently these identical clines have appeared in groups between which an exchange of genetic factors is no longer possible, because of either biological, geographical, or temporal isolation. This criterion may be formulated: *Related forms possess the potentiality for the development of identical clines, and under similar environmental conditions identical clines may develop at different times or in separate geographical areas.*

Paradromism and Multiprotoformity

Evolution usually involves the simultaneous modification of a number of characters. Although some characters may change more rapidly than others, over a long period the gross effect of evolution is a composite change. During such evolutionary changes various phyletic lines diverge. This divergence will perforce involve a segregation of the chrono-clines present. Some clines will pass imperceptibly from the prototype into the derived series where they remain identical by changing in the same direction, and result in identical or nearly identical derived extremes. Occasionally, following the initial divergence of two phyletic lines, one may encounter an environment which will not tolerate the expression of mutants necessary for the continued development of the cline. An intervenient of such a cline may remain static for long periods. Subsequently the environment might again shift, permitting the interrupted sequence of changes to continue. This sequence, appearing much later in time, will nevertheless be identical to the uninterrupted cline of the alternate phyletic series. It is considered possible, then, that complexes of clines persist through the divergence of newly evolved taxonomic categories and remain identical thereafter. Even if a cline should pause in its evolution and then continue subsequent to the isolation of the new group within which it occurs, its direction and identity will coincide with clines of similar characters in related groups. The divergence of phyletic series may be followed by the complete cessation of various clines in each series. In this event the prototypic series will share with each derived series a number of clines which are not mutually shared by the derived series. These clines are of problematic value in establishing taxonomic relationships because the missing clines in either series may subsequently reappear and resume their orderly phyletic changes, resulting in derived extremes identical to those which have already evolved in the divergent stocks.

New clines usually appear at the time of the divergence of phyletic series or soon after. These clines, in contrast to those with antecedents in the prototype, are of considerable taxonomic significance because they are of more certain diagnostic value. But the polarity of these clines which have no antecedents is often difficult to determine. This difficulty is especially acute in morpho-clines, if many of the primitive intervenients are lacking, because this leaves only an isolated fragment of a cline showing no affinity with characters or morpho-clines occurring in related groups. If the principles discussed under the heading of identical clines are valid, it follows that whenever a chronocline is initiated, its subsequent development will be simultaneous with that of unrelated clines in the same phyletic series. In other words, the new cline will parallel the pre-existing clines. Furthermore, the direction of change in two such paradromic clines will be the same.

Oliver (1948) in his study of the Neotropical tree snakes of the genus *Thalerophus* concluded from consideration of a number of characters which formed morpho-clines, that *T. depressirostris* was the most primitive member. This species possesses endpoints which most closely approach a generalized condition typical of several closely related genera. This is an example of the use of the first principle of divergence. But Oliver thereupon (p. 261) carefully states that "In so far as can be determined without detailed genetic analysis, most of the evolutionary trends indicated for the characters considered in the preceding section vary independently of the trend for any other character. Still, a number of these trends may be distributed geographically in a parallel manner and thus facilitate the determination of phylogenetic lines." While Oliver primarily used zoogeographic criteria in determining relationships he recognized the occurrence of paradromism and used it in determining phylogenies.

Smith (1942), however, makes a much more specific use of this principle in his discussion of the genus *Adelphicos* which has already furnished an illustration of the principle of specialization. In this Central American snake genus the chin-shields tend to become tremendously enlarged. Smith showed this to be a specialization, with large chin-shields representing the derived extreme of this morpho-cline. But Smith points out that there are within this genus other trends which involve:

1. A reduction in the size of the third infra-labial scale.

2. An increase in the number of pala-
tine and pterygoid teeth.

3. A reduction in the number of hemipenial spines.

4. A reduction in the number of ventral and subcaudal scales.

5. A loss of pigment on the belly.

6. A restriction of pigment of the subcaudal region to a midventral line.

7. A shortening and broadening of the
frontal plate of the crown.

8. A general decrease in size.

Most of these characters are unrelated
to the condition of the chin-shields and
yet all of these morpho-clines are paradromic to each other. Furthermore, they
are paradromic to the morpho-cline concerned with the enlargement of the chin-
shields. Smith concluded that the primitive extremes of these clines are those
extremes which appear in species with
unmodified chin-shields. He has, in effect,
determined the polarity of one morpho-
cline and then assumes that the direction
of change in all paradromic clines is the
same.

A striking example of this use of the
principle of paradromism can be found in
Dunn's monograph (1942) of the American coecilian amphibians. Here Dunn
(p. 440) categorically stated that "A primitive coecilian should, theoretically, have
the following characteristics." He then
lists 13 characters, but gives no specific
reasons here as to why he considers these
characters as primitive. But in the text
of his monograph it becomes apparent
that these characters form the endpoints
of morpho-clines. Furthermore, most of
the characters listed are similar to those
found in the more generalized Caudata;
for example, the presence of "a definite
tail," "teeth of any given tooth row uniform in size"; "eye well developed and in
an open orbit." He established the polarity of a number of morpho-clines by invoking the first principle of divergence.
The polarity of the remaining morpho-
clines was determined by the principle of
paradromism. One of the latter morpho-

clines is of particular interest. Superficial
annular grooves occur in coecilians, marking the boundaries of the body segments.
These grooves, called primaries, are comparable anatomically to the costal grooves
of salamanders. But in coecilians secondary grooves, never present in salamanders,
may also occur. These range in number
from none to as many secondaries as
there are primaries, forming a clear gradient. If recourse were made to the principle of divergence alone it might be concluded that a lack of secondaries was the
primitive extreme of this morpho-cline.
But Dunn points out that there is a full
complement of secondaries only in those
species in which also occur the primitive
extremes of the morpho-clines listed
above. He concluded that the primitive
extreme of this morpho-cline is the presence of a full complement of secondaries,
that lack of secondaries is derived.

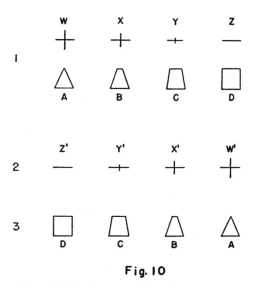

Fig. 10

FIG. 10. Paradromism. Series 1 depicts four
taxonomic categories arranged according to
the known morpho-cline $WXYZ$. Paradromic
to this morpho-cline is an unknown cline
$ABCD$. Series 2 represents the correct arrangement of the morphocline from the primitive extreme, Z', to the derived extreme, W',
as it is found in a related group. Series 3
then represents the correct arrangement of
the morphocline found in series 1; D is primitive and A derived.

These examples fit a common pattern which may be diagrammed. In Figure 10, series 1, there are two paradromic morpho-clines. Cline $ABCD$ is in this instance unfamiliar to the taxonomist. Cline $WXYZ$ is a familiar cline which appears in several phyletic series. In series 2 it has been established from other criteria that the primitive extreme of the cline $W'X'Y'Z'$ is Z'. The direction of change should, therefore, be indicated by designating the cline as $Z'Y'X'W'$. But this cline is identical to the one appearing in series 1, hence the direction of change of the cline in series 1, according to the principle of identity, is also $ZYXW$, the primitive extreme being Z. According to the principle of paradromic clines the primitive extreme of cline $ABCD$ is D, not A, and the cline should be arranged as in series 3, extending from the primitive extreme D through various intervenients to the derived extreme A. This principle of paradromism may be formally stated: *If an unknown cline is paradromic to a known one the direction of change in the unknown cline is the same as that of the known.*

The simultaneous study of many more than two clines in a group of organisms is frequently possible. If the direction of change of one cline can be established, then, using the principles of paradromism and parallelism, the polarity of the others can be determined. Many clines will be incomplete and many others will be exhibited in but a few of the species, yet it will generally be true that the primitive extremes of the various clines will occur in the same species. In some clines the primitive extremes will simply be static phases in which the changes apparent in the more advanced members of the group have not begun. This principle is so frequently used that it practically amounts to an axiom. Bailey (1928) found that a number of paradromic morpho-clines occur in the American lizard genus *Ctenosaura*. The primitive extremes of all of these occur in *C. acanthura*. Then, partly on this basis and partly on its zoo-

geographic position, he concluded that it is the most primitive species in the genus. Smith (1942) suggested that in the Central American snake genus *Adelphicos*, *A. veraepacis veraepacis,* and other forms which similarly exhibit the greatest number of primitive extremes, should be considered the most primitive members. This principle may be stated in the form of a criterion: *If two or more clines occur in a number of forms, that form which has the greatest number of primitive extremes is the most primitive.*

Antipodal Characters

If it is valid to identify chrono- and morpho-clines, then a morpho-cline although represented only by its extremes may be treated as a cline. Hence, most of the criteria applied to clines apply equally to antipodal characters. The analysis of two pairs of antipodal characters is a common taxonomic problem. For example, if the intervenients of two paradromic clines are lacking and only galaxies of extremes are available, the forms are easily classified, but the interpretation of their phyletic relationships is difficult. In Figure 11 eight species are shown which exhibit the extremes and near-extreme intervenients of a morpho-cline $A-N$. These species may be arranged into two contrasting groups, $ABCD$ and $KLMN$, which most taxonomists would consider natural groups. The extremes of another cline, $Q-Z$, are also present (see Fig. 12). This cline is a familiar one, and has proved of value in determining generic relationships in related organisms. If the extremes of cline $A-N$ are used in an antipodal arrangement of the forms, the extremes of cline $Q-Z$ segregate, as shown in Figure 12, into two unnatural groups, $QRWX$ and $STYZ$. If, on the other hand, the latter cline is used in establishing relationship, an arrangement as depicted in Figure 13 is derived. But now the first cline is disrupted into the apparently unnatural groups $ABKL$ and $CDMN$. The only alternative arrangement is the erec-

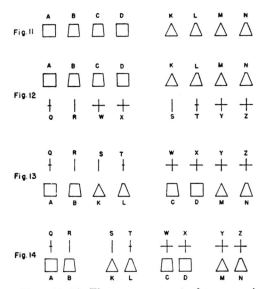

FIGS. 11–14. The arrangement of taxonomic categories on the basis of two sets of antipodal characters. Eight species exhibit portions of two paradromic clines, *A–N* and *Q–Z*. On the basis of the *A–N* cline alone the species fall into two natural groups (Fig. 11). With this arrangement, however, the grouping of the *Q–Z* characters is unnatural (Fig. 12). On the other hand the characters of the *A–N* cline fall into unnatural groups if the *Q–Z* characters are ordered (Fig. 13). All eight species will fall into four pairs in which the groupings of both sets of characters are natural (Fig. 14).

tion of four groups of equal rank in which the extremes of each cline are segregated in an orderly way (Fig. 14).

This complex could have arisen in two ways. The prototype of all four forms might have given rise to four separate phyletic series within each of which the two characters evolved independently but in a parallel fashion. Or a dichotomous divergence affecting only one character might have occurred, followed by another dichotomy within each phyletic line. Which chrono-cline appeared first, or did they arise simultaneously? Two antipodal characters alone do not provide an answer. But if additional paradromic clines are analyzed, it is often possible to ascertain the presence or absence of dichotomy and the chronological order in which the various chrono-clines arose. Thus did Noble

(1922) conclude that the loss of teeth in anuran amphibians was a derived character.

Multiple divergence of one prototype into a large series of phyletic lines occurs rarely, if at all, to judge from both zoogeographic and morphological criteria. What may appear as radial evolution may actually be a condensation of sequential dichotomies, trichotomies, or other simple divisions. Groups of independent chrono-clines are presumed to characterize the initial line. If only two groups of paradromic clines appear subsequently, the divergence must have been dichotomous. Even if these divergent paradromic fasces of chrono-clines are represented by a series of escapes, a dichotomy of groups of discontinuous paradromic morpho-clines would still be apparent. Should isolated but identical characters appear in these two divergent groups, they must represent parallelisms which originated subsequent to the initial dichotomy. The more frequent the appearance of such parallelisms the more difficult it is to decide whether the characters are derived or primitive. The difficulty is increased when the paradromic morpho-clines are poorly defined.

Occasionally phyletic relations can be determined for a complex of species in which the extremes of only two clines are available. This may be seen in Clark's analysis (1944) of two morpho-clines in the hemipenial structure of snakes. In one the primitive extreme is represented by an organ ornamented proximally with flounces. The derived condition is represented by an organ in which this ornamentation is replaced by spines. The second morpho-cline involves the conversion of a primitively forked sulcus spermaticus to an unforked one. Various combinations of these extremes occur in specialized families of snakes. The problem was to decide which cline appeared first, for then whichever group exhibited the primitive extreme of the earlier cline would be the most primitive. It happens that only three of the four possible combi-

79

nations of the antipodal extremes of these two clines are realized. These are spinous, unforked; spinous, forked; and flounced, forked. The flounced, unforked class is lacking. Clark concluded, then, that the earlier cline was that concerned with the replacement of flounces by spines. On this basis the Xenodontinae with forked hemipenes which are spinous at the base would be the most primitive subfamily of the Colubridae.

This reasoning may be represented diagrammatically. Figure 15 shows a complex of six species in which the extremes or near-extremes of two clines are distributed at random. The species can be arranged as in Figure 16, using the extremes of the morpho-cline "plus" to "bar" to segregate two groups of species. On the other hand, the extremes of the morpho-cline "square" to "triangle" can be used to segregate two groups, in which the extremes of the cline "plus" to "bar" are randomly distributed (Fig. 17). In either arrangement the groups are of unequal size, one group being 100 per cent larger than the other. This is a clue to the possible number of classes in the complex. It has been seen previously that in the consideration of the extremes of only two clines, four classes can be recognized (Fig. 14). Here, however, only three classes are present. This can be seen by arranging the extremes of the two clines in a checkerboard graph (Fig. 18). The class in which the extremes "triangle" and "bar" should be associated is missing. If, now, the direction of change in the two morpho-clines is known, it should be possible to construct by extrapolation a phyletic diagram of the group. The starting point will be, according to the principle of multiprotoformity, that species in which the primitive extremes of both clines are present (Fig. 19). Two lines diverged from this species, in one of which the chrono-cline "bar" to "plus" never expressed itself. In this line the existent species represent escapes which have remained in an environment with which they are in equilibrium with respect to

Fig.15 | □   +△ | □   +□   +△   +□
Fig.16 +□   +□   +△   +△ | □ | □
Fig.17 | □ | □   +□   +□   +△   +△

	□	△
		□	
+	+□	+△

Fig.18

Fig.19

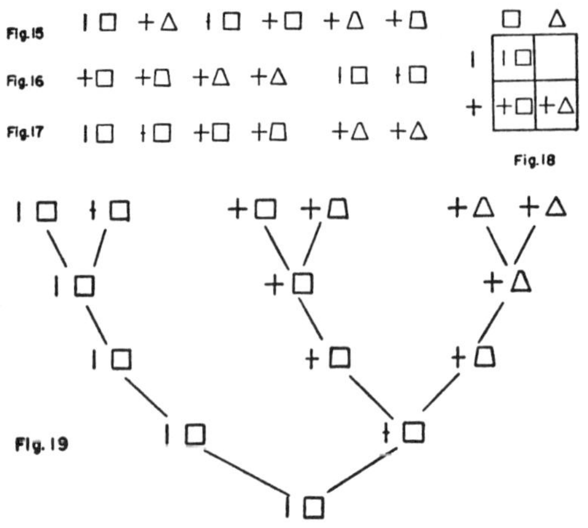

FIGS. 15–17. The arrangement of taxonomic categories on the basis of two sets of antipodal characters. A series of six species exhibits the apparently random associations of extremes and near-extreme intervenients of the clines "plus" to "bar" and "square" to "triangle" (Fig. 15). Whether arranged on the basis of the "plus" to "bar" cline (Fig. 16) or "square" to "triangle" (Fig. 17) the species fall into two groupings of unequal size.

FIG. 18. Checkerboard graph of the classes found in the series of forms depicted in Fig. 15 which shows that one class, bar-triangle, is missing.

FIG. 19. A possible phylogeny for the species of Fig. 17.

both the "bar" and "square" clines. The other series is immediately exposed to an environment which will not tolerate the expression of the character "bar" and it becomes modified towards "plus." Thereafter all descendants of this line will show varying degrees of modification of this character and will form a natural group. But environmental conditions are soon encountered which will not tolerate the primitive condition "square." There is a second divergence with one line escaping into an environment which will tolerate "square" but not "bar." The other line remains exposed to the selective action of the altered environment in which both "square" and "bar" are intolerable, resulting finally in a class having the derived extremes, "plus" and "triangle." No other phyletic diagram can be constructed from the data unless it is assumed that the class "bar"-"triangle" did evolve but subsequently became extinct. The probability is good that this analysis is correct, but this kind of analysis is possible only when the missing class possesses the derived extreme of one cline and the primitive extreme of the other. This type of analysis is only rarely applicable. Furthermore, the same conclusion can often be reached by using the principle of paradromic clines and analyzing morpho-clines in which adequate intervenients are present.

Convergence was mentioned earlier but without any solution of the problems involved. The term convergence as used in this paper refers to the appearance in two or more taxonomic categories of identical or nearly identical characters which are derived phylogenetically from different prototypes. Convergent characters, no matter how dramatic and conspicuous, can be ignored if they appear in relatively few members of two taxonomic groups. Relationships can be worked out on the basis of paradromic characters or morpho-clines. The same procedure may be employed where convergence affects only one organ. But occasionally two different species converge in so many characters that their differences long go unrecognized. This is especially true if the convergent forms are phylogenetically isolated, but even here analysis of less superficial characters usually reveals their true identity. The lesson to be drawn from this discussion of convergence is never to rely on too few characters in working out relationships.

Although morpho-clines were treated as characters in the discussion of the principle of precurrence, they are usually of little taxonomic value because they frequently occur in only a few members of a group. A morpho-cline listed as a diagnostic feature of a group would hardly assist in determining the taxonomic position of a new form which did not exhibit any intervenient of this particular morpho-cline. But occasionally morpho-clines are of real value when used as characters. An expression comparable to the following frequently appears in the literature: "Group *A* in contrast to group *B* has a tendency or trend towards the development of character *X*." Here the principle of heterochronic parallelism is subconsciously invoked. Character *X* does not appear in every form but the potentiality for its development is assumed to be inherent in all members of the group. The character used is not the presence of a particular structure but the presence of certain genetic factors which are capable under the proper circumstances of producing a particular sort of phenotype.

Morpho-clines could be more frequently employed with profit in the description of various groups. Trends are often reluctantly omitted from the diagnosis of a taxonomic category because manifestations of these trends do not occur in all members of the group. But a series of overlapping clines could give as accurate a diagnostic description of a group as a diagnosis relying only upon those characters and intervenients of clines which appear in every member of a group.

The problem of allocating annectent groups or forms to their proper taxonomic categories is frequently encountered. Annectent forms are almost invariably primitive in a number of characters. Furthermore, these characters frequently constitute the primitive extremes of clines which do not run through all members of either of the groups to which the annectent might be allocated. If clines are used in the description of these groups the position of the annectent form can often be determined. It is possible that an annectent form might not typically display any of the diagnostic characters used in the formal description of a group, yet its position could be definitely established by the use of clines.

Summary

Some of the principles employed in determining phylogenetic relationships from morphological evidence have been described in this paper, and formulated into criteria. In summary these are:

1. *Principle of similarity.* Similar organisms are related.
2. *Principle of identity.* Morpho-clines are partially or entirely identical to the chrono-clines from which they are derived.
3. *Principles of divergence.*
   a. If one extreme of a morpho-cline resembles a condition found in the less modified members of related groups of the same rank, this extreme is primitive.
   b. If two or more morpho-clines appear in different groups of organisms, and an extreme of each morpho-cline occurs in the same taxonomic unit; then these extremes are the primitive extremes.
   c. If two extremes of two clines are identical and are found in the same taxonomic unit, the identical extremes are primitive and the dissimilar extremes are derived.
4. *Principle of precurrence.* If an intervenient of any cline is identical to an extreme of another, then the first cline is primitive and the intervenient which forms an extreme of the second cline is its primitive extreme.
5. *Principle of relicts.* If a morphologically isolated character which occurs in one member of a group of related forms also appears in the primitive members of closely related groups of comparable rank, then the character under consideration is primitive and the form possessing it is a relict.
6. *Principle of specialists.* If a morphologically isolated character is unique or appears only sporadically in the derived forms of related groups of comparable rank, then the character is specialized and the form possessing the character is a specialist.
7. *Principle of parallelism.* If the same morpho-cline appears in two related

groups the direction of change of the clines in both groups is identical.
8. *Principle of heterochronic parallelism.* Related forms possess the potentiality for the development of identical clines, and under similar environmental conditions identical clines may develop at different times or in separate geographical areas.
9. *Principle of paradromism.* If an unknown cline is paradromic to a known one, the direction of change in the unknown cline is the same as that in the known.
10. *Principle of multiprotoformity.* If two or more clines occur in a number of forms, that form which has the greatest number of primitive extremes is the most primitive.

REFERENCES

ADAMS, LOWELL. 1942. The natural history and classification of the Mount Lyell salamander, *Hydromantes platycephalus. Univ. Calif. Publ. Zool.,* 46, 179-204.

BAILEY, JOHN W. 1928. A revision of the lizards of the genus *Ctenosaura. Proc. U. S. Nat. Mus.,* 73, 1-55.

BLANCHARD, FRANK N. 1923. The snakes of the genus *Virginia. Papers Michigan Acad. Sci., Arts & Let.,* 3, 343-365.

BLANCHARD, FRANK N. 1942. The ring-neck snakes, genus *Diadophis. Bull. Chicago Acad. Sci.,* 7, 1-144.

BOGERT, C. M., and B. L. MATALAS. 1945. Results of the Archbold expeditions. No. 53. A review of the elapid genus *Ultrocalamus* of New Guinea. *Am. Mus. Novit.,* No. 1284.

CLARK, HUGH. 1944. The anatomy and embryology of the hemipenes of *Lampropeltis, Diadophis* and *Thamnophis* and their value as criteria of relationship in the family Colubridae. *Iowa Acad. Sci.,* 51, 411-445.

DUNN, E. R. 1917. The salamanders of the genus *Desmognathus* and *Leurognathus. Proc. U. S. Nat. Mus.,* 53, 393-433.

DUNN, E. R. 1942. The American Coecilians. *Bull. Mus. Comp. Zool.,* 91, 439-540.

HUXLEY, J. S. 1939. Clines: an auxiliary method in taxonomy. *Bijdr. Dierk.,* 27, 491-520, 624, 626, 627.

MICHENER, C. D. 1949. Parallelisms in the evolution of the saturnid moths. *Evolution,* 3, 129-141.

NOBLE, G. K. 1922. The phylogeny of the Salientia. I. The osteology and the thigh musculature; their bearing on classification

and phylogeny. *Bull. Am. Mus. Nat. Hist.,*
46, 1-87.

OLIVER, JAMES A. 1948. The relationships and
zoogeography of the genus *Thalerophis*
Oliver. *Bull. Am. Mus. Nat. Hist., 92,* 157-
280.

SCHMIDT, K. P., and C. M. BOGERT. 1947. A new
fringe-footed sand lizard from Coahuila,
Mexico. *Am. Mus. Novit.,* No. 1339.

SMITH, HOBART M. 1936. The lizards of the
torquatus group of the genus *Sceloporus*
Wiegmann, 1828. *Univ. Kansas Sci. Bull.,*
24, 539-693.

SMITH, HOBART M. 1942. A review of the
snake genus *Adelphicos. Proc. Rochester
Acad. Sci., 8,* 175-195.

SMITH, HOBART M., and LEONARD E. LAUFE.
1945. Notes on a herpetological collection
from Oaxaca. *Herpetologica, 3,* 1-32.

STEBBINS, R. C., and C. H. LOWE. 1949. The
systematic status of *Plethopsis* with a discussion of speciation in the genus *Batrachoseps. Copeia, 1949,* 116-129.

STULL, OLIVE G. 1940. Variations and relationships in the snakes of the genus *Pituophis. Bull. U. S. Nat. Mus.,* No. 175,
225 pp.

STURTEVANT, A. H., and E. NOVITSKI. 1941. The
homologies of the chromosome elements in
the genus *Drosophila. Genetics, 26,* 517-
541.

STURTEVANT, A. H. 1948. The evolution and
function of genes. *Am. Scientist, 36,* 225-
236.

WOOD, A. E. 1947. Rodents—A study in evolution. *Evolution, 1,* 154-162.

WOOD, A. E. 1950. Porcupines, paleogeogra-
phy, and parallelisms. *Evolution, 4,* 87-98.

T. PAUL MASLIN is Assistant Professor of
Zoology at the University of Colorado. The
author wishes to acknowledge the kindness
of Drs. William S. Creighton, Hugo G. Ro-
deck, and William A. Weber, who have read
portions of this paper and offered suggestions
and criticisms.

6

Copyright © 1965 by the Society of Systematic Zoology
Reprinted from *Syst. Zool.* **14**:221–236 (1965)

Similarity *versus* Relationship In *Drosophila*

LYNN H. THROCKMORTON

Introduction

When we consider numerical and orthodox taxonomies, it is immediately apparent that the seriousness of the differences between them, and the possibilities for their mutual agreement, depend finally on the ultimate goals of taxonomy. These goals are by no means as well established as one might think from reading the declarations of leading spokesmen (Simpson, 1961; Mayr, Linsley, and Usinger, 1953) for orthodoxy. If there were unanimous agreement on the aims of taxonomy, the applicability of any proposed method would be readily apparent, or at least unequivocally testable, in which case there would be no controversy over methods. Basically, the real problem that taxonomy faces is that of defining its aims. I would like to defer the problem of aims for later consideration, however, and to devote attention now to various aspects of method. It is at this level that the primary attributes of numerical taxonomy are most apparent, and it is here that orthodox taxonomy is said to be conspicuously deficient (Sokal and Sneath, 1963).

To some degree, the bitterness of the controversy between numerical and orthodox taxonomists has resulted from opposing views as to the aims of taxonomy. Another large part of the disagreement has arisen from a confusion of the issues involved. Up to the present, the lines of argument seem mostly to have been drawn as if it were necessary to choose *between* methods, as if we must have orthodox methods *or* numerical methods. No areas of mutual assistance have been recognized or sought. In reality, we are not faced with the problem of choosing one methodology over the other.

We have instead to ascertain the present state of our science and hence to determine the areas open to improvement. In doing this we must ask what, or which, methods are best for the problems we face. We cannot assume that orthodox methods are perfect and simply defend them. Neither can we accept the numerical taxonomists' evaluation of their wares and blindly adopt their methods. We must instead evaluate *both* methods in terms of the taxonomic problems we need to solve.

One of the obvious means for evaluation of methods is the testing of alternatives, and this brings us to a curious impasse. To me, and perhaps to others, one of the strangest aspects of the discussions concerning numerical taxonomy has been the virtual absence of concrete statements of methods alternative to it. For some reason there is no detailed method, comprised of theoretically justified, operational steps, that can stand as *the* method of orthodox taxonomy. Too often, formal explanations bog down with references to art or to experience at precisely the points where clarity is most needed. One cannot help but wonder why. Surely it is not because the method is too complicated. A method that is too complicated to explain is too complicated for reliable use. Adequate methods must be explainable, but inadequate methods may not be. Actually, most taxonomists have learned their science by osmosis or absorption from adjacent taxonomists. Primarily, they have learned or been taught attitudes, concepts, and customs, but not methods. These they have had to produce intuitively, or by imitation of their predecessors. Judging from the shifting patterns of synonymy and homonymy in the literature, not all taxonomists have been equally successful in generating or imitating methods. At least, all methods have not had equivalent results in the eyes of all taxonomists. Perhaps, then, we should not be too surprised at the diffidence of individual taxonomists when it comes to publishing their private methods. All too easily, this could be disastrous. Unfortunately, silence is neither constructive nor convincing, and we can hardly test numerical methods against orthodox methods if the latter are not explicitly stated. Until orthodox methods are clearly stated, we have only the orthodox taxonomists' word for it that their methods are adequate and that they accomplish what is claimed for them. This is an unhealthy state of affairs. It makes it possible for orthodox methods, however admirable, to remain unassailable simply by remaining obscure. We can neither evaluate nor improve methods that are hidden in the minds of systematists, and most taxonomists would probably agree that a clear statement of method could materially benefit taxonomy. It would certainly make recognition of the differences and similarities between orthodox and numerical methods easier. And with it the possible interactions between the two approaches should become more apparent. For this reason we must explore the problem of method. But since orthodox methods are neither well-explained nor followed consistently by all taxonomists, we cannot attempt to describe what *is* done. We must make a different approach, and I would like to do this by indicating the way in which taxonomic method might be developed by a problem-oriented, as distinct from a method-oriented, scientist. In this way we may be able to view both numerical and orthodox positions from a vantage point independent of either.

Phylogeny

Before we can investigate method we must determine its objective. Here we have several alternatives, but the traditional goal of taxonomy has been the production of a

classification that reflects phylogeny, and this can serve our present purpose. Now, if one looks at this taxonomic goal as a problem-oriented scientist *must* look at it, the first thing the taxonomist must do is produce phylogeny. Then he must devise a classification that will reflect it. By any other procedure the correlation between the classification and the phylogeny will be accidental, and there is no virtue in being right by accident.

As it happens, phylogeny itself is a compound problem, and the approach to it is determined by where one starts. For clarity, it is simplest to start by assuming that the organisms being studied have never been treated taxonomically. There are then three steps. These are: (1) species recognition, (2) discovery of primary groups, and (3) analysis of relationships among primary groups. Each of these steps is a separate and distinct problem, and the methods appropriate to one are not appropriate to the others. This has been a continuing source of confusion among some taxonomists. I have spoken with individuals who maintained that one uses the same method to group specimens into species, species into genera, and genera into higher categories. If this were true, we would (if we were rational) welcome phenetic–numerical methods enthusiastically, since that is exactly what they purport to do. Unfortunately, real biological problems cannot be solved so easily.

Basically, the problem we have to solve is that of classifying the products of organic evolution. This means that we deal with a unique array of objects, an array that has peculiar properties because of the manner of its production. We cannot proceed by putting extant species into groups. They already exist in groups by descent. Our objective must be to *discover* the groups to which organisms belong. This is a simple distinction, but it has a critical effect on perspective. It signifies that the theory that underlies taxonomic practice today is evolution theory, not some general theory of classification, and our problem is to discover what has been produced by evolution. Taxonomists should not think of themselves as classifiers. They should instead consider themselves to be students of evolution, with a classification one useful by-product of their activities.

I do not wish to discuss all three of the steps to phylogeny—species recognition, discovery of groups, and the discovery of groups of groups—in detail here. I am most interested in the third step and will devote more space to it than to the others. However, I would like to mention briefly each of the first two steps in order to indicate how problem-orientation affects the way in which we go about solving them. The effects of problem-orientation can be seen most readily in considering the difficulties of species recognition. If we accept the biological species concept, we are interested in recognizing populations that are evolutionarily independent of each other. In the vast majority of cases, however, it is impossible to test reproductive isolation, and we are forced to draw inferences from characteristics of the samples available to us. Formerly, before the biological species concept was widely accepted, taxonomists might approach the problem of species recognition by asking: How *different* are they? Now, the comparable questions are: Do the characteristics of the samples provide evidence for genetic discontinuity between the populations from which the samples are drawn? And, if there is evidence for genetic discontinuity, is the interruption of gene flow temporary or permanent? This does not make the problem of species recognition any easier. It simply places it in biological perspective. As has been emphasized many times by many people, at the species level we are not classifying objects, we are evaluating the properties of gene pools. At the operational level, the species problem is the problem of detecting genetic discontinuity. And one does not use the same analytical approach to discover genetic discontinuity that he uses to answer the question, How different are they? If one asks this last question, and

applies it to the problem of species recognition, he will be right, by accident, in a certain proportion of cases. He would be right then, not because of the scientific excellence of his procedures but because the evolutionary system produces, in a certain fraction of the total cases, species that match his preconceptions. We might contrast these two methods as "accidental" and "intentional" taxonomy.

At the species level, the intentional approach has been developed carefully over many years. There has been much space devoted to it, and I do not wish to add anything, either by way of definition or by way of method, to what has already been said on the subject. For my purposes here, the species question, and its ultimate resolution, is an excellent example of the way in which the recognition and formulation of a problem *determines* the method that will solve it. It is also the first problem that must be solved on the road to phylogeny, and eventually to classification.

The second step that must be carried out is that of discovering what I will call "primary" groups. This step and the next are very usefully considered against the background of evolution in the genus *Drosophila,* and hence the title of this paper. The genus *Drosophila* is particularly helpful here because it provides some of the few instances where the results of taxonomic practice, or the validity of taxonomic inference, can be evaluated by independent, non-taxonomic criteria. The case to which I wish to draw attention is that of the species groups and subgroups that have been recognized within the different subgenera of *Drosophila.* Judging from conversations I have had with others, it is apparently not fully realized that these species groups are, for the most part, established on purely phenetic grounds. They are phenetic groups in the same sense as are the phenetic groups of the numerical taxonomists. They differ only in being a restricted type of phenetic group, representing just the ultimate terminal clusters of phyletic lineages. The recognition of these clusters is based, not on selected or key characters, but on analysis of the total phenotype that is accessible to the investigator. Species are recognized as belonging together in primary groups when phenetic resemblances are so high that the probability of their monophyletic origin, from a single ancestor unique to them, amounts almost to a certainty. Concomitantly, there is a very low probability that any other known species or group originated from among the species in question. There are readily detectable gaps between groups, and the species within the groups are so similar that in many cases they can be distinguished from each other only with difficulty. It has been suggested (Stone, 1962) that these groups are the equivalent of the superspecies of some authors (e.g., Simpson, 1961). Attempts to subdivide these groups would have to be based on such a low number of characters that the implied relationships would be very uncertain. Some persons, not accustomed to such fine discrimination in their detection of phenetic differences, might question the validity or reality of such groups of species, even though one might predict that evolution should produce such clusters. Fortunately, these groups have been validated, unintentionally for the most part, by genetic and cytogenetic studies (Patterson and Stone, 1952; Stone, 1962). Not all groups have been investigated so thoroughly, of course, but with few exceptions among the studied groups, genetic and cytogenetic evidence has only served to confirm the conclusions that taxonomists have drawn from phenetic criteria. Thus, on genetic, cytogenetic, and phenetic grounds, primary groups in *Drosophila* appear to be real groups. They can be recognized by discriminating phenetic studies without the necessary assistance of genetic or cytogenetic techniques, and there is no reason to think that such groups cannot be, or are not, recognizable by taxonomists working with any groups of organisms. Primary groups such as these have a certain utility of their own in classification. I wish to emphasize them now,

however, not as a category but as real units, the recognition of which is a very useful if not an essential step in the discovery of phylogeny. If these clusters have reality as the terminal branches of phyletic lineages, and this is what the genetic and cytogenetic evidence indicates, then important deductions and inferences can be drawn from characteristics distributed among them.

Before we continue, we must digress briefly and consider factors involved in taxonomic inference when phylogenetic conclusions are to be drawn. These develop from the fact that phylogeny is a consequence of genetic continuity, and any phylogeny is accurate only to the extent that it indicates the genetic history of the groups involved. Implicitly or explicitly, taxonomic inference, if it is to support phylogenic conclusions, must be genetic inference. It may not be genetic inference to the level where specific anatomical or other traits are correlated with specific alleles, but there must at least be the inference that more similar phenotypes indicate more similar genotypes. For this purpose it is convenient to use the concept of genotypic homology, and to view the problem of phylogeny as that of tracing the genotypic changes that have occurred during the evolution of many individual characteristics.

Genotypic homology is a consequence of the interaction between the replicative capacity of the genetic system and natural selection. If replication were perfect, we might not have any evolution, but in any case we would speak of genotypic identity rather than homology. Homology is a concept that has utility only when we are dealing with attributes that are similar but different. We infer genetic similarity, and genetic similarities are a consequence of genetic replication. Hence they are *evidence* of common descent. In this sense, genotypic homology is a more useful operational tool than conventional homology, since that is usually defined as *due to* descent. Since replication is not always perfect, but nearly

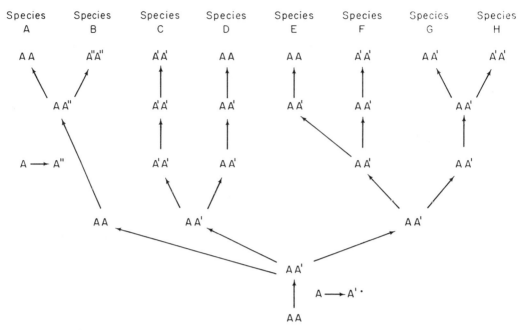

Fig. 1. Evolutionary consequences of heterozygosity.

so, and since natural selection acts on changed products (mutations) to produce slow and gradual replacement of a genotype in time, we must deal with degrees of genotypic homology rather than with the absolutes of gene identity. Since most characteristics are determined or influenced by many loci, we are concerned with evolutionary changes in complex genotypes. These can, however, be simplified for descriptive and predictive purposes.

Some predictions regarding the probable consequences of genotypic change in evolutionary lineages are shown in Figure 1. The emphasis there is on the evolutionary consequences of heterozygosity. In sexually reproducing diploid organisms it is, so far as we know, impossible to go from one character state of an organism to a more derivative state without passing through some genetic intermediate stage. If the two alternative character states are visualized as homozygous (which they need not be), then the transitional stages would include genetic elements necessary to produce *both* the new and the old phenotypes, plus,

perhaps, some intermediate phenotypes, depending on the way in which the "new" and the "old" genotypes interacted. Since evolution is opportunistic, the subdivision of a population can occur any time, and not necessarily only when there is little or no genetic variability in the gene pool. In fact, we might expect that gene pools that were rich in genetic variability might more often produce successful descendent species than gene pools with little or none. Hence, persistent polyallelism (heterozygosity) in evolutionary lineages might be predicted as a very probable state of affairs, in which case, there will be, for perhaps a considerable period of time, the possibility for alternative character states to be produced by segregation within descendent gene pools. In the genetic sense, the genotypes responsible for the character states will be *completely homologous*, if not virtually identical; but the character will *appear* many times independently, which is to say that it can occur independently in species derived from species that *did not* show the characteristic, even

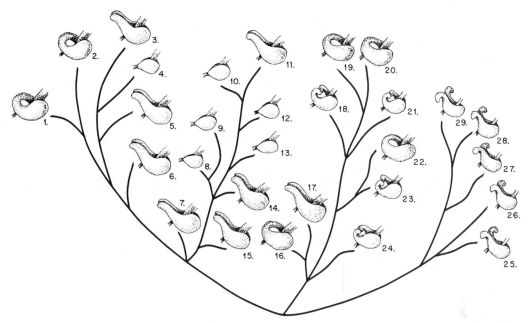

Fig. 2. The distribution of types of ejaculatory bulbs related to the cytological phylogeny of the *repleta* group of *Drosophila*. After Throckmorton, 1962.

though the genotype for the character was potential within their gene pools. This is a critical difference between genotypic homology and conventional homology, since in the latter case one generally insists that a character be present in both ancestor and descendent species before it can be considered homologous. If the character cannot be inferred to have been present in the ancestors and is not traceable back to one common ancestor, it is customarily referred to as a parallelism, or even mistaken for convergence. For genotypic homology, the genotype existed "unassembled," so to speak, in the *gene pool* of the ancestor. Few if any individuals would have had the genotype at that time, and this ancestor probably would not have shown the character. Later the genotype could have been "assembled" many times independently in separate descendent lineages, but it would not have been any the less homologous for that reason. Genotypic homology is determined by the derivation of the genetic *elements* from a common ancestral gene pool, not by the time of assembly of the

genotype (Throckmorton, 1962). Consideration of the potentialities of the genetic-evolutionary system, and particularly of the evolutionary consequences of persistent polyallelism in gene pools, emphasizes the essential equivalence of so-called homologies and parallelisms. They can both be used, with about the same degree of confidence, when one infers genotypic histories, which is to say, when one infers phylogeny.

The genus *Drosophila* has many examples of this particular evolutionary phenomenon. Figure 2 shows details of the ejaculatory bulb, a part of the male reproductive system, in relationship to the cytological phylogeny of the *repleta* group. There is one type of bulb (Fig. 2: 1, 2, 16, 19, 20, 22) that is restricted to about six of the more than three hundred species of the genus I have examined to date. All of these species are in the *repleta* group, and four subgroups; the primary groups in this case, are shown in the figure. The species sharing this type of ejaculatory bulb belong to two different cytological phylads, but it would be highly unrealistic to argue that

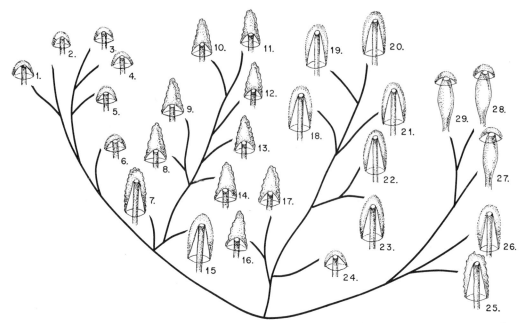

FIG. 3. The distribution of types of spermathecae related to the cytological phylogeny of the *repleta* group of *Drosophila*. After Throckmorton, 1962.

they do not reflect genotypic homology. Their detailed resemblance almost certainly is a reflection of a detailed similarity between their genotypes.

These ejaculatory bulbs also represent parallelism, as that term is commonly used. They do indicate relationship, as we can see from the evidence of cytology, but the relationships they indicate are not the immediate relationships of the species that share them. Rather, they are the relationships of the primary groups to which the species belong. Figure 3 shows much the same thing for another structure, the dorsal seminal receptacle of the female. Here we have several different types (e.g., Figs. 2: 1, 24, and 29; 14 and 16; 7 and 25; 15, 23, and 26), again appearing to segregate (to use this term rather loosely) to several different species in several different phyletic lineages. Here again the presence of a given character state tells us not that the species that share it are very close relatives but that the close relationships are between the primary groups to which these species belong. This provides us with the informa-

tion on which to base the last procedure in the phylogenetic analysis. It also indicates that we must be very cautious in inferring phylogenetic relationships between closely related species.

In most instances in *Drosophila*, closely related species are complex mosaics of the characteristics of their nearest relatives. They show, individually, very little that is unique to themselves. They show, instead, unique *combinations* of the characters found among other close relatives. When they do share a few distinctive traits with other species, the possibilities of parallelisms are such that these few common traits cannot be given weight as determining the phylogeny *of the species*. It is true that some unique characters are indications of relationship, but we have no *a priori* way of determining which character states undergo parallel development and which do not. Thus, on an *a priori* basis, we cannot determine which characters represent conventional homology and which represent parallelism. If we would have an objective analysis, we must treat all character states as if they were parallelisms. If we pretend to recognize homologies and group our organisms accordingly, we bias our procedures by our preconceptions regarding the evolution of the groups with which we deal. The operational assumption that characters reflect parallelism is the least restrictive assumption that can be made during a phylogenetic analysis, and it also proves to be a perfectly workable assumption, as shown below.

Parallelism is the rule rather than the exception for *individual characters* in *Drosophila*. There is no evidence from *Drosophila* that parallelism has ever involved extensive character complexes, however, and the genetic and cytogenetic tests are uniquely suited to uncover such a situation, if it exists. For the individual characters, however, "reversals" of character state are common, and ancestral character states recur often. "Reversals" over about three character states are frequent. In genetic terms, this suggests that, at any level in time, the average gene pool could generate genotypes for the character states existing at that time, for those that had been supplanted, and for some that had not yet emerged. Since we deal with complex genotypes, this need not imply polymorphism, although polymorphism could be a manifestation of this phenomenon. The data supporting these conclusions have been published elsewhere (Throckmorton, 1962).

These considerations tend to emphasize, first, that the phylogeny of individual species may be highly uncertain and, second, that the phylogeny of primary groups need not be. Recall that this is the substance of the conclusions from comparing character states within a known phylogeny (Figs. 2 and 3). It is this that requires the detection of primary groups as the second step in the analysis of phylogeny. This step can be carried out very readily, as long as one depends on the analysis of the total accessible phenotype and not on selected or key characters for delimiting the group. Recognition of the need for this step in the analysis of phylogeny could probably come

only from a problem-oriented approach. Essentially we have investigated the properties of the evolutionary system, and the characteristics of its products, by prediction (Fig. 1) and by example (Figs. 2 and 3). We have asked what *is* produced. Then we ask what one must do to discover this product. The existence of parallelisms cautions against an approach to phylogeny through the species, and at the same time it indicates great promise for an approach that utilizes primary groups consciously to exploit the phylogenetic implications of parallelisms.

What we need to do next is to derive the gene pools that have existed in the past and from which our existing primary groups are derived. This may sound formidable, but operationally it turns out to be almost alarmingly simple. This type of analysis is described elsewhere (Throckmorton, 1962) in greater detail, and only the bare essentials will be presented here. Technically, one can use all characters available, and the characters are not weighted. I have found by experience that the process is made much easier if one uses the characters in a certain order. Those to be used first are the ones that bring the largest number of primary groups together into groups of groups. The largest possible groups are formed from the character having the fewest states. The steps themselves can be almost mechanical, although some consideration may be given to the direction of evolution when that can be inferred.

The general steps involved are shown in Figure 4. This is a great simplification and shows an analysis based on nine characters and nine primary groups. The primary groups are numbered across the top, and each group includes two species. Each species is indicated by its character state in brackets under the number of its primary group. For this example, I have selected the primary groups and the two species each from those available among the close relatives of *Drosophila*. This example is, therefore, a real but abbreviated case. The steps in the analysis are numbered down the left side, and the characters used are indicated beneath, together with a notation for the character states and the direction of evolution.

The first step involves the use of the character states of the Malpighian tubules. Actually there are three character states, but practically there are only two since the two more derivative states exist side by side in most of the primary groups. This is the first benefit of working from primary groups. We can often see at a glance which character states show promise of giving phylogenetic information and which do not, simply by their distribution among primary groups. In the present example, the wide distribution of the two more derivative character states of the Malpighian tubules implies that the genotypes for these character states have continued to segregate over a rather considerable period of evolutionary time. Quite literally, it appears that the ancestral gene pools, from which primary groups 1 through 8 were derived, were segregating for the genotypes for the derivative states of this character. It is not unreasonable to assume that these eight gene pools were themselves derived from a gene pool that was also segregating for these genotypes. It is probable that a series of gene pools was involved, judging from the number of groups involved, but the assumption of a single gene pool is most parsimonious at this stage. It presumes the shortest possible time period during which the evolutionary segregation of character states could occur. That is to say, the period during which a gene pool could segregate genotypes for two or more character states is assumed to be short unless and until evidence from other characters requires the assumption of a longer period. Hence, a single division is indicated by these character states, and this division is shown by extending downward the line between groups eight and nine. Primary groups are, of course, indivisible, and the disposition of one species from the group fixes the disposition of all species of the group.

The next character also exists in three states (Fig. 4, step 2), and the sequence of the primary groups must be changed. This change is made by considering the probable contents of the gene pools from which the primary groups were derived and by considering the direction of evolution for the character under consideration, when that can be inferred. In this case, groups 1 and 4 appear to be derived from gene pools that were segregating two character states. Specifically, the species in these groups show either the presumed derivative state of the character or a state of the character that is intermediate between the primitive and derivative states. A segregating gene pool is taken to be evidence (operational evidence, not absolute and unequivocal evidence) of intermediacy in the evolutionary sequence. Hence, groups 1 and 4 are placed between the groups that show only the primitive character state and those that show only the derivative character state. Also, primary groups 1 and 4 include species having a presumed intermediate character state, which tends to support the intermediate disposition of these two groups. The genetic inference here would be that the intermediate character state had been established genetically prior to the origin of the gene pool from which primary groups 1 through 7 were derived. The genotypes for character state C had also become available by this time, but they co-existed in the gene pool with those for state B. Therefore, both character states could (and did) segregate from a gene pool that existed at that time. Groups 1 and 4 were established from this gene pool, and their original gene pools themselves continued to segregate for these two character states. Primary groups 2, 3, 5, 6, and 7 probably also came from this segregating gene pool, but fixation of one genotype (for C) was involved in their derivation from it. I need hardly point out again that these are operational interpretations, but they are working assumptions that are justified in view of what we know of the properties of

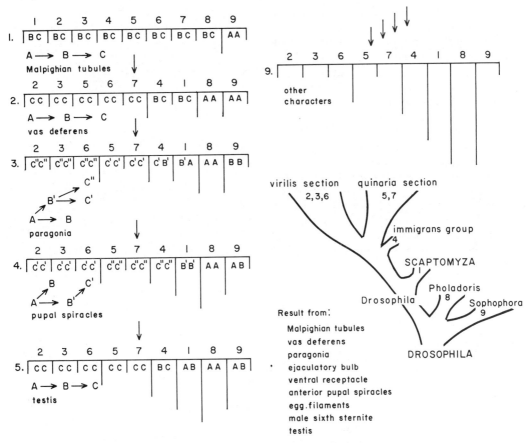

FIG. 4. General method for deducing phylogeny.

the genetic system. The new sequence is therefore adopted, and the additional group divisions (between seven and four, and between one and eight) are indicated.

In the third step, several character states exist, and they do not seem to have a linear relationship with each other. Two different character states appear to have been derived from the primitive. One is found in primary group 9, the other in primary groups 1 and 4. This may be an indication of diverging lineages, but it does not alter the sequence. These groups are already adjacent to the primitive group (8) on the basis of the earlier characters. Again, primary groups 1 and 4 are intermediate when judged on the basis of segregating gene pools. Group 1 is transitional in having members with either the primitive or the

intermediate character state. Group 4 is also transitional, but it has the intermediate state together with one of the derivative states (C'), suggesting that it originated somewhat later than did group 1. This, too, is consistent with the existing sequence. Finally, the positions of groups 5 and 6 need to be reversed, to keep the character states grouped as much as possible. The new divisions are marked in as before (between 5 and 6), and when additional evidence for a previous separation is obtained, the line indicating that division is extended (lines between 1 and 8, and between 8 and 9).

By the fourth step no changes in sequence were necessary. Character 4 confirmed the results from the previous characters. The fifth step likewise required no change in sequence, and it again gave a pattern consistent with the previous interpretations. This analysis was continued up to nine characters, not all shown in Figure 4. After the first three characters, few changes were required by the addition of new characters. For the most part, additional characters reinforced the pattern produced by the first ones. Hence, the order in which characters are used does not seem to be too important. Also, evidence from the additional characters supported the *inferences* on which the earlier dispositions of groups were based. Had the earlier *operational* assumptions for one character been improper, one would not have expected such a marked concordance among all the characters used. The general phylogeny for the groups in the example is shown to the lower right of Figure 4 to indicate how the operational figure may be converted to a typical dendrogram. This type of analysis is designed to produce only sequence (cladistics), so distance and angle in the figure have no significance.

One of the most important features of this approach to phylogeny is that it requires no inferences regarding probable characters of ancestors, and it requires no reconstruction of ancestral types. Technically, we infer only the probable contents of gene pools, and since we have fewer preconceptions about gene pools than about morphology, we gain in objectivity by this procedure. In this approach the most important evidence for intermediate position is the possession, by two or more species in one primary group, of character states that indicate which other primary groups were derived with them from a common and unique ancestral gene pool. In *Drosophila* it has been possible to resolve almost all questions of sequence on the basis of the intragroup variation, and not on individual evaluations of what constitutes morphological intermediacy. The evidence from intragroup variation and that from presumptive morphological intermediacy were consistent with each other

in most instances. When there were conflicts it was because some character states were overlooked, in prospect, that appeared as perfectly obvious intermediate states, in retrospect. In short, this method has proved to be a rather powerful one for discovering direction of evolution of different characters.

Phylogenetic Method and Numerical Taxonomy

Up to this point we have been discussing method as it should be developed with a single problem in mind. The problem was to derive the sequence of genetic change that occurred during the evolution of a group of organisms. This is the ostensible aim of taxonomy so long as it makes any pretense of having a phylogenetic basis. The method that has been outlined may seem foreign to some, and the genetic interpretation and justification of the steps may distress others. This last, the genetic interpretation of taxonomic data, is a subject that might profitably be discussed at some later time. At this point it is well to call attention to the ways in which the method just outlined reflects procedures used by orthodox taxonomists. We can note also the extent to which the technical procedures for each level of analysis might be refined by contributions by numerical taxonomists. That is, we can suggest potential interactions and cooperations between orthodox and numerical taxonomists, at least insofar as their methods are concerned.

The first point that needs to be re-emphasized is that, from the problem-oriented viewpoint, the preliminary steps to classification are three, and a different method is required for each. No single procedure, numerical or otherwise, will adequately reduce taxonomic data for classification. The three steps outlined above are the minimum number that can be employed prior to assigning groups to categories. The first step, that of species recognition, requires little comment except to suggest that there is here a fertile field that might be exploited by numerical methods. To date, this level of taxonomic research has been disdained by most numerical taxonomists, and some have even defined it out of existence (Sokal and Sneath, 1963; Sokal, 1964; Ehrlich, 1961). The species problem remains, nonetheless, and the possibilities for refining phenetic techniques specifically for detecting genetic discontinuities between gene pools seems an obvious and potentially fruitful outlet for energetic computers. Such methods, if developed, would be primarily for resolving special and troublesome cases, or for demonstrating as conclusively as possible that in a given instance phenetic analysis could not uncover evidence for species distinction. It would not be expected that computers could give us unequivocal answers to all species problems. Error and uncertainty are inherent in our system so long as we are constrained to depend on phenetic methods to resolve problems that can only be resolved unequivocally by reproductive tests. Still, our responsibilities as systematists must be toward resolving species questions as accurately as possible in as many cases as possible, and numerical methods should help us increase the number of cases where accurate species recognition is assured. At this level of taxonomic investigation, therefore, numerical methods should be supplemental to orthodox methods, just as cytological and biochemical methods are. We do not choose between these methods. We use all of them as skillfully as possible on the problems for which they are best suited.

The second level of investigation, that of the discovery of primary groups, is eminently suited to the numerical approach. Unfortunately, many numerical methods undertake to do much more than this. It is this step that most taxonomists are thinking of when they speak of "classification." It is an obvious and necessary step in any taxonomic procedure, and it is quite obviously a step that is based on estimates of degree of resemblance. Regrettably, many taxonomists have not realized that their only valid inference at this level is that a very high degree of similarity indicates very recent common ancestry and a low degree does not. This allows one to identify the primary groups to which species belong, but it does not allow one to rank species by degree of similarity and infer therefrom their probable phylogeny. The existence of an evolutionary system in which heterozygosity plays a part, however small, automatically entails parallelism and hence precludes the assumption that degree of resemblance is equivalent to recency of common ancestry, except as already noted, when extremely close relatives are involved. The more extensive the parallelism, the less tenable this assumption will be. It is true that in some cases, even in many cases, this assumption may be substantially correct, but if we trust in this we will, again, be right only by accident. Taxonomy, as a science, must have more lofty aims than this! Hence, the need for a third method for evaluating relationships outside of the primary groups.

Numerical methods can make great contributions at this second level, in two different ways: they can refine methods for discovering primary groups, and they can develop methods for evaluating the properties (homogeneity, for example) of already recognized groups. This is an area that should be most actively explored by orthodox and numerical taxonomists alike since it promises to place on a firm and uniform basis the taxonomic category (the primary group, whatever its category designation) that has the greatest practical usefulness, whose phylogenetic implications are least equivocal, and whose predictive qualities are highest. If orthodox and numerical taxonomists would collaborate just to the extent of evaluating some existing groups, we ought soon to have a much better understanding both of what numerical methods can do and of what orthodox methods have done. Some assertions of numerical taxonomists might prove to be incorrect, and some of the skepticism of orthodox taxonomists might prove to be unjustified. It is often remarkable the extent to which simple tests may contribute to the abatement of controversy, and the present case may prove to be no exception.

The last step to phylogeny, that of grouping of groups, is somewhat less utilitarian but of great biological interest. The system outlined here may need some modification before it could be treated as a general method, but in its present form it is objective and repeatable, and it can also serve as a method for the *a posteriori* weighting and evaluating of characters. These are achievements that are often referred to as attainable only through art or through experience. The outlined procedure is a simple way by which one gains experience quickly and objectively. When this is done, taxonomy becomes not an art but a science. The role of computer methods at this level of analysis remains to be determined. The work of Camin and Sokal (1964, and in publication) suggests that there is real hope for progress in this direction.

Classification

Figure 5 shows a phylogeny of *Drosophila* and its close relatives, derived basically in the manner indicated earlier. It provides an example of the extent to which orthodox methods have achieved a phylogenetic classification, and the concordance between phylogeny and the existing classification is not impressive. This figure also provides an example of the magnitude of the problem one faces in attempting to reflect phylogeny in a classification. A major difficulty with this phylogeny is that it shows a pronounced vertical development, rather than the pattern of sequential divergent branching presumed by the Linnean hierarchy. Most of the diversification in this family has occurred by divergence from a single lineage that was itself changing slowly in time. At a given level, several groups may have evolved from this basic type, but they have not all diverged or diversified to the same degree. Thus, species, species groups, and sub-genera separate from nearly common points in the phylogeny. Those that have diverged in their external and traditionally diagnostic features are classified in other genera. Where these same features have remained unchanged, and in spite of other changes,

the forms are classified as *Drosophila*. Any attempt systematically and consistently to impose a hierarchical arrangement on such a pattern will result in a wildly asymmetrical product that rapidly exhausts the category and subcategory names available.

There is, then, an observation to note and a point to be emphasized. The *observation* is that orthodox methods have not produced what they claim to produce, that is, a phylogenetic classification. The point to be emphasized is that the hierarchical Linnean system is severely limited as to what phylogenetic information it can carry. The phylogenetic *detail* in Figure 5 simply cannot be reflected by a Linnean hierarchy. This last is not a new point (see Hull, 1964), but it certainly has not been given the serious attention it deserves from taxonomists.

One would like to know the reason for the dichotomy between the principles of orthodox taxonomy and the practices of orthodox taxonomists, as these are revealed by the situation shown in Figure 5. Under ordinary circumstances, in most scientific disciplines, this would be easy. We would simply ask whether the error resulted from improper basic premises, from improper methods, or from the incompetent application of methods that were themselves adequate. As noted in the introduction to this paper, taxonomists have denied themselves this standard procedure of self-evaluation and improvement by the simple expedient of never explicitly defining aims and never explicitly stating their methods. This has led to a great deal of confusion and inconsistency, and this confusion is probably the clearest message that can be read from Figure 5. Whatever the criticisms that orthodox taxonomists may have of numerical taxonomy and their methods, they certainly cannot claim for themselves that they have clear objectives or more consistent and defensible practices. We can readily agree that some taxonomists do have proper aims, and proper methods to achieve these aims. We must at the same time recognize that all taxonomists do not share the same aims or use the same methods, even when they do seem to have the same objectives.

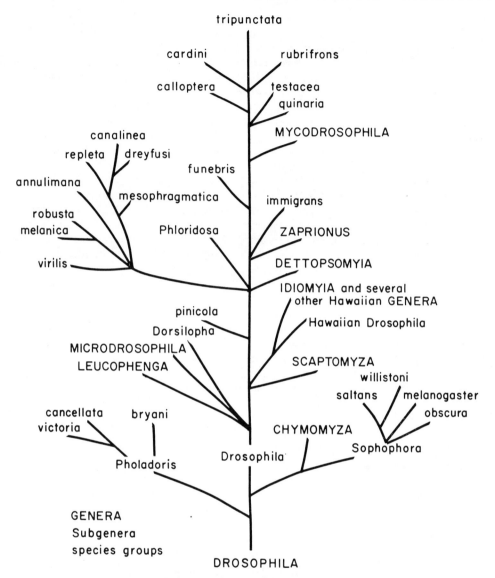

FIG. 5. Phylogeny of *Drosophila* and related genera.

In the last analysis, the major question faced by taxonomy today is not, as implied by the controversy over numerical methods, whether classification should be phylogenetic or phenetic. It is probable that the great majority of biologists agree that a phylogenetic system is preferred. Many taxonomists seem to think that this settles the problem of aims. Unfortunately, the problem still remains as to what constitutes a phylogenetic classification, granting that we will use the Linnean hierarchy to express it.

Considering the problem of phylogenetic classification, there are two basic formats that must be distinguished. One is a classification that is a literal translation of phylogeny. This is what most biologists have been taught to expect of a classification (see for example Grant, 1963:34), even though such a translation may in fact exist only rarely. It is this form of classification that cannot be impressed on the groups shown in Figure 5. The second major format is one that simply shows groups and diversification within groups. Phylogenetic sequence is not implied here, except in that groups are intended to be monophyletic in a very strict sense. This form of classification would not, and could not, show divergence. It is the only classification that can show, unambiguously, the little phylogenetic information that can actually be contained in a hierarchical classification. The Linnean hierarchy can show gross phylogenetic relationships (if its groups are rigidly monophyletic) and diversification (the number of different monophyletic groups within a larger group). It is suited to show little else.

The existing system of classification seems to be an unwholesome hybrid between the two basic systems, or something else that is neither. The classification of the groups in Figure 5 shows divergence in some parts (apparently in the vain hope that divergence *per se* would have some phylogenetic significance), and diversification in others. Hence, it shows little about phylogeny. It would show more about phylogeny if there had been a reasonable degree of equivalence between degree of difference and recency of common ancestry. As noted earlier, we would not expect such to be the case, and, in the Drosophilidae at least, it certainly is not the case.

Where are we then with regard to the interactions between numerical and orthodox taxonomy? The numerical taxonomists recognized long ago, along with other less vocal taxonomists, that there were inconsistencies in the practices of orthodox taxonomists. They set out to change this state of affairs and have played a major role in stimulating a re-examination of taxonomic methods and goals. This in itself is no mean achievement. Unfortunately, they seem to have interpreted the confusion in taxonomic practice as resulting from a conflict between phenetics and phylo-

genetics. Hence, they have pressed methods without seriously evaluating problems. It is more probable, however, that the basic confusion in taxonomy is a consequence of discord resulting from a misunderstanding among taxonomists regarding what a phylogenetic classification can and should do, and of how phylogenetic data is to be converted into a phylogenetic classification. This confusion remains. Numerical taxonomy has done nothing to dimish it (phenograms are no more amenable to hierarchical classification than are phylogenies), and it must be eliminated before further constructive attacks can be made on method. Taxonomy today is a hodgepodge of goals and methods. We must settle firmly and conclusively the problem of the aims of taxonomy by specifying the format by which phylogeny will be expressed in classification. If we do this, the problem of methods should settle itself since, as I have shown above, appropriate methods exist for all the critical steps in the taxonomic procedures that precede classification. We must decide what should be done, how it is best done (and what should *not* be done), and then do it. Within this emerging system it is almost inevitable that numerical methods will attain increasing importance for particular problems at certain stages of taxonomic analysis, but we cannot expect them to completely supplant procedures that now exist.

REFERENCES

CAMIN, J. H., and R. R. SOKAL. 1964. A method for deducing phylogenies. Symposium: Interactions between numerical and orthodox taxonomies. Knoxville meeting of the Society of Systematic Zoology, December 27–30, 1964.

EHRLICH, P. R. 1961. Has the biological species concept outlived its usefulness? Systematic Zool. 10:167–176.

GRANT, V. 1963. The origin of adaptation. Columbia University Press, New York. 606 p.

HULL, D. L. 1964. Consistency and monophly. Systematic Zool. 13:1–11.

MAYR, E., E. G. LINSLEY, and R. L. USINGER. 1953. Methods and principles of systematic zoology. McGraw Hill, New York. 336 p.

PATTERSON, J. T., and W. S. STONE. 1952. Evolution in the genus *Drosophila*. Macmillan, New York. 610 p.

SIMPSON, G. G. 1961. Principles of animal taxonomy. Columbia University Press, New York. 247 p.

SOKAL, R. R. 1964. The future systematics. *In* C. A. Leone [ed.], Taxonomic biochemistry and serology. Ronald, New York. 728 p.

SOKAL, R. R., and P. H. A. Sneath. 1963. Principles of numerical taxonomy. W. H. Freeman, San Francisco. 359 p.

STONE, W. S. 1962. The dominance of natural selection and the reality of superspecies (species groups) in the evolution of *Drosophila*. Studies in Genetics II. University of Texas Publ. 6205: 507–537.

THROCKMORTON, L. H. 1962. The problem of phylogeny in the genus *Drosophila*. Studies in Genetics II. University of Texas Publ. 6205: 207–343.

Part III

HENNIGIAN ARGUMENTATION METHOD

Editors' Comments
on Papers 7 and 8

Two papers have been chosen to illustrate early applications of Hennig's original description of his method. Koponen (Paper 7) illustrates the logic for determining plesiomorphy and apomorphy. An attempt is made to define monophyletic groups in Mniaceae. Included is a discussion of various groupings and competing arrangements of groups at various ranks. This is one of the earliest applications of Hennig's method and the first in botany.

Nelson (Paper 8) discusses the need for wider application of Hennig's method in ichthyology. He argues that common ancestry be the basis for classification or grouping, that ranking be undertaken by time of origin, and that sister groups be given equal rank. Such classifications will provide a general reference system for biologists and will result in the rejection of all nonmonophyletic groups. Many of these issues are still being actively debated. Nelson also provides a comparison of Hennigian, Mayrian, and Simpsonian meanings of the concept of relationship.

7

Copyright © 1968 by Societas Biologica Fennica Vanamo
Reprinted from *Ann. Bot. Fenn.* **5**:117–151 (1968)

Generic revision of Mniaceae Mitt. (Bryophyta)

Timo Koponen

Department of Botany, University of Helsinki

Preface

The need for monographic studies in the taxonomy of bryophytes has been stressed in so many connections (V. F. Brotherus according to Kotilainen 1950, Malta 1926, Verdoorn 1934, 1950, Meijer 1951, Steere 1955, Tuomikoski 1958, Anderson 1963, Watson 1967) that there is no need to discuss it further. Another field which is no less important but has been almost completely neglected is the revision of the supraspecific taxa, such as families and genera. One of the pioneers in this field was Loeske who paid attention to generic revision in a number of papers (e.g. 1907 a, 1907 b, 1911, 1932) and even reviewed the delimitation of several European genera (1910). Later Steere (1947) and Tuomikoski (1958), in particular, underlined the importance of such studies. However, rather few revisions have been made (e.g. Hilpert 1933, Lawton 1957, Robinson 1962).

The writer began to study *Mnium* with some intensity in 1964. The original aim of the research was to make a monograph of the genus *Mnium* covering all the species. From the very beginning special attention was paid to characters which, though neglected earlier, might be of value not only for the specific but also for the supraspecific taxonomy of the group. In the field of hepatic taxonomy, K. Müller (1948, 1951, cf. also Richards 1959) met with considerable success in corresponding studies. In the genus *Mnium* the neglected stem characters (p. 122) appeared to be useful. In the course of the study it became more and more obvious that a generic revision could not be avoided. In addition to the genus *Mnium* in its customary delimitation, the other genera of the traditional *Mniaceae* were included in the revision. It was thought most suitable to publish this revision as the first part of the study.

Acknowledgements. – The present work has been carried out in the Department of Botany, University of Helsinki. To the Director of that institute, Dr. AARNO KALELA, Prof. of Botany, I wish to express my sincere thanks for providing working facilities and arranging financial assistance, as well as for the comments on my manuscript.

Dr. RISTO TUOMIKOSKI, Prof. of Biotaxonomy, originally roused my interest in bryology and suggested the family *Mniaceae* as the subject of research. I wish to thank him for his stimulating teaching and generous help in all phases of this study.

Dr. JAAKKO JALAS, Associate Prof. of Botany, and Docent TEUVO AHTI, Ph.D., read the manuscript and made valuable suggestions. They also took part in discussions on nomenclatural problems, in which they were joined by Dr. PEKKA ISOVIITA, who was particularly generous in offering his help. I wish to express my grateful thanks to all of them.

Working facilities for my herbarium studies were largely arranged through the kind offices of Dr. HEIKKI ROIVAINEN, the Emeritus Curator, Botanical Museum of the University of Helsinki. Help in acquiring the necessary literature was given by Mrs. ELSA NYHOLM, Keeper of the Paleobotanical Department Riksmuseet, Stockholm, and Dr. HERMAN PERSSON, the Emeritus Keeper of the same Institute, Dr. HARUMI OCHI, Prof. of Botany, Tottori University, Tottori, Japan, and Dr. ZENNOSKE IWATSUKI, the Hattori Botanical Laboratory, Nichinan, Japan. Mrs. AUNE KOPONEN, Phil. Cand., drew the final copies of the figures. Mrs. ANNA A. DAMSTRÖM, M.A., kindly revised the English language of the manuscript and Mrs. MARJA KAILA, Phil. Mag., translated the Latin diagnoses. I wish to express my appreciation to all these persons.

I should also like to thank *Emil Aaltosen Säätiö, Betty Väänäsen Stipendirahasto, Suomen Kulttuurirahasto, Suomalainen Tiedeakatemia,* the *British Council,* and the University of Helsinki for the grants given in aid of this study.

I. Introduction

A. Genus Mnium

The name *Mnium* dates back to DILLENIUS's »Historia Muscorum» (1741). LINNAEUS (1753) also included three hepatics in his genus *Mnium* (cf. EVANS 1907) and this caused some nomenclatural confusion (see p. 140). HEDWIG's (1801) genus *Mnium* consisted of thirteen species, only five of which are included in *Mnium* to-day, viz. *M. hornum, M. stellare, M. cuspidatum, M. punctatum,* and *M. undulatum.* The others now belong to the genera *Aulacomnium, Pohlia, Bryum, Rhodobryum,* and *Philonotis.* The current delimitation of the genus as far as the European species are concerned is derived from BRUCH, SCHIMPER and GÜMBEL (1838) who, however, also included *Cinclidium* as a subgenus, and later as a section (1846). The delimitation of *Mnium* in respect to the non-European taxa advanced more slowly. C. MÜLLER (1848 – 1849, cf. also 1901) had a wide generic concept and included the current genera *Aulacomnium* and *Rhizogonium.* LINDBERG (1868) discussed MÜLLER's concept and separated two genera, *Trachycystis* and *Leucolepis,* frc *Mnium.* It may be mentioned that ANDREWS still included them in *Mnium* as late as in 1940. The works of FLEISCHER (1902 – 1904), BROTHERUS (1909, 1924), and KABIERSCH (1936) contributed to the establishment of the current delimitation. However, HOLMEN separated a new genus, *Cyrtomnium,* from *Mnium* as late as in 1957.

From time to time the possibility of splitting the genus *Mnium* into independent genera has been discussed. LOESKE (1910) was the first who showed that the section *Polla* (Brid.) Mitt. differs in many characters from the other species of *Mnium* and constitutes a separate genus, *Polla* (Brid.) Loeske. LOESKE also discussed the relationships of *Cinclidium* and *Mnium* and stated that the species of the section *Rhizomnium* are the nearest relatives of *Cinclidium.* He concluded that the best solution would be to include *Cinclidium* as a section in *Mnium.* He suggested, on the other hand, that if it was wished to treat *Cinclidium* as a separate genus, *Mnium* should be divided into three genera, viz. *Polla, Mnium* proper (*Eu-Mnium*) and *Cinclidium,* the latter including also the section *Rhizomnium.* LOESKE's proposal has been discussed by several authors. KABIERSCH (1936) and ANDREWS (1940) did not accept LOESKE's ideas but retained the traditional concept of *Mnium.* STEERE (1947) seems inclined to accept the separation of *Polla,* and LAZARENKO (1955) and SHLYAKOV (1961; cf. also DOMBROVSKAJA & SHLYAKOV 1967) accepted LOESKE's concept and dealt with the »biserrate» species of *Mnium* as a separate genus using, however, the now illegitimate generic name *Polla* (Brid.) Loeske (see p. 140).

B. Subdivision of Mnium

The first attempt to carry out a subdivision of *Mnium* was made by BRUCH, SCHIMPER & GÜMBEL (1838). They based it on the characters of the leaf margin:

Subgen. *Cinclidium*
Subgen. *Mnium*
    A. Foliis marginatis
       a. integris
       b. Foliis dentatis
    B. Foliis immarginatis
       a. serratis
       b. Foliis subintegris

Categories other than subgenera were not given definite rank and were not properly named. In 1846, the same authors called their subdivisions sections but named only *Cinclidium* validly. C. MÜLLER (1848 – 1849) introduced the section *Eumnium* which, however, comprised the current genus *Mnium*; his other sections were *Aulacomnium* and *Rhizogonium* and thus no real progress had occurred at the subdivisional level. LIMPRICHT's (1895) subdivisions *Biserratae*, *Integerrimae*, and *Serratae* were also without definite rank but they already roughly corresponded to the genera proposed by LOESKE (cf. above). A more detailed subdivision of *Mnium* was presented by KINDBERG (1897).

The rank of the subdivisional names in KINDBERG's (1896, 1897) papers is unclear. He uses two categories of subdivisions. The more comprehensive ones are marked with Roman numerals (I, II, III, etc.) and the others with Arabic numerals (1, 2, 3, etc.). In the preface to the first part of his synopsis (1896) KINDBERG gives only a vague explanation: »I believe that the greatest importance is to attach[e] to the natural affinity of such species as could be joined to common types (subgenera or groups)». However, he does not give any examples of what he means by »subgenera». On the other hand, in the footnote to the subdivision *Bryum* VII. *Pohlia* (1897, p. 347) he writes: »This section could be related as a proper genus to fam. *Meeseaceae*, allied to *Plagiobryum* and *Orthodontium*». It is evident that the word »section» in this context is not used in the present meaning of the Code but as a synonym of the word »group». Later KINDBERG (1909) published some corrections and alterations which were necessitated by BROTHERUS's (1909) work. Throughout this paper, apart from two exceptions, KINDBERG accorded the rank of subgenus to those categories which were marked with Roman numerals in 1896 and 1897, and the rank of section to those with Arabic numerals. KINDBERG (1909) did not list all the names mentioned in the earlier papers and the subdivisions of *Mnium* are among those not considered.

The question of names published without a clear indication of taxonomic rank has been recently discussed by ISOVIITA (1966, pp. 215 – 216) and BRUMMIT & CHATER (1967). The latter authors submitted a proposal for modifying the Code, Article 35, suggesting that »such names published before 1953 should be treated as validly published but inoperative in questions of priority. If taken up later in a definite rank the new name should be regarded as a new combination with the original name as the basionym». If the Code and this suggestion are followed, the rank of KINDBERG's (1897) subdivisional names was determined by KABIERSCH (1936) in the case of *Mnium*. He regarded the names marked with Roman numerals (cf. above) as sections and those with Arabic numerals as subsections. His view thus differed from that of KINDBERG (1909) himself.

WIJK et al. (1964) did not follow KABIERSCH (1936). For instance, they considered KINDBERG'S II. *Pseudo-Bryum* as a subgenus (cf. below). Thus, according to the Code, Article 35, they in fact introduced a number of new combinations. Since as a rule they introduced the necessary new combinations in separate papers (cf. WIJK & MARGADANT 1958, WIJK et al. 1959) the new combinations introduced in »Index Muscorum» can be considered as incidental and invalidly published (the Code, Art. 34).

KINDBERG's (1897) and KABIERSCH's (1936) subdivisions are to some extent used as a basis for the present study and they are referred to frequently later. It therefore seems convenient to reproduce them here:

KABIERSCH (1936)	KINDBERG (1897)
	II. *Eu-Mnium*
	1. *Hymenophylloidea*
sect. I. *Pseudoleucolepis*	
sect. II. *Polla*	
subsect. 1. *Horna*	10. *Horna*
2. *Spinosa*	8. *Spinosa*
3. *Stellariformia*	3. *Stellariformia*
4. *Serrata*	9. *Serrata*
sect. III. *Eumnium*	
subsect. 1. *Cuspidatiformia*	4. *Cuspidatiformia*
2. *Venusta*	
3. *Rosulata*	5. *Rosulata*
4. *Rostrata*	6. *Rostrata*
5. *Sublimbata*	
6. *Undulata*	7. *Undulata*
sect. IV. *Rhizomnion*	2. *Punctatiformia*
V. *Pseudobryum*	I. *Pseudo-Bryum*
VI. *Pinnaticosta*	

C. Family Mniaceae

Two different concepts have prevailed in the suprageneric taxonomy of *Mnium*. Some of the authors of the 19th century (BRIDEL-BRIDERI 1826 – 1827, BRUCH, SCHIMPER & GÜMBEL 1838, SCHIMPER 1860, 1876, KINDBERG 1882, 1897) considered the genus *Mnium* as a part of the family (or »tribe») Bryaceae. Some other authors (C. MÜLLER 1848 – 1849, MITTEN 1859, LINDBERG 1868, 1878, 1879, LIMPRICHT 1895) accepted the family (or »tribe») Mniaceae. Which genera the family Mniaceae comprised largely depended on each author's own famliy concept and the material dealt with. For instance, C. MÜLLER's (op. cit.) tribe »Mnioideae» consists of 12 genera, including *Cinclidium*, *Georgia*, *Timmia*, and *Dawsonia*. MITTEN's (1859) Mniaceae contains six genera, *Fissidens*, *Rhizogonium*, *Mnium*, *Timmia*, *Mniadelphus*, and *Daltonia*. LINDBERG (1878, 1879) included five genera in the family, *Cinclidium*, *Astrophyllum* (= *Mnium* proper) *Timmia*, *Mnium* (= *Aulacomnium* androgynum), and *Sphaerocephalos* (= *Aulaco-*

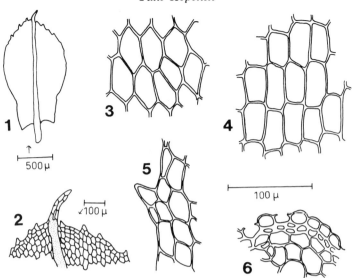

Figs. 1 – 6. *Bryomnium solitarium* Card. 1 = leaf. 2 = apex of leaf. 3 = cells at centre of lamina. 4 = cells at leaf base
5 = margin of leaf. 6 = transverse section of stem epidermis.

mnium s.str.), and LIMPRICHT's (1895) *Mniaceae* comprises only *Cinclidium* and *Mnium*. Since the appearance of BROTHERUS's (1909, 1924, cf. also KABIERSCH 1936) works, the concept of the family *Mniaceae* has remained rather stable. The following nine genera have been included:

Bryomnium Card.	*Orthomnion* Wils.
Cinclidium Sw.	*Orthomniopsis* Broth.
Cyrtomnium Holmen	*Roellia* Kindb.
Leucolepis Lindb.	*Trachycystis* Lindb.
Mnium Hedw.	

PODPĚRA (1954) also included the genus *Rhodobryum* in *Mniaceae*.

The aim of this research is not to revise the limits of the family *Mniaceae* (cf. pp. 134, 137). However, two of the genera listed above, *Bryomnium* and *Roellia*, are excluded from the family. *Roellia* is not a near relative of the genera under consideration but rather of *Rhodobryum* or *Bryum* (cf. ANDREWS 1938, CRUM 1967). The type specimen of *Bryomnium solitarium* Card., the only known *Bryomnium* species, was studied, and it also shows closer affinities to *Rhodobryum* than to *Mnium* (Figs. 1 – 6). However, the taxonomic status of *Bryomnium* will be uncertain until the sporophyte is found.

II. Taxonomy

A. Material and methods

The main source of material used for this research is the moss herbarium of the University of Helsinki (H). The herbarium of V. F. BROTHERUS (H-BR) was used especially frequently when specimens of Asiatic or tropical species were needed. Fresh samples collected by the writer (in Finland, N. Norway, Central Sweden, Bavaria in Germany, and England) were used when possible. Specimens have been studied in the case of each of the species listed (p. 140 onwards), if not otherwise stated.

Before examination the dry moss was soaked in ethyl alcohol and cleared in KOH solution. The transverse sections were mounted in gum arabic and sealed immediately with varnish. The figures not representing transverse sections were drawn from specimens in water preparations.

Toluidin blue was used in staining to show up the rhizoid initials (cf. BUCH 1947).

The specimens from which the figures were drawn are listed below with the numbers of the figures.

Mnium hornum Hedw. – Finland. Ahvenanmaa. Sund, Gesterby, Tjännan, moist wood, 1914 Collander (H). – Figs. 7, 8, 46, 64, 70, 76, 90, 91, 101, 102.

Trachycystis flagellaris (Sull. & Lesq.) Lindb. – Japonia centralis, in montibus, 3 – 5 000 f. s.m. 1890 Bayr (H-BR). – Fig. 17.

T. microphylla (Doz. & Molk.) Lindb. – Japan. Prov. Rikuzan, Matsushima, 1915 Yasuda (H-BR). – Figs. 16, 62, 67, 77.

Leucolepis menziesii (Hook.) Steere. – Oregon. Multnomah Co. Base of cliffs west of Bridal Veil Falls, 1962 Hale no 21 504 (H). – Figs. 12, 45, 63, 68, 75, 87, 88, 89.

Cinclidium arcticum (Schimp.) C. Müll. – Norvegia. Sör-Tröndelag. Opdal, near Kongsvoll, in springs just above the timber-line, ca 1 300 m, 1908 Buch (H). – Fig. 79.

C. stygium Sw. – Finland. Kemi Lapland, fen at Purnuoja, 1940 Fageström (H). – Figs. 9, 38, 42, 74.

C. stygium Sw. – Finland. North Bothnia. Tervola. Pisavaara Nature Reserve, area 260. Peräletto, eutrophic fen, 1966 Koponen & Koponen 8639 (H). – Figs. 52, 56.

Rhizomnium minutulum (Besch.) Kop. – Japan. Oita Pref. Naruko-gawa valley, on decayed wood, 1959 Miyamoto (Musci Japonici no 791, H). – Figs. 50, 84.

R. punctatum (Hedw.) Kop. var. *elatum* (Schimp.) Kop. – Finland. Rovaniemi. Pisavaara Nature Reserve, area 55, by Rajapuro brook, 1965 Koponen & Koponen 8319 (H). – Figs. 10, 11, 33, 34, 37.

R. punctatum (Hedw.) Kop. var. *punctatum.* – Finland. Ålandia. Eckerö, Storby, at ditch edge, 1899 Axelson (H). – Figs. 49, 61, 71, 83, 95.

Cyrtomnium hymenophyllum Holmen. – Norway. Finnmark. Alta, Kaafjord, canyon of Mathisfos, on calcareous cliffs, 1964 Koponen 10 249 (H). – Figs. 13, 48, 57, 73, 80.

Orthomnion bryoides (W. Griff.) Norkett. – Sikkim Himalaya, Sureil, 6000 ft. Gammie 112/2 (H-BR). – Figs. 18, 40, 51, 72, 85, 93.

O. loheri Broth. – Philippines. Prov. Benguet. Luzon. On small trees, alt. 4 000 ft., 1904 Elmer 6 487 (H-BR). – Figs. 92, 105, 106.

O. loheri Broth. – Philippines. Prov. Benguet, Baguio, 1907 Elmer 8446 (H-BR). – Figs. 94, 99.

Orthomniopsis dilatata (Mitt.) Nog. – Japan. Nara Pref. Ikenomine, Shimokitayama, on trunks of trees, ca. 400 m alt., 1962 Kodama (Musci Japonici no 990, H). – Figs. 19, 66, 82, 97.

O. dilatata (Mitt.) Nog. – Japan. Kyushu. Kumamoto Pref. Yatsushirogun, Mt. Ryuho, ca 500 m alt., 1963 Iwatsuki 673 (H). – Figs. 103, 104.

Plagiomnium affine (Funck) Kop. – Finland. Uusimaa, Helsinki. Vanhakaupunki, grass-herb wood on N. shore of River Vantaa, 1935 Buch (H). – Figs. 31, 32.

P. confertidens (Lindb. & Arn.) Kop. – U.S.S.R. Irkutsk Dist. Kirensk, fl. Angara-Ilim, pr. pagum Kotschenga, in ripa fl. Ilim, 1909 Ganeschin (H-BR). – Fig. 81.

P. cuspidatum (Hedw.) Kop. – Finland. Uusimaa, Helsinki. Villa Haga, 1923 Buch (H). – Figs. 14, 15, 41, 43, 47, 60, 69, 78, 98.

P. rostratum (Schrad.) Kop. – Finland. Varsinais-Suomi. Lohja, Torhola cave, on limestone, 1964 Koponen 4298 (H). – Fig. 20.

P. rugicum (Laur.) Kop. – Finland. North Bothnia. Tervola, Pisavaara Nature Reserve, area 260, slightly sloping eutrophic fen, 1966 Koponen & Koponen 8606 (H). – Fig. 44.

P. elat∷n: (B.S.G.) Kop. – Sweden, Vestmanland, Viker. limestone area N of Älvhyttan, in grass-herb wood, 1966 Koponen 10474 (H). – Figs. 35, 36.

P. venustum (Mitt.) Kop. – British Columbia. Vancouver Island. Miracle Beach Provincial Park (S of Oyster River), on forest floor in dense coastal forest, 1961 Ahti & Ahti 15 151 (H). – Fig. 100.

Pseudobryum cinclidioides (Hüb.) Kop. – Finland. South Häme, Iitti. Säyhde, swamp forest at the edge of Lake Punalampi, 1963 Lounamaa (H). – Figs. 22, 23, 24, 29, 30, 39.

P. cinclidioides (Hüb.) Kop. – Finland. North Savo, Kuopio. Suovu, in moist wood by Lake Lummelammi, 1906 Linkola (H). – Figs. 55, 59, 65, 96.

P. speciosum (Mitt.) Kop. – Japan. Nagano Pref. Mt. Yatsu, on humus under snow, alt. 2300 m., 1954 Iwatsuki (Musci Japonici no 476, H). – Figs. 21, 53, 54, 58, 86.

Bryomnium solitarium Card. – Congo Belge. Dibele, 1903 Laurent & Laurent (Holotype, PC). – Figs. 1 – 6.

Bryum alpinum Brid. – Finland. Varsinais-Suomi, Karjalohja. Karkali Nature Reserve, area 27, sloping cliff with trickling water by shore, 1962 Koponen 3553 (H). – Fig. 25.

Rhizogonium bifarium (Hook.) Schimp. – Nova Seelandia. 1888 Dall, ex. herb. Melbourne no 719 (H-BR). – Fig. 28.

Rhodobryum roseum (Hedw.) Limpr. – Finland. South Karelia, Vehkalahti. Pyhältö, in wood by N shore of Lake Pyhältöjärvi, 1967 Fagerström (H). – Fig. 27.

Roellia roellii (Röll) Crum – British Columbia. Wells Gray Provincial Park, sample plot 45, 1961 Ahti & Ahti 14 781 (H). – Fig. 26.

In the descriptions on pp. 139 – 147 the leaf shape is given, as far as possible, according to the proposal of the Systematics Association Committee for Descriptive Biological Terminology (1962), and the shape of the apex and base of the leaves according to LAWRENCE (1956). The size of the lamina cells is roughly indicated by the following terms: small = > 15 μ, medium-sized = 15 – 40 μ, and large = < 40 μ. In solving the nomenclatural problems the International Code of Botanical Nomenclature (LANJOUW et al. 1966, called »the Code» in this paper) was used. The abbreviations of the herbaria are according to LANJOUW & STAFLEU (1964).

B. Description of some characters used

Some characters of the genera are indicated in Figures 7 – 106. Some of them, especially those seldom or never used in the past, perhaps deserve a fuller discussion and more accurate description. To avoid confusion the nomenclature presented on pp. 139 – 147 is used from this point onwards, if not otherwise stated.

1. Growth-form and branching of the stem

The growth-form of a moss depends on the way in which it branches and the direction of growth of the branches. CORRENS (1899 a, b) found that *Plagiomnium undulatum* appears to have a bud (»ein ruhendes Auge») at the axil of each leaf, from which the branches may originate. He called this type of bud distribution the Bryum type. The type was found in all genera of *Mniaceae* studied by the present writer. The bud develops from the same segment of the apical meristem as the leaf immediately above (LORCH 1931, PARIHAR 1965) and is, therefore, often situated laterally at the axil of the leaf (cf. also HUBER 1967). MEUSEL (1935) included the European *Mnium* species (s. lat.) in the type in which the branches originate at the base of the stem (»Typ der basitonen Innovationen»). According to him, *Plagiomnium undulatum* represents a form intermediate between »basitone» and »acrotone» mosses. In addition to occurring in *P. undulatum* (and the whole section *Undulata*), subapical branching takes place in *Leucolepis* and in *Trachycystis microphylla*. *T. flagellaris* has subapical flagellae which originate from the buds and thus are modified branches (cf. CORRENS 1899 b). A similar dendroid growth-form was also occasionally found in some species of *Mnium*. In other genera some upper buds may also develop into branches if the apical meristem has been destroyed or if the 'inflorescence' has developed. MISIURA (1964) showed that cuttings from the stems of *Rhizomnium punctatum* may regenerate from dormant buds (cf. also AINSWORTH 1956).

Since the more or less dendroid growth-form is restricted to the genera *Leucolepis*, *Trachycystis*, *Mnium*, and *Plagiomnium* sect. *Undulata*, although the species of the other genera have practically unlimited branching potentiality (cf. above), the presence or absence of subapical branches under normal conditions must be genetically controlled. On the whole, subapical branching seems to be present in all the species of a given genus, except in the case of *Plagiomnium* where it is useful as a sectional character.

Another point where frequent branching takes place is the base of the stem (cf. above). In the tribes *Mnieae* and *Cinclidieae* and in the genus *Pseudobryum* these branches either develop into fertile erect stems or remain more or less erect sterile shoots. These sterile shoots may have an arched somewhat plagiotropic apex, such as is often seen, for example, in *Mnium stellare*. However, they do not reach the substratum and »root» at the apex. In the genus *Plagiomnium* in addition to erect branches there are also more or less horizontal ones. In some species, e.g. in *P. rostratum*, the whole length of these plagiotropic shoots is adpressed against the substratum, in some others, e.g. in *P. affine*, they are arcuate and »rooting», especially at the apex of the shoot. According to MEUSEL (1935), the plagiotropic shoots may continue to grow horizontally for several years but finally develop an erect fertile stem.

The genera *Orthomnion* and *Orthomniopsis* differ from the other genera in that the »main» stem is adpressed and the branches erect. WIJK (1958) used this as a character for separating these genera from *Mnium* (s. lat.). In *Orthomnion* the sporophytes develop on these erect branches. In *Plagiomnium*, e.g. in *P. rostratum*, the sporophytes may also develop on the branches of plagiotropic shoots, but these branches mostly have at least a short plagiotropic

basal part. I was not able to check whether *Orthomnion* really has a continuously monopodial plagiotropic growth-form or whether a terminal sporophyte finally develops on the horizontal »main» stem, as is the case in *Orthomniopsis*. The horizontal stems in *Orthomnion* and *Orthomniopsis* are not as complanate as in some *Plagiomnium* species, and they are possibly an adaptation to a special type of substratum. The species in question are mostly corticolous, or grow on rotten wood.

As stated above, the occurrence of horizontal plagiotropic shoots or stems with a »rooting» ability is confined to the genera *Plagiomnium*, *Orthomnion*, and *Orthomniopsis*. Thus it seems to be useful as a generic character.

2. Stem anatomy

The stem of the species of the family *Mniaceae* consists of an epidermis which is one or several cell layers thick, the parenchymatous cortex and the central conductive strand. In the cortex there are smaller strands of conductive tissue called false leaf traces. They are downward continuations of the conductive strands of leaves which end blindly without reaching the central conductive strand (cf. LORCH 1931).

The stem epidermis shows characters which may have taxonomic significance.

In the tribes *Mnieae* and *Cinclidieae* the epidermis consists mainly of one cell layer with strongly thickened cell walls (Figs. 9 – 13). In the mature stem parts the cell wall is thickened by an inner, often clearly defined, secondary layer, which nearly fills the lumen of the cell. In *Rhizomnium*, *Cinclidium*, and *Cyrtomnium* the secondary layer usually turns bright yellow when treated with KOH solution; in *Mnium*, *Leucolepis*, and *Trachycystis* it often assumes a brownish or reddish colour. In *Cinclidium stygium* the radial walls are seen to be more strongly thickened than the tangential ones and the lumen has a characteristically narrow shape (cf. also LORCH 1931). In other *Cinclidium* species and in other genera all the walls thicken equally, or the outermost wall is the thinnest. This is especially characteristic in *Trachycystis* (Fig. 17), in which the outermost cell wall may disappear in old stem parts producing a distinctive structure. According to LORCH (op. cit., pp. 56 – 57) the yellow secondary wall layer of *Cinclidium stygium* consists of almost pure cellulose.

In *Plagiomnium*, *Orthomnion*, and *Ortho-*

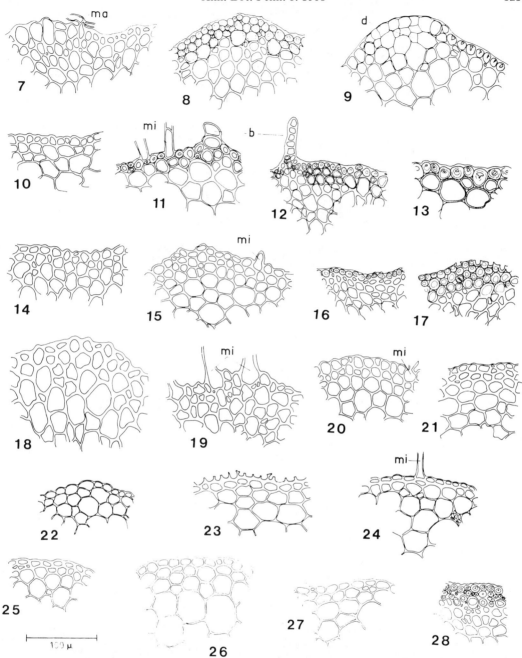

Figs. 7 – 28. Structure of stem epidermis, transverse sections. b = decurrent base of leaf, d = dormant bud, mi = micronema, ma = macronema. 7, 8 = young and old *Mnium hornum*. 9 = *Cinclidium stygium*. 10, 11 = young and old *Rhizomnium punctatum* var. *elatum*. 12 = *Leucolepis menziesii*. 13 = *Cyrtomnium hymenophyllum*. 14, 15 = young and old *Plagiomnium cuspidatum*. 16 = *Trachycystis microphylla*. 17 = *T. flagellaris*. 18 = *Orthomnion bryoides*. 19 = *Orthomniopsis dilatata*. 20 = *Plagiomnium rostratum*. 21 = *Pseudobryum speciosum*. 22, 23, 24 = young, old, and middle-aged *P. cinclidioides*. 25 = *Bryum alpinum*. 26 = *Roellia roellii*. 27 = *Rhodobryum roseum*. 28 = *Rhizogonium bifarium*.

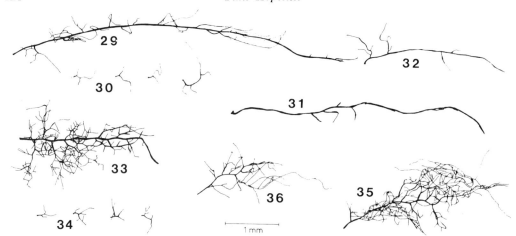

Figs. 29–36. Structure of rhizoids. 29, 30 = macronema and micronemata of *Pseudobryum cinclidioides*. 31, 32 = macronema and micronema of *Plagiomnium affine* from tip of arcuate shoot. 33, 34 = macronema and micronemata of *Rhizomnium punctatum* var. *elatum*. 35, 36 = macronema and micronema of *Plagiomnium elatum*.

mniopsis the epidermis is composed of two or more layers of cells. In typical cases the cell walls show no or only slight secondary thickening (Figs. 14, 15, 18 – 20). Such thickening was seen in some mature stem parts, e.g. in *Plagiomnium cuspidatum*, but in no case did it reach the degree observed, for instance, in *Mnium*. The same type of thickening was found to be present in *Bryum*, *Rhodobryum*, and *Roellia* (Figs. 25 – 27).

In the stems of *Pseudobryum cinclidioides* and *P. speciosum* the main thickening takes place in the tangential walls between the first and the second cell layers (Figs. 21 – 24). In addition, the outermost wall of the epidermis is extremely thin, especially in *P. cinclidioides*. In old stems the thin wall disappears (Fig. 23) in the same way as in *Trachycystis* (cf. above). Similar thickening of the tangential walls is present in *Bryomnium* (see Fig. 6).

In the following, the three different types of epidermis are called the Bryum, Mnium, and Pseudobryum types. Each type seems to occur throughout a given genus and they can therefore be included among the generic characters. The Mnium type is most striking and seems to separate the tribes *Mnieae* and *Cinclidieae* from the other groups. The three types described are also present in other groups of mosses. For instance, Figs. 20, 30, 32, 33, and 34 in LORCH (1931) represent the Mnium type, and Figs. 25, 27, and 35 the Pseudobryum type.

3. Rhizoids and rhizoid topography

Although rhizoids and rhizoid topography have long been used as taxonomic characters in hepatics (cf. e.g. SCHUSTER 1966), they were neglected in mosses until fairly recently. However, WARNSTORF (1906), probably influenced by CORRENS (1899 b), used them as specific characters in the genus *Drepanocladus* (cf. also TUOMIKOSKI 1949). Later their significance was discovered by TUOMIKOSKI (1958) although he did not publish any detailed study on the subject. In *Mniaceae* the abundance or absence of rhizoids on the stem has been used as a distinguishing character at specific level (cf. e.g. KABIERSCH 1936).

The rhizoids on the stem in *Mniaceae* can be divided into two groups according to their place of origin, size and structure. They are here referred to as macronemata and micronemata (macronema and micronema in the singular).

The macronemata are large, often profusely branching organs (Figs. 29, 33, 35). They always originate from the large initial cells surrounding buds (cf. Figs 41 – 43). The micronemata originate from initial cells which are scattered along the stem without any definite location in relation to the leaves and buds (Figs 37 – 39). They are much thinner and shorter than the macronemata and the branching is less profuse and often pseudodichotomous (Figs. 30, 32, 34). The micronematous initials become visible on the stem at a rather early phase and before the

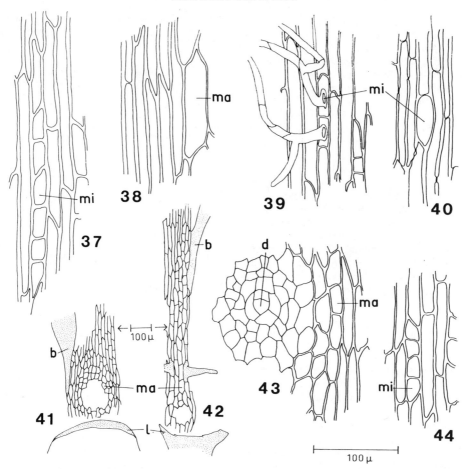

Figs. 37–44. Rhizoid topography and initials of rhizoids, surface view. b = decurrent base of leaf, d = dormant bud l = leaf removed, ma = macronematous initial, mi = micronematous initial. 37 = micronematous initials of *Rhizomnium punctatum* var. *elatum*. 38 = macronematous initial of *Cinclidium stygium*. 39 = micronematous initials and micronemata of *Pseudobryum cinclidioides*. 40 = micronematous initial of *Orthomnion bryoides*. 41 = macronematous initials surrounding the bud of *Plagiomnium cuspidatum*. 42 = elongated area of macronematous initials in *Cinclidium stygium*. 43 = dormant bud and macronematous initials in *Plagiomnium cuspidatum*. 44 = micronematous initials in *P. rugicum*.

thickening of the walls of the epidermis cells. In most cases they differ markedly from the other cells of the stem epidermis in being shorter and broader, and having thinner walls. They seem to originate from young epidermis cells by transverse cell divisions. The number of initials corresponding to one normal epidermis cell may have taxonomic significance, for instance, in *Plagiomnium* there are often four initials or more but in *Orthomnion* one or two (Figs 40, 44). In some species, e.g. *P. undulatum* and *P. cuspidatum*, the micronematous initials are less distinctly differentiated morphologically, being,

however, shorter than the normal epidermis cells (cf. also CORRENS 1899 b).

Although the micronematous initials are present on the stem at a rather early phase the formation of micronemata seems to some extent to be controlled by the habitat. In wet localities or when the stem is submerged the micronemata appear at an early stage and often are numerous. In drier conditions the micronemata sprout later and may be few in number. There also seem to be specific differences, for instance, even in rather dry localities *Plagiomnium medium* is often covered with micronemata up to the 'inflorescence' while the stems of the closely related *P. insigne* are almost bare. In mixed stands of *P. rugicum* and *P. elatum* (= *Mnium seligeri* auct.) it is often possible to observe that the stems of *P. rugicum* are covered with micronemata up to the apex, while those of *P. elatum*, growing in the same conditions, are still bare. On plagiotropic shoots micronemata (and also macronemata) may be present on the very apex of the shoot and in such cases often develop on the ventral side of the shoot only, although there are initials on the dorsal side as well.

The structure and distribution of the macronemata and micronemata seem to some extent to be distinguishing generic characters. Since buds are present in all the genera, the macronematous initials and macronemata surrounding the buds are also found in every genus. *Orthomnion* and *Orthomniopsis* differ from the other genera in that the micronemata and macronemata are not differentiated morphologically but are of about the same size and shape. In some species of *Plagiomnium*, e.g. in *P. cuspidatum*, the micronemata and macronemata are also roughly of the same size and similarly branched, the main difference being that the micronemata are thinner. In *Mnium*, *Trachycystis*, *Leucolepis*, *Cinclidium*, and *Cyrtomnium* the micronemata, if present at all, are confined to the very base of the stem and micronematous initials are not found on mature stem parts or on the apex. In these genera only macronemata may be present on the upper stem parts. In *Cinclidium* the area of the marconematous initial cells is distinctively elongated (Figs. 38, 42) and thus the macronemata are arranged in longitudinal rows on the stem, a character which is peculiar to this genus. In *Plagiomnium*, *Pseudobryum*, *Orthomnion*, and *Orthomniopsis* the micronemata are not confined to the base of the stem but are generally also present on the upper part. As a rule they sprout earlier than the macronemata. On plagiotropic shoots, however, both types of rhizoids may begin to grow simultaneously.

The genus *Rhizomnium* is the only one in which there are species both with and without micronemata. In *R. glabrescens*, *R. minutulum*, *R. nudum*, and *R. striatulum* and possibly also in *R. andrewsianum* micronemata (and micronematous initials) are absent, but they are frequent in *R. pseudopunctatum*. *R. punctatum* is the most problematic species in this respect, since it comprises both types of rhizoid topography.

When the inconsistency in the rhizoid topography of *R. punctatum* was noticed, more material was studied to solve this problem. It was observed that the stems which lacked micronemata did not have micronematous initials either, which is in accordance with the observations presented above. According to the labels, most of the specimens with micronemata were collected in wet habitats, by springs, brooks etc. They are dioicous, rarely possess sporophytes, and are mostly large specimens with broadly elliptic – oblong or obovate, obtuse leaves. The leaf border is multi- or bistratose at the base only and always unistratose at the apex. All these characters are diagnostic of the taxon called *»Mnium» punctatum* var. *elatum* Schimp. (cf. SCHIMPER 1860, PERSSON 1947, 1962, PERSSON & WEBER 1958). By contrast, the specimens devoid of micronemata (and initials) seem to be smaller, their leaves are narrower, the border being mostly multi- or bistratose throughout, and their habitats are drier. Most of the micronematous specimens were collected in northern districts, while the taxon without micronemata shows a southern distribution. Several mixed specimens of the two types were found in the herbaria. The rhizoid topography seems to offer a character for separating these two taxa. The question is dealt with further in a separate paper (in print).

In addition to the types of rhizoids described above, rhizoids similar to the micronemata on the stem may originate from the leaves. The rhizoid initials in the lamina, which alternatively may produce secondary protonemata, are thinwalled, and usually smaller than the other lamina cells. They are single cells or groups of a few cells. They generally occur mainly near the border and costa but may also be more dispersed. A preliminary study to discover any possible taxonomic significance gave a negative result, which was what might have been expected, in view of CORRENS's (1899 b) study (cf. also LERSTEN 1961, MISIURA 1964). In addition to the lamina, the costa may produce rhizoids, and does so in great abundance in *Orthomnion* and *Orthomniopsis*. Even rhizoids originating from the decurrent bases of the leaves were occasionally found.

The rhizoid topography, and especially the presence or absence of micronemata, seems to represent a valuable taxonomic character in Mniaceae. It must, however, be used with caution until it has been ascertained to what extent the presence or absence of micronematous initials is genetically controlled and what influence is exerted by environmental factors, such as moisture. The micronemata are most frequent in species growing in wet habitats but, on the other hand, they are absent in *Cinclidium* which inhabits bogs and fens, and present in *Plagiomnium* species growing in rather dry habitats. The absence of micronemata could also be connected with the thickness of the epidermis since

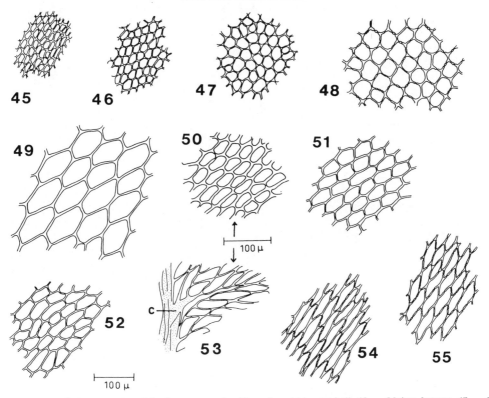

Figs. 45 – 55. Areolation at centre of lamina. c = costa. 45 = *Leucolepis menziesii*. 46 = *Mnium hornum*. 47 = *Plagiomnium cuspidatum*. 48 = *Cyrtomnium hymenophyllum*. 49 = *Rhizomnium punctatum* var. *punctatum*. 50 = *Rhizomnium minutulum*. 51 = *Orthomnion bryoides*. 52 = *Cinclidium stygium*. 53, 54 = branch of costa and areolation in *Pseudobryum speciosum*. 55 = *P. cinclidioides*.

they are mostly absent in species with an epidermis of the Mnium type (see p. 124). However, they occur in some species of *Rhizomnium*, and, moreover, the micronematous initials originate before the thickening of the cell walls begins. Further experimental work is needed to find out whether it is possible, for instance, to induce the development of micronematous initials in the upper stem parts of *Mnium*, *Leucolepis* etc. by culture in a wet habitat or by auxin treatments.

4. Leaf characters

The shape of the leaf and the size and the shape of the lamina cells (Figs. 45 – 55) represent roughly distinguishing characters in each of the genera. However, in the taxonomy of *Mniaceae*, more significance attaches to the structure of the leaf margin and the costa.

Differences are found in the serration of the margin and in the degree of the differentiation of the border (Figs. 56 – 74). The margin may be toothed or entire, the border may be well differentiated, multistratose – unistratose, or undifferentiated. The double serrature in *Mnium* and *Trachycystis* is perhaps not a taxonomically significant, independent character since it may be connected with the multistratose border; a double serrature can hardly be present in leaves with a unistratose border.

The structure of the midrib offers useful characters (cf. LIMPRICHT 1895, LOESKE 1910). In the transverse section of the costa (cf. e.g. Fig. 76) four different elements can be distinguished: dorsal and ventral epidermis, large central guide cells, conductive strand, and a maximum of two stereid bands, one dorsal and one ventral. The most conspicuous character is the absence or presence of stereid bands. In

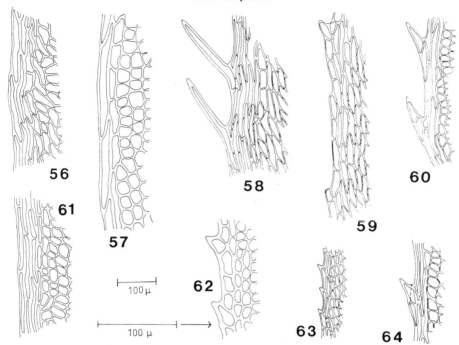

Figs. 56 – 64. Leaf margin, surface view. 56 = *Cinclidium stygium.* 57 = *Cyrtomnium hymenophyllum.* 58 = *Pseudo-bryum speciosum.* 59 = *P. cinclidioides.* 60 = *Plagiomnium cuspidatum.* 61 = *Rhizomnium punctatum* var. *punctatum.* 62 = *Trachycystis microphylla.* 63 = *Leucolepis menziesii.* 64 = *Mnium hornum.*

Mnium, Trachycystis, and *Leucolepis* as a rule both bands of stereids are present. In *Cyrtomnium, Cinclidium,* and *Plagiomnium* (and *Orthomniopsis*) only the dorsal is developed, and in *Rhizomnium* sect. *Rhizomnium, Orthomnion,* and *Pseudobryum* typical stereids are lacking. The differentiation of the epidermal cells and the length of the costa (Figs 75 – 86) offer further characters. The costa is often toothed at the back in the tribe *Mnieae.*

The shape and size of the leaves may be different in different parts of the stem. The basal leaves in some species of the tribe *Mnieae* differ from the fully developed ones in being triangular and acuminate, though this may be connected with the fact that the leaves in this tribe are generally rather narrow. However, in *Leucolepis* at least, a true difference definitely exists. The short basal leaves have a hyaline apex and are narrowly acuminate, and characteristically serrate. In the other tribes and genera there is not such a contrast between the shapes of the basal and normal leaves. Differences are also found in the number and in the degree of differentiation of the perichaetial leaves (Figs. 87 – 92), and the number of leaves per stem can be a distinguishing character, as was shown statis-

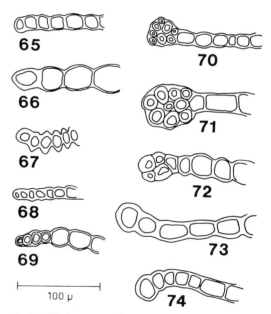

Fig. 65 – 74. Structure of border, transverse sections. 65 = *Pseudobryum cinclidioides.* 66 = *Orthomniopsis dilatata.* 67 = *Trachycystis microphylla.* 68 = *Leucolepis menziesii.* 69 = *Plagiomnium cuspidatum.* 70 = *Mnium hornum.* 71 = *Rhizomnium punctatum* var. *punctatum.* 72 = *Orthomnion bryoides.* 73 = *Cyrtomnium hymenophyllum.* 74 = *Cinclidium stygium.*

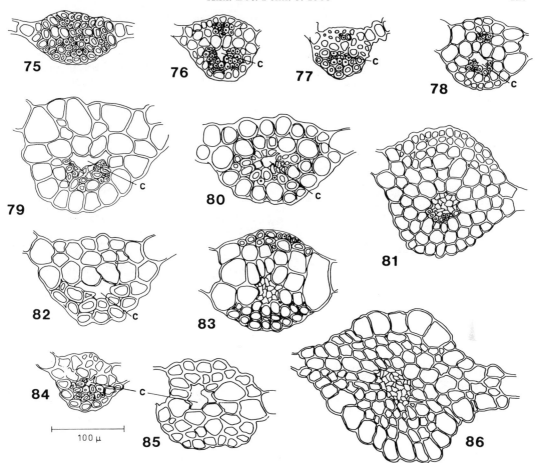

Figs. 75 – 86. Structure of costa, transverse sections. c = easily damaged conductive strand. 75 = *Leucolepis menziesii*. 76 = *Mnium hornum*. 77 = *Trachycystis microphylla*. 78 = *Plagiomnium cuspidatum*. 79 = *Cinclidium arcticum*. 80 = *Cyrtomnium hymenophyllum*. 81 = *Plagiomnium confertidens*. 82 = *Orthomniopsis dilatata*. 83 = *Rhizomnium punctatum* var. *punctatum*. 84 = *R. minutulum*. 85 = *Orthomnion bryoides*. 86 = *Pseudobryum speciosum*.

tically by MacLeod (1917) e.g. it separates *Mnium hornum* from *Rhizomnium punctatum*.

The length and form of the decurrent leaf bases also offer generic in addition to specific characters (cf. Tuomikoski 1936, Koponen 1967 b). Most of the genera have decurrent leaves but there are differences in the length and width of the decurrent parts. *Mnium* (*M. hornum* excluded), *Trachycystis*, and *Leucolepis*, do not differ greatly from each other in this respect, but *Rhizomnium* diverges from the other genera in having very long and extremely narrow decurrent bases (cf. Tuomikoski 1936, pp. 19 – 20). In the latter genus they are often formed by a

single cell row and therefore easily overlooked, especially when the leaves have been removed from the stem. The width and length of the decurrent parts to some extent depend on the vitality of the stem. As a rule they are not as well defined in plagiotropic parts of the shoots as in erect stems. Of course, the length of the decurrent bases depends on the distance between the leaves. If leaves are close to each other the decurrent parts are short. Therefore, in principle, comparison should not be based on the absolute lengths of the decurrent parts but on the ratio of their length to the length of a segment.

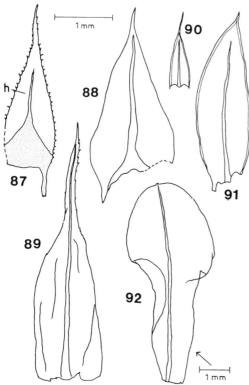

Fig. 87 – 92. Differentiation of leaves. h = hyaline apical part. 87, 88, 89 = basal, normal, and perichaetial leaves in *Leucolepis menziesii*. 90, 91 = basal and normal leaves in *Mnium hornum*. 92 = perichaetial leaf in *Orthomnion loheri*.

5. Characters of the sporophyte

The separation of the genera of *Mniaceae* has in many cases been based on the differences in the sporophyte. This is the case in the genera *Cinclidium*, *Orthomnion*, *Orthomniopsis*, and *Cyrtomnium*.

The peristome is different from the common type (see p. 140) in *Cinclidium*, where the inner peristome is dome-shaped, and in *Orthomnion* and *Orthomniopsis*, in which the peristome is more or less rudimentary.

The stomata are as a rule cryptopore and confined to the neck. *Cyrtomnium* with phaneropore stomata and *Plagiomnium* sect. *Rostrata* with more widely dispersed stomata are exceptions.

The ovate shape of the capsule is rather uniform but its erect position in *Orthomnion* and *Orthomniopsis* represents a distinctive character. An annulus seems to be present in all the genera except *Orthomnion*, although the author was not able to check all the species in this respect. The presence or absence of the rostrum is useful at generic or sectional levels. The structure of the seta proved to be rather uniform except in *Orthomnion* and *Orthomniopsis* in which the epidermis is not differentiated and may be mamillate (Figs. 93 – 100); in addition, the difference between the structure of the epidermis of the seta and the exothecium of the capsule is not so pronounced as in the other genera (Figs. 101 – 106).

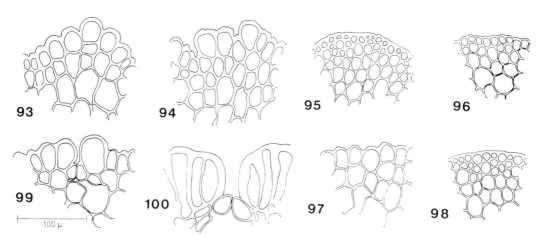

Figs. 93 – 98. Structure of epidermis of seta. Figs. 99 – 100. Cryptopore stomata, transverse sections. 93 = *Orthomnion bryoides*. 94 = *O. loheri*. 95 = *Rhizomnium punctatum* var. *punctatum*. 96 = *Pseudobryum cinclidioides*. 97 = *Orthomniopsis dilatata*. 98 = *Plagiomnium cuspidatum*. 99 = *Orthomnion loheri*. 100 = *Plagiomnium venustum*.

The study of the spores might perhaps reveal some further characters (cf. McCLYMONT 1955); for instance, the spores of *Mnium hornum* may easily be seen to differ from those of *Plagiomnium cuspidatum* in respect of the size and morphology of the papillae of the sporodermis. As regards the size of the spores, those of *Orthomnion* are larger (up to 100 μ) than the spores of the other genera (15 – 60 μ), while those of *Orthomniopsis* have an intermediate position, being about 75 μ. By and large, the sporophyte characters in the genera *Orthomnion* and *Orthomniopsis* differ very distinctly from those of the common type.

6. Colour substances and other chemical characters

Although the division of the family *Mniaceae* presented on pp. 139 – 147 is based mainly on the »orthodox» morphological characters, increasing attention is being paid to chemical characters in the taxonomy of *Mniaceae* and mosses in general (cf. McCLURE & MILLER 1967). The presence or absence of a kind of reddish colour in the gametophyte appears to be very useful. The colour substance is situated in the cell walls and is most easily observed in the stem, midrib, and leaf border. In some species the reddish colour is not visible in old stem parts which then appear brown or black. The reddish colour is present in the tribes *Mnieae* and *Cinclidieae*, with the exception of the genus *Cyrtomnium*, but is absent in *Plagiomnieae* and *Orthomniopsis*. In *Orthomnion* the colour is present in one of the two species. In species which lack the red substance, the young cell walls are colourless and the plants thus give the impression of being green. The old stem parts and the walls of dead lamina cells are brownish or black. It must be mentioned, however, that a reddish colour is present in the seta of many *Plagiomnium* species and of *Pseudobryum cinclidioides*. Though here the colour substance may be different from that in *Mnieae* and *Cinclidieae*, since it is most intense in young sporophytes but is absent or turns to brown in old herbarium specimens.

SENFT (1924 b) studied these differences in the colour substances with the aid of ferrichloride and found that each of the three sections of LIMPRICHT (1895) took on a characteristic colour. *Mnia biserrata* (= *Mnium* s. str.) turned red, *Mnia serrata* (= roughly the genus *Plagiomnium*) turned brown, and *Mnia integerrima* (= roughly the genus *Rhizomnium*) assumed a more or less

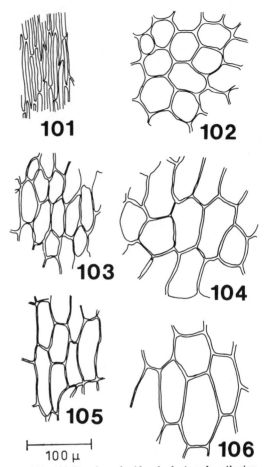

Fig. 101 – 106. Structure of epidermis of seta and exothecium, surface view. 101, 102 = *Mnium hornum*. 103, 104 = *Orthomniopsis dilatata*. 105, 106 = *Orthomnion loheri*.

violet appearance. This result may indicate that the reddish substances in *Mnium* and *Rhizomnium* are not quite identical. There also seems to be some difference in the distribution of the colour substance in the tissues. In stems of *Rhizomnium* and *Cinclidium*, the walls of the cortex cells are coloured as well as the epidermis; in other genera the colour is conspicuous mainly in the epidermis. The unique bluish colouring of the leaves in *Cyrtomnium* is also worth noting.

In addition to the colour substances situated in the cell walls there are colour substances in the cytoplasm and cell sap. The most striking is mnioindigon which is formed postmortally in *Mnium stellare* (cf. SENFT 1924 a). The blue colour is present in the other species of the section *Stellariformia* as well (*M. blyttii*, *M. heterophyllum*) but is not as strong as in *M. stellare*. The colour is known only in this section and thus has taxonomic significance. BROTHERUS (1923) even used this character in his key for *Mnium* species (cf. also LIMPRICHT 1895, CORRENS 1899 b, p. 293, AMANN 1921). In the section *Rosulata* of *Plagiomnium* and in *Orthomnion* there are yellow colour substances which like mnioindigon become visible after the death of the cells. The old herbarium specimens of *Plagiomnium elatum* (= *Mnium seligeri* auct.) are often distinctly yellowish and if shoots from such a specimen are boiled in water to which KOH is then added, the solution turns a strong yellow.

Modern research on the colour substances and other chemical substances in *Mniaceae* has been slight and largely accidental. MELCHERT & ALSTON (1965) found eight different flavone C-glycosides in »*Mnium affine*» (obviously not identical with *Plagiomnium affine* in the present paper, but some other species of the section *Rosulata*, cf. KOPONEN 1967 a, 1967 b). Some other flavonoids were found in *Mnium arizonicum*, and saponaretin in »*Mnium cuspidatum*». KOZLOWSKI (1921) reported saponarin in *P. cuspidatum*, but did not find it in *P. affine*, *P. undulatum*, *Mnium hornum*, or *Rhizomnium punctatum*. BENDZ et al. (1966) did not find proanthocyanids in *Plagiomnium undulatum*. TACHIBINA & MEEUSE (1960) found trans-aconitic acid in *Plagiomnium affine* (cf. above) but not in *Rhizomnium glabrescens* and *Leucolepis menziesii*, and thus it may have some taxonomic value, though the authors were not of this opinion (they used ANDREWS's (1940) classification). Even such an early worker as CZAPEC (1899) found differences in the constituents of the cell walls of *Mnium hornum* and *Rhizomhium punctatum*.

On the basis of the above discussion and examples, it seems evident that the colour substances and their distribution in the genera of *Mniaceae* are of taxonomic significance and should be studied further by modern biochemical methods (cf. MELCHERT & ALSTON op. cit., McCLURE & MILLER 1967). For instance, it would be interesting to know whether the red colour in the sporophytes of *Plagiomnium* is the same as in the gametophytes of, for example, *Mnium*, and the relation of the red colour substances of *Bryum* and some *Rhodobryum* species to those of *Mnium* and related genera (cf. BENDZ et al. 1962, BENDZ & MÅRTENSSON 1963).

7. Karyological evidence

The family *Mniaceae* is one of the best known among the mosses as regards karyological data. The chromosome numbers of some 35 species have been counted. According to HEITZ (1942), LOWRY (1948), TATUNO & ONO (1966), BOWERS (1966), and several other authors (cf. below), the basic chromosome number of the genus *Mnium* (s. lat.) is n = 6 or 7. *Leucolepis* is the only genus with the basic number 5.

However, the study of karyotypes has been shown to be more significant in taxonomy than the chromosome number alone. In his classical study, LOWRY (1948) showed that KABIERSCH's (1936, cf. p. 119) division of *Mnium* (s. lat.) into sections is well founded from this point of view. LOWRY's conclusions are based on the differences in the chromosome morphology and in the length of corresponding chromosomes. His most important conclusions from the standpoint of the present study are that *Leucolepsis* should be excluded from the genus *Mnium* (cf. also STEERE, ANDERSON & BRYAN 1954), that, as regards karyotype, *Cinclidium* is more similar to *Rhizomnium* than to the other groups, and that *Mnium cinclidioides* (*Pseudobryum*) occupies an independent position having longer and thinner chromosomes than the other groups studied. TATUNO and ONO (1966) confirmed that not only *Trachycystis* and *Mnium* (s. lat.) but also the subdivisions of the latter genus have karyotypes which differ from each other. It even seems that there are differences between the karyotypes of some sections of *Mnium*, in the sense of the present work. This can be seen in Table 3 in TATUNO's and ONO's (op. cit.) paper. LOWRY (op. cit.) stated that *Mnium hornum* differs from other »biserrate» species in its chromosome morphology.

Thus, the chromosome morphology and other karyological data revealed by two studies based on different materials accord with each other and also support the division of *Mnium* s. lat. into smaller well-delimited genera based mainly on morphological differences (pp. 139–147). Further research in the field of cytotaxonomy is needed in *Mniaceae*. The study of the karyotypes of *Pseudobryum speciosum*, *Mnium immarginatum*, and the *Orthomnion* and *Orthomniopsis* species, in particular, would provide supplementary information on their relationships.

In the following the known chromosome numbers in *Mniaceae* are listed with the references to the literature. Some older records listed by LOWRY (1948), WYLIE (1957) and TATUNO & ONO (1966) are excluded from the list.

	n	
Mnium		
sect. *Mnium*		
M. hornum	6	Several authors, cf. TATUNO & ONO 1966
	7	TATUNO & ONO 1966
	12	HOLMEN 1958
sect. *Spinosa*		
M. ambiguum	6	BOWERS 1966
M. arizonicum	6	BOWERS 1966
	12	BOWERS 1966
M. laevinerve	7	TATUNO & ONO 1966
M. lycopodioides	6	BOWERS 1966
M. marginatum	12	LOWRY 1948, HOLMEN 1958, ANDERSON & CRUM 1958, BOWERS 1966
M. orthorrhynchum	6	HEITZ 1942, LOWRY 1948, BOWERS 1966
M. spinosum	6	HEITZ 1942, LAZARENKO & LESNYAK 1966
M. spinulosum	8	LOWRY 1948, ANDERSON & CRUM 1958, IRELAND 1967
sect. *Stellariformia*		
M. blyttii	7	BOWERS 1966
M. stellare	7	HEITZ 1942, LOWRY 1948
Trachycystis		
T. flagellaris	7	TATUNO 1951, TATUNO & YANO 1953, TATUNO & ONO 1966
T. microphylla	6	KURITA 1939
	7	TATUNO 1951, TATUNO & YANO 1953, TATUNO & ONO 1966
Leucolepis		
L. menziesii	5	LOWRY 1948, STEERE et al. 1954, IRELAND 1967
Cinclidium		
C. stygium	14	LOWRY 1948
C. subrotundum	14	STEERE 1954
Rhizomnium		
sect. *Rhizomnium*		
R. andrewsianum	7	BOWERS 1966
R. glabrescens	6	LOWRY 1948
R. pseudopuncta- tum	13	HEITZ 1942, VAARAMA 1950,
	14	LOWRY 1948, BOWERS 1966
R. punctatum	7	Several authors, cf. TATUNO & ONO 1966, LAZARENKO et al. 1965, 1967, BOWERS 1966
R. punctatum var. elatum	7	BOWERS 1966
Cyrtomnium		
C. hymenophyl- loides	7	HEITZ 1942

Plagiomnium		
sect. *Plagiomnium*		
P. cuspidatum	6	LOWRY 1948
	12	HEITZ 1942, LOWRY 1948, HOLMEN 1958, LAZARENKO et al. 1965, BOWERS 1966
P. drummondii	6	LOWRY 1948
P. japonicum	7	TATUNO & ONO 1966
P. trichomanes	6	KURITA 1939, BOWERS 1966
	7	TATUNO 1951, TATUNO & ONO 1966
sect. *Rosulata*		
P. »affine» (cf. KOPONEN 1967 b)	6	LOWRY 1948
P. affine	6	HOLMEN 1958
P. (affine var.) ciliare	6	LOWRY 1948
P. elatum (= M. seligeri auct.)	6	HEITZ 1942, HOLMEN 1958
P. insigne	6	LOWRY 1948, IRELAND 1967
P. medium	12	LOWRY 1948, VAARAMA 1950, HOLMEN 1958, BOWERS 1966
P. rugicum	6	HOLMEN 1958, BOWERS 1966
sect. *Undulata*		
P. undulatum	6	HEITZ 1942, HAMANT 1950, HOLMEN 1958
	7	TATUNO & ONO 1966
sect. *Rostrata*		
P. maximoviczii	7	Several authors, cf. TATUNO & ONO 1966
P. rostratum	12	HEITZ 1942, HOLMEN 1958
P. rostratum? (as »Mnium longi- rostrum»)	7	TATUNO & ONO 1966
P. succulentum	7	TATUNO & ONO 1966
Pseudobryum		
P. cinclidioides	6	LOWRY 1948
P. speciosum	7	TATUNO & YANO 1953

C. Grounds for the classification

The last few years have seen the development of two classificatory systems based on essentially different principles, the numeral-phenetic system (e.g. SOKAL & SNEATH 1963) and the cladistic-phyletic (e.g. HENNIG 1950, 1966, ZIMMERMANN 1960). Any numerical-phenetic classification is out of the question in the present case because the number of the characters needed is high, according to MICHENER & SOKAL (1957) 60 at least. The organization of mosses is rather simple and, accordingly, the number of useful characters is low especially when, as in the present study, attention is mainly limited to taxa above the specific rank. Biochemical data may, however, later raise the number of characters (cf. p. 131) so that a numerical classification may be attempted. The low number of characters also hinders classification on the cladistic-phyletic basis. In addition, the traditional family *Mniaceae* seems to be too limited to form the sole object of a phylogenetic research; better results can perhaps be attained by extending the study to include some related families. Although no final classification can yet be established, the phylogenetic method may contribute some information on such questions as the monophyly of the taxa proposed in this study. This would be a better basis for the classification than mere »general similarity» or outdated traditional opinions.

1. Phylogenetic considerations

Before the possible phyletic relationships of the taxa are considered, an attempt must be made to classify the characters as original (primitive, plesiomorphous) or derived (advanced, apomorphous). This cannot always be decided and is especially difficult when *Mniaceae* alone is considered. Therefore, some comparative studies were carried out on the related families *Rhizogoniaceae* and *Bryaceae*. The methods employed were those presented by HENNIG (1950, 1966).

Research on the phylogenetic relationships of bryophytes is hampered by the scarcity of fossils (see GAMS 1932, STEERE 1946, 1958, SAVICZ–LJUBITZKAJA & ABRAMOV 1959). From the point of view of *Mnium*, the most important writings on this subject may be NEUBURG's (1956, 1960) reports from Permian deposits in the area of the ancient Angara land (roughly comprising northern Siberia and a part of European Russia). Among the material she describes are several species of the genus *Intia* Neub. The leaf characters, areolation and serrulation, resemble those of the present family *Bryaceae* or some genera of *Mniaceae*, especially *Plagiomnium* and *Pseudobryum*. DIXON (1916) mentions *Mnium antiquorum* Card. & Dix. (*Trachycystis* according to KABIERSCH, 1936) dating from the Pliocene, discovered by the River Maas, at the boundary between the Netherlands and Germany, and SZAFRAN (1950) describes *Trachycystis szaferi* found in the Miocene deposits in Poland. The finds are remarkable because the genus *Trachycystis* no longer occurs in Europe.

Leaf characters. – The character multistratose versus unistratose leaf border divides *Mniaceae* into two parts. The unistratose leaf border is the common type in bryophytes, including *Bryaceae*, although multistratose borders are present in some *Bryum* species, too (cf. BROTHERUS 1924, NYHOLM 1958). It would seem probable that the multistratose border, being more complicated, is in general a derived character and that the unistratose border is original. Regressive evolution from multistratose to unistratose has probably taken place in both *Mnieae* and *Cinclidieae*. *Mnium stellare* provides rather good evidence

that regression really has taken place (see p. 140). The regression has been accompanied by the disappearance of a definite border (*M. stellare*, *Leucolepis*, *Mnium immarginatum*). It also seems probable that the toothed margin is a primitive character in *Mniaceae* and the lack of serratures a derived one. Toothed borders are present in all the phyletic branches except *Cinclidieae* and *Orthomnieae*. The presence of double teeth in *Mnium* and *Trachycystis* is perhaps not an independent character, in so far as it is connected with the multistratose leaf border (cf. p. 127). A large number of stereids in the costa is probably a more primitive character than a small number or their absence. In several genera of *Bryaceae* they are numerous, practically filling the costa. Thus *Leucolepis* is the most »primitive» genus within *Mniaceae* in this respecr. The two main phyletic branches, *Mnieae– Cinclidieae* and *Plagiomnieae* (Fig. 107), include species with both bands of stereids which suggests that this situation is the original one. Otherwise this rather complicated structure must have originated on two separate occasions. *Mnium stellare* and *Rhizomnium* sect. *Micromnium* (see p. 143) offer evidence of the reduction of stereids. Although the leaf and cell shapes and size are rather similar within the genera considered, they may represent parallel developments caused by similar environmental conditions. For instance, in *Rhizomnium punctatum* var. *elatum*, *Plagiomnium rugicum* and *Pseudobryum cinclidioides* the shapes of the leaves and the size of the lamina cells are rather similar, although they are not necessarily close relatives. Thus these characters cannot be taken into consideration in the phyletic discussion.

Stem characters. – As regards the structure of the stem epidermis, the Bryum type is most probably the original one. The Mnium type is more complicated, and although slight secondary thickening was found in some *Bryum* and *Plagiomnium* species it has not advanced so far in them as in the tribes *Mnieae* and *Cinclidieae*. The Mnium type was also present in all the *Rhizogonium* species studied. The Pseudobryum type differs less distinctly from the Bryum type, but is, however, characteristic, and, being more complicated than the Bryum type, is probably derived. The types of rhizoids and their distribution on the stem will have to be studied in other groups of mosses before general conclusions can be drawn. However, the stem with both types of rhizoids distributed along its length is probably the original one. *Orthomnion* and *Orthomniopsis* thus appear to possess the most primitive type and this view is supported by the fact that the two rhizoid types are not morphologically differentiated in these genera. The presence of plagiotropic shoots as well as rich subapical branching are probably derived characters. Since the significance of the colour substances is still practically unknown, it can only be pointed out that in the genera studied the presence of reddish colouring was found to be positively correlated with a multistratose leaf border, a stem epidermis of the Mnium type, and the absence of micronematous initials on the stem.

Sporophyte characters. – The characters which unite most of the taxa under discussion are the shape of the capsule and the structure of the peristome and seta. However, a rather similar peristome is present in some related families, such as *Bryaceae* and *Rhizogoniaceae*. The dome-shaped inner peristome in *Cinclidium* is clearly a younger derivative. It is more difficult to estimate the significance of the rudimentary peristome in *Orthomnion* and *Orthomniopsis*, because it may be connected with the erect capsule (cf. LOESKE 1910). The phaneropore stomata in *Cyrtomnium* are probably not a derived feature since they are the common type in *Bryaceae* (cf. PATON & PEARCE 1957). The scattered distribution of the stomata in *Plagiomnium* sect. *Rostrata* is possibly a relic of the original type. The complete absence of the rostrum of the lid may be a derived character.

Tribe Mnieae. – The genus *Mnium* does not seem to have any exclusive characters (autapomorphous characters sensu HENNIG 1966). However, within *Mnium* there is at least one section which would seem to be monophyletic, viz. the section *Stellariformia*. The extreme reduction of the rostrum and the stereids in the costa have taken p'ace in other phyletic branches, too, but the blue colour substance (p. 131) is present exclusively in this section. Derived characters also exist in the other two genera, viz., the extreme differentiation of the basal leaves in *Leucolepis*, and the mamillate lamina cells, and perhaps also the unique structure of the stem epidermis, in *Trachycystis* (p. 122). A difficult problem is the relation of the genus *Rhizogonium* to *Mnieae*. In some *Rhizogonium* species the stem epidermis, the absence of micronemata, the toothed costa,

the bi- or multistratose leaf border, and the presence of reddish colour are similar to the characters found in *Mnium*. It has already been suggested by KABIERSCH (1936) that *Rhizogonium* shares so many characters with *Mnium* (s. str.) and with *Trachycystis* that these genera probably constitute a monophyletic group. As mentioned above, the characters of the peristome are also similar. The sporophyte of *Rhizogonium* is, however, »lateral»; it develops from a short branch originating from a bud. Since the fertile stems of *Mnium* and related genera are in fact branches (pp. 121 – 122), the main difference between the »lateral» and acrocarpic sporophyte in this case seems to be that the branch in *Rhizogonium* is shorter, and that it may also originate on the upper part of the stem. For instance, in the case of *Trachycystis microphylla* and *Rhizogonium bifarium* the phenetic similarity is obvious, and the main differences are constituted by the mamillate lamina cells of *Trachycystis* and the lateral sporophyte of *Rhizogonium*.

There seem to be two derived and two original characters in *Mnieae* which separate it from *Cinclidieae* (Fig. 107). However, the tribe *Mnieae* is possibly not monophyletic in the strict sense of the word (cf. HENNIG 1966) because one of the groups which may belong to this branch, *Rhizogonium*, has been excluded. Further study will probably lead to alterations in the present treatment.

Tribe Cinclidieae. – It seems fairly certain that the genus *Cinclidium* is monophyletic. It is hardly possible that two such unconnected characters as the peculiar structure of the inner peristome and the elongated area of the macronematous initials (p. 126) can have originated together more than once. The genus *Rhizomnium* is not as uniform as *Cinclidium*. The presence of micronematous initials seems to some extent to correlate positively with the partly unistratose leaf border (p. 143). The unique long and narrow decurrent leaf bases and the complete reduction of stereids in the costa (except in *R. minutulum*) appear to be derived characters when compared with those of *Cinclidium*. The genus *Cyrtomnium* has retained the possibly original phaneropore stomata. The absence of reddish colouring and the bluish leaves are perhaps derived characters. The thick outermost cell row of the lamina in *C. hymenophyllum* may indicate the reduction of a multistratose leaf border.

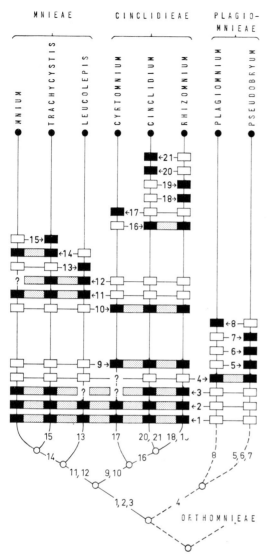

It seems rather probable that *Cinclidium* and *Rhizomnium* together constitute a monophyletic branch, to which possibly also *Cyrtomnium* belongs.

Tribe Orthomnieae. – The two species of *Orthomnion* have so many unique sporophyte characters in common (cf. p. 130) that their monophyly is fairly certain. The only *Orthomniopsis* species examined in this study differs from them in several sporophyte characters, e.g. the peristome is not as rudimentary, the calyptra is not hairy, an annulus is present and the seta is shorter. It is hardly possible to decide which of these characters are original and which are derived. The growth-form, the rhizoid topography, and the stem epidermis are similar in both genera and seem to unite them with *Plagiomnium*. KABIERSCH (1936; cf. also Fig. 108) suggested that they have developed from *Plagiomium* sect. *Rostrata*. However, the bistratose, entire leaf border, the structure of the costa, and the reddish colour present in *Orthomnion bryoides* seem to point in the direction of *Rhizomnium*, as was supposed by BROTHERUS (1905).

Although *Orthomnion* is probably monophyletic, any conclusions regarding its phylogenetic relationships can hardly be drawn on the basis of the present data. The possibility cannot be ruled out that the tribe suggested, although in part phenetically uniform, is polyphyletic in origin. The characters of the sporophyte, not only those of the capsule but also those of the seta, are very different from the common type in *Mniaceae*. This may indicate that the two genera are not very close phyletic relatives of the other groups now under discussion. Accordingly, the tribe *Orthomnieae* is not included in the dendrogram in Fig. 107.

Tribe Plagiomnieae. – The other main phyletic branch (Fig. 107), *Plagiomnium* – *Pseudobryum*

Fig. 107. The assumed phylogeny of *Mniaceae*. The black squares indicate derived (apomorphous) characters and the open squares the corresponding original (plesiomorphous) characters present in the basic type. Question-marks are used to show the deviations in this scheme, and thus may also indicate characters which appear to be derived when compared with those of the sister group. The tribe *Orthomnieae*, whose location is somewhat uncertain, has been left out. The numbers indicate the following derived characters (in the column of *Rhizomnium* at the bottom, after 18, read 19):

1 = stem epidermis of Mnium type
2 = absence of micronemata, except in certain species of *Rhizomnium*
3 = bistratose or multistratose leaf border, except in *Cyrtomnium* and *Leucolepis*
4 = absence of reddish colouring in the gametophyte (also in *Cyrtomnium*)
5 = complete reduction of stereids in costa
6 = branched midrib

7 = stem epidermis of Pseudobryum type
8 = plagiotropic shoots
9 = complete reduction (entral stereid band
10 = entire border
11 = costa toothed at the back apically
12 = frequent subapical branching (rare in *Mnium*)
13 = differentiation of basal leaves
14 = complete or partial reduction of ventral stereid band
15 = mamillate leaf cells
16 = cryptopore stomata
17 = bluish colour in the leaves
18 = extremely narrow and long decurrent leaf bases
19 = complete reduction of stereids, except in sect. *Micromnium*
20 = dome-shaped inner peristome
21 = elongated area of macronematous initials

is distinguished from *Mnieae – Cinclidieae* by one possibly derived character, the absence of the reddish colouring, and by such original characters as a stem of the Bryum (or Pseudobryum) type, a unistratose leaf border, and the presence of micronematous initials. The presence of plagiotropic shoots in *Plagiomnium* is possibly a derived character, and other derived characters exist in some of its sections, for example, the complete reduction of stereids in *Venusta*, and the subapical branching in the section *Undulata*. The section *Rostrata* has retained the possibly original characters, rostrate operculum and scattered distribution of stomata. Its possible phyletic relationship with *Orthomnieae* may eventually necessitate alterations in the present concepts. The sections *Plagiomnium* and *Rosulata* do not seem to have any clearly exclusive characters. However, these sections are phenetically rather uniform (cf. p. 145). The genus *Pseudobryum* seems to be rather definitely monophyletic. The similarity of the structure of the stem epidermis, and the areolation (cf. however p. 134) of the lamina in the two species, and the structure of the branching costa support this opinion.

The scheme of the possible phylogeny of *Plagiomnieae*, presented in Fig. 107, may have to be revised when some related groups are taken into account. The genera *Rhodobryum*, *Roellia*, *Bryomnium*, and *Bryum* could perhaps be located between the two main branches in the present concept. For instance, *Rhodobryum* partly shares the characters of *Plagiomnium*, viz., the absence of reddish colouring, unistratose leaf border, serrate leaf margin, reduction of stereid bands to one or nil, stem epidermis of Bryum type, presence of micronemata, and even the plagiotropic shoots (underground organs in *Rhodobryum*). On the other hand, some *Rhodobryum* species may be related to *Bryum*, e.g. they possess the reddish colour and reflexed leaf border. Thus the suggested tribe *Plagiomnieae* may also be a paraphyletic group.

When the classification presented on pp. 139 – 147 is considered in the light of the above discussion and the possible phyletic relationships (Fig. 107), it is seen that the classification has been largely drawn up on the evolutionary basis. One of the most central problems in phyletic systematics is to what degree the claim of monophyly must be applied in the delimitation of the taxa (cf. HENNIG 1950, 1966, SIMPSON 1961, SOKAL & SNEATH 1963, MAYR 1965, BRUNDIN

1966, TUOMIKOSKI 1967). Most of the genera now proposed have yielded evidence of probable monophyly. On the other hand, *Cinclidieae* may be the only tribe which is monophyletic in the strict sense of the word; *Mnieae* and *Plagiomnieae* are possibly paraphyletic, and *Orthomnieae* may even be polyphyletic. The classification presented here is thus not strictly cladistic-phyletic but also takes account of phenetic discontinuities and practical considerations.

In the course of the present study a preliminary comparative research was carried out on some related groups (cf. also p. 120). This investigation suggested that the line of demarcation between the families *Rhizogoniaceae* and *Mniaceae* is not well drawn. For instance, the genus *Rhizogonium* is perhaps more closely related to *Mnium* and *Trachycystis* (cf. p. 135) than are the other genera of *Mniaceae*, e.g. *Plagiomnium* and *Pseudobryum*. The boundary between *Mniaceae* and *Bryaceae* also seems to be open to question. For instance, ENGLER's Syllabus (MELCHIOR & WEDERMAN 1964) gives as separating characters only the differences in the form of the lamina cells (which does not apply in the case of *Rhodobryum* and *Pseudobryum*), and in the shape of the male inflorescence and paraphyses. Some features of *Rhodobryum* have already been mentioned. The stem, rhizoid, and colour characters, together with the growth-form and some characters of the sporophyte, such as the structure of the seta and stomata, might possibly be useful in the delimitation of these families. However, I do not want to enter further into this question in this connection and on the basis of an incomplete preliminary research. The unsatisfactory delimitation of the families is the main reason that taxa higher than tribes are not presented in this paper.

When the present concept of the possible phylogenetic relationships is compared with the earlier classifications, it is found that the three main phyletic branches in the dendrogram roughly correspond to the subdivisions presented by such workers as LIMPRICHT (1895) and LOESKE (1910; cf. also p. 118). Fig. 108 gives the phylogenetic scheme of *Mniaceae* presented by KABIERSCH (1936, redrawn). The main differences between his system and the present concept (Fig. 107) are that KABIERSCH derived *Rhizomnium* and *Cinclidium* from *Plagiomnium*, and that he considered forms like *Mnium* and *Rhizogonium* as primitive ones. KABIERSCH's scheme agrees with the present in giving a rather in-

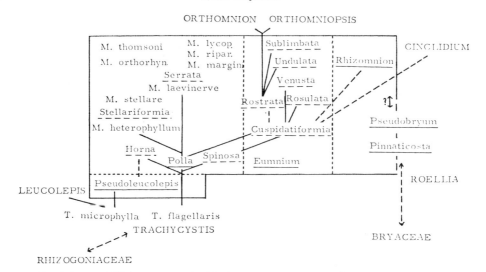

Fig. 108. The phylogeny of *Mniaceae* as conceived by Kabiersch (1936, redrawn). Solid lines indicate rather probable phyletic lines, the broken ones less probable. The boundaries of the genus *Mnium* are indicated by solid lines, or when some uncertainty exists, with a broken line. The boundaries of the sections are marked by dotted lines.

dependent position to *Pseudobryum*. His dendrogram even gives the impression that he considers the family *Mniaceae* as possibly polyphyletic in origin.

2. Categories used

In the light of the above discussion (cf. also pp. 139 – 147), the ten taxa which I call genera seem to be rather »natural». They are phenetically uniform, and in addition, most of them are probably monophyletic. The generic rank was selected because the ten taxa fulfil the criteria which as a rule have been used in the delimitation of genera (cf. e.g. Davis & Haywood 1963), and in particular those presented by Steere (1947), in connection with mosses. For instance, the characters of both the sporophyte and the gametophyte have been taken into account, each of the genera are separated from the nearest related ones by at least two unique characters, etc.. Altogether, Steere (op. cit.) listed thirteen different points and criteria and, on the whole, these have been accepted in the present study. An additional advantage of the present concept is that the largest genera *Mnium* and *Plagiomnium* can be divided further into natural subdivisions, named sections in this paper.

If generic rank is not accepted for the taxa concerned there are two other possibilities: to

retain the large traditional genus *Mnium* with ten sections, two of them with subsections, or to consider as genera the four larger taxa which I call tribes. In the latter case there would be the genera *Mnium* and *Cinclidium* with three sections, *Orthomnion* with two (or one) section and *Plagiomnium* with two sections. The first alternative would take us back to the classification of C. Müller (1848 – 1849, 1901), and the resulting genus would be incoherent phenetically and possibly also phylogenetically. For instance, such commonly accepted genera as *Leucolepis* and *Cinclidium* would have to be reduced in rank. The second possibility, to consider the present tribes as genera, is more acceptable. The genera thus formed would be phenetically rather uniform, although they would probably not fulfil the condition of monophyly (see p. 137). In addition, this choice would divide the traditional genus *Mnium* s. lat., which in the light of the present study is rather incoherent. It is not phenetically uniform and each of the old sections (see p. 119) *Polla*, *Eumnium*, *Rhizomnion*, and *Pseudobryum* seem to belong to different phyletic branches The nearest relatives of the section »*Polla*» are not the sections »*Eumnium*» or »*Rhizomnion*» but the genera *Leucolepis* and *Trachycystis*. Similarly, »*Rhizomnion*» seems to be one of the nearest relatives of *Cinclidium* and *Cyrtomnium* and, if these are accorded generic rank, it must also be given to *Rhizomnion* (cf. also p. 118). The section »*Eumnium*» (= the present *Plagiomnium*) shows a greater resemblance to *Pseudobryum* than to »*Rhizomnion*» or »*Polla*», and has perhaps closer phyletic connections with *Pseudobryum*.

The present arrangement causes numerous alterations in the nomenclature, although an attempt was made to avoid the creation of new names by raising old sectional names to generic rank. If the taxa referred to here as tribes are considered as genera, the nomenclatural changes will be almost equally numerous.

III. Classification

Analytic key to the genera

1. Leaf margin serrate
  2. Back of midrib toothed apically
   3. Lamina cells not mamillate
    4. Border unistratose, teeth single
     5. Dendroid growth-form, basal leaves differentiated .... *Leucolepis*
     5. Subapical branches few, basal leaves not so clearly differentiated .. *Mnium immarginatum*
    4. Border multistratose, with double teeth .. *Mnium*
   3. Lamina cells mamillate .... *Trachycystis*
  2. Back of midrib not toothed
   6. Micronemata on upper stem parts, no blue colour, red colour in sporophyte at most
    7. Lamina cells rhomboid, no plagiotropic shoots .... *Pseudobryum*
    7. Lamina cells rounded, hexagonal or elongated, plagiotropic shoots generally present .... *Plagiomnium*
   6. Only macronemata on upper stem parts, blue postmortal discoloration, reddisch colour in stem and leaves .... *Mnium* sect. *Stellariformia*
1. Leaf margin entire
  8. Leaf border, at least partly, bi- or multistratose
   9. No plagiotropic shoots, capsule horizontal or pendulous
    10. Only macronemata present, their initial region elongated, strong red colour in the gametophyte, inner peristome fused, dome-shaped .. .. *Cinclidium*
    10. Macronematous initial region elliptic or rounded, micronemata may be present, red colour weaker, inner peristome not dome-shaped .... *Rhizomnium*
   9. Plagiotropic shoots present, capsule erect *Orthomnion*
  8. Border unistratose throughout
   11. Reddish colour present in the gametophyte, no plagiotropic shoots
    12. Macronematous initial region elongated, micronemata not present, inner peristome dome-shaped .... *Cinclidium*
    12. Macronematous initial region rounded or ovate, micronemata may be present, peristome not dome-shaped .... *Rhizomnium*
   11. Reddish colour absent in the gametophyte, plagiotropic shoots may be present
    13. Micronemata and plagiotropic shoots present
     14. Capsule erect, micro- and macronemata not morphologically differentiated
      15. Seta 0.5 cm long at most, calyptra not hairy .... *Orthomniopsis*
      15. Seta about 2 cm long, calyptra hairy .... *Orthomnion*
     14. Capsule horizontal – pendulous, macro- and micronemata differentiated .... *Plagiomnium*
    13. Micronemata and plagiotropic shoots absent, leaves bluish green .... *Cyrtomnium*

Familia Mniaceae Mitt. 1859

Typus: *Mnium* Hedw.

[Fam.] XIV. *Mniaceae* Mitten, Journ. Linn. Soc. Suppl. Bot. 1: 137. 1859.

Subtribus I. *Mniaceae* C. Müller, Syn. Musc. Frond. 152. 1848.

Fam. *Bryaceae* 11. *Mnioidae* a) *Mniaceae* Rabenhorst, Deutschl. Krypt. Fl. 2 (3): 223. 1848 (nom. inval.).

Fam. *Cinclidiaceae* Kindberg, Gen. Eur. N. Am. Bryin. 35. 1897 (nom. nud.).

Note. The ranks of the taxa in MITTEN's (op. cit.) paper are not clearly indicated. However, on p. 1 the sentence: »– The Cleistocarpous Order has been suppressed, because its component groups are readily referable to families of higher development: thus *Sphagnaceae* have been removed from the side of *Leucobryaceae*, ...» makes it possible to accept his taxa marked with Roman numerals and *-aceae* suffixes as families. The »*Mniaceae*» of RABENHORST (op. cit.) contravenes Art. 33 of the Code. I have not followed the example in Art. 61, but that of Art. 19, according to which the authors would be *Mniaceae* (C. Müll.) Mitt.

Since the delimitation of the family *Mniaceae* is possibly not final no diagnostic description for separating *Mniaceae* from the other families is given here. For practical purposes, however, the characters common to the taxa below the rank of family are listed. They are not repeated in the descriptions of the lower taxa and are valid unless otherwise stated.

Acrocarpic mosses with buds at axil of every leaf. Shape of macronematous initial region rounded or ovate. Fertile erect and sterile arcuate stems develop mostly from basal buds. Colour substances in stem mostly restricted to epidermis. Leaves mainly decurrent, with distinctly differentiated border. The majority of basal leaves undifferentiated. Costa mostly with stereids, unbranched. Cells of lamina rounded quadrate or hexagonal and elongated, not mamillate. Polysety common. Seta long. Capsule horizontal – pendulous. Stomata cryptopore, confined to the neck. Cells of the epidermis of seta and exothecium distinctly different from each other. Operculum mostly rostrate, calyptra not hairy. Annulus differentiated. Outer peristome yellowish, papillose, teeth 16, lanceolate, lamellae numerous, inner peristome with basal membrane, papillose, yellowish or reddish brown, segments perforated, cilia 2 – 4, nodulose. Spores 15 – 60 μ wide.

A. Tribus Mnieae C. Müll. 1848.

Typus: *Mnium* Hedw.
Trib. XIV. *Mnioideae* C. Müller, Syn. Musc. Frond. 152. 1848.

Subapical branching common especially in fertile erect stems. Stem epidermis of Mnium type. Micronemata absent on the upper stem parts. Leaves narrowly elliptic, or elliptic, acute or acuminate, toothed, border mostly bi – multistratose. Reddish colour present. Costa mostly with two stereid bands, usually toothed at the back apically.

1. Genus Mnium Hedw. 1801.

Mnium Hedwig, Spec. Musc. 188. 1801 nom. cons., non Linnaeus Sp. plant. 1109. 1753. – Typus: *M. hornum* Hedw. (typ. cons.).
Astrophyllum [Neck.] Lindberg, Musc. Scand. 13. 1879 (nom. illeg. pro gen. *Mnium*). – [*Astrophyllus* Necker, Elem. Bot. 3: 326. 1790]. – *Mnium* subg. *Astrophyllum* Gams, Kl. Krypt. Fl., ed. 4., 151. 1957 (nom. inval. et illeg. pro *Mnium* subg. *Mnium*). Lectotypus: *Astrophyllum hornum* (Hedw.) Lindb. (selected here).
non *Astrophyllum* Torrey & A. Gray, Pac. Railr. Rep. 2: 161. 1854 (Phan.) not seen, cf. WIJK et al. 1959).
Polla (Brid.) Loeske, Stud. Morph. Syst. Laubm. 129. 1910 (nom. illeg. pro gen. *Mnium*). – *Bryum Polla* Bridel-Brideri, Bryol. Univ. 688. 1826. – *Mnium* sect. *Polla* (Brid.) Mitten, Trans. Linn. Soc. London Bot. ser. 2 (3): 169. 1891. – Lectotypus: *Polla horna* (Hedw.) Loeske (selected here).

Note. PROSKAUER (1963) showed that the name *Mnium* Hedw. in fact is a later homonym of the name *Mnium* L. and proposed that *Mnium* Hedw. should be conserved against *Mnium* L. This proposal was accepted by the Edin-

burgh Botanical Congress in 1964 (cf. WIJK 1964) and the genus *Mnium* Hedw. was included in the list of »Nomina generica conservanda» in the Code, 1966.

Subapical branching rare. Lamina cells medium-sized. Leaf margin usually with double teeth, border multistratose. Costa ending mostly in the apex.

Taxonomic remarks. – The main difference between the present concept and KINDBERG's (1897) and KABIERSCH's (1936) subdivision is the union of the (sub)sections *Spinosa* and *Serrata*. The separation was based on differences in size and in the shapes of leaves and lamina cells. An additional character which unites *M. spinosum* and *M. spinulosum* is the brown outer peristome, which in other *Mnium* species is yellow. The differences between *M. spinosum* and *M. spinulosum* and the other species of the section *Spinosa* are not as definite as the differences between *Spinosa* and the sections *Mnium* and *Stellariformia*. *M. stellare* in the latter section is most exceptional, having a unistratose leaf margin without any border, and teeth which are usually single, and lacking stereid bands. The stem characters, the chromosome morphology (LOWRY 1948), and the blue colour formed in the dead cells (see p. 131) show that it belongs to the section *Stellariformia* and to *Mnium*.

Sectio Mnium

Holotypus: *Mnium hornum* Hedw.
Mnium b. *serratifolia* Ångström in E. Fries, Summ. Veg. Scand. 1: 87. 1846 (nom. nud.).
Mnium b. *serratifolia* *marginata* Ångström in E. Fries, Summ. Veg. Scand. 1: 87. 1846 (nom. nud.).
Mnium sect. *Eumnium* C. Müller, Syn. Musc. Frond. 1: 155. 1848 (nom. illeg. pro sect. *Mnium*). – Lectotypus (selected here): *M. hornum* Hedw.
Mnium A. *Biserratae* Limpricht, Laubm. Deutschl. 2: 452. 1892. – Lectotypus (selected here): *M. hornum* Hedw.
Mnium sect. *Polla* (Brid.) Mitten, cf. above.
Mnium sect. *Biserrata* Amann & Meyl., Fl. Mouss. Suisse 1: 137 1912 (nom. illeg. pro sect. *Mnium*). – Lectotypus (selected here): *M. hornum* Hedw.
Mnium sect. *Polla* subsect. *Horna* (Kindb.) Kabiersch, Hedwigia 76: 17. 1936 (nom. illeg pro subsect. *Mnium*). – *Mnium* II. *Eu-Mnium* 10. *Horna* Kindberg, Eur. N. Am. Bryin. 2: 339. 1897. – Holotypus: *M. hornum* Hedw.

Leaves not decurrent, costa ending below apex. Capsule larger than in other sections, cernuous. Operculum not rostrate. Monosetous.

The only species of this section is *Mnium hornum* Hedw.

Sectio Spinosa (Kindb.) Koponen, comb. nova

Mnium II. *Eu-Mnium* 8. *Spinosa* Kindberg, Eur. N. Am. Bryin. 2: 339. 1897. – *Mnium* sect. *Polla* subsect. *Spinosa* (Kindb.) Kabiersch, Hedwigia 76: 17. 1936. – Lectotypus (selected here): *Mnium spinosum* (Voit) Schwaegr. *Mnium* sect. *Polla* subsect. *Serrata* (Kindb.) Kabiersch, op. cit. p. 22. – *Mnium* II. *Eu-Mnium* 9. *Serrata* Kindberg, op. cit. p. 339. 1897. – Lectotypus (selected here): *Mnium marginatum* (With.) P. Beauv.

Note. It may be mentioned that if it is desired to follow the section concept of KABIERSCH (1936), the name *Serrata* cannot be used at sectional level because AMANN's & MEYLAN's (1912) section *Serrata* has priority (cf. p. 145).

Leaves decurrent, costa ending in apex, capsule horizontal – cernuous, operculum rostrate. Polysety often present.

Species:

Mnium arizonicum Amann
M. laevinerve Card.
M. lycopodioides Schwaegr.
M. marginatum (With.) P. Beauv.
M. orthorrhynchum Brid.
M. spinosum (Voit) Schwaegr.
M. spinulosum B.S.G.
M. thomsonii Schimp.

Sectio Stellariformia (Kindb.) Koponen, comb. nova

Mnium II. *Eu-Mnium* 3. *Stellariformia* Kindberg, Eur. N. Am. Bryin. 2: 338. 1897. – *Mnium* sect. *Polla* subsect. *Stellariformia* (Kindb.) Kabiersch, Hedwigia 76: 19. 1936. – Lectotypus (selected here): *Mnium stellare* Hedw. *Mnium* b. *serratifolia* **immarginata* Ångström in Fries, Summ. Veg. Scand. 1: 87. 1846 (nom. nud.). *Mnium* Gruppe 2. *Mnia Stellaria* C. Jensen, Danm. Moss. 2: 448. 1923. – Holotypus: *Mnium stellare* Hedw.

Costa lacking stereid bands or with dorsal band, usually ending below apex, its back not toothed. Leaf border uni- or bistratose, or without border. Operculum not rostrate. Postmortal blue colour (mnioindigon) present. Monosetous.

Species:

Mnium blyttii B.S.G.
M. heterophyllum (Hook.) Schwaegr.
M. stellare Hedw.

2. Genus Trachycystis Lindb. 1868

Trachycystis Lindberg, Not. Sällsk. F. Fl. Fenn. Förh. 9 (N. S. 6): 80. 1868. – *Mnium* sect. *Trachycystis* (Lindb.) Mitten, Trans, Linn. Soc. London ser. 2 (3): 169. 1891. – *Mnium* 4. *Trachycystis* (Lindb.) C. Müller, Gen. Musc. Frond. 138. 1901. – Holotypus: *T. microphylla* (Doz. & Molk.) Lindb.

Subapical branching frequent. Stem epidermis more or less intermediate between Mnium and Pseudobryum types. Lamina cells small, mamillate. Leaf margin with double or single teeth, border uni- or partly multistratose. Costa with dorsal stereid band, single ventral stereids sometimes present. Operculum with a short point.

Species:

Trachycystis flagellaris (Sull. & Lesq.) Lindb.
T. microphylla (Doz. & Molk.) Lindb.

Taxonomic remarks. – The double teeth and bordered leaves show that *Trachycystis flagellaris* has closer phenetic affinities to *Mnium* than *T. microphylla*. The small size of the lamina cells and the subapical branching of the stem (the flagellae in *T. flagellaris* are, of course, branches, see p. 122) connect the two species with *Leucolepis*. According to TATUNO & ONO (1966), *Trachycystis* and *Mnium* (sect. »*Polla*») have different karyotypes.

3. Genus Leucolepis Lindb. 1868.

Leucolepis Lindberg, Not. Sällsk. F. Fl. Fenn Förh. 9 (N. S. 6): 80 – 81. 1868. – *Mnium* 3. *Leucolepis* (Lindb.) C. Müller, Gen. Musc. Frond. 137. 1901. – Holotypus: *Leucolepis acanthoneura* (Schwaegr.) Lindb. = *L. menziesii* (Hook.) Steere.

Subapical branching frequent. Leaves ovate, narrowly elliptic or triangular, acute or acuminate. Basal leaves distinctly differentiated with hyaline, serrate point. Leaf margin with single teeth, border unistratose, slightly differentiated. Lamina cells small. Operculum with a short point.

Leucolepis menziesii (Hook.) Steere is the only species of the genus. However, *Mnium immarginatum* Broth. may belong here (see below).

Taxonomic remarks. – The regularly dendroid stem of *Leucolepis menziesii* may suggest that it cannot be a near relative of *Trachycystis* and *Mnium*. However, rather clear evidence of relationship is furnished by the rhizoid topography, the colour and structure of the stem, the differentiation of the basal leaves (still more pronounced than in *Trachycystis* and in some species of *Mnium*), and the toothed costa. The shape and unistratose margin of the leaves separate it clearly from *Mnium*. The stereid bands are large filling practically the whole midrib.

The problem proper in the case of the genera *Mnium*, *Trachycystis*, and *Leucolepis* concerns the position of *Mnium immarginatum* Broth. KABIERSCH (1936) stated that *M. immarginatum* is not closely related to the other *Mnium* species but is perhaps related to *M. hornum*. On the other hand, he maintained that *Leucolepis* has no near relatives but *M. immarginatum*. He did not, however, include *M. immarginatum* in *Leucolepis*, but proposed a new section, *Pseudoleucolepis*, in the genus *Mnium* for this species (cf. also ANDREWS 1938).

Mnium immarginatum is really an intermediate species between *Leucolepis*, *Trachycystis*, and *Mnium*. The structure of the epidermis of the stem is intermediate between the Bryum and Pseudobryum types and thus nearer to that of *Trachycystis* than to those of the other genera. The leaf shape is about the same as in *Mnium* and *Trachycystis*, the size of the smooth lamina cells the same as in *Leucolepis* and *Trachycystis*. The leaf border is unistratose and not differentiated, the teeth are single and subapical branching is present as in *Leucolepis* and *Trachycystis microphylla*. On the other hand, branching similar to that of *Mnium immarginatum* also occurs, although rarely, in other *Mnium* species. In the midrib only the dorsal band of stereids is developed. The capsule is similar to those of *Leucolepis* and *Mnium hornum*, larger than in the other *Mnium* species and *Trachycystis*. Unfortunately, the chromosome number and the karyotype are unknown, and, therefore, it may be better to leave the question open until they have been studied.

B. Tribus Cinclidieae Koponen, tribus nova

Pars superior caulis ramos raro emittens. Epidermis caulina Mnio similis. Micronemata in partibus superioribus caulis raro adsunt. Color ruber etiam in cellulis corticalibus plerumque adest. Folia late elliptica – late ovata vel obovata, apiculata – acuta vel obtusa, margine plerumque bi- vel multistrato, integro. Costa sine stereidis vel fasciculo dorsali stereidarum praedita, edentata. Typus: *Cinclidium* Sw.

Subapical branching rare. Stem epidermis of Mnium type, red colour mostly present in cortex, as well. Micronemata mostly absent on the upper stem parts. Leaves broadly elliptic –

broadly ovate, or obovate, apiculate – acute, or obtuse, border mainly bi- or multistratose, margin entire. Costa with dorsal stereid band at most, not toothed.

4. Genus Cinclidium Sw. 1803

Cinclidium Swartz, Schraders Journ. f. Bot. 1801 (1): 27. 1803. – *Mnium* subg. *Cinclidium* (Sw.) Bruch, Schimper & Gümbel, Bryol. Eur. 4: 178. 1838 (fasc. 5, Mon. 14). – *Mnium* sect. *Cinclidium* (Sw.) Bruch, Schimper & Gümbel, Bryol. Eur. 4: 203. 1846 (fasc. 5, Mon. Suppl. 1, 1). – Holotypus: *C. stygium* Sw.

Extremely strong red colour. . Initial cell region of macronemata long and macronemata thus distributed in rows along the stems. Base of leaves attenuate, not or indefinitely decurrent. Lamina cells irregularly polygonal, elongated. Border mostly bi- or unistratose. Capsule with a distinct neck, pendulous. Operculum not rostrate. Outer peristome teeth short and blunt, papillose, yellowish, with few lamellae, inner peristome not divided into segments and cilia but forming a dome-like structure, with somewhat irregular openings at the side, corresponding in number and position with the outer peristome teeth.

Species:

Cinclidium arcticum Schimp.
C. latifolium Lindb.
C. stygium Sw.
C. subrotundum Lindb.

Taxonomic remarks. – *Cinclidium* is a clearly delimited genus. In addition to the characters of the peristome, the longitudinal rows of macronemata and the strong red colour separate it from the related genera *Cyrtomnium* and *Rhizomnium*. The leaf border of *Cinclidium stygium* was found to be unistratose but in transverse section the outermost cell row is seen to be characteristically higher than the other lamina cells (Fig. 74).

5. Genus Rhizomnium (Broth.) Koponen, status nova

Mnium sect. *Rhizomnium* Brotherus, Fl. Fenn. 1: 332–333. 1923. – *Mnium* subg. *Rhizomnium* Gams, Kl. Krypt. Fl. ed. 4. 153. 1957 (nom. inval.). – Lectotypus (selected here): *Mnium punctatum* Hedw. = *Rhizomnium punctatum* (Hedw.) Koponen.

Red colour present. Micronemata also sometimes present on the upper part of the stem. Lamina cells mainly large, polygonal, often hexagonal, elongated. Leaves with long and extremely narrow decurrent base. Costa mostly without stereid bands. Border bi – multistratose. Capsule horizontal – pendulous. Operculum rostrate.

Taxonomic remarks. – Obovate – elliptic leaves with a bi- or multistratose border and a narrowly decurrent base, and the characteristic shape (Fig. 83) and structure of the midrib are repeated in all the species of *Rhizomnium* except *R. minutulum*, in which stereids are present. The capsule of the latter is, however, similar to the capsules of the other *Rhizomnium* species and different from that of *Cinclidium*. In *R. nudum*, *R. punctatum* var. *elatum*, *R. pseudopunctatum*, and *R. andrewsianum* the leaf border is unistratose in the apical part, but even then is generally at least bistratose at the base. In *R. striatulum*, *R. glabrescens*, *R. minutulum*, and *R. punctatum* (s.str.) the border is bi - – multistratose throughout. The micronemata are absent in species with completely bi - – multistratose leaves and in *R. nudum*.

Sectio Rhizomnium

Lectotypus: *Rhizomnium punctatum* (Hedw.) Koponen.
Mnium C. *Integerrimae* Limpricht, Laubm. Deutschl. 2: 482. 1893. – Lectotypus (selected here): *Mnium punctatum* Hedw.
Mnium II. *Eu-Mnium* 2. *Punctatiformia* Kindberg, Eur. N. Am. Bryin. 2: 338, 339. 1897. – Lectotypus (selected here): *Mnium punctatum* Hedw.
Mnium sect. *Rhizomnion* Brotherus, Nat. Pflanzenfam. 1 (3): 612. 1904 (nom. illeg. pro sect. *Rhizomnium*). – *Mnium* sect. *Rhizomnion* Mitten, Trans. Linn. Soc. Lond. 2. Ser. Bot. 3 (3): 167. 1891 (nom. nud.). – Lectotypus (selected here): *Mnium punctatum* (Hedw.) Kop.
Mnium sect. *Integrifolia* Amann & Meyl., Fl. Mouss. Suisse 1: 135. 1912 (nom. illeg. pro sect. *Rhizomnium*). – Lectotypus (selected here): *Mnium punctatum* Hedw. – *Mnium* a. *integrifolia* Ångström in E. Fries, Summ. Veg. Scand. 1: 87. 1846 (nom. nud.).
Mnium sect. *Rhizomnium* Brotherus, cf. above.
Mnium Gruppe 5. *Mnia Punctata* C. Jensen, Danm. Moss. 2: 454. 1923. – Lectotypus (selected here): *Mnium punctatum* Hedw.

Large species, with large lamina cells. Costa without stereid bands.

Taxa:

Rhizomnium punctatum (Hedw.) Koponen, comb. nova. – Basionym: *Mnium punctatum* Hedwig, Spec. Musc. 193. 1801.
Rhizomnium punctatum var. *elatum* (Schimp.) Koponen, comb. nova. – Basionym: *Mnium punctatum* var. *elatum* Schimper, Syn. Musc. Eur. 398. 1860.
Rhizomnium pseudopunctatum (Bruch & Schimp.) Koponen, comb. nova. – Basionym: *Mnium pseudo-punctatum* Bruch & Schimper, London Journ. Bot. 2: 669. 1843.

Rhizomnium andrewsianum (Steere) Koponen, comb. nova. – Basionym: *Mnium andrewsianum* Steere, Bryologist 61: 175. 1958.
Rhizomnium nudum (Britt. & Williams) Koponen, comb. nova. – Basionym: *Mnium nudum* Britton & Williams, Bryologist 3: 6. 1900.
Rhizomnium glabrescens (Kindb.) Koponen, comb. nova. – Basionym: *Mnium glabrescens* Kindberg, Ottawa Natur. 7: 18. 1893.
Rhizomnium striatulum (Mitt.) Koponen, comb. nova. – Basionym: *Mnium striatulum* Mitten, Journ. Linn. Soc. Ser. 2 (3): 167 – 168. 1891.

Sectio Micromnium Koponen, sect. nova

Species parvae, cellulis laminarum parvis. Costa fasciculo dorsali stereidarum praedita. –
Typus: *Rhizomnium minutulum* (Besch.) Koponen

Small species with small- – medium-sized lamina cells, costa with dorsal stereid band.

The only species of the section is:

Rhizomnium minutulum (Besch.) Koponen, comb. nova. – Basionym: *Mnium minutulum* Bescherelle, Ann. Sc. Nat. Ser. 7 (17): 346. 1893.

6. Genus Cyrtomnium Holmen 1957

Cyrtomnium Holmen, Bryologist 60: 138. 1957. – Holotypus: *C. hymenophyllum* (B.S.G.) Holmen.
Mnium II. *Eu-Mnium* 1. *Hymenophylloidea* Kindberg, Eur. N. Am. Bryin. 2: 338. 1897. – Holotypus: *C. hymenophylloides* (Hüb.) Koponen.

Reddish colour not present. Leaves bluish, with or without decurrent base. Lamina cells rounded quadrate or rounded hexagonal. Border unistratose. Capsule arcuate or ovate. Stomata phaneropore. Operculum not rostrate, with short point only. Peristome may be rudimentary.

Species:

Cyrtomnium hymenophylloides (Hüb.) Koponen, comb. nova. – Basionym: *Mnium hymenophylloides* Hübener, Musc. Germ. 416. 1833.
C. hymenophyllum (B.S.G.) Holmen.

Note. Since no page number is quoted for the combination of the epithet *hymenophylloides* with the genus *Cyrtomnium* given by NYHOLM (1958), the combination is invalid and I prefer to introduce a new one.

Taxonomic remarks. – The species of the genus resemble each other in having phaneropore stomata (cf. HOLMEN 1957, NYHOLM 1958), bluish green colour in the leaves and similar leaf areolation. Bi- or multistratose leaf borders were not found. In the border of the *C. hymenophyllum* leaf there is only one definitely differentiated cell row. Its cells are, however, characteristically shaped, being higher than the other cells of the lamina. Thus the border is thicker than the remainder of the lamina, as in *Cinclidium stygium* (see Figs. 73, 74). This character is lacking in *C. hymenophylloides*, in which the leaf border is 2 – 4 cells broad. The capsules were not studied by the author, but according to HOLMEN (op. cit.) the capsule is arcuate in *C. hymenophyllum* and ovate in *C. hymenophylloides* (PERSSON 1914).

C. Tribus Orthomnieae Koponen, tribus nova

Pars superior caulis ramos raro emittens. Surculi adpressi plerumque adsunt. Epidermis caulina Bryo similis. Macro- et micronemata forma inter se non differunt. Micronemata usque ad apicem caulis adsunt. Color pallide rubellus adest vel deest. Folia elliptica, ₌piculata vel obtusa, margine uni- vel bistrato, integro. Costa plerumque sine stereidis, dorso edentata. Capsula erecta. Cellulae exothecii et epidermis setae haud plane inter se differunt.
Typus: *Orthomnion* Wils.

Subapical branching rare. Plagiotropic stems present. Stem epidermis of Bryum type. Macro- and micronemata not morphologically differentiated. Micronemata also on the upper stem parts. Weak reddish colour sometimes present. Leaves elliptic, apiculate, or obtuse, margin entire, border uni- or bistratose. Costa mostly without stereid bands, not toothed at the back. Capsule erect. Cells of exothecium and epidermis of seta not distinctly different.

7. Genus Orthomnion Wils. 1857

Orthomnion Wilson in Mitten, Kew Journ. Bot. 9: 368. 1857. – *Mnium *Orthomnion* (Wils.) Mitten, Journ. Linn. Soc. London, Suppl. Bot. 1: 142. 1859. – Typus: *Orthomnion bryoides* (W. Griff.) Norkett (cf. NORKETT 1958).

Note. WIJK et al. (1964) considered the description of the genus *Orthomnion* to be invalidly published and accepted the name *Orthomnium* (Mitt.) Broth. However, it is quite clear that WILSON (op. cit.) intended to introduce a description of a genus (cf. also NORKETT 1958). The description begins with »658 Nov. genus» and, in addition to »*O. crispum* Wils. MSS.», another species »*Orthomnion trichomitrium* Wils. MSS.» is clearly mentioned in the paragraph following the description. Thus, according to the Code (Art. 34, Note 1), the hesitation expressed by the question-mark does not invalidate the description in this case. The original spelling must be retained (Art. 73, cf. also DIXON 1909).

Seta about 2 cm long, definitely or slightly mamillate. Annulus absent. Calyptra hairy.

Outer peristome white, papillose, teeth lanceolate, blunt, without lamellae, inner peristome rudimentary, formed by the basal membrane only.

Species:
Orthomnion bryoides (W. Griff.) Norkett
O. loheri Broth. (ut »*Orthomnium*»)

Taxonomic remarks. – The two species differ from each other in several characters. *O. bryoides* has decurrent leaves with a bistratose border, while the leaves of *O. loheri* are unistratose and not decurrent. In *O. bryoides* a clear reddish colour is often present in the stem and leaves, which is lacking in *O. loheri*. Thus, the suggestion of NORKETT (1958, p. 446) that the taxa in question possibly deserve only subspecific rank does not seem well founded. The specific difference is also reinforced by different distribution patterns. *O. bryoides* inhabits the SW Asiatic continent and *O. loheri* mainly occurs in the Philippines and East Indies.

8. Genus Orthomniopsis Broth. 1907

Orthomniopsis Brotherus, Öfv. Finska Vet. Soc. Förh. 49 (10): 1. 1907. – Holotypus: *O. japonica* Broth. = *O. dilatata* (Mitt.) Nog. (cf. NOGUCHI 1966).

Seta 0.5 cm long at most. Capsule erect or slightly arcuate. Annulus present. Calyptra not hairy. Outer peristome yellowish brown, lanceolate, papillose, lamellae numerous, inner peristome with basal membrane, papillose, yellow brown, segments not perforated, cilia 2, short, not nodulose.

Species:
Orthomniopsis dilatata (Mitt.) Nog.
(*O. elimbata* Nog.)

Taxonomic remarks. – The writer had at his disposal material of *Orthomniopsis dilatata* only. The main differences between *Orthomnion* and *Orthomniopsis* are the shorter seta and hairless calyptra of the latter and differences in the peristome. The gametophytic characters, the structure of the stem epidermis, the rhizoid topography and the growth-form are similar. The resemblance between the gametophytes of *Orthomniopsis* and *Orthomnion loheri* is especially marked (cf. also KARIERSCH 1936). Karyological research is needed to throw further light on the extent of the similarity between the two genera. According to CHEN (1955), *Mnium dilatatum*

Mitt. has an erect capsule and elongated seta but the calyptra is not hairy. He included it in *Orthomnion*. Recently NOGUCHI (1966) suggested that it belongs to *Orthomniopsis*. The same opinion was held earlier by REIMERS (cf. KABIERSCH op. cit. p. 56) on the basis of the gametophytic characters. According to NOGUCHI (1953) and BARTRAM (1965), a second species, *O. elimbata* Nog., exists.

D. Tribus Plagiomnieae Koponen, tribus nova

Pars superior caulis ramos raro emittens. Surculi arcuati vel plus minusve adpressi e basi caulium erectorum prodeunt. Epidermis caulina Bryo vel Pseudobryo similis. Macronemata et micronemata forma magnitudineque non semper plane inter se differunt. Micronemata etiam in partibus caulium apicalibus adsunt. Color rubellus caulibus et foliis deest, in seta et capsula nonnumquam adest. Folia forma varia, margine unistrato, dentibus singulis praedito. Costa plerumque fasciculo dorsali stereidarum praedita, dorso edentata.

Typus: *Plagiomnium* Kop.

Subapical branching rare. More or less plagiotropic shoots from the base of orthotropic stems. Stem epidermis of Bryum or Pseudobryum type. The difference in the shape and size of the macro- and micronemata not distinct in all cases. Micronemata also on the apical stem parts. Reddish colour absent in the gametophyte but sometimes present in the sporophyte. Leaves of various shapes, border unistratose, margin with single teeth. Costa mostly with dorsal stereid band, not toothed at the back.

9. Genus Plagiomnium Koponen, genus novum

Surculi arcuati vel adpressi in speciebus plurimis. adsunt. Epidermis caulina Bryo similis. Costa plerumque fasciculo unico dorsali stereidarum praedita. Operculum plerumque sine rostro.

Typus: *Plagiomnium cuspidatum* (Hedw.) Koponen.

Plagiotropic shoots present in most species. Stem epidermis of Bryum type. Costa generally with one stereid band. Operculum mostly not rostrate.

Taxonomic remarks. – There are two points of difference between KABIERSCH's (1936) sub-division and the present division into sections. *Plagiomnium maximoviczii* is not included in the section *Undulata* but in the section *Rostrata* (cf. also ANDERSON 1954, HORIKAWA & ANDO 1964), and the species of the subsection *Sublimbata* are included in the section *Rostrata*. On the whole, the sections *Plagiomnium*, *Venusta*, *Rosulata*, and *Undulata* (cf. SMIRNOVA 1964, NOGUCHI 1966) are rather well delimited, but the section *Rostrata* is more problematic (cf. also p. 136). A large number of species related to *P. rostratum* have been described from different parts of the world, but their taxonomic value can only be solved by monographic research. These names, many of which are probably superfluous are not mentioned here (cf. e.g. NOGUCHI 1952, INOUE 1956, ANDERSON op. cit., HORIKAWA & ANDO op. cit.). KABIERSCH's (1936) separation of the subsections *Rostrata* and *Sublimbata* is based on the differences in the gametophytes, but there are sporophyte characters which join these subsections. All the species which I was able to study had a rostrate operculum. The scattered distribution of the stomata was seen in *P. rostratum*, *P. maximoviczii*, and *P. succulentum* (cf. also NYHOLM 1958).

Two bands of stereids have been recorded in *P. cuspidatum* and *P. medium*.

The most exceptional and divergent species is *Mnium handelii* Broth., which may belong to this genus. It was described by BROTHERUS (1929) on the basis of two specimens only. KABIERSCH (1936) included it in the subsection *Sublimbata* but I am not sure of its relationships. The shape of the leaves is different from that of the species of *Plagiomnium*, and the absence of plagiotropic shoots and stereids in the costa might constitute sufficient grounds for its separation as a section at least. However, as the sporophyte has not been seen and the karyotype is unknown, the best solution is perhaps to leave the matter open, until more material has been collected.

Sectio Plagiomnium

Typus: *Plagiomnium cuspidatum* (Hedw.) Koponen

Mnium B. *Serratae* Limpricht, Laubm. Deutschl 2: 467. 1893. – Lectotypus (selected here): *Mnium cuspidatum* Hedw.

Mnium II. *Eu-Mnium* 4. *Cuspidatiformia* Kindberg, Eur. N. Am. Bryin 2: 339. 1897. – *Mnium* sect. *Eumnium* subsect. *Cuspidatiformia* (Kindb.) Kabiersch, Hedwigia 76: 33. 1936 (nom. illeg. pro subsect. *Plagiomnium*). – Lectotypus (selected here): *Mnium cuspidatum* Hedw.

Mnium sect. *Serrata* Amann & Meyl., Fl. Mouss. Suisse

1: 136. 1912. – Lectotypus (selected here): *Mnium cuspidatum* Hedw.

   Mnium Gruppe 3. *Mnia cuspidata* C. Jensen, Danm. Moss. 2: 449. 1923. – Lectotypus (selected here): *Mnium cuspidatum* Hedw.

Leaves obovate-elliptic, acute, with broad and long decurrent base. Teeth present on apical half of the leaves only. Teeth sharp, formed by 1 – 2 cells. Lamina cells mostly rounded hexagonal, small – medium-sized.

Species:

Plagiomnium cuspidatum (Hedw.) Koponen, comb. nova. – Basionym: *Mnium cuspidatum* Hedwig, Spec. Musc. 192. 1801.

   Plagiomnium trichomanes (Mitt.) Koponen, comb. nova. – Basionym: *Mnium trichomanes* Mitten, Kew. Journ. Bot. 8: 231. 1856.

   Plagiomnium japonicum (Lindb.) Koponen, comb. nova. – Basionym: *Mnium japonicum* Lindberg, Acta Soc. Sc. Fenn. 10: 226. 1872.

   Plagiomnium drummondii (Bruch & Schimp.) Koponen, comb. nova. – Basionym: *Mnium Drummondii* Bruch & Schimper, London Journ. Bot. 2: 669. 1843.

Sectio Venusta (Kab.) Koponen, comb. nova

   Mnium sect. *Eumnium* subsect. *Venusta* Kabiersch, Hedwigia 76: 39. 1936. – Holotypus: *Mnium venustum* Mitt.

Plagiotropic shoots absent. Leaves toothed down to base, stereids lacking in costa. Cells surrounding stomata mamillate.

Species:

Plagiomnium venustum (Mitt.) Koponen, comb. nova. – Basionym: *Mnium venustum* Mitten, Kew Journ. Bot. 8: 231. 1856.

Sectio Rosulata (Kindb.) Koponen, comb. nova

   Mnium II. *Eu-Mnium* 5. *Rosulata* Kindberg, Eur. N. Am. Bryin. 2: 339. 1897. – *Mnium* sect. *Eumnium* subsect. *Rosulata* (Kindb.) Kabiersch, Hedwigia 76: 41. 1936. – Lectotypus (selected here): *Mnium affine* Funck.

Leaves elliptic – oblong, acute – apiculate or mucronate, mostly with decurrent base, toothed down to base. Teeth often long, formed by as many as 4 cells. Lamina cells hexagonal, elongated, large.

Taxa:

Plagiomnium affine (Funck) Koponen, comb. nova. – Basionym: *Mnium affine* Funck, Crypt. Gew. Fichtelgeb. 17: 3. 1810.

   Plagiomnium ciliare (C. Müll.) Koponen, comb. nova. – Basionym: *Mnium affine* var. *ciliare* C. Müller, Syn. Musc. Frond. 159. 1849.

   Plagiomnium elatum (B.S.G.) Koponen, comb. nova. – Basionym: *Mnium affine* var. *elatum* Bruch, Schimper & Gümbel, Bryol. Eur. 4: 195 (fasc. 5, Mon. 31). 1838 (= *Mnium seligeri* auct.).

Plagiomnium insigne (Mitt.) Koponen, comb. nova. – Basionym: *Mnium insigne* Mitten, Kew Journ. Bot. 8: 230. 1856.

   Plagiomnium medium (B.S.G.) Koponen, comb. nova. – Basionym: *Mnium medium* Bruch, Schimper & Gümbel, Bryol. Eur. 4: 196 (fasc. 5, Mon. 31). 1838.

   Plagiomnium medium ssp. *curvatulum* (Lindb.) Koponen, comb. nova. – Basionym: *Astrophyllum curvatulum* Lindberg, Bot. Centralbl. 6 (10): 363. 1881.

   Plagiomnium rugicum (Laur.) Koponen, comb. nova. – Basionym: *Mnium rugicum* Laurer, Flora 19: 292. 1827.

   Plagiomnium tezukae (Sak.) Koponen, comb. nova. – Basionym: *Mnium Tezukae* Sakurai, Journ. Japan. Bot. 29 (4): 114 – 115. 1954.

Sectio Undulata (Kindb.) Koponen, comb. nova

   Mnium II. *Eu-Mnium* 7. *Undulata* Kindberg, Eur. N. Am. Bryin. 2: 339. 1897. – *Mnium* sect. *Eumnium* subsect. *Undulata* (Kindb.) Kabiersch, Hedwigia 76: 58. 1936. – Holotypus: *Mnium undulatum* Hedw.

Subapical branching present. Leaves narrowly oblong – oblong, acute – apiculate or mucronate, with long and broad decurrent base, toothed down to base. Teeth sharp, formed mostly by one cell. Lamina cells rounded hexagonal or elongated, small-sized.

Species:

Plagiomnium undulatum (Hedw.) Koponen, comb. nova. – Basionym: *Mnium undulatum* Hedwig, Spec. Musc. 195. 1801.

   Plagiomnium arbusculum (C. Müll.) Koponen, comb. nova. – Basionym: *Mnium arbusculum* C. Müller, Nuovo Giorn. Bot. Ital. n. ser. 5: 161. 1898.

   Plagiomnium confertidens (Lindb. & Arn.) Koponen, comb. nova. – Basionym: *Astrophyllum confertidens* Lindberg & Arnell, K. Sv. Vet. Akad. Handl. 23 (10): 17. 1890.

Sectio Rostrata (Kindb.) Koponen, comb. nova

   Mnium II. *Eu-Mnium* 6. *Rostrata* Kindberg, Eur. N. Am. Bryin. 2: 339. 1897. – *Mnium* sect. *Eumnium* subsect. *Rostrata* (Kindb.) Kabiersch, Hedwigia 76: 43. 1936. – Holotypus: *Mnium rostratum* Schrad.

   Mnium sect. *Eumnium* subsect. *Sublimbata* Kabiersch, Hedwigia 76: 50. 1936. – Lectotypus (selected here): *Mnium succulentum* Mitt.

Subapical branching not present. Leaves oblong – elliptic, mostly apiculate, with or without decurrent base, toothed down to base, sometimes entire. Teeth blunt, mostly formed by one cell. Border generally not distinctly differentiated, sometimes absent. Lamina cells large and elogated, or small and rounded hexagonal – elongated. Stomata not confined to the neck but also present in the exothecium. Operculum rostrate.

Species:

Plagiomnium maximoviczii (Lindb.) Koponen, comb. nova. – Basionym: *Mnium Maximoviczii* Lindberg, Acta Soc. Sc. Fenn. 10: 224–225. 1872.

Plagiomnium rostratum (Schrad.) Koponen, comb. nova. – Basionym: *Mnium rostratum* [Schrad.] Schrader, Bot. Zeit. Regensburg 1: 79. 1802.

Plagiomnium succulentum (Mitt.) Koponen, comb. nova. – Basionym: *Mnium succulentum* Mitten, Journ. Linn. Soc. Bot. Suppl. 1: 143. 1859.

Plagiomnium vesicatum (Besch.) Koponen, comb. nova. – Basionym: *Mnium vesicatum* Bescherelle, Ann. Sc. Nat. Bot. ser. 7 (17): 345. 1893.

In addition, the following species probably belong here, though this point remained uncertain owing to lack of sufficient material: *Mnium carolinianum* Anderson, *M. elimbatum* Fleisch., *M. formosicum* Card., *M. integrum* Bosch. & Lac., and *M. luteolimbatum* Broth.

10. Genus Pseudobryum (Kindb.) Koponen, status nova

Mnium I. *Pseudo-Bryum* Kindberg, Eur. N. Am. Bryin. 2: 338. 1897. – *Mnium* sect. *Pseudobryum* (Kindb.) Kabiersch, Hedwigia 76: 68. 1936. – Holotypus: *Mnium cinclidioides* Hüb.

Mnium Gruppe 4. *Mnia cinclidioidea* C. Jensen, Danm. Moss. 2: 453. 1923. – Holotypus: *Mnium cinclidioides* Hüb.

Mnium sect. *Pinnaticosta* Kabiersch, Hedwigia 76: 68. 1936. – Holotypus: *Mnium speciosum* Mitt.

Plagiotropic shoots absent. Stem epidermis of Pseudobryum type. Costa without stereid bands, branched. Lamina cells prosenchymatous, polygonal, rhomboidal. Operculum not rostrate.

Species:

Pseudobryum cinclidioides (Hüb.) Koponen, comb. nova. – Basionym: *Mnium cinclidioides* Hübener, Musc. Germ. 416. 1833.

Pseudobryum speciosum (Mitt.) Koponen, comb. nova. – Basionym: *Mnium speciosum* Mitten, Trans. Linn. Soc. London Bot. ser. 2 (3): 166 – 167. 1891.

Taxonomic remarks. – The leaf areolation (Figs. 54, 55) and the structure of the stem epidermis are similar in these two species and different from those of all the other genera. The »branched» costa of *P. speciosum*, according to which KABIERSCH (1936) named his section *Pinnaticosta*, is also present in *P. cinclidioides* but is less distinct. According to LOWRY (1948), the chromosomes of *Pseudobryum cinclidioides* are exceptionally long and thin and, therefore, different from those of the other groups studied by him.

IV. Summary

1. A revision of the genera and sections currently included in the family *Mniaceae* is presented. The family consists of ten genera: *Mnium* Hedw., *Trachycystis* Lindb., *Leucolepis* Lindb., *Cinclidium* Sw., *Rhizomnium* (Broth.) Kop., *Cyrtomnium* Holmen, *Orthomnion* Wils., *Orthomniopsis* Broth., *Plagiomnium* Kop., and *Pseudobryum* (Kindb.) Kop. The genera *Bryomnium* Card. and *Roellia* Kindb. are excluded from the family. The genus *Mnium* of earlier authors is divided into five genera, *Mnium*, *Rhizomnium*, *Cyrtomnium*, *Plagiomnium*, and *Pseudobryum*.

2. Both gametophytic and sporophytic characters have been used in the delimitation of the genera. Genera earlier separated mainly on the basis of the sporophyte characters also showed differential characters in the gametophyte. The structure of the stem, rhizoid topography, colour and growth-form were found to be useful characters in the delimitation of the genera and had also sometimes diagnostic value at specific level. The generic division suggested is well founded karyologically.

3. An attempt has been made to base the classification on evolutionary concepts. The genera proposed are possibly monophyletic and also phenetically uniform. They also fulfil the conditions which are generally used and accepted in the delimitation of genera. On the other hand, it was not possible to apply the criterion of monophyly in the case of taxa above the rank of genus. Only one of the tribes suggested is probably monophyletic; the others may be paraphyletic while one is perhaps even polyphyletic. The delimitation of the families *Bryaceae*, *Mniaceae*, and *Rhizogoniaceae* appears to be in need of revision.

4. A sectional division of the genera is proposed. It follows KABIERSCH's (1936) subdivision with minor amendments. The species studied are listed under each genus or section and the necessary new combinations are introduced.

5. Three new tribes, *Cinclidieae* Kop., *Orthomnieae* Kop., and *Plagiomnieae* Kop., three new genera, *Rhizomnium* (Broth.) Kop., *Plagiomnium* Kop., and *Pseudobryum* (Kindb.) Kop., and one new section, *Rhizomnium* sect.

133

Micromnium Kop. are introduced. The independent position of the section *Pinnaticosta* Kab. is not maintained and instead it is united with the genus *Pseudobryum*. Similarly, the subsection *Sublimbata* Kab. is united with *Plagiomnium* sect. *Rostrata* (Kindb.) Kop. A number of new combinations are introduced at sectional and specific level in the genera *Mnium* (p. 140), *Rhizomnium* (p. 142), *Plagiomnium* (p. 145), and *Pseudobryum* (p. 147). Numerous typifications are given at generic and sectional level.

References

AINSWORTH, W., 1956: Axillary shoots of the stolons of Mnium cuspidatum. – Bryologist 59, 187 – 191.

AMANN, J. & C. MEYLAN, 1912: Flore des Moussesde la Suisse I. Tableaux synoptiques pour la détermination des mousses européennes. – 215 pp. Lausanne.

– » 1921: Nouvelles additions et rectifications à la Flore des Mousses de la Suisse 3. – Bull. Soc. Vaud. Scienc. Nat. 200 (Vol. 54), 33 – 66.

ANDERSON, L. E., 1954: A new species of Mnium from the southern Appalachians. – Bryologist 57, 177 – 188.

– » – 1963: Modern species concepts: Mosses. – Ibid. 66, 107 – 119.

ANDERSON, L. E. & H. CRUM, 1958: Cytotaxonomic studies on mosses of the Canadian Rocky Mountains. – Nat. Mus. Can. Bull. 160, 1 – 89.

ANDREWS, A. LEROY, 1938: Review: W. KABIERSCH, Studien über die ostasiatischen Arten einiger Laubmoosfamilien (Mniaceae – Bartramiaceae). Hedwigia 76, 1 –94, (1936); 77, 71 – 136, (1937). – Bryologist 41, 95 – 98.

– » – 1940: Bryaceae II. Mniaceae. – In A. J. GROUT, Moss flora of North America north of Mexico 2: 4, 211 – 285.

BARTRAM, E. B., 1965: Mosses of the eastern highlands, New Guinea, from the 6th Archbold expedition, 1959. – Contr. U. S. Nat. Herb. 37: 2, 43 – 67.

BENDZ, G., O. MÅRTENSSON & L. TERENIUS, 1962: Moss pigments. 1. The anthocyanins of Bryum cryophilum O. Mårt. – Acta Chem. Scand. 16, 1183 – 1190.

BENDZ, G. & O. MÅRTENSSON, 1963: Moss pigments. 2. The anthocyanins of Bryum rutilans Brid. and Bryum weigelii Spreng. – Ibid. 17, 266.

BENDZ, G., O. MÅRTENSSON & E. NILSSON, 1966: Moss pigments. 4. An investigation of the occurrence of proanthocyanids in mosses. – Ibid. 20, 277 – 278.

BOWERS, M. C., in A. LÖVE, 1966: IOPB chromosome number reports 8. – Taxon 15, 279 - 284.

BRIDEL–BRIDERI, S. E., 1826 – 1827: Bryologia universa seu systematica ad novam methodum dispositio, historia et descriptio omnium muscorum frondosorum hucusque cognitorum cum synonymia ex auctoribus probatissimis. – I – XLVI + 860 pp. Lipsiae.

BROTHERUS, V. F., 1909, 1924: Unterklasse Bryales. II. Spezieller Teil. – In A. ENGLER: Die natürlichen Pflanzenfamilien 1: 3, 277 – 700. – 1924: 2. Aufl. Ibid. 10: 1, 143 – 478.

– » – 1905: Contributions to the bryological flora of the Philippines I. – Öfv. Finsk. Vet. Akad. Soc. Förh. 47: 14, 1 – 12.

– » – 1923: Die Laubmoose Fennoskandias. – Flora Fenn. 1, I – XIII + 1 – 635.

– » – 1929: Musci. – In HANDEL – MAZETTI, Symbolae Sinicae IV, 1 – 147 + IV pls.

BRUCH, P., W. PH. SCHIMPER & TH. GÜMBEL, 1838; 1846: Bryologia Europaea seu genera muscorum Europaeorum monographice illustrata. – 1838: Bryaceae: Mnium, Vol. IV, 165 – 201 (Fasc. 5, 1 – 37, 13 pls.). – 1846: Mnium. Supplementum I, Vol. IV, 203 – 208 (Fasc. 31, Mon. Suppl. 1, 1 – 6, 5 pls.).

BRUMMIT, R K. & A. O. CHATER, 1967: On names published without clear indication of rank. – Taxon 16, 403 – 406.

BRUNDIN, L., 1966: Transantarctic relationships and their significance as evidenced by chironomid midges, with a monograph of the subfamilies Podonominae and Aphroteniinae and the austral Heptagyiae. – K. Sv. Vetensk. Akad. Handl. Ser. 4. 11: 1, 1 – 472, 29 pls.

BUCH, H., 1947: Über die Wasser- und Mineralstoffversorgung der Moose II. – Soc. Scient. Fenn. Comm. Biol. 9: 20, 1 – 61.

CHEN, P. C., 1955: Bryophyta Nova Sinica. – Feddes Repert. 58, 23 – 52.

CORRENS, C., 1899 a: Über Scheitelwachsthum, Blattstellung und Astanlagen des Laubmoosstämmchens. – Schwendener, Festschrift. 28 pp. Berlin.

– » – 1899 b : Untersuchungen über die Vermehrung der Laubmoose durch Brutorgane und Stecklinge. – 472 pp. Jena.

CRUM, H., 1967: Studies in North American Bryaceae I – II. – Bryologist 70, 106 – 110.

CZAPEK, F., 1899: Zur Chemie der Zellmembranen bei den Laub- und Lebermoosen. – Flora 86, 361 – 381.

DAVIS, P. H. & V. H. HEYWOOD, 1963: Principles of angiosperm taxonomy. – 556 pp. Edinburgh and London.

DILLENIUS, J. J., 1741: Historia muscorum in qua circiter sexcentae species veteres et novae ad sua genera relatae describuntur et iconibus genuinis illustrantur: cum appendice et indice synonymorum. – XVI + 579 pp. + LXXXV pls. Oxonii.

DIXON, H. N., 1909: An undescribed structure in Mnium, with notes on the genus Orthomnion. – Rev. Bryol. 36, 141 – 147.

– » – 1916: Mnium antiquorum Cardot and Dixon, an extinct moss. – Bryologist 19, 51 – 52.

DOMBROVSKAJA, A. V. & R. I. SHLYAKOV, 1967: Lishajniki mkhi severa evropejskoj chasti SSSR. – 182 pp. Le ningrad.

EVANS, A. W., 1907: The genus Calypogeia and its type species. – Bryologist 10, 24 – 30.

FLEISCHER, M., 1902 – 1904: Die Musci der Flora von Buitenzorg (zugleich Laubmoosflora von Java). II. – Pp. 381 – 643. Leiden.

GAMS, H., 1932: Quaternary distribution. – In FR. VERDOORN, Manual of Bryology, pp. 297 – 322. The Hague.

HAMANT, C., 1950: Le noyau dans le sporogone et au cours de la division hétérotypique dans le genre Mnium et en particulier chez M. undulatum L. – Compt. Rend. Acad. Sci., Paris 231, 301 – 302.

HEDWIG, J. (†, ed. F. SCHWAEGRICHEN), 1801: Species muscorum frondosorum. – VI + 352 pp. + LXXVII pls. Lipsiae.

HEITZ, E., 1942: Über die Beziehung zwischen Polyploidie und Gemischtgeschlechtlichkeit bei Moosen. – Arch. Klaus-Stift. Vererb. Forsch. 17, 444 – 448.

HENNIG, W., 1950: Grundzüge einer Theorie der phylogenetischen Systematik. – 370 pp. Berlin.

– » – 1966: Phylogenetic systematics. – 263 pp. Urbana.

HILPERT, F., 1933: Studien zur Systematik der Trichostomaceen. – Beih. Bot. Centralbl. 50: 2, 585 – 706.

HOLMEN, K., 1957: The sporophyte of Mnium hymenophyllum. – Bryologist 60, 135 – 138.

– » – 1958: Cytotaxonomical studies in some Danish mosses. – Bot. Tidsskr. 54, 23 – 43.

HORIKAWA, Y. & H. ANDO, 1964: Contributions to the moss flora of Thailand. – Nature and life in Southeast Asia. III, 1 – 44.

HUBER, H., 1967: Über die Anordnung der Zellen in der Astrinde der Torfmoose (Sphagnum). – Bauhinia 3, 209 – 216.

INOUE, T., 1956: Variations in Mnium maximoviczii. (In Japanese.) – Misc. Bryol. Lich. 1: 7, 1.

IRELAND, R. R., 1967: Chromosome studies on mosses from the state of Washington. II. – Bryologist 70, 335 – 338.

ISOVIITA, P., 1966: Studies on Sphagnum L. I. Nomenclatural revision of the European taxa. – Ann. Bot. Fenn. 3, 199 – 264.

KABIERSCH, W., 1936: Studien über die ostasiatischen Arten einiger Laubmoosfamilien (Mniaceae – Bartramiaceae). – Hedwigia 76, 1 – 94.

KINDBERG, N. C., 1882: Die familien und Gattungen der Laubmoose (Bryineae) Schwedens und Norwegens hauptsächlich nach dem Lindbergschen Systeme. – Bih. K. Sv. Vet. Akad. Handl. 6: 19, 1 – 25.

– » – 1896, 1897: Species of European and Northamerican Bryineae (Mosses). Part 1. (1896) Pleurocarpous; Part 2. (1897) Acrocarpous. – 410 pp. Linköping.

– » – 1909: Notes on the synonymy of European and North-american Bryineae. – Rev. Bryol. 36, 115 – 117.

KOPONEN, T., 1967 a: The typification of Mnium affine Funck and M. medium Bruch, Schimp. & Gümb. – Ann. Bot. Fenn. 4, 64 – 66.
– » – 1967 b: Biometrical analysis of a mixed stand of Mnium affine Funck and M. medium B.S.G. – Ibid. 4, 67 – 73.
KOTILAINEN, M. J., 1950: In memory of the 100th anniversary of V. F. Brotherus' birthday. – Arch. Soc. 'Vanamo' 4, 107 – 109.
KOZLOWSKI, A., 1921: Sur la saponarine chez le Mnium cuspidatum. – Compt. Rend. Acad. Sci., Paris 173, 429 – 431.
KURITA, M., 1939: Chromosomen bei einigen Laubmoosen. – Bot. Zool. Tokyo, 7, 385 – 388. (Not seen, referred according to TATUNO & ONO 1966.)
LANJOUW, J. et al. 1966: International Code of botanical nomenclature adopted by the Tenth International Botanical Congress Edinburgh, August 1964. – Regnum Veget. 46, 1 – 402.
LANJOUW, J. & F. A. STAFLEU, 1964: Index herbariorum. I. The herbaria of the world. 5th ed. – Ibid. 31, 1 – 251.
LAWRENCE, G. E., 1956: An introduction to plant taxonomy. 2nd. ed. – 179 pp. New York.
LAWTON, E., 1957: A revision of the genus Lescuraea in Europe and North America. – Bull. Torrey Bot. Club. 84, 281 – 307.
– » – 1964: The structure and distribution of Mnium nudum. – Bryologist 67, 44 – 47.
LAZARENKO, A. S. 1955: Opredelitel' listvennyh mkhov Ukrainy. – 468 pp. Kiev.
LAZARENKO, A. S. & E. I. VISOTSKAYA, 1965: Contribution to the study of chromosome numbers in Ukrainian mosses. – Zitol. i Genet. 1965, 174 – 178.
LAZARENKO, A. S. & E. M. LESNYAK, 1966: Khromosomni chisla i mokhiv z Kazakhstanu i Tadzhikisktanu. – Dopovide Acad. Nauk. Ukrains'koi RSR 1966, 407 – 411.
LAZARENKO, A. S., O. I. VISOTSKA & E. M. LESNYAK, 1967: Chromosome numbers in mosses of the Black Sea coast of the Caucasus. – Ibid. 1967, 85 – 88.
LERSTEN, N. R., 1961: A comparative study of generation from isolated gametophytic tissues in Mnium. – Bryologist 64, 37 – 47.
LIMPRICHT, K. G., 1895: Die Laubmoose Deutschlands, Oesterreichs und der Schweitz. II. Bryineae. – 853 pp. Leipzig.
LINDBERG, S. O., 1868: Observationes Mniaceis europaeis. – Not. Sällsk. F. Fl. Fenn. 9 (N. S. 6), 41 – 88.
– » – 1878: Utkast till en naturlig gruppering af Europas bladmossor med toppsittande frukt (Bryineae acrocarpae). – 39 pp. Helsingfors.
– » – 1879: Musci Scandinavici in systemate novo naturali dispositi. – 50 pp. Upsaliae.
LINNAEUS [= VON LINNÉ], C., 1753: Species plantarum, exhibetes plantas rite cognitas, ad genera relatas, cum differentiis specificis, nominibus trivialibus, synonymis selectis, locis natalibus, secundum systema seuale digestas. II. – Pp. 561 – 1 200. Holmiae.
LOESKE, L., 1907 a: Zur Systematik der europäischen Brachytheciaceae. – Flora 1907: 1 – 2, 1 – 4.
– » – 1907 b: Drepanocladus, eine biologische Mischgattung. – Hedwigia 46, 300 – 321.
– » – 1910: Studien zur vergleichenden Morphologie und phylogenetischen Systematik der Laubmoose. – 224 pp. Berlin.
– » – 1911: Kritische Bemerkungen über Lesquereuxia S. O. Lindb. – Hedwigia 50, 311 – 328.
– » – 1932: Kritik der europaeischen Anomobryen. – Rev. Bryol. Lich. N. S. 5: 4, 169 – 201.
LORCH, W., 1931: Anatomie der Laubmoose. – In K. LINSBAUER, Handbuch der Pflanzenanatomie VII: 1, 1 – 358.
LOWRY, R. J., 1948: A cytotaxonomic study of the genus Mnium. – Mem. Torrey Bot. Club. 20: 2, 1 – 42.
MACLEOD, J., 1917: Quantitative description of ten British species of the genus Mnium. – Journ. Linn. Soc. Bot. 44, 1 – 58.
MALTA, N., 1926: Die Gattung Zygodon Hook. et Tayl. Eine monographische Studie. – Latvijas Univ. Bot. Därza Darbi 1, 1 – 185.
MAYER, E., 1965: Numerical phenetics and taxonomic theory. – Syst. Zool. 14, 73 – 97.
McCLURE, J. W. & H. A. MILLER, 1967: Moss chemotaxonomy. A survey for flavonoids and the taxonomic implications. – Nova Hedwigia 14, 111 – 125.
McCLYMONT, J. W. 1955: Spore studies in the Musci, with special reference to the genus Bruchia. – Bryologist 58, 287 – 306.
MEIJER, W., 1951: The genus Orthodontium. – 80 pp. Amsterdam.
MELCHERT, T. E. & R. E. ALSTON, 1965: Flavonoids from the moss Mnium affine Bland. – Science 150, 1170 – 1171.
MELCHIOR, H. & E. WEDERMANN, 1954: A. Engler's Syllabus der Pflanzenfamilien. I Band. 12. Aufl. – 267 pp. Berlin.
MEUSEL, H. 1935: Wuchsformen und Wuchstypen der europäischen Laubmoose. – Nova Acta Leopoldina N. F. 3, 123 – 277.
MICHENER, C. D. & R. R. SOKAL, 1957: A quantitative approach to a problem in classification. – Evolution 11, 130 – 162.
MISIURA, M., 1964: Regeneracja gametofitu i rozmnazanie wegetatywne Mnium punctatum (Schreb.) Hedw. – Acta Soc. Bot. Poloniae 33, 451 – 459.
MITTEN, W., 1959: Musci Indiae Orientalis; an enumeration of the mosses of the East Indies. – Journ Proc. Linn. Soc. Suppl. Bot. 1, 1 – 171.
MÜLLER, C., 1848 – 1849: Synopsis muscorum frondosorum omnium hucusque cognitorum. I. Musci vegetationis acrocarpicae. – VIII + 812 pp. Berolini.
– » – 1901: Genera Muscorum Frondosorum. – 474 pp. Leipzig.
MÜLLER, K., 1948: Morphologische und anatomische Untersuchungen an Antheridien beblätterter Jungermanien. – Bot. Not. 1949, 71 – 80.
– » – 1951: Neue für die Lebermoostaxonomie wichtige Merkmale. – Mitt. Bad. Landsver. Naturk., Freiburg in Br. N. F. 5, 182 – 185.
NEUBURG, M. F., 1956: Discovery of Scale mosses in Permian deposits of the USSR. – Doklady Akad. Nauk SSSR. 107, 321 – 324.
– » – 1960: Listostebel'nye mkhi iz permskikh otlozhenij Angaridy. – Trud. Geol. Inst. Akad. Nauk. SSSR. 19, 1 – 104, 78 pls.
NOGUCHI, A., 1952: Mnium vesicatum Besch. and its allies. – Journ. Jap. Bot. 27, 27 – 32.
– » – 1953: Mosses of Mt. Sarawaket, New Guinea. – Journ. Hattori Bot. Lab. 10, 1 – 23.
– » – 1966: Mniaceae. – In H. HARA, The Flora of eastern Himalaya, 559 – 565. Tokyo.
NORKETT, A. H., 1958: Himalayan moss notes. I. Orthomnion Wils. – Trans. Brit. Bryol. Soc. 3, 441 – 447.
NYHOLM, E., 1958: Illustrated moss flora of Fennoscandia. II. Musci. – Fasc. 3, 189 – 288. Malmö.
PARIHAR, N. S., 1965: An introduction to Embryophyta I. Bryophyta. – 375 pp. Allahabad.
PATON, J. A. & J. V. PEARCE, 1957: The occurrence, structure and functions of the stomata in British bryophytes. – Trans. Brit. Bryol. Soc. 3, 228 – 259.
PERSSON, H. 1914: Bladmossfloran i sydvästra Jämtland och angränsande delar af Härjedalen. – Ark. Bot. 14: 3, 1 – 70.
– » – 1947: Further notes on Alaskan – Yukon bryophytes. – Bryologist 50, 279 – 310.
– » – 1962: Bryophytes from Alaska collected by E. Hultén and others. – Sv. Bot. Tidskr. 56, 1 – 35.
PERSSON, H. & W. A. WEBER, 1958: The bryophyte flora of Mt. McKinley National Park, Alaska. – Bryologist 61, 214 – 242.
PODPÉRA, J., 1954: Conspectus muscorum europaeorum. – 697 pp. Praha.
PROSKAUER, J., 1963: Later »starting points» and the genera Mnium L., Mnium Hedwig, and Calypogeja Raddi. – Taxon 12, 200 – 201.
RICHARDS, P. W., 1959: Bryophyta. – In W. B. TURRILL, Vistas in Botany. – A Volume in Honour of the Bicentenary of the Royal Botanic Gardens, Kew, 387 – 420. London.
ROBINSON, H., 1962: Generic revision of North American Brachytheciaceae. – Bryologist 65, 73 – 146.
SAVICZ-LJUBITZKAJA, L. I. & I. I. ABRAMOV, 1959: The geological annals of Bryophyta. – Rev. Bryol. Lich. N. S. 28, 330 – 342.
SCHIMPER, W. P., 1860, 1876: Synopsis muscorum europaeorum praemissa introductione de elementis bryologicis tractante. – CLIX + 735 pp. + VIII pls. Stuttgartiae; 2nd ed. CXXX + 886 pp. + VII pls. Stuttgartiae.
SCHUSTER, R. M., 1966: The Hepaticae and Anthocerotae of North America east of the hundredth meridian. I. – 802 pp. New York and London.
SENFT, E., 1924 a: Mnioindikan, ein neuer pflanzlicher Farbstoff. – Stud. Plant. Phys. Lab. Charles Univ. Prague 2, 95 – 100.
– » – 1924 b: Über gefärbte Membranen der Moose aus der Familie der Mnia. – Ibid. 2, 100 – 106.
SHLYAKOV, R. N., 1961: Flora listostebel'nyh mkhov Khibinskikh gor. – 249 pp. Murmansk.
SIMPSON, G. G., 1961: Principles of animal taxonomy. – 247 pp. New York.

SMIRNOVA, Z. N., 1964: Mnium confertidens (Lindb. et Arn.) Par. et species ei affines. – Novit. Syst. Pl. non Vascul. 1964, 301 – 317.

SOKAL, R. R. & P. H. A. SNEATH, 1963: Principles of numerical taxonomy. – 359 pp. San Francisco.

STEERE, W. C., 1946: Cenozoic and mesozoic Bryophytes of North America. – Amer. Midland Natural. 36, 298 – 324.

– » – 1947: A consideration of the concept of genus in Musci. – Bryologist 50, 247 – 258.

– » – 1954: Chromosome number and behavior in arctic mosses. – Bot. Gaz. 116, 93 – 133.

– » – 1955: Bryology. – A Century of Progress in the Natural Sciences, 1853 – 1953, 267 – 299.

– » – 1958: Evolution and speciation in mosses. – Amer. Natural. 92: [862], 5 – 20.

STEERE, W. C., L. E. ANDERSON & V. S. BRYAN, 1954: Chromosome studies on California mosses. – Mem. Torrey Bot. Club 20: 4, 1 – 75.

Systematic Association Committee for Descriptive Biological Terminology, 1962: II. Terminology of simple symmetrical plane shapes. – Taxon 11, 145 – 156, 1 chart.

SZAFRAN, B., 1950: Trachycystis Szaferi a new species of moss from the miocene of Poland. – Acta Soc. Bot. Poloniae 20: 1, 247 – 250.

TACHIBANA, S. & B. J. D. MEEUSE, 1960: Isolation of transaconitic acid from the moss Mnium affine. – Science 132, 1671.

TATUNO, S., 1951: Über die Chromosomen der Laubmoose, mit besonderer Rücksicht auf ihre Heterochromosomen. I. – La Kromosomo 8, 305 – 310. (Not seen, referred according to TATUNO & ONO 1966.)

TATUNO, S. & ONO, 1966: Zytologische Untersuchungen über die Arten von Mniaceae aus Japan. – Journ. Hattori Bot. Lab. 29, 79 – 95.

TATUNO, S. & K. YANO, 1953: Geschlechtschromosomen bei vier Arten von Mniaceae. – Cytologia 18, 36 – 42.

TUOMIKOSKI, R., 1936: Über die Laubmoosarten Mnium affine, Mnium rugicum und Mnium seligeri. – Ann. Bot. Soc. 'Vanamo' 6: 5, 1 – 45.

TUOMIKOSKI, R., 1949: Über die Kollektivart Drepanocladus exannulatus (Br. eur.) Warnst. – Ibid. 23: 1, I – V + 1 – + I – IV pls.

– » – 1958: Über den heutigen Stand der Laubmoosesystematik. – Uppsala Univ. Årsskr. 1958: 6, 65 – 69.

– » – 1967: Notes on some principles of phylogenetic systematics. – Ann. Entomol. Fenn. 33, 137 – 147.

VAARAMA, A., 1950: Studies on chromosome numbers and certain meiotic features of several Finnish moss species. – Bot. Not. 1950, 239 – 256.

VERDOORN, F., 1934: Bryologie und Hepaticologie, ihre Methodik und Zukunft. – Ann. Bryol. 4: Suppl., 1 – 39.

– » – 1950: The future of exotic cryptogamic botany. – Bryologist 53, 1 – 9.

WARNSTORF, C., 1906: Laubmoose. – Kryptogamenflora der Mark Brandenburg und angrenzender Gebiete II. Laubmoose. – 1160 pp. Leipzig.

WATSON, E. V., 1967: The structure and life of bryophytes. 2nd ed. – 192 pp. London.

WIJK, R. VAN DER, 1958: Precursory studies on Malaysian mosses. A preliminary key to the moss genera. – Blumea 9, 143 – 186.

– » – 1964: Reports Nomenclature Committees, Committee for Bryophyta. – Taxon 13, 136 – 137.

WIJK, R. VAN DER & W. D. MARGADANT, 1958: New combinations in mosses I. – Ibid. 7, 287 – 290.

WIJK, R. VAN DER, W. D. MARGADANT & P. A. FLORSCHÜTZ, 1959, 1964: Index Muscorum I, III. – I (A – C), 1959, Regnum Veget. 17, 548 pp. Utrecht.; III (Hypnum – O), 1964, ibid, 33, 529 pp. Utrecht.

WYLIE, A. P., 1957: The chromosome numbers of mosses. – Trans. Brit. Bryol. Soc. 3, 260 – 278.

ZIMMERMANN, W., 1960: Geschichte und Methode der Evolutionsforschung. – Arbeitstagung zu Fragen der Evolution zum Gedenken an Lamarck – Darwin – Haeckel, 15 – 36.

Index to supraspecific names

The pages listed in the indexes are those which contain some discussion of the taxa in question (as opposed to a mere reference), and those which deal with its nomenclature. References to the figures are provided in the list of specimens (pp. 120 - 121). References to taxa outside *Mniaceae* are given in every case.

Index to specific and subspecific epithets

8

Reprinted from *Syst. Zool.* **21**:364–374 (1972)

COMMENTS ON HENNIG'S "PHYLOGENETIC SYSTEMATICS" AND ITS INFLUENCE ON ICHTHYOLOGY

GARETH J. NELSON

For better or worse, creators of biological classifications have traditionally been unwilling to explain, discuss and weigh the relative merits of their purposes and achievements. But during the last ten years or so, the niche of expositor-discussant-arbiter has not remained empty, being filled (one might even say overpopulated) by theoreticians, experimentalists and mathematicians (Darlington, 1971:341). One result, perhaps ultimately beneficial, is that the taxonomist is now under some pressure to be explicit about his purposes and precise about his methods (Hull, 1970:49–50).

Among other taxonomists, Hennig (1966, 1969, 1971; also Schlee, 1971) has written on the subject of methods and purposes. His view is that many different classificatory systems are possible, and perhaps useful in one way or another, but that among them there is one best suited to serve as a general reference system for the science of biology. He defined the concept of phylogenetic relationship (kinship, or genealogical relationship) on which this system of classification is based and termed it the "phylogenetic system." It demands, axiomatically, that species related by common ancestry be classified together. Whereas some commentators consider Hennig's work

a modern expression of the Darwinian tradition in biology (e.g., Crowson, 1970:95; Nelson, 1971; see also Crowson, 1958), and value it as such, others consider it absurdly extreme or too rigid, to be of general applicability (Darlington, 1970). Some persons contest its claim to be "phylogenetic" (e.g., Mayr, 1965a:167, 1968:547, 1969:70), and some warn of the "disastrous" results that would follow its application to classification (e.g., Ashlock, 1971:64).

Critics of Hennig's system of classification have attacked, among others, the principles that species (or groups of them) are to be grouped according to the branching sequence of their ancestry, and ranked (1) according to their relative time of origin or (2) such that sister-groups are given equal rank. Each of these three principles may appear in itself rather arbitrary and unconnected with the other two, in short, a convenient target for hostile criticism. All three, however, are simple corollaries of the axiom that species related by common ancestry be classified together (grouped in the same taxon), and logically follow from it. So far there have been few explicit attempts to refute directly the axiom itself, probably because Hennig's concept of relationship is, to one degree or another, generally acceptable within the Darwinian tradition. Also, to accept Hennig's concept of relationship and to reject his axiom of classification would be to argue (1) that related species (or groups of them) should not be classified together, or (2) that unrelated species should be classified together. To date, few if any biologists have seemed consciously willing to so argue. A number of them have, however consciously or unconsciously, adopted concepts of relationships broader than that of Hennig. These broader concepts are of two sorts, here termed "Mayrian" and "Simpsonian" after the persons who, perhaps most clearly, defined them (no other significance is implied by the use of these terms in the following discussion of concepts of relationships).

Mayrian relationships. "The argument of . . . [Hennig] fails to recognize that the term *relationship* has two distinct meanings, genetic relationship and genealogical relationship." ". . . let me emphasize the essential equivalence of 'genealogic' with 'cladistic' ['phylogenetic' in the sense of Hennig], and the close correlation between genetic and phenetic. When an evolutionary taxonomist speaks of the relationship of various taxa, he is quite right in thinking in terms of genetic similarity, rather than in terms of genealogy." "When a biologist speaks of phylogenetic relationship, he means relationship in gene content rather than cladistic genealogy" (Mayr, 1965b:79).

Whatever else Mayrian relationships might consist of, their essential ingredient is the concept of overall genetic similarity. Although very little is known about the genetic similarity of any species or groups of them, there have been devised various measures of phenotypic similarity—sometimes assumed to be a rough estimate of genetic similarity (e.g., Sokal and Sneath, 1963).

The concept of Mayrian relationships is widespread in biology, and there is general agreement that it is useful for purposes

other than classification. For example, Simpson (1961:7) stated that "*Systematics is the scientific study of the kinds and diversity of organisms and of any and all relationships among them.*" "In this definition, the word 'relationships' is to be understood not in any particular, narrow sense (for instance, in the sense of phylogenetic connections), but in a fully general way, including specifically all associations of contiguity and of similarity." One may agree with Simpson that systematics is a broadly based science concerned with Mayrian relationships, but does this mean that classifications of organisms have been, are, or ever will be expressive, in any precisely definable sense, of this broad kind of relationships? Evidently not, for in the actual practice of classification, Simpson and others have adopted narrower concepts of relationships probably because, as Hennig and others have pointed out (e.g., Hennig's 1966, remarks on "syncretistic systems," and Griffiths', in press, remarks on "combined classification"; also Jardine and Sibson, 1971:139; Brundin, in press), Mayrian relationships pose unnecessary and insoluble problems, which arise from confounding, whether deliberately or not, the concepts of similarity and kinship. A practical example provided by Simpson (1960:123) concerns the classification of two contemporaneous genera of fossil mammals, which so far as one might suppose, were very similar in overall "gene content" and on that basis might have been classified together in accordance with a Mayrian concept of relationship. Simpson, nevertheless, classified them in different suborders in accordance with their presumed kinship: "The ancestral form . . . was *Hyracotherium*. . . . It is supposed to be slightly closer [in kinship] to horses than to tapirs and so is classified as a member of the horse suborder and family. The contemporary *Homogalax* . . . may be slightly closer to tapirs and for that reason is now classified in a different suborder and family from *Hyracotherium*. Nevertheless, *Hyracotherium* and *Homogalax* are almost identical in structure, to such an extent that the most skilled paleontologists long failed to distinguish them correctly and even now are likely to mistake specimens of one for the other."

Simpsonian relationships. "Is a man more closely related to his father, son or brother? The actual genes involved may be quite different, but the degree of genetical relationship to father and to son is invariably the same (0.5 in terms of proportion of shared chromosomes). Genetical relationship to a brother is variable—from 1.0 to 0.0 in terms of chromosomes, although the probability of those extremes is exceedingly low—but the mean value is the same as for father or son. Unfortunately relationships among taxa do not have such fixed a priori expectations, and they cannot be precisely measured. The two kinds of relationships nevertheless exist: among successive taxa in an ancestral-descendant lineage, and among contemporaneous taxa of more or less distant common origin. In accordance with the usual coordinates of tree representation, the former relationships are called *vertical* and the latter *horizontal*." "One kind of relationship is obviously just as objective as the other and may be just as close or distant. Moreover, it seems obvious that the two are equally phylogenetic" (Simpson, 1961:129).

The "vertical" and "horizontal" relationships of Simpson pertain to kinship, and he calls them "phylogenetic." His doing so differentiates them from the other categories of "relationships" ("associations of contiguity and of similarity"—the various Mayrian relationships), included by Simpson in his broad definition of systematics (see above). Of all these categories of relationships, Simpson's "phylogenetic" relationships are central to his discussion of phylogeny and classification. In classifying, according to Simpson, one must "compromise" between extreme vertical and extreme horizontal classification. In practice this means always classifying phylogenetically related species together, but choosing sometimes to operate with the concept of vertical relationships and sometimes with the concept of horizontal relationships. The actual taxa formed may be different. Following Simpson's analogy of fathers, sons, and brothers, one may choose to classify one brother with another brother rather than his father (expressing a horizontal relationship), or a brother with his father rather than another brother (a vertical relationship).

The concept of kinship between species is a corollary of the theory of evolution, and has influenced much of post-Darwinian biology. Simpsonian relationships are not concepts of mere kinship, but kinship partitioned into vertical and horizontal aspects. They are, furthermore, distinctive in arising from different sources: "In dealing with recent animals or with contemporaneous faunas of fossils, only horizontal relationships are *directly* involved." "In temporal sequences of fossils vertical relationships are directly presented" (Simpson, 1961:129).

Simpson's discussion of relationships has caused some comment. For example, Brundin (1966:28) stated that "all phylogenetic relationships are by necessity vertical. Brother relationships are quite irrelevant," a point contested by Tuomikoski (1967:143). But as implied by Brundin, the kinship between two contemporaneous species (Simpson's "horizontal" relationship) may be partitioned into two vertical relationships: a vertical relationship between an empirically unknown, common ancestral species (hypothetical) and each of the two descendant species observed to be contemporaneous. One may justifiably conclude that horizontal relationships do not exist as a separate type of kinship. Rather, they are patterns of kinship involving the vertical relationships of empirically unknown ancestral species, and are reducible thereto.

Here Simpson blurs the distinction between phylogenetic relationships (kinship) and similarity "relationships." Kinship between species is, of course, never "presented" to the investigator, but can only be assumed by him following an analytical appraisal of what species there are, and how they might be related. What are presented are specimens and, in a loose sense, their similarities and differences (Colless, 1967: 294). In any case, Simpson's partition of kinship into horizontal and vertical aspects gives a hint of, but is not itself, the essential ingredient of Simpsonian relationships. Indeed, Simpson might well grant that horizontal relationships can all be reduced to vertical relationships of unknown (hypothetical) ancestors—in the sense of kinship as opposed to similarity—without modifying the main thrust of his argument: to emphasize vertical relationships of ancestors presumed to be directly known in, and revealed by, the fossil record.

It may be widely believed that problems of relationship (kinship) are ultimately solvable only when adequate temporal (stratigraphic) sequences of fossils are at hand. Then, when temporal sequences of fossils are seemingly "continuous," one might suppose that the various ancestral and descendant species and groups—and their "vertical relationships"—could be recognized as such, apprehended empirically by means of the "control" introduced by the stratigraphic (temporal) sequence: "The great drawback of the comparative method and of contemporaneous evidence is that they are not in themselves historical in nature. The drawing of historical conclusions from them is therefore full of pitfalls unless it can be adequately controlled by *directly* historical evidence" (Simpson, 1960:122). However, "proponents of this view have failed to grasp an essential point: an actual phylogeny is not capable of outright discovery; it is a system of relationships that needs to be analyzed" (Schaeffer, Hecht and Eldredge, in press). Throughout his career Simpson has, nevertheless, been a strong spokesman for this view, and has repeatedly stressed the "true time dimension" (e.g., 1961:83) provided by fossils, and their unique ability to "prove" (1961:86) or "confirm" (1961:99) what otherwise would be purely hypothetical.

According to this view, "the fossil record . . . alone provides a relative time scale and it alone separates what actually happened from what might have been" (McKenna, 1969:217).

Such pronouncements, which evoke healthy skepticism from some persons, have been countered in a variety of ways, e.g.: "By maintaining that biostratigraphic [temporal] data should be ignored when evaluating relationships, we are simply arguing that phylogenies must be based on comparative morphology. Data concerning relative stratigraphic position necessarily bias the results by narrowing the range of possible relationships held by the taxa in question. Obviously Taxon A cannot be ancestral to Taxon B, if the latter is some two million years older than Taxon A. But a penchant for recognizing ancestral-descendant sequences may result in ignoring the possibility that the taxa A and B, rather than possessing an ancestor-descendant relationship, actually possess a common ancestor, i.e., neither is ancestral to the other" (Schaeffer et al., in press). To assume that the relationships between all known species and groups of species, whether fossil or recent, involve hypothetical common ancestral species, which are empirically unknown and unknowable, divorces problems of relationships from data concerning stratigraphic distribution of fossils. To so assume is to abandon, at least temporarily, the hope of discovering actual ancestral-descendant relationships between species.

Hennigian relationships, indeed, imply a distinctive research strategy. They eliminate what Hennig (1966) and Brundin (1966) claim to be conceptual and pragmatic weaknesses inherent in Mayrian and Simpsonian relationships, the influence of which is pervasive enough in biology to be identified with tradition: "an alternative to the traditional method—'an attempt first to define orders and other higher taxa and then to speculate upon their origin, albeit in the light of the known fossils' (Greenwood et al., 1966:346) is to use the criteria recommended by Hennig (1966:88, 120) and to search . . . for the sister group (Hennig, 1966:139; see also Brundin, 1966: 17)" (Patterson, 1968:95).

One may compare Mayrian, Simpsonian and Hennigian relationships. Mayrian relationships are broad and vague concepts of overall genetic and phenotypic similarity; kinship is deemphasized or ignored. Simpsonian relationships are more narrow and precise concepts of kinship, which is partitioned into horizontal and vertical aspects, defined in relation to ancestors presumed to be unknown (hypothetical) and known (non-hypothetical) respectively. Hennigian (sister-group) relationships, concepts more narrow still, are common ancestral relationships (patterns of vertical relationships involving unknown, hypothetical common ancestors), and correspond to Simpson's "horizontal relationships"— insofar as his pertain to kinship rather than to similarity.

Hennigian relationships are, by definition, relative. Contemporaneous or not, two species—however different—may be assumed to have evolved from a common hypothetical ancestral species—however remote. With consideration of a third species —either fossil or recent—one may enquire whether it is more closely related to, and shares a more recent hypothetical common ancestry with, either of the two species. Those species sharing a more recent common ancestry are, by Hennig's definition, more closely related among themselves than they are to other species.

That Hennigian relationships may be worked out generally—for fossil as well as recent species—is now clear (Eldredge, 1972; Gaffney, 1972): "a phylogeny is of exactly the same form whether it involves fossils, Recent organisms, or a mixture of both" (Schaeffer et al., in press). Hennigian relationships embody the principles that all common ancestral species are necessarily hypothetical, and that ancestral species, although they may be reconstructed (e.g., Fitch, 1971), will forever remain unknown and unknowable in a directly empirical sense (e.g., in the sense that species

are "known" by way of inference from observation of study material). To work out Hennigian relationships (kinship) presupposes rejection of (1) Simpson's concept of "vertical relationships," "directly presented" by the fossil record and "known . . . to be true from fossil evidence" (Simpson, 1961: 103), for the reason that the geological ages of the species in question (Simpson's "true time dimension") are irrelevant for the purpose of determining Hennigian relationships; and with it (2) the premise (e.g., Cain, 1959:211; Sokal and Sneath, 1963: 227; Jardine and Sibson, 1971:160) that fossils somehow have been, are, or ever will be necessary, on a priori grounds, to establish kinship between contemporaneous species. In other words, to determine kinship with the use of biostratigraphic (temporal) data provided by fossils "is simply wrong" (Schaeffer et al., in press). If only because of the general inadequacy in preservation of fossil material, kinship of species represented by fossils is in principle no more, and in practice often less, firmly determinable than kinship of recent species. In any case, for working out Hennigian relationships (kinship) the same methods— and their limitations—apply equally to all species, fossil or recent (e.g., Schlee, 1971: 60; cf. Colless, 1967:294). These methods are those that allow (1) characters to be recognized as being relatively primitive (plesiomorphous in Hennig's terminology) or advanced (apomorphous in Hennig's terminology), regardless of the stratigraphic positions of the species in which they occur; and (2) relationships to be based only on shared advanced characters (synapomorphies in Hennig's terminology), regardless of the stratigraphic positions of the species or groups related.

To some extent Hennigian relationships have been commented on by ichthyologists active in research on higher classification of both fossil and recent fishes: "Ridewood (1904), Hennig (1966), and a host of others have pointed out that relationships cannot satisfactorily be determined on the basis of ancestral characters" (Gosline, 1969:

214); "Hennig (1965) has correctly pointed out that ' . . . it is not the extent of resemblance or difference . . . that is of significance (in determining) phylogenetic relationship, but the connection of the agreeing or divergent characters with earlier conditions' " (Rosen and Patterson, 1969:417); "We here accept a basic tenet of the 'phylogenetic cladists' (Hennig, . . .) that assessments of affinity are most reliably based on shared, derived characters" (Schaeffer et al., in press); "The analysis of relationships used here follows . . . the phylogenetic approach suggested by Hennig (1966) and Brundin (1966)" (Weitzman, in press).

From the above, it seems that Hennig's definition of relationship is consistent enough with ichthyological tradition to be acceptable as such. So also seems his principle that relationships can be recognized only by means of shared advanced characters, i.e., synapomorphies in Hennig's terminology (Schlee, 1971:5), and most if not all of his methods of recognizing advanced characters. As regards his system of classification, however, among ichthyologists there is some disagreement with his axiom that related species be classified together.

In recent years there have been repeated attempts to provide an updated, generally applicable and useful, classification of teleostean fishes (e.g. Greenwood, Rosen, Weitzman and Myers, 1966; McAllister, 1968; Gosline, 1971; Rass and Lindberg, 1971; for other fish groups see e.g., Nelson, 1969; Lindberg, 1971; Moy-Thomas and Miles, 1971). These classifications, as might be expected, vary considerably. To one degree or another, most of the concerned authors made an attempt to explain what they were doing and why. If one compares their efforts with earlier ones (e.g., Regan, 1929; Berg, 1940), there is discernible a trend toward a more critical and explicit treatment of relationships and an increasing awareness that Hennigian relationships are what most investigators are trying to discover regardless of how the relationships

might subsequently be expressed in classification.

With the vagaries of formal classification left aside for the moment, a preliminary assessment of the situation seems to confirm the apparently extravagant prediction of Tuomikoski (1967:147): "if the recent works of Hennig and Brundin are carefully studied by the systematic monographers and biogeographers and strict methodical reasoning in phylogenetic matters is accepted and put into practice, a new era of fruitful research will dawn, greatly contributing to a better understanding of the history of life on earth." Why this seems so is perhaps best stated by Jarvik (1968: 522): Hennig's method "enforces a phylogenetic way of thinking and a careful consideration of the evidence." In other words, Hennig's work is an attempt "to put the practice of phylogeny (perhaps more an art, at present, than a science) on a sound philosophic base" (Thomson, 1971:141).

Some persons are dubious, however, that Hennig's system of classification must be adopted in order that science might fully benefit from the use of his precise conceptual tools for determining relationships: "Brundin (1966) . . . has shown that a knowledge of the relative ages of various lineages is essential for work in historical zoogeography. But for teleost fishes, . . . fossil forms will provide more reliable information on this subject than any attempt to reclassify modern teleosts ever could" (Gosline, 1971:94). Unfortunately, "The fossil record is clearly no panacea to students of phylogeny [or historical biogeography]. The reason for this is not really that the record is 'incomplete,' but that many neontologists and paleontologists have assumed that if enough rocks are split the record of past life will reveal itself" (Schaeffer et al., in press). Clearly, fossils by themselves do not solve problems of phylogeny or historical biogeography, but like recent specimens can only be the objects of historical analysis.

For the purpose of historical analysis, what are the benefits to be derived from phylogenetic classification, both of fossil and recent species? For a definitive evaluation the minimal demands are (1) a phylogenetic classification of all organisms, or at least most or all organisms of a major group, and (2) a demonstration of its benefits relative to other types of classifications of the same organisms.

Pending the announcement of some such "definitive evaluation," some persons who profess a phylogenetic approach argue that classification need not precisely embody detailed patterns of relationships, which might be better left represented by diagrams (Tuomikoski, 1967:147, Lindeberg, 1971). But if classification need never embody, nor indeed express precisely, detailed patterns of relationships, will a definitive evaluation ever be achieved? And meanwhile, by what principles are classifications to be constructed—even on an interim basis? It seems pointless to insist, as many have done, that classifications be "convenient" or "practical" without specifying what they might be convenient or practical for, and the principles by which one might judge a classification more convenient or practical than another. Similarly, to appeal to the concept of "status quo" or "tradition" begs the question of whose status or tradition to embrace.

Hennig's system of classification requires that species related by common ancestry be classified together to form monophyletic groups. Non-monophyletic groups (paraphyletic or polyphyletic) are not permitted. The result is what some persons recognize as a "genealogical" classification in the tradition of Darwin (1859:486: "Our classifications will come to be, as far as they can be so made, genealogies"): "we believe that a classification should reflect propinquity of descent, a view first formulated by Darwin (see Ghiselin, 1969) and later expanded by Hennig and Brundin (see Brundin, 1966) and, most elegantly, by Crowson (1970)" (Greenwood and Rosen, 1971:39). But neither Darwin nor anyone else has ever announced a "definitive evaluation" in the above sense. Darwin simply

argued that "all true classification is genealogical; that community of descent is the hidden bond which naturalists have been unconsciously seeking, and not some unknown plan of creation, or the enunciation of general propositions, and the mere putting together and separating objects more or less alike" (1859:420). There is little basis for equivocation about the meaning of his argument (cf., Ghiselin, 1969:84; Mayr, 1969:70; Nelson, 1971:375; Darlington, 1972): "If it could be proved that the Hottentot had descended from the Negro, I think he would be classed under the Negro group, *however much he might differ in colour and other important characters* from negroes (Darwin, 1859:424, italics added; the reader is invited to substitute "bird," "reptile" and "reptiles" for the appropriate quoted terms).

To some persons Darwin's argument was, and still is, convincing, at least on an interim basis—pending the announcement and confirmation of a "definitive evaluation." If his argument is still valid today, it means that phylogenetic (genealogical) classification has been carried along implicitly or unconsciously in the work of numerous systematists over the years. It means, also, that the classificatory traditions associated with, for example, the names of Linnaeus, Owen and Adanson are ultimately meaningful only to the extent that the principles of phylogenetic classification can be seen to be implicit in them. It means, further, that Hennig's (1966, 1969) effort, together with Brundin's (1966) account of the austral midges, has put the case for phylogenetic classification on an explicit and detailed basis, supported further by Crowson (1970), Griffiths (in press) and others. Indeed, their work may be the best demonstration of the nature and benefit of phylogenetic classification since Darwin first enunciated the concept. But only time will tell whether a phylogenetic classification of all organisms, or even those of a major group (see, however, Hennig's 1969 classification of insects) will ever be achieved so as to permit a "definitive evaluation," rather than an indefinitely long interim period of relative indecision, and the confusion and controversy that attends it. There are still at work "traditions" other than the Darwinian-Hennigian.

An example of a non-monophyletic group in Hennig's sense is the group "Reptilia" of common parlance (which on occasion has been made monophyletic by the inclusion of birds and the exclusion of "mammal-like reptiles"). Why does a group as notoriously non-monophyletic as the "Reptilia" tend to persist in classification (despite repeated efforts to eliminate or emend it)? Is it because of some requirement of Linnaeus, Owen or Adanson, or traditions associated with them? Or is the decisive factor only the historical inertia provided by accumulated and uncritical usage, an inertia that finds as "impractical" change of any sort? Indeed, why else except for its clash with this "tradition" (which except as inertia seems unspecifiable) does the Darwin-Hennig system of classification on occasion meet with such disfavor (Ashlock, 1971:64)?

Several commentators have argued that certain non-monophyletic groups have traditionally been found practical or convenient and, therefore, should be maintained (see above). Others have argued that they "*must*" (Mayr, 1965a:168, 1969: 75) be maintained so that classification can embody, however vaguely and imprecisely, aspects of overall similarity and divergence (Mayrian relationships), confounded or not with kinship (e.g., by McAllister, 1970:210: "Hennigian views run directly counter to those of some classical and many numerical taxonomists who consider divergence as an important index of relationship"). Still others have argued that they are unavoidable, because of the refractory nature of ancestor-descendant relationships (the vertical aspect of Simpsonian relationships, presumed to be directly presented by the fossil record): "if any two lineages are traced back far enough, they will be found to originate among brothers who cannot be distinguished in biologically meaningful

terms; that is, the brothers may not show the characteristics of the descendant lineages at all. In practice, one follows lineages back to a basal group and then switches from a father-son (vertical) to a fraternal (horizontal) classification for that basal group" (Gosline, 1971:93; cf., above, Simpson's discussion of *Hyracotherium* and *Homogalax*). It may be interjected here that the concept of vertical relationships between contemporaneous groups (i.e., recent groups of teleostean fishes) seems, at best, a misleading oversimplification. Indeed, how, except by way of unknown (hypothetical) common ancestral species, might lines of kinship (actual "lineages") be traced between contemporaneous groups? Even when fossils and their stratigraphic data are available, one may contest that it is ever really possible on an empirical basis to "follow lineages" back to "brothers" that "do not show the characteristics of the descendant lineages at all." In such a case, what would be the "biologically meaningful" evidence that each "brother" is the ancestor of a distinct lineage? The minimal requirement for each such relationship is the demonstration that the species (the "brother" and the species that compose the lineage descending from it) share one or more advanced characters indicative of common ancestry. "The notion of ancestry and descent [the vertical aspect of Simpsonian relationships] is, of course, implicit in the concept of phylogeny and is a logical concomitant to the entire idea of evolution. But there exists a large information gap between what we know *must have* happened and knowledge of what actually *did* happen" (Schaeffer et al., in press). A willing ability to negotiate this information gap by the explicit use of the concept of hypothetical (unknown and unknowable) ancestors characterizes the Hennigian approach; in the absence of this ability the principles of phylogenetic systematics cannot be exploited, and the significance of Hennig's contribution cannot be fairly criticized.

Perhaps some of the current adverse

criticism of Darwin-Hennig classification has arisen from conceptual inertia and the lingering traditions of Linnaeus, Owen and Adanson; as such it may be dismissed without further comment. More significant, in quantity at least, is the criticism arising from seemingly naive acceptance of concepts of Mayrian or Simpsonian relationships (e.g., Cain, 1967; Darlington, 1970; Howden, 1972); biased through neglect of a worthy alternative (the concept of Hennigian relationships, especially as it may be applied to the study of fossils), this criticism can only confuse rather than clarify the issues.

There remains in the minds of some persons the "vexed question" (Colless, 1969: 142) of non-monophyletic groups (the paraphyletic and polyphyletic groups of Hennig) and the extent to which they should be tolerated—if at all—for whatever reasons. In general, ichthyologists seem intolerant of them: "the objectives of a phylogenetic classification are not met by maintaining a paraphyletic group (Hennig, 1966, p. 146)" (Rosen and Greenwood, 1970:22). Indeed, recent work in teleostean classification gives every indication, because of the repeated elimination of non-monophyletic groups, of progress toward a completely phylogenetic classification in the Darwin-Hennig sense. As implied by Cavender (1970:2), the appearance of the revised classification of teleosts of Greenwood et al. (1966), "in selecting not primitive but derived character states as the more useful for inferring monophyletic groups," was a decisive, major step toward Darwin-Hennig classification. Of interest is that it was taken not because of Hennig's influence, but because of progressive forces within ichthyology itself, as may be well appreciated, for example, in the remarks of Garstang (1931). In view of the popularized assertions that Darwin-Hennig classification is "totally distorted" and "absurd" (Mayr, 1968:547, 1969:231) or "absolutely impossible" (Simpson, 1961: 130), the tendency toward its accomplishment in ichthyology may have a significant

influence on biological systematics in general.

ACKNOWLEDGMENTS

I am indebted to Profs. Lars Brundin, Roy Crowson, Ernst Mayr and Willi Hennig, and other colleagues too numerous to mention singly, who have read and commented on this manuscript (their agreement with any of the ideas expressed therein is not implied).

REFERENCES

ASHLOCK, P. D. 1971. Monophyly and associated terms. Syst. Zool. 20:63–69.
BERG, L. S. 1940. Classification of fishes, both recent and fossil. Trav. Inst. Zool. Acad. Sci. URSS, 5:87–517.
BRUNDIN, L. 1966. Transantarctic relationships and their significance. K. Svenska Vetensk.-akad. Handl., ser. 4, vol. 11, no. 1.
BRUNDIN, L. In press. Evolution, causal biology, and classification. Zool. Scripta 1:107–120.
CAIN, A. J. 1959. Deductive and inductive methods in post-Linnaean taxonomy. Proc. Linn. Soc. Lond. 170:185–217.
CAIN, A. J. 1967. One phylogenetic system. Nature, London 216:412–413.
CAVENDER, T. 1970. A comparison of coregonines and other salmonids with the earliest known teleostean fishes. (In) Lindsey, C. C., and C. S. Woods (eds.), Biology of coregonid fishes, 1–32. The University of Manitoba Press, Winnipeg.
COLLESS, D. H. 1967. The phylogenetic fallacy. Syst. Zool. 16:289–295.
COLLESS, D. H. 1969. The interpretation of Hennig's "phylogenetic systematics"—a reply to Dr. Schlee. Syst. Zool. 18:134–144.
CROWSON, R. A. 1958. Darwin and classification. (In) Barnett, S. A. (ed.), A century of Darwin, 102–129. Harvard University Press, Cambridge.
CROWSON, R. A. 1970. Classification and biology. Atherton Press, Inc., New York, and Heinemann Educational Books, Ltd., London.
DARLINGTON, P. J., JR. 1970. A practical criticism of Hennig-Brundin "phylogenetic systematics" and Antarctic biogeography. Syst. Zool. 19:1–18.
DARLINGTON, P. J., JR. 1971. Modern taxonomy, reality, and usefulness. Syst. Zool. 20:341–365.
DARLINGTON, P. J., JR. 1972. What is cladism? Syst. Zool. 21:128–129.
DARWIN, C. 1859. On the origin of species by means of natural selection. John Murray, London.

ELDREDGE, N. 1972. Systematics and evolution of *Phacops rana* (Green, 1832) and *Phacops iowensis* Delo, 1935 (Trilobita) from the Middle Devonian of North America. Bull. Amer. Mus. Nat. Hist. 147:45–114.

FITCH, W. M. 1971. Toward defining the course of evolution: minimum change for a specific topology. Syst. Zool. 20:406–416.

GAFFNEY, E. S. 1972. The systematics of the North American family Baenidae (Reptilia, Cryptodira). Bull. Amer. Mus. Nat. Hist. 147: 241–320.

GARSTANG, W. 1931. The phyletic classification of Teleostei. Proc. Leeds Phil. Lit. Soc., sci. sect. 2:240–260.

GHISELIN, M. T. 1969. The triumph of the Darwinian method. The University of California Press, Berkeley and Los Angeles.

GOSLINE, W. A. 1969. The morphology and systematic position of the alepocephaloid fishes. Bull. Brit. Mus. (Nat. Hist.), Zool. 18:183–218.

GOSLINE, W. A. 1971. Functional morphology and classification of teleostean fishes. The University of Hawaii Press, Honolulu.

GREENWOOD, P. H., AND D. E. ROSEN. 1971. Notes on the structure and relationships of the alepocephaloid fishes. Amer. Mus. Novitates 2473:1–41.

GREENWOOD, P. H., D. E. ROSEN, S. H. WEITZMAN, AND G. S. MYERS. 1966. Phyletic studies of teleostean fishes, with a provisional classification of living forms. Bull. Amer. Mus. Nat. Hist. 131:345–455.

GRIFFITHS, G. C. D. In press. The phylogenetic classification of Diptera Cyclorrhapha, with special reference to the male postabdomen. Junk, The Hague, Ser. entomologica, no. 8.

HENNIG, W. 1965. Phylogenetic systematics. Ann. Rev. Ent. 10:97–116.

HENNIG, W. 1966. Phylogenetic systematics. The University of Illinois Press, Urbana.

HENNIG, W. 1969. Die Stammegeschichte der Insekten. Senckenbergischen Naturforschenden Gesellschaft, Frankfurt am Main.

HENNIG, W. 1971. Zur Situation der biologischen Systematik. Erlanger Forsch., ser. B., Naturwissensch. 4:7–15.

HOWDEN, H. F. 1972. Systematics and zoogeography: science or politics. Syst. Zool. 21: 129–131.

HULL, D. L. 1970. Contemporary systematic philosophies. (*In*) Johnston, R. F., et al. (ed.), Ann. Rev. Ecol. Syst. 1:19–54.

JARDINE, N., AND R. SIBSON. 1971. Mathematical taxonomy. John Wiley & Sons Ltd., London, New York, Sydney and Toronto.

JARVIK, E. 1968. Aspects of vertebrate phylogeny. (*In*) Ørvig, T. (ed.), Current problems of lower vertebrate phylogeny, 497–527. Almqvist and Wiksell, Stockholm.

LINDBERG, G. U. 1971. Families of the fishes of the world. Leningrad.

LINDEBERG, B. 1971. Comments on the branching of phyletic lineages and the formation of taxa. Ann. Ent. Fenn. 37:54–57.

MCALLISTER, D. E. 1968. Evolution of branchiostegals and classification of teleostome fishes. Natl. Mus. Canada Bull. 221:i–xiv, 1–239.

MCALLISTER, D. E. 1970. Gill arches and the phylogeny of fishes. Quart. Rev. Biol. 45:209–210.

MCKENNA, M. C. 1969. The origin and early differentiation of therian mammals. Ann. New York Acad. Sci. 167(1):217–240.

MAYR, E. 1965a. Classification and phylogeny. Amer. Zool. 5:165–174.

MAYR, E. 1965b. Numerical phenetics and taxonomic theory. Syst. Zool. 14:73–97.

MAYR, E. 1968. Theory of biological classification. Nature, London 220:545–548.

MAYR, E. 1969. Principles of systematic zoology. McGraw-Hill Book Co., New York.

MOY-THOMAS, J. A., AND R. S. MILES. 1971. Palaeozoic fishes. W. B. Saunders Co., Philadelphia and Toronto.

NELSON, G. J. 1969. Gill arches and the phylogeny of fishes, with notes on the classification of vertebrates. Bull. Amer. Mus. Nat. Hist. 141:475–552.

NELSON, G. J. 1971. "Cladism" as a philosophy of classification. Syst. Zool. 20:373–376.

PATTERSON, C. 1968. The caudal skeleton in Mesozoic acanthopterygian fishes. Bull. Brit. Mus. (Nat. Hist.), Geol. 17:47–102.

RASS, T. S., AND G. U. LINDBERG. 1971. Modern concepts of the natural system of recent fishes. Voprosy Ikhtiol. 11:380–407; Jour. Ichthyol. 11:302–319 (translation).

REGAN, C. T. 1929. Fishes. (*In*) Encyclopaedia Britannica, 14th. ed. 9:305–328.

RIDEWOOD, W. G. 1904. On the cranial osteology of the fishes of the families Elopidae and Albulidae. Proc. Zool. Soc. Lond. 1904:35–81.

ROSEN, D. E., AND P. H. GREENWOOD. 1970. Origin of the Weberian apparatus and the relationships of the ostariophysan and gonorynchiform fishes. Amer. Mus. Novitates 2428:1–25.

ROSEN, D. E., AND C. PATTERSON. 1969. The structure and relationships of the paracanthopterygian fishes. Bull. Amer. Mus. Nat. Hist. 141:357–474.

SCHAEFFER, B., M. K. HECHT, AND N. ELDREDGE. In press. Paleontology and phylogeny. (*In*) Dobzhansky, T., et al. (ed.), Evolutionary biology, 6:31–46.

SCHLEE, D. 1971. Die Rekonstruktion der Phylogenese mit Hennig's Prinzip. Aufsätze Red. Senckenberg. Ges. 20:1–62.

SIMPSON, G. G. 1960. The history of life. (*In*) Tax, S. (ed.), The evolution of life, 1:117–180. The University of Chicago Press, Chicago and London; The University of Toronto Press, Toronto.

SIMPSON, G. G. 1961. Principles of animal taxonomy. Columbia University Press, New York and London.

SOKAL, R. R., AND P. H. A. SNEATH. 1963. Principles of numerical taxonomy. W. H. Freeman and Co., San Francisco.

THOMPSON, K. S. 1971. The adaptation and evolution of early fishes. Quart. Rev. Biol. 46:139–166.

TUOMIKOSKI, R. 1967. Notes on some principles of phylogenetic systematics. Ann. Ent. Fenn. 33:137–147.

WEITZMAN, S. H. In press. Osteology and evolutionary relationships of the stomiatoid fish family Sternoptychidae with a new classification of stomiatoid families. Bull. Amer. Mus. Nat. Hist.

(*Received March, 1972*
Revised August, 1972)

Part IV

PARSIMONY METHODS

Editors' Comments
on Papers 9 Through 13

The most widely applied methodology for the estimation of evolutionary history is parsimony analysis, in various forms. The preferred cladogram is the one that requires the fewest character state changes to explain the distribution of character states among the taxa analyzed. Different types of parsimony analysis are obtained depending on the types of character state change allowed. Character states may change from primitive to advanced, or may reverse from advanced to primitive. Either the primitive or an advanced state will be found at interior nodes of the cladogram. Certain methods allow interior nodes to be polymorphic. The cladogram that minimizes the number of polymorphic interior nodes is preferred.

The development of parsimony dates from modifications of Wagner's Groundplan-divergence method. Paper 9 outlines the method and states the step-by-step procedures involved. Wagner also discusses the applications of this method. The use of parsimony is not explicit in this paper (nor in Paper 2). Parsimony is used in Groundplan-divergence to minimize the character state change in branches of the cladogram defined by derived character states that occur uniquely in the taxa on that branch. Camin and Sokal (Paper 10) use similar arguments to describe a parsimony method that allows only parallelisms. They advocate the use of characters that agree and in which states would arise uniquely on a cladogram for the group being studied. These characters are then used to develop a "procladogram," with parsimony applied to minimize the length of the branches defined by the procladogram.

Farris (Paper 12) offered a parsimony method to allow both parallelisms and reversals. This method, known as Wagner parsimony, was developed as a numerical implementation of the Groundplan-divergence method. It has been one of the most widely used methods, and the basic algorithm given in this paper has been adopted in the development of several computer programs for this type of analysis. Kluge and Farris (Paper 11) also discuss this type of method in the context of the evolution of anurans. Particularly important is the concept of additive binary coding to transform a qualitative multistate data matrix into an equivalent matrix of binary characters. Solbrig (Paper 13) discusses evolutionary relationships in *Gutierrezia* using a variety of phenetic and cladistic techniques. This paper illustrates the application of parsimony techniques and the value of comparative phenetic and cladistic studies.

9

Reprinted from *Syst. Bot.* **5**:173-193 (1980)

Origin and Philosophy of the Groundplan-divergence Method of Cladistics

W. H. Wagner Jr.

ABSTRACT. The Groundplan-divergence Method of constructing phylogenetic trees was created during the 1950's for illustrating systematic principles, but it was taken up widely by researchers carrying out monographic research. Based upon a study of certain Hawaiian ferns, the method attempts to deduce pathways of genetic change on the basis of phenetic evidence. It tries to estimate the amounts (grades), directions (clades), and sequences (steps) of phylogenetic divergence, using the concept of generalized or groundplan characteristics as the basis of judging primitiveness. Groundplan-divergence analysis is concerned with pathways of actual biological changes rather than with their chronology (when the changes took place) or their geography (where they took place). Obvious parallelisms, reticulations, and co-existences of ancestors and derivatives are accepted and embodied in the cladistics. The method involves the interplay of phenetic classification, the detection of taxa of hybrid origin, analysis of character trends, synthesis of generalized character states, estimation of divergence levels, grouping according to divergence formulas, and the connecting of lines with hybrid reticulations. The method is not able to overcome problems that are caused by such factors as major gaps in the phylogenetic record, important missing characters, excessive hybridization, and evolutionary patterns that involve massive randomness or overwhelming amounts of parallelism. Examples of various uses of this method are cited.

In view of recent resurgence in cladistic theory and methodology, it seems appropriate to review the history and philosophy of the Groundplan-divergence Method, one of the pioneering attempts to place the construction of phylogenetic trees on an objective basis. The method remains as useful today as it was when it was introduced over a quarter of a century ago for purposes of teaching principles of systematic botany. As will be seen, the approach is a conservative one, utilizing time-honored ideas of phylogenetic induction and bringing them together in a particular way. The method is based upon traditional concepts of the relationships of classification to phylogeny. Although both are involved, it makes no claims for either perfect parsimony or complete compatibility. It allows for a certain amount of parallelism as well as reticulation. The last, the origin of taxa by hybridization, has not previously been discussed in connection with Groundplan-divergence, nor for that matter has the entire method itself been reviewed in any detail, although various authors to be cited below have discussed various aspects of it.

Until the 1950's, phylogenetic trees were based on subjective analysis and drawn without definite reference to evolutionary theories. There was no effort to quantify relationships and each author drew his tree intuitively. Little or no explanation went along with it. There was no statement of how the tree was drawn. How were the characters used? The ancestor determined? The branching sequence? Nevertheless, some of these early cladograms became classics. In the ferns we used F. O. Bower's family tree, and in the flowering plants C. E. Bessey's ordinal tree (actually more of a cactus). Many of these trees did seem to make sense, being expressions of ideas of morphological trends or of assumed "dicta," although sometimes this was hard to put into words or explain to students. By using the mind and the eye, authors were able to generate trees that satisfied our intuitive sense of phylogenetic relationships.

During the late 1940's I began my study of phylogeny of the ferns, a group of plants that provides morphologists a paradise of parallelisms. Practically every character of stem, leaf, sporangium, spore, and gametophyte shows multiple parallelisms, causing chaos in taxonomies based upon few characters. Sorus structure, the classical basis of taxonomy in ferns, was shown on the grounds of extensive outside evidence to be misleading, and a broad spectrum of character comparisons became necessary for understanding phylogenetic affinities.

The fern genus *Diellia*, which I chose for my doctoral thesis, is a small set of Hawaiian ferns, the species of which are extremely rare or local, or extinct. As it turned out, this genus was ideal for developing an organized rationale for cladistics. Its taxonomic position was undecided and it had been associated with several different fern families. With only five species in all plus a few varieties, and only a small number of well-marked character trends, it was possible to construct a simple cladogram that efficiently accounted for the facts. The steps in arriving at it were obvious: classify the plants, find the character trends, estimate an ancestor on the basis of the most generalized character states, find the relative amount of advancement for each species and variety and form the tree by shared divergent characters. (Wagner 1952, pp. 19–136). Fortunately, origin of fertile taxa by interspecific hybridization apparently does not occur in *Diellia*, and only one possible sterile hybrid was detected (op. cit., p. 163). Thus there was no mixture of several evolutionary lines.

The cladogram that resulted from this study showed one of the species (*D. falcata*) to be the most evolved or divergent, and another (*D. erecta*) to be the most primitive or generalized. The latter showed the most generalized character states according to both ingroup and outgroup comparisons. At the base of the tree was an hypothetical ancestor similar to the latter, but with the soriation of the parent genus *Asplenium*. The hypothetical ancestor was presumed to be extinct. However, during a post-doctoral year at Harvard following the completion of my thesis, while working over unidentified specimens at the Gray Herbarium, I discovered a fern that conformed to my predicted ancestor. To be true, the specimen was only a parched herbarium mummy of a collection made in 1879 in Makawao, Maui. This plant had been confused with a common and widespread Hawaiian *Asplenium*, and its true nature had been overlooked by pteridologists. The important point is that this plant had the predicted features of the hypothetical ancestor of *Diellia* (Wagner 1953), and although it evidently became extinct (for it was never found again), it can, for all practical purposes, be counted as truly a "living ancestor."

The concepts and methods used in the study of *Diellia* were formalized and codified for teaching Systematic Botany classes at the University of Michigan, Ann Arbor. They proved to be valuable for getting over certain basic ideas of systematics, especially for the non-systematists in my classes such as those then frightening adversaries, the physiologists and molecular biologists, who demanded an objective method. The steps were simple and easy to understand; the students could discuss them, find fault with them, and evaluate the pros and cons of their use. I did not anticipate that this so-called Groundplan-divergence Method would be picked up by researchers but that is what happened. It caught on, especially in plant taxonomy at Michigan and a number of other American universities. In one form or another it has been used by more than three dozen researchers and it remains as popular today as it was in the 1950's. The method and its variations played a role in the creation of other, quite different techniques.

PHILOSOPHICAL CONSIDERATIONS

The Groundplan-divergence Method is not concerned only with branching sequences and recency of common ancestry. It is aimed at finding and describing all of the pathways of genetic change. The method attempts to explain taxonomic diversity in the form of a phylogenetic tree but without respect to time or place. The tree is based upon taxonomy, not the reverse (taxonomy being defined here as phenetic classification, clustering by relative amounts of resemblance and difference).

Evolution and speciation are the processes, and phylogeny is the result. What evolution accomplishes through mutation, recombination, selection, drift, and isolation, involves changes in the genotypes of populations. The alterations in the gene pools of different phylogenetic lines express themselves in phenotypic changes at all levels from the molecular to the organographic, and the changes take place rapidly or slowly, depending on the organisms and the circumstances. Phylogenetics attempts to record the genetic changes themselves, not the rate of these changes. A complete phylogeny, then, involves the amount (grades or levels), directions (clades or lines), and sequences (steps or series of shared advancements) of divergences in genotypes, all involving biological changes. It is neither concerned with chronological time (as represented in fossil strata) nor space (as represented in geographical variation). The cladogram of Groundplan-divergence shows the branchings in respect entirely to estimated amount of evolution, i.e., grades, levels, or patristic distances. Chronological and geographical plots of the same clades yield very different patterns, as shown in the diagrams in figure 1. Chronocladistics distorts the actual phylogenetic distances to conform to times

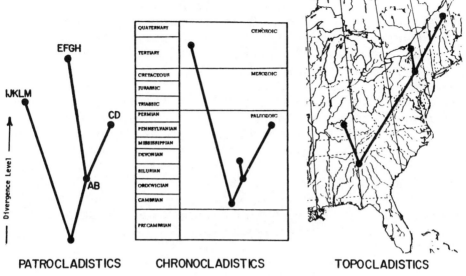

FIG. 1. Comparison of forms of cladistic trees when different measures are used for branch lengths. All branchings are based on shared divergent character states. *Patrocladistics*: Groundplan-divergence tree with branch lengths based on divergence levels. *Chronocladistics*: Chronological tree lengths based on time of appearance in the fossil record. *Topocladistics*: Geographical tree based on where the taxa appeared in their migrations.

of branching, thus showing rates of evolution. Topocladistics emphasizes changes of populations in their migrations. Problems of when and where evolutionary changes took place are interesting in themselves, of course, but they cannot be determined until after the cladogram based upon patristics has been established.

Our measure of amount, direction, and sequence of biological changes is calculated by using successive divergences (advancements, specializations, derivations) in characters—attributes of plant habit, leaf shape, corolla color, or any other intrinsic morphological, cytological, physiological, or chemical character. In one phylogenetic line we may change from the generalized or groundplan woody tree or shrub habit to an herbaceous habit (e.g., Cornaceae), from blade shape sagittate and basally attached to orbiculate and centrally attached (Nymphaeaceae), or from corolla color yellow to pink (Hypericaceae). One isozyme may be transformed to another. We take these changes as evidences of changes that must have occurred in the gene pool and use them accordingly as measures. Phylogeny is simply not the same as the genealogy of a human family, even though we have adopted many of the same terms. Genealogical family trees may occur with or without substantial genetic change. Phylogeny, on the other hand, is fundamentally concerned with the divergence that is involved in the descent of populations and taxa. The true phylogenetic tree ideally is the one that expresses the pathways of change, where they start, where they branch, and how long they are. If there are cross-connections, as there commonly are in higher plants, these should be mapped as well.

Not only are evolutionary rates different in different groups but a given group may become stagnant and remain essentially unchanged for hundreds of millions of years, such divergence as might occur being trivial or undetectable. In ferns it seems probable that families like Gleicheniaceae, Schizaeaceae, Osmundaceae, and Dipteridaceae may represent such conservatives. In contrast, other groups at present are apparently undergoing very rapid evolution. We think, for example, in ferns of Adiantaceae and Aspleniaceae (s.str.), families with such a welter of species formation at all levels that they are extremely difficult taxonomically. Rapidly evolving flowering plant groups include such taxa as Asteraceae, Myrtaceae, Apiaceae, Euphorbiaceae, and Brassicaceae. Their myriad closely related elements make them difficult objects for cladistic analysis.

If conditions permit, both the ancestor, or a type close to it, and its derivative type, may co-exist. Like the "wild types" of many cultigens, the generalized ancestor continues to exist in the original home while derived cultivars flourish at the same time in garden, orchard, or farm. Although most biologists today accept this concept, the acceptance is not universal, and in the extremist view a wild type of an horticultural variety cannot be considered its ancestor because an ancestor, according to this view, may not in genealogical terms be present at the same time as its descendant. In purely genetical terms, however, this is not a problem— the genetic ancestor, or a population like it or little different from it, i.e., the non-divergent gene pool, still exists together with its derivative. This is discussed further below under 4. Determination of the Common Ancestor. In terms of the Groundplan-divergence Method, the ancestor is the genetic prototype, whatever its genealogical relationship happens to be.

For the most part our recognition of relationships of gene pools has to be on the basis of phenetic resemblances and differences. Identity, for practical purposes, means identical gene pools as in certain clones and apomicts. Slight phenetic differences indicate slight genetic differences. Strong phenetic differences between taxa indicate divergent gene pools, and in strictly genetic terms the last has the most different common ancestry, remote in a qualitative sense as well as quantitative.

With this in mind, the ideal, most comprehensive, most predictive, and informative phylogenetic cladogram should possess the following attributes: (1) The tree should fit the phenetic classification of the organisms in question. (2) It should utilize all characters that show phylogenetic change. (3) The ancestral condition for the group as a whole should be given. (4) Estimates of overall divergence for each taxon should appear. (5) The direction (= combined divergences) should be formulated, not only by uniquely derived divergences but parallelisms or reversals too. (6) The cladogram should include the complete sequence of changes, including all the branches, whether the nodes be hypothetical or real. And (7) it should reveal the reticulations—grafts between branches resulting from intervarietal, interspecific, or intergeneric hybridizations.

I realize that what has just been outlined is a big order, but the items included are precisely the things that modern systematic botanists engaged in critical monographic work are concerned with—taxonomy, character trends, ancestors, primitive and specialized taxa, non-parallel and parallel divergences, branching patterns, and reticulations. It seems to me that the ideal method of phylogenetic cladistics is the one that allows for the induction of such comprehensive cladograms.

The Groundplan-divergence Method attempts to do this, and an outline of the procedure is as follows:

a. Systematic analysis, based upon taxonomy.
   Step 1. Phenetic classification.
   Step 2. Detection and removal of hybrid taxa.
b. Determination of the common ancestor.
   Step 3. Analysis of individual character trends.
   Step 4. Combination of groundplan character states.
c. Phylogenetic synthesis based upon divergences.
   Step 5. Calculation of divergence levels.
   Step 6. Grouping by shared divergences.
   Step 7. Insertion of hybrid connections.

Table 1 presents an hypothetical example; the resultant cladogram is plotted on a target graph in figure 2. Each of the steps is discussed below, followed by an overall evaluation of some of the problems of phylogenetic cladistics encountered in using this and other methods.

1. *Classification.*—Correct phenetic classification underlies any phylogenetic analysis. The first step, therefore, is to order the data, which amounts to defining and assembling the taxa according to their similarities and differences, using time-honored procedures of taxonomy. It cannot be repeated enough that to carry out cladistic analysis, one "must have a thorough knowledge of the group under study," as was recently emphasized by Gardner and LaDuke (1978). This knowledge cannot be acquired without a rigorous taxonomic treatment. A cladogram is no better than the quality of the taxonomy upon which it is based. If several truly distinct biological species are subsumed under one, or if one species is arbitrarily separated into several species, or if a species is placed in the wrong genus, or if a hybrid is confused with a divergent species—then the cladogram will be erroneous. As a rule, competent monographers come up with highly repeatable taxonomies, especially if they base them on a wide range of character comparisons. It is remarkable how often classifications made by traditional methods of careful comparison and subjective assortment match closely those made by computer methods such as those of the so-called neo-Adansonian pheneticists, using what today is referred to as Numerical Taxonomy. A highly critical classifi-

TABLE 1. Character states, taxon types, divergence formulae, and divergence levels for ten species of an hypothetical genus. Each character has theoretically been studied and the trend for each determined; 1 = divergent state; ancestor has 0 value for all states. DS = divergent species; HY = hybrid. In divergence formulae, capital letter = divergent state, each letter being a different character. Divergence level totals all changes, irrespective of what particular characters are involved. Characters F, K, L, and M show parallelism.

Taxon	A	B	C	D	E	F	G	H	I	J	K	L	M	N	O	P	Q	R	S	T	Taxon type	Divergence formula	Divergence level
alba		1				1					1			1					1		DS	BFKNS	5
brunnea		1									1	1		1	1		1	1			DS	BKLNOQR	7
caerulea		1			1				1	1		1	1			1					DS	BEIJLMP	7
coccinea		1										1		1	1			1		1	DS	BLNORT	6
fulva		1									½	1		1	1		½	1		½	HY	B(K/2)LNO(Q/2)R(T/2)	6½
nigra	1			1		1						1	1								DS	ADFLM	5
purpurea		1			1		½	½	½	½		½	½			1					HY	BE(GHIJLM/2)P	6
pallida		1	1											1					1		DS	BCNS	4
rubra		1			1		1	1								1					DS	BEGHP	5
viridis		1												1							DS	BN	2

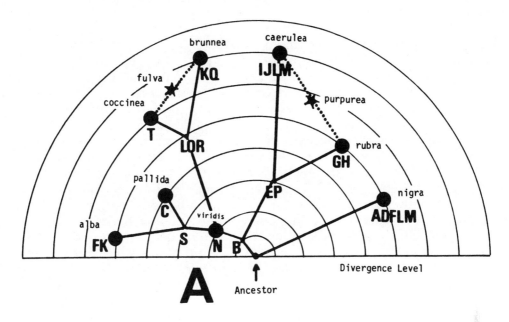

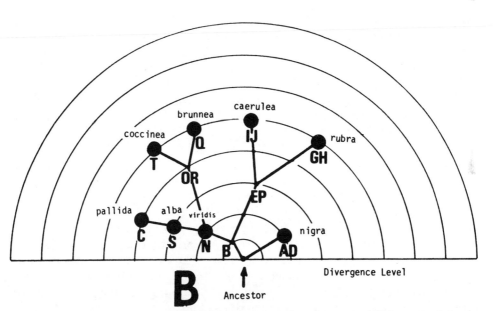

FIG. 2. Hypothetical cladograms. A. Groundplan-divergence cladogram of the data in table 1, including both parallelisms and reticulations. B. Same branching sequence as A but eliminating parallelisms and reticulations. Note that *alba* appears to be directly ancestral to *pallida*.

159

cation is not concerned with the numbers of species as a basis of grouping: One section of a genus might contain eight species, another 35, and another only one. For comparative purposes in estimating character trends, the subgenus with only a single species is equivalent to the one with 35. It is the discreteness of similarity coefficients that counts. Many little variations on a theme are not equivalent to a major differentiation.

2. *Detection and removal of hybrids.*—Equating taxa of hybrid origin with taxa of divergent phylogenetic origin will cause serious errors in cladistic analysis. Hybrids by their nature tend to break down the complexes of divergences that have arisen in separate lines of evolution and thus cloud the phylogenetic picture. Typically hybrids are intermediate in all or most of their characters. They arise from simple combinations of parental character states rather than by complex phylogenetic pathways of mutation, combination, selection, isolation, and drift. The wonder is that so many interspecific and intergeneric hybrid plants survive so successfully in nature, when we consider the theoretically "adaptive" nature that so many biologists ascribe to divergences. A distinctive new hybrid taxon may be produced by a single fertilization, where a normal divergent species requires systems of populations in which to evolve. The characteristics of the hybrid spring from the simple coming together of only two individuals. Furthermore, the same hybrid can be polyphyletic and be produced repeatedly, in nature and in the laboratory, unlike true species that can originate only once. Although they are usually sterile, some hybrids are sexual homoploids (rarely), apomicts (commonly), or polyploids (very commonly), and are thus able to form propagules with which they build up large populations, sometimes with wide geographical ranges. Such fertile hybrids behave in the wild like normal species, and may have equal importance floristically and ecologically. Although this is probably rare, some hybrids, the homoploids especially, may proceed to evolve on their own and even develop new phylogenetic clades. Whether high polyploids ever do this is questionable. And apomicts likewise tend to undergo little or no evolutionary change. The fact is that most interspecific and intergeneric crosses in nature are sterile deadends, even though some of them are re-formed frequently enough to merit taxonomic attention. Any modern taxonomy must include the significant taxa of reticulate origin, but before any cladistic analysis can be carried out they must be removed from the data set so that they will not obscure the basic skeleton of divergent branching patterns. All hybrids should be removed, including their direct derivatives.

The removal of hybrids is based upon the fact that most hybrid character states in plants tend to be intermediate between those of the parents, due presumably to additive polygenic effects. If X = a hybrid character state, and A and B are the homologous parental character states, then $X = (A + B)/2$. So closely do the quantitative data usually correspond to this formula (at least in higher plants, though not necessarily certain animals), as the great botanist Edgar Anderson pointed out long ago, we can have in hand one parent and its hybrid with an unknown species, and predict with surprising accuracy the characters of the unknown part, using the formula $2X = A + B$. Generally, a taxonomist who has a thorough knowledge of the group under study will recognize his hybrids rapidly and assign parentage to them with no difficulty. Besides intermediacy, he will also use other evidence for hybridity, such as co-existence with parents, occurrence in disturbed or hybrid habitats or geographical areas, allopolyploidy, aneuploidy, pollen abortion, and (or) apomixis. Sometimes it is easy to confirm the nature of the hybrid artificially by synthesizing it in the laboratory or greenhouse. The trouble comes when there are too many different hybrid combinations within a group, as in certain rosaceous genera like *Crataegus* and *Rubus,* where the underlying framework of the phylogenetic cladogram may become so obscured as to be almost undetectable.

3. *Analysis of character trends.*—Each of the individual characters must be evaluated for trends. If a given character shows two distinct trends, then each must be considered separately as separate genetic divergences. For example, if the estimated primitive condition of the leaf margin is dentate, and it undergoes two changes, one to an entire margin and the other to a bidentate margin, then we have to treat these as two different divergences, e (the primitive condition) to E (entire margin) and b (same primitive condition) to B (bidentate margin). The point is that we are concerned with two different pathways, even though we started from the same base and it involves the same character.

Character states being compared must be homologous. The detection of what represents homology is rarely a problem in plants at the levels of order, family, and below. A sepal of a member of the Rosaceae is surely homologous with the sepal of a member of the Saxifragaceae. Stamens of the Magnoliales are homologous with those of the Aristolochiales. In contrast, no systematic botanist would consider the cyathium inflorescence of *Euphorbia* to be homologous to the flower of *Celastrus,* no matter how flower-like the *Euphorbia* cyathium might be or how closely related we may conclude the Euphorbiaceae and Celastraceae may be. At the levels of class and above, however, problems of homology do set in. The leaf of Lycopodiopsida may not be homologous to the leaf of Equisetopsida. The same applies to leaves of the Pinopsida vs. Magnoliopsida (cf. Wagner 1974). As to secondary chemical products of plants, we may not be able to establish homologies until we understand their biosynthetic pathways.

To establish homology we use the classical tests of similarities in (a) position of the character in question with respect to other organs or processes, (b) ontogenetic sequences or biosynthetic processes, (c) mature states, and (d) extent of differences of all these attributes from those of non-homologous characters. In plants that are taxonomically close enough to hybridize, homology can be tested by the ability of the char-

161

acter states in question to combine. Thus stems of hybrids will not form intermediates with leaves, stipules with sepals, or cortex with pith. On the other hand, if two character states, e.g., one leaf shape and another leaf shape or one type of pith and another, are able to combine additively in hybrids, those states involve homologous characters.

Probably many so-called parallelisms and reversals are actually convergences to some extent, depending upon the degree of relationships of the plants being compared and whether or not these changes involve like or unlike genetic changes. The same attributes can be acquired along different pathways. Pinnately constructed leaves are found in some monocots and many dicots. Conversely, leaves with parallel main veins are found in many monocots and some dicots. Both of these types of blade construction represent convergences in the two groups, even though originally all angiosperms probably had the same basic type of leaf. We can expect all degrees of intergradation between, on the one hand, true parallelism and reversal involving strict homology, and, on the other, convergence involving analogy. The "leaves" of *Ruscus* (Ruscaceae) are actually modified shoots. We cannot always deduce this unless we have complete sequential evidence from ontogeny, heteroblasty, and stages in a character trend. We usually have no idea what is the actual genetic basis, so we can only use our phenetic evidence. Strict convergence is usually readily evident, as in the case of climbing ferns (*Lygodium*) of which the stem-like rachises of the leaves twine indeterminately and resemble strikingly the stems of certain angiospermous vines.

Assuming that we can affirm homology, we can usually tell whether a given trend is a parallelism rather than uniquely derived using a foundation of other correlated evidence. To illustrate, if four divergent lines are described respectively by the divergence formulas (common divergences underlined) ABCDEFGH, ABCDEIJKLX, KLMNOPQR, and KLMNOSTUX, then X is quantitatively clearly a parallel development in the two separate clades based on ABCDE and KLMNO. The latter two divergence formulas strongly define two separate phylogenetic lines. X has happened once in each.

To determine which of two alternative states of a character is the primitive one we use mainly the age-old concept of generalized or groundplan character state (cf. Danser 1950) and its corollary that primitive attributes tend to coincide more often with other primitive attributes than primitive with divergent or divergent with divergent (unless there is some obligatory functional coincidence). The most generalized character states usually coincide with those of the prototype. The probability that groundplan attributes are primitive is an obvious consequence of the structure of evolutionary trees, namely that the majority of individual specialized character states are found in the minority of taxa. The situations are reciprocal: Found in one or a few taxa = divergent; found in all or most taxa = primitive. Ingroup and outgroup comparisons are merely different aspects of the same thing—the former dealing within the group in question only, and the latter with all of its immediately related and coordinate groups. Although within a group, common or generalized character states tend on the average to be primitive (cf. Estabrook 1977), outside comparisons should always be made if possible. For example, the phenomenon of absence of vessels in the wood is exceptional in flowering plants, and one might be tempted to regard this condition as a specialization, as did the Harvard anatomist Jeffrey. Outgroup comparison in this case, however, shows that it is actually the vessels that are divergent in the flowering plants (as well as in all other tracheophytes). Further comparisons, like the classical ones made by I. W. Bailey, showed that vessel-less wood coincides nicely with numerous other, very distinctive primitive characters of various parts of the plant, as in certain Magnoliidae, especially the woody Magnoliales.

It is, of course, entirely possible to go wrong in any specific character trend interpretation. We must therefore be cautious. Dubious character trends should be rejected if there is any substantial room for doubt. If a given trend is questionable as to its directionality, it is best to await more evidence relative to it and not use it in a Groundplan-divergence analysis. Because of this, in many plant monographs, such as a number of those listed below, we may end up with ten or even fewer characters that possess substantial evidence of directionality and can be used.

4. *Determination of the common ancestor.*—The most logical way to estimate the characteristics of the common ancestor is to amalgamate all of the generalized states of the characters. To use a simple case, if we have six taxa with divergence formulas of clades based on their advanced character states as follows: ABC, ABD, EF, EFG, EFHI, and JK, then, obviously, the most parsimonious estimate of the common ancestor would be a compound of the primitive states of each of the characters, as indicated by the small letters for each, viz, abcdefghijk. The prototype thus logically possesses all of the groundplan states from which the six taxa have diverged.

It should be pointed out that there are numerous cladistic methods available today for creating what are called "networks" or "rootless trees" in which the taxa may be said to float in all directions and are connected only by branchings with no designated ancestor to give them directions. Such trees may be "rooted" by merely selecting a taxon that in the judgement of the writer seems to be the most primitive. The defect in this procedure arises from the effects of unequal evolutionary changes.

Even the most primitive plants, such as *Lycopodium, Osmunda, Cycas,* and *Drimys,* tend to have at least some probable divergences. Only exceptionally does a known taxon have all of the primitive characters that were presumably associated with the ancestral gene pool. For this reason, I believe that the ancestor should be estimated entirely on the basis of the generalized or groundplan character states. The root thus determined may on rare occasions be a real plant or near it (see above, in the discussion of *Diellia*), either a fossil or a living form. This is illustrated

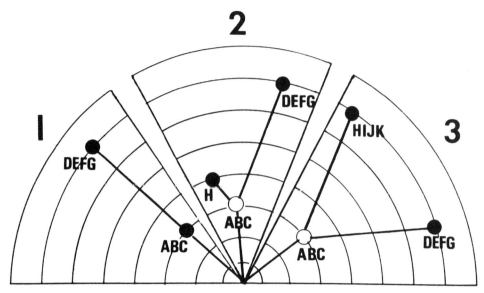

FIG. 3. Clades showing different branching patterns. 1. Ancestor ABC persisting after giving rise to derivative ABCDEFG. 2. Ancestor ABC disappearing after changing slightly to ABCH and giving rise to derivative ABCDEFG. (Although more primitive in total divergence level, ABCH could have arisen after ABCDEFG.) 3. Ancestor ABC disappearing after giving rise to two equally and strongly divergent lines, ABCHIJK and ABCDEFG.

in figure 3, Clade 1, in which one of the sister taxa remains static. This so-called "straight-line evolution" can arise in various ways, such as the ancestral stock remaining as a massive continental population with strong evolutionary entropy and the derivative developing as a small insular population; the ancestral stock remaining ecologically unchanged, and the derivative becoming ecologically specialized; or the ancestral stock becoming a high polyploid or apomict and thus having its evolutionary potential arrested. Clades 2 and 3 show more common situations in which the sister taxa both undergo divergence from the prototype. Clade 2, however, is not very different from Clade 1.

  5. *Phylogenetic synthesis: divergence level.*—Divergence levels or patristic distances can be expressed in two ways: If the ancestor is designated and its derivative taxon arises from it in "straight-line evolution" (see above), the divergence is shown entirely in terms of the estimated amount of evolution that has taken place, i.e., the grade or level (fig. 3, Clade 1). Here we can be concerned with patristic distance only, as calculated by a divergence distance or level, the total of all the changes that have occurred. We count each change as equivalent to each other, since we do not have any quantitative way of assessing the relative amount of genetic change in each. The divergence level is arrived at simply by adding up the divergences, giving each of them a value of 1, unless the

change has only partially occurred. Stamens that are unfused would be 0 (no divergence, groundplan condition), partially fused 0.5, and entirely fused 1.0 (trend completed, fully divergent). If the groundplan condition for the whole group were abcdefgh, and there were four derivatives, ABC, AB$\frac{D}{2}$E, ABDEF, and GH, their divergence levels would be respectively 3, 3½, 5, and 2.

In the 1940's and 1950's, when the Groundplan-divergence Method was developed, botanists were much concerned with relative evolutionary advancements of the families and orders of higher plants. This interest in relative advancement seems to have subsided somewhat in recent years, but it should be resurrected in my opinion for its potential biological significance in phylogeny. Ophioglossales and Marattiales are considered to be primitive ferns (i.e., in spite of numerous specializations, having a preponderance of primitive states), and Polypodiales are considered to be advanced (i.e., having a majority of specialized or divergent states). Flowering plant families like Winteraceae and Schizandraceae, along with other woody Magnoliales are closer genetically to the proangiosperm stocks than are Caprifoliaceae, Rubiaceae, and other Asterales. Sporne (1954, and references) developed a measure he called "Advancement Indexes" for the purpose of quantifying amounts of divergence. His index was a percentage of a total made up of the maximum advancement.

In the Groundplan-divergence Method, the amount of advancement is expressed on the cladogram by divergence levels of the taxa, the lengths of the branches equalling the total divergences. I have found it useful to use a target graph in which successive concentric levels are placed, but this is not necessary (cf. fig. 1, patrocladistics, and figs. 2A and B). To arrive at this sum which we define as the divergence level, all clear-cut changes, including parallelisms, reversals, and convergences should be used, as well as non-parallel or uniquely derived character states. In my opinion, to leave them out would be misleading, and this is illustrated by comparing figures 2A and B. Cladogram B is distorted. The clade leading to *alba* becomes a straight line rather than branched, and *alba* assumes the position of a direct ancestor to *pallida*.

It cannot be over-stressed that at least in plants the study of parallelism and convergence is essential and is an integral part of phylogenetic cladistics. Not only do parallelisms give botanists evidence on the nature and biological significance of certain adaptations, but many parallelisms occur so often and in such bewildering profusion that to ignore them in evaluating phylogeny would be fallacious indeed. I need only mention such trends as reduction/amplification of organs, fusions/separations of appendages, woody/herbaceous habit, triporate/polyporate pollen, dehiscent/indehiscent fruits, pubescent/glabrous epidermis, and so on. In the ferns, past classification was chaotic because of confusions arising from parallel and convergent evolution of soral types. In algae, parallelisms of body types created the artificial taxonomies of the 1800's that were rectified only after consideration of a host of other variables, including biochemical characters.

6. *Phylogenetic synthesis: shared divergent character states.*—Once taxa are positioned in a distance relationship to the common ancestor, the stage is set for tying lines together parsimoniously. This is accomplished through the use of divergence formulas and finding the successive groundplans that tie them together. The branching points are determined by the shared derived character states. If two taxa have the formulas ABCDEFG and ACFHIJ, they are naturally connected by the shared ancestor ACF. The indexes for the two derived taxa are 7 and 6 respectively, but the index for their hypothetical ancestor is only 3. Except for a few refinements, this is the classic method of phylogenetics, adopted more or less intuitively by all students of plant evolution. As pointed out by Faegri (1979), the logic of grouping taxa that share derived characters is "what every phylogeneticist has used as a leading principle since phylogenetics started."

7. *Phylogenetic synthesis: placing the hybrids.*—Taxa of reticulate origin will usually have all of the elements of the divergence formulas of the parents, although the values of most character states that are not shared by both of the parents will be altered. The divergence index of a hybrid should be the same as both of its parents if their indexes are equal. Accordingly, taxa ABCDE = 5 and ABFGH = 5, and their hybrid $AB\frac{CDEFGH}{2} = 5$. The divergence level of a hybrid should be intermediate if the parents have different divergence levels. Thus ABCDE = 5, ABFGHIJKL = 9, and the hybrid $AB\frac{CDEFGHIJKL}{2} = 7$. The hybrid taxon will plot at a distance or level that depends on the divergence levels of the parents. However, the precise position of the hybrid, unlike that of the divergent parental taxa, does not depend upon branching sequences defined by ensembles of divergent character states but rather on cross connections between the parents. It is not permissible for a single divergent taxon to have two different phylogenetic parents, for obvious reasons. On the contrary, as discussed above, hybrids always arise from different lines. Good examples of the method of plotting hybrids on a Groundplan-divergence cladogram are found in Brown (1964, pl. 1) and Mickel (1962, pl. 11).

DISCUSSION

Probably no phylogenetic cladogram, no matter how it is constructed, can be totally "proved" or "falsified." Cladograms can only fit the facts better or worse; they attempt to correlate the character patterns in the most logical way that the author can devise. The validity inherent in a Groundplan-divergence cladogram may be challenged in a number of ways. It must, for example, be in agreement with the phenetic taxonomy; if it is not, then the validity of the cladogram must be questioned. Likewise, the cladogram should fit known facts about stratigraphy, geography, and ecology. The more correspondence that exists, the more likely the cladogram is to be correct. Another test of a Groundplan-divergence cladogram is the extent to which it is confirmed by using other cladistic methods, such as the modified Wagner Tree methods of Whiffin and Bierner (1972) and Nelson and Van Horn (1975), and the Character Compatibility Method of LeQuesne (1974) and Estabrook et al. (1975). In a cladistic problem that involves mainly "clean data" (i.e., lacks excessive parallelisms, reversals, and reticulations), the different methods seem to give essentially the same results.

There is increasing suspicion that reliable phylogenetic trees probably cannot be constructed for taxonomic groups that have so-called "messy data." Mere numbers of characters and (or) taxa do not necessarily by themselves cause the problem. Hauke (1978), in dealing with only eight species and eight hybrids of *Equisetum* subg. *Equisetum,* was unable to construct a phylogenetic cladogram because of what he called "reticulate distribution of characters." Although hand methods of working out relationships are not as convenient as computer methods when the numbers of species involved are very large, this is not often a problem in plants. Most monographic studies involve only 10–35 taxa. Walters (1961) determined that the average genus in the 20 largest families of flowering plants has 15 species. In ferns I have calculated from the data of Lovis (1977) that the average genus has 30 species, but this is somewhat misleading, because more than three-fourths of the genera have fewer than 20. Mickel (1962), using the Groundplan-divergence Method, constructed a cladogram from 39 species and 5 hybrids of the genus *Anemia* (Schizaeaceae). At levels higher than the family, the problems become exacerbated because the numbers of 2-state characters increase with the increase in rank.

One of the major reasons that reliable cladograms cannot be constructed for certain taxonomic groups is simply the existence of overwhelmingly large gaps in the record of taxa, both past and present. In living plants a good example is found in the Psilotales, the order of whiskferns. So distant are the two genera and several species of this order from all other known plants, fossil and living, that it has been impossible with any assurance to associate them with any other known group. Their relationships have been interpreted to be with Trimerophytes (Devonian fossils), Lycopodiales (clubmosses), and Polypodiales (true ferns), three widely disparate assemblages. The consensus of a recent symposium (White 1977) was that Psilotales are without demonstrable relatives and are the survivors of an ancient line that lacks a fossil record. Under such circumstances we cannot determine character trends. For each pair of contrasting character states there is equal evidence for either.

Missing characters can cause serious problems. In ancient plants we usually deal with mere fragments—a leaf, a piece of petrified wood, a pollen grain. There are no correlated characters to aid in insuring homologies or phylogenetic affinities. The same is true in modern plants of the loss of organs, as illustrated by the flowers of the so-called "Amentiferae," and the plant bodies of certain parasites such as Orobanchaceae and Rafflesiaceae. In very rare circumstances the whole plant becomes lost, as the sporophyte generation of the Appalachian gametophyte, a vittarioid fern in which the diploid generation has been lost and there remains no fern as such. All that is left are the ribbon-like branching sterile prothallia (Wagner and Sharp 1963; Farrar 1974).

A third serious problem has to do with the phylogenetic patterns themselves, the outlines of which may have become too confused or even have been lost. As indicated previously, where hybridization has been very extensive, the basic clades may be so obscured as to be practically undetectable. In very actively evolving groups, the evolutionary processes may be so randomized and unpatterned that they resist analysis. We can imagine, for example, a population so broken up into separate colonies by insularity in an extensive archipelago that random genetic drifting in all directions may predominate. Some plant groups may not even require an extreme framework such as that. *Senecio* (Asteraceae) possesses a bewildering array of repetitive forms. This cosmopolitan genus has been estimated to contain as many as two or three thousand species. Other well-known, "explosive" plant groups with hundreds of "difficult" species include *Eugenia* (Myrtaceae) and *Euphorbia* (Euphorbiaceae). It is possible that cladograms in such groups may be less reliable, in general, than in such groups as, say, the Magnoliaceae, which is a small family that appears to have discrete lines. In the latter case the gaps may materially benefit our ability to make cladograms.

The Groundplan-divergence Method has been described in more or less detail a number of times (Wagner 1961, 1962, 1966, 1969; Radford et al. 1974; and in many of the references cited below). Various applications and modifications have been adopted for flowering-plants (Bacon 1978; Delisle 1963; Fryxell 1971; Graham 1963; Hardin 1957; Huft 1979; Iltis 1959; Keener 1967; Myint 1966; Olsen 1979; Richardson 1977; Scora 1966, 1967; Solbrig 1969; Stern 1961, 1970; Southall and Hardin 1974; Turner and Morris 1976), in pteridophytes (Blasdell 1962; Brown 1964; Evans 1964; Duek 1975; Hoshizaki 1972; Lellinger 1965, 1967; Mickel 1962; Miller 1965; Hauke 1959), in mosses (Buck 1979; Vitt 1970), and in fungi (Mazzer 1972; Peterson 1971). Perhaps the major use of the technique has been for teaching concepts in systematic botany (e.g., Wagner in Benson 1962a; Cox and Miller 1972). Other manual methods have grown out of the Groundplan-divergence Method, and others have affinities with it. These include the Wagner Trees, derived from the original computer programs of Kluge and Farris (1969) that use Manhattan Distance for connections in order to achieve the most parsimonious cladograms; the tree may be rooted by either a taxon chosen to be the most primitive or by an hypothetical ancestor derived essentially as described in the present article. The computerized programs of Kluge and Farris (1969) and Farris (1970) were modified for manual use by Whiffin and Bierner (1972) and Nelson and Van Horn (1975). The independently developed "Argumentation Plan" Method of Hennig (1966), which contains many of the elements of the Groundplan-divergence Method, has been used by a few European botanists (Ehrendorfer et al. 1977; Bremer 1976; and Bremer and Wanntorp 1978). The Character Compatibility Method (LeQuesne 1974; Estabrook et al. 1975, 1976), although originally computerized, is also now available for manual use (Estabrook and students, unpublished). Funk and Stuessy (1978) have reviewed the salient features of many of these methods. All of them have a number of basic precepts in common, but each of them stresses a different approach. It provides a salutary exercise in the teaching of systematic botany to compare these methods with one another and assess the differences in their underlying philosophies. More often than not, what appear to be differences are actually similarities. Even different words used turn out to mean the same things: groundplan = generalized = common = mutual, trend = transformation series, and level = grade = patristic distance. A symplesiomorphy = common groundplan; and advanced = specialized = divergent = derived. Future progress in cladistics will come only when the relative values of different approaches are evaluated and when we can decide at which time to use one or the other.

In conclusion, then, I believe that phylogenetic cladistics in the sense exemplified by the Groundplan-divergence Method described here is sufficiently well established as a technique to warrant its use by systematists, either in its original form or some modification thereof. I am keenly aware of the problems involved in constructing accurate phylogenetic trees, and at least some of these problems should be apparent from the discussion given herein. The Groundplan-divergence Method is one of the oldest, if not the oldest, formalized method for making cladograms and has probably been used by more monographers to date than any other method.

Classification that is supported by cladistic analysis is of considerably greater value than classification by itself. Using a method such as is described here, or some of those listed above, forces us to investigate the nature of character states and to evaluate all of the available characters. Once we have a phylogenetic tree founded upon the most probable divergences of the gene pools of the organisms involved we are in a position to correlate the resultant clades and grades with paleobotany, phytogeography, and ecology. The tree is predictive in that it estimates hypothetical forms yet to be discovered. Above all, by using cladistic methods and explaining why and how each branch was obtained, we place the cards on the table. Others can understand what we did and how we reached our conclusions, thus making it possible for other workers to evaluate them or to modify them. As to whether to use manual or computer techniques, one advantage to the former is that they are less expensive and may actually take less time. In general, in contrast to certain ones of the computer techniques, the method described here is readily understood, even if the researcher lacks a technical background, and, if necessary, the same data set can be run through a computer program that will usually give the same results.

In any event, the systematist should not simply "plug in" his data set and allow the computer to come up with the cladogram. He should think it out for himself, and this, scientifically, may be one of the most useful rewards of following each of the procedures of the Groundplan-divergence Method.

ACKNOWLEDGMENTS. Without committing any of them to agreement with all or part of the above, I express my thanks for suggestions and inspiration from my students, most recently T. F. Daniel, M. J. Huft, J. M. Beitel, and D. M. Johnson, and my colleagues, especially G. F. Estabrook, A. G. Kluge, T. F. Stuessy, and F. S. Wagner. I am especially indebted to the individuals who ran the data given in table 1 using different cladistic methods, namely Mark W. Bierner, Charles W. Nelson, Gene S. Van Horn, Leslie R. Landrum, and Christopher A. Meacham.

LITERATURE CITED

BACON, J. D. 1978. Taxonomy of *Nerisyrenia* (Cruciferae). Rhodora 80:154–227.

BLASDELL, R. F. 1962. A monographic study of the fern genus *Cystopteris*. Mem. Torrey Bot. Club 21(4):1–102.

BREMER, K. 1976. The genus *Relhania* (Compositae). Opera Bot. 40:1–85.

——— and H. E. WANNTORP. 1978. Phylogenetic systematics in botany. Taxon 27:317–329.

BROWN, D. F. M. 1964. A monographic study of the fern genus *Woodsia*. Beih. zur Nova Hedwigia 16:i–x; 1–154; pl. 1–40.

BUCK, W. R. 1979. A generic revision of Endodontaceae (Bryophyta: Musci). Ph.D. thesis, Univ. Michigan.

COX, J. W. and C. N. MILLER. The evolution of flowers: a laboratory exercise. Science Activities 7(2):22–25.

DANSER, B. H. 1950. A theory of systematics. Bibl. Biotheoret. 4(3):1–20. (cf. p. 4).

DELISLE, D. G. 1963. Taxonomy and distribution of the genus *Cenchrus*. Iowa State J. Sci. 37:259–351.

DUEK, J. J. 1975. Osmundaceae, Schizaeaceae, Gleicheniaceae (Pteridophyta). *Flora de Venezuela*. Merida, Venezuela: Univ. de los Andes.

EHRENDORFER, F., D. SCHWEIZER, H. GREGOR, and C. HUMPHRIES. 1977. Chromosome banding and synthetic systematics in *Anacyclus* (Asteraceae-Anthemideae). Taxon 26:387–394.

ESTABROOK, G. F. 1978. Some concepts for the estimation of evolutionary relationships in systematic botany. Syst. Bot. 3:146–158.

———, C. S. JOHNSON, and F. R. McMORRIS. 1975. An idealized concept of the true cladistic character. Math. Biosci. 23:263–272.

———, ———, and ———. 1976. A mathematical foundation for the analysis of cladistic character compatibility. Math. Biosci. 29:181–187.

EVANS, A. M. 1964. Interspecific relationships in the *Polypodium pectinatum-plumula* complex. Ann. Missouri Bot. Gard. 55:193–293.

FAEGRI, K. 1979. The emperor's new taxonomic dress. Taxon 28:168.

FARRAR, D. R. 1974. Gemmiferous fern gametophytes—Vittariaceae. Amer. J. Bot. 61:146–155.

FARRIS, J. S. 1970. Methods for computing Wagner trees. Syst. Zool. 19:83–92.

FRYXELL, P. A. 1971. Phenetic analysis and the phylogeny of the diploid species of *Gossypium* L. (Malvaceae). Evolution 25:554–562 (cf. pp. 555–561, fig. 1).

FUNK, V. A. and T. F. STUESSY. 1978. Cladistics for the practicing plant taxonomist. Syst. Bot. 3:159–178.

GARDNER, R. C. and J. C. LADUKE. 1978. Phyletic and cladistic relationships in *Lipochaeta* (Compositae). Syst. Bot. 3:197–207.

GRAHAM, S. A. 1963. Systematic studies in the genus *Cuphea* (Lythraceae). Ph.D. thesis, Univ. Michigan.

HARDIN, J. W. 1957. A revision of the American Hippocastanaceae. Brittonia 9:145–171.

HAUKE, R. L. 1959. A taxonomic monograph of the genus *Equisetum* subgenus *Hippochaete*. Beih. zur Nova Hedwigia 8:1–123, tables 1–12, graphs 1–3, pl. 1–21.

———. 1978. A taxonomic monograph of *Equisetum* subg. *Equisetum*. Nova Hedwigia 30:385–455.

HENNIG, W. 1966. *Phylogenetic systematics*. Urbana: Univ. Illinois Press.

HOSHIZAKI, B. J. 1972. Morphology and phylogeny of *Platycerium* species. Biotropica 4:93–117.

HUFT, M. J. 1979. A monograph of *Euphorbia* section *Tithymalopsis*. Ph.D. thesis, Univ. Michigan.

ILTIS, H. H. 1959. Studies in the Capparidaceae-VI. *Cleome* sect. *Physostemon*: Taxonomy, geography, and evolution. Brittonia 11:123–162.

KEENER, C. S. 1967. A biosystematic study of *Clematis* subsection *Integrifoliae* (Ranunculaceae). J. Elisha Mitchell Sci. Soc. 83:1–42.

KLUGE, A. G. and J. A. FARRIS. 1969. Quantitative phyletics and the evolution of anurans. Syst. Zool. 18:1–32.

LELLINGER, D. B. 1965. A quantitative study of generic delimitation in the adiantoid ferns. Ph.D. thesis, Univ. Michigan.

———. 1967. *Pterozonium* (Filicales: Polypodiaceae). Pp. 2–23 in *The Botany of the Guayana Highlands*, ed. Bassett Maguire. Mem. New York Bot. Gard. 17.

LEQUESNE, W. J. 1974. The uniquely evolved character concept and its cladistic application. Syst. Zool. 23:513–517.

LOVIS, J. D. 1977. Evolutionary patterns and processes in ferns. Pp. 229–440 in *Advances in botanical research*, ed. R. D. Preston and H. W. Woolhouse. London: Academic Press.

MAZZER, S. J. 1972. A monographic study of the genus *Pouzarella*. Ph.D. thesis, Univ. Michigan.

MICKEL, J. T. 1962. A monographic study of the fern genus *Anemia*, subg. *Coptophyllum*. Iowa State J. Sci. 36:349–383.

MILLER, C. N., JR. 1965. The evolution of the fern family Osmundaceae. Ph.D. thesis, Univ. Michigan.

MYINT, T. 1966. Revision of the genus *Stylisma* (Convolvulaceae). Brittonia 18:97–117.

NELSON, C. H. and G. S. VAN HORN. 1975. A new simplified method for constructing Wagner Networks and the cladistics of *Pentachaeta* (Compositae, Astereae). Brittonia 27:363–373.

OLSEN, J. S. 1979. Systematics of *Zaluzania* (Asteraceae: Heliantheae). Rhodora 81:449–501.

PETERSEN, R. H. 1971. Interfamilial relationships in the clavarioid and cantherelloid fungi. Pp. 345–371 in *Evolution in the higher basidiomycetes*, ed. R. H. Petersen. Knoxville: Univ. Tennessee Press.

RADFORD, A. E., W. C. DICKISON, J. R. MASSEY, and C. R. BELL. 1974. *Vascular plant systematics.* New York: Harper and Row. (Groundplan-divergence Method described, pp. 562–563.)

RICHARDSON, A. T. 1977. Monograph of the genus *Tiquilia* (*Coldenia,* sensu lato), Boraginaceae: Ehretioideae. Rhodora 79:467–572.

SCORA, R. W. 1966. The evolution of the genus *Monarda* (Labiatae). Evolution 20: 185–190.

———. 1967. Divergence in *Monarda.* Taxon 16:499–505.

SOLBRIG, O. T. 1969. The phylogeny of *Guttierrezia:* an eclectic approach. Brittonia 22:217–229.

SOUTHALL, R. M. and J. W. HARDIN. 1974. A taxonomic revision of *Kalmia* (Ericaceae). J. Elisha Mitchell Sci. Soc. 9:1–23.

SPORNE, K. R. 1954. Statistics and the evolution of the dicotyledons. Evolution 7:55–64.

STERN, K. R. 1961. Revision of *Dicentra* (Fumariaceae). Brittonia 13:1–57.

———. 1970. Pollen aperture variation and phylogeny in *Dicentra* (Fumariaceae). Madroño 20:354–359.

VITT, D. H. 1970. The infrageneric evolution, phylogeny, and taxonomy of family Orthotrichaceae (Musci) in North America. Nova Hedwigia 21:683–711.

TURNER, B. L. and M. A. MORRIS. 1976. Systematics of *Palafoxia* (Asteraceae: Heleniae). Rhodora 78:567–628. ·

WAGNER, W. H., JR. 1952. The fern genus *Diellia:* Structure, affinities, and taxonomy. Univ. Calif. Publ. Bot. 26:1–212, pl. 1–21.

———. 1953. An *Asplenium* prototype of the genus *Diellia.* Bull. Torrey Bot. Club 80:76–94.

———. 1961. Problems in the classification of ferns. Pp. 841–844 in *Recent advances in botany,* Vol. 1. (Lectures and symposia from IX Internatl. Bot. Congr., Montreal, 1957.) Toronto: Univ. Toronto Press.

———. 1962a. The synthesis and expression of phylogenetic data. Pp. 273, 276, 277 in *Plant taxonomy. Methods and principles,* by L. Benson. New York: Ronald Press.

———. 1962b. A graphic method for expressing relationships based upon group correlations of indexes in divergence. Pp. 415–417, Ibid.

———. 1966. Modern research on evolution in the ferns. Pp. 164–184 in *Plant biology today: Advances and challenges,* 2nd ed., ed. W. A. Jensen and L. K. Kavaljian. Belmont, California: Wadsworth Publ. Co.

———. 1968. Hybridization, taxonomy, and evolution. Pp. 113–138 in *Modern methods in plant taxonomy,* ed. V. H. Heywood. London and New York: Academic Press.

———. 1969a. The construction of a classification. Pp. 67–90 in *Systematic biology.* Natl. Acad. Sci. U.S.A. Publ. 1692.

———. 1969b. The role and taxonomic treatment of hybrids. BioScience 19:785–789, 795.

———. 1974. The classification of leaf types of land plants. (Abstract). Amer. J. Bot. 61(5, supplement):67.

——— and A. J. SHARP. 1963. A remarkably reduced vascular plant in the United States. Science 142:1483–1484.

WALTERS, S. M. 1961. The shaping of angiosperm taxonomy. New Phytologist 60:74–84.

WHIFFIN, T. and M. W. BIERNER. 1972. A quick method for computing Wagner trees. Taxon 21:83–90.

WHITE, R. A. (convener). 1977. Taxonomic and morphological relationships of the Psilotaceae. A symposium. Brittonia 29:1–68.

10

Reprinted from *Evolution* **19**:311–326 (1965)

A METHOD FOR DEDUCING BRANCHING SEQUENCES IN PHYLOGENY[1,2]

JOSEPH H. CAMIN AND ROBERT R. SOKAL[3]

With the advent of relatively objective classifications, such as the phenetic classifications produced by the operational techniques of numerical taxonomy (Sokal and Sneath, 1963), it was inevitable that biologists would wonder what phylogenetic conclusions could be drawn from them and with what reliability. If these phenetic taxonomies did not reflect all of the elements of phyletics (Sokal and Camin, 1965), could techniques be devised for deducing the latter? For example, could operational methods be devised for deducing the cladistic relationships among taxa, so that, given the same initial information, different investigators would obtain the same results? By cladistic relationships we mean the evolutionary branching sequences among taxonomic units without regard to phenetic similarities among them or to an absolute time scale.

There is no question that phylogenies could probably be reconstructed without error for any taxonomic group if complete fossil sequences for that group were available. However, can cladistic reconstructions be carried out with any degree of

reliability if only characters of recent forms are considered? Several recent studies have also considered this question from different points of view (Doolittle and Blombäck, 1964; Edwards and Cavalli-Sforza, 1964; Simpson, 1963; Throckmorton, 1965; Wilson, 1965).

Since 1962 a group at the Entomology Department of The University of Kansas has been examining the principles by which phylogenies are constructed conventionally, as well as the relation between the principles and practices of phylogeny and those of taxonomy, both orthodox and numerical. In addition to the authors, the group includes G. W. Byers and C. D. Michener and several graduate students. The study was based on a group of imaginary animals possessing a number of morphological characteristics generated by one of us (JHC) according to rules known so far only to him, but which are believed to be consistent with what is generally known of transspecific evolution. Genetic continuity was accomplished by tracing the drawings of the animals from sheet to sheet, permitting the preservation of all characters except for such modifications as were desired. Although the study is still in progress, it has already led to an empirical method which we believe capable of deducing probable cladistics from the characters of existing organisms.

Detailed studies of subsets of the assemblage of hypothetical animals by orthodox phylogenetic methodology resulted in differing, but internally consistent, cladistic schemes, the choice among which was not apparent to those uninitiated in the true phylogeny. Comparison by Camin of these various schemes with the "truth" led him to the observation that those trees which most closely resembled the true cladistics

[1] This paper was presented on December 29, 1964, at a symposium entitled "Interactions between numerical and orthodox taxonomies" at Knoxville, Tennessee, before the Society of Systematic Zoology.

[2] Contribution No. 1261 from the Department of Entomology, The University of Kansas, Lawrence. Research for this paper was supported by NSF Grant G 21011.

This investigation also was supported in part by a Public Health Service research career program award (No. 3-K3-GM-22, 021-01S1) from the National Institute of General Medical Sciences to Robert R. Sokal.

[3] We are indebted for constructive comments and criticisms to W. A. Clemens, P. R. Ehrlich, C. D. Michener, F. J. Rohlf, P. H. A. Sneath, the Biosystematics Discussion Group and the Evolutionists at The University of Kansas.

invariably required for their construction the least number of postulated evolutionary steps for the characters studied. Subsequently we examined the possibility of reconstructing cladistics by the principle of evolutionary parsimony. The following technique seems capable of doing this.

Technique

The technique requires a conventional data matrix as used for numerical taxonomy (Table 1). The columns of this matrix represent the *operational taxonomic units* (OTU's), which can stand for any taxonomic unit from individual through species up to higher categories. The rows of the matrix represent characters scored into different *character states* as qualitative or quantitative subdivisions of each character, differing among the OTU's.

The basic assumptions underlying character coding are fundamental to the entire technique and must be carefully examined.

1. We assume that characters can be expressed in discrete states differing among at least some of the OTU's of the study.
2. The character states can be arrayed in some logical order. If the characters are quantitative (*e.g.*, counts of bristles, segments, or leaves, or increments of size) linear order for the states is easily accomplished. Qualitative characters (shapes, colors, etc.) may require some ingenuity as well as some arbitrariness in coding states. If a logical order cannot be found for a qualitative character, the states may have to be recoded as several two-state characters. For details of this procedure see Sokal and Sneath (1963, p. 74ff).

From our knowledge of evolutionary processes the following three asumptions are not valid for all cases, although they are probably true for the majority of characters and taxa. Thus, they are only working assumptions which, as we shall show, can themselves be tested by the technique and may be relaxed in certain instances.

3. It is assumed that we have knowledge of the direction of the evolutionary trends within characters, and therefore the character states can be arrayed in a presumed evolutionary sequence from primitive (ancestral) to derived. In the linear sequence of character states, 1, 2, 3, 4, 5, 6, the presumed evolutionary sequence may be from 1 to 6, or from 6 to 1, or from 3 in two directions (toward 1 as well as toward 6). The primitive character state in these three examples would be 1, 6, or 3, respectively. For convenience, the primitive state is coded zero, derived states

positively or negatively, as required (see Table 1).
4. The ancestral state arose only once in the taxa at hand. Wilson (1965) has called such character states *unique*. Derived character states may, however, have arisen repeatedly in different branches of the group studied.
5. Evolution is irreversible, *i.e.*, a line, having attained a derived character state, cannot return to a character state ancestral to the derived one.

Under these assumptions the minimum number of evolutionary steps necessary to evolve c states of a character is $c - 1$. The number of character states and the minimum number of evolutionary steps necessary for each character are shown in Table 1. The most parsimonious cladistic dendrograms for characters 5 and 6 are shown at the bottoms of Tables 2A and 2B. We suggest the term *cladogram* to distinguish a cladistic dendrogram from a phenetic one which might be called a *phenogram*.[4] The principles of our approach are best illustrated during a preliminary step in the computations, the calculation of the so-called compatibility matrix.

All characters in the study are fitted to the pattern of the cladogram of each character. By this we mean that we compute the number of evolutionary steps required to arrive at the correct character states for all the OTU's in the study via the various cladistic patterns provided. The pathways of a pattern cladogram are unidirectional from the base of the dendrogram to the tips. Since changes in character states are irreversible (assumption 5) an evolutionary step in a character affects all pathways beyond that step. Provisionally any time several branches come off a stem at the same place a single evolutionary step suffices to produce the same change in any or all branches. Evolutionary steps increasing character state codes are shown graphically as short lines crossing the stems, while those decreasing the character state codes are shown as X-marks across the stems. They are marked with the num-

---

[4] Ernst Mayr (1965) has independently suggested the same terms with identical meanings. Arnett (1963) has used *phenogram* for a profile-type summary of characters.

TABLE 1. *Data matrix.*

				t OTU's					Character states (*c*)	Minimum steps (*c* – 1)
		7	8	13	14	15	25	28		
	1.	1	0	1	0	1	1	1	2	1
	2.	1	1	0	0	0	2	0	3	2
	3.	1	1	1	2	0	2	2	3	2
	4.	3	2	0	0	3	1	0	4	3
n Characters	5.	1	1	2	1	0	1	3	4	3
	6.	1	0	0	0	0	–1	0	3	2
	7.	0	0	0	0	–1	0	1	3	2
										15

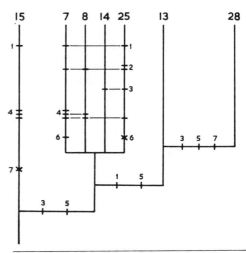

FIG. 1. Pattern cladogram of character 5. This cladogram follows the pattern shown in Table 2A. Numbers at the tips of the branches are code numbers of OTU's. The evolutionary steps for each character are marked on the branches. Evolutionary steps increasing character state codes are shown as lines across branches, those decreasing character state codes as X-marks. The number of the character represented is next to each mark. Single steps affecting several members of a cluster are diagrammed as heavy cross lines which indicate changes for the stems concerned, connected by thin lines to indicate that these comprise a single step.

ber of the character which they represent (Fig. 1). Each mark represents a character state increment or decrement of 1 only. To indicate a change from state 3 to 5, for example, two marks must be shown on a stem. The state code at the base of the dendrogram is assumed to be 0 (primitive) for all characters. It is convenient to set up the data in a *pattern table* as shown in Table 2. This arranges the OTU's by the state codes of the characters that form the bases for the patterns (*e.g.*, characters 5 and 6). The character state codes inside the table are the columns of the data matrix in Table 1 rearranged in the new order. The cladogram at the bottom of the table may be helpful, although we have not generally found it necessary for our computations.

Fitting character 1 to the pattern of character 5 (Table 2A) we find that it possesses state 1 in OTU 15. A mark cannot be placed at the base of the cladogram because OTU's 8 and 14 are coded 0 and would be changed by such an evolutionary step. A positive mark is therefore placed on the branch to OTU 15. A single step suffices to change OTU's 7 and 25 to character state code 1. The mark for this step cannot be placed on the stem leading to the cluster (7–25) as it would raise all members of the cluster to state 1. We therefore place the marks over the cluster as shown in Fig. 1, heavy cross lines indicating changes for the stems concerned and thin lines connecting the heavy ones to indicate that these comprise a single step. Such clusters may then be resolved later in the

procedure. Finally a mark before the point where OTU's 13 and 28 separate will raise these to code 1. These marks and those for fitting the other characters to pattern 5 are shown graphically in Fig. 1.

Character 2 needs a single positive mark to raise OTU's 7, 8, and 25 to state code 1 and a second for OTU 25 only, to raise it to code 2. Character 3 has a step prior to cluster (7–25) raising all OTU's other than 15 to code 1. A single positive mark raises OTU's 14 and 25 to code 2 and another raises OTU 28 to code 2. In character 4, because OTU's 14, 13, and 28 are at state 0, three separate positive steps are required to change OTU 15 to code 3. One step will raise OTU's 7, 8, and 25 to state 1, a second will raise OTU's 7 and 8 to state 2, and a third will raise OTU 7 to state 3.

Character 5, being the pattern character,

TABLE 2. *Sample pattern tables.*

To demonstrate the method of computing the number of necessary evolutionary steps and the measure of compatibility between the various characters and patterns constructed according to other characters. The cladograms at the bottoms of the tables are the most parsimonious pathways for evolution of characters 5 and 6, respectively, whose states are underlined in the tables.

A. Pattern table of character no. 5.

Characters	OTU's 15	7	8	14	25	13	28	Total steps	Extra steps
1.	1	1	0	0	1	1	1	3	2
2.	0	1	1	0	2	0	0	2	0
3.	0	1	1	2	2	1	2	3	1
4.	3	3	2	0	1	0	0	6	3
5.	0	1	1	1	1	2	3	3	X
6.	0	1	0	0	–1	0	0	2	0
7.	–1	0	0	0	0	0	1	2	0
								21	6

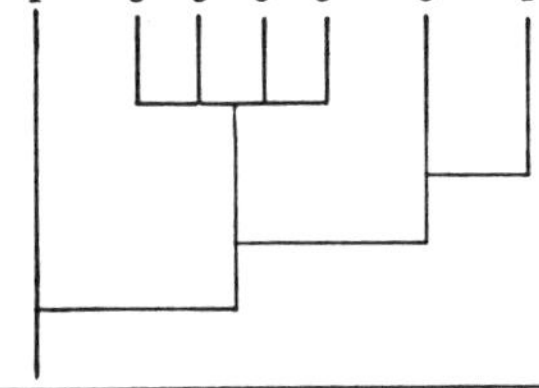

B. Pattern table of character no. 6.

Characters	OTU's 25	8	13	14	15	28	7	Total steps	Extra steps
1.	1	0	1	0	1	1	1	2	1
2.	2	1	0	0	0	0	1	3	1
3.	2	1	1	2	0	2	1	4	2
4.	1	2	0	0	3	0	3	6	3
5.	1	1	2	1	0	3	1	4	1
6.	–1	0	0	0	0	0	1	2	X
7.	0	0	0	0	–1	1	0	2	0
								23	8

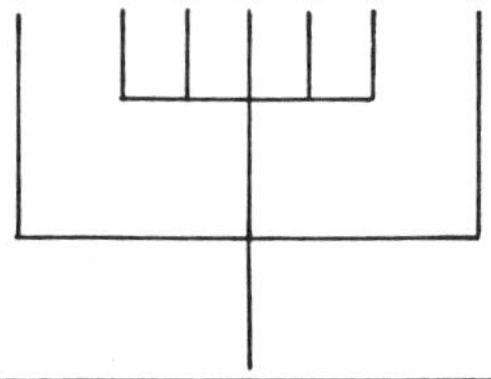

requires only the minimum number of steps as indicated. In character 6 a single positive and a single negative step change OTU's 7 and 25 to codes 1 and –1, respectively. Finally, character 7 needs a positive and a negative step to change OTU's 28 and 15 similarly. The total number of steps required for any one character to fit a given pattern is listed to the right of the pattern table (Table 2A). After a little practice the number of steps required can be written down simply by inspection of the pattern table.

Some additional problems in evaluating the number of evolutionary steps are illustrated by Table 2B which is a pattern table based on character 6, showing a V-shaped evolutionary trend. Evolutionary steps now have to be calculated in both directions from the pivotal stem. Parsimony may result by following the provisional rule on branches arising from one place on the stem. For example, a single step may turn the left and right arms or the pivotal stem and one of the arms in the same direction. For character 1 in Table 2B a single step turns OTU's 25 and 7 to character state 1 and a second step turns OTU's 13, 15, and 28 to state 1, leaving OTU's 8 and 14 at state 0. Thus two evolutionary steps are required to fit character 1 to the pattern of character 6.

Subtracting the minimum number of steps $(c-1)$ for each of the characters (see Table 1) from the total number of steps yields the extra number of steps necessary to fit a given character to a pattern. These values are shown in the last column of each pattern table (Table 2). Whenever this value is 0, the character is compatible with the pattern provided. The number of extra steps is a measure of the incompatibility of the character to any given pattern. A check on the computation is provided since the sum of the total necessary steps minus the sum of the extra steps must give the sum of the minimum number of steps, $\Sigma(c-1)$. When all (n) characters have been fitted to the n patterns (one for each character) the numbers of extra steps for each pattern are assembled in a *compatibility matrix* (Table 3). The diagonal elements of the compatibility matrix are zeros since, obviously, every character is compatible with its own pattern. They are indicated by X's and excluded from the computation. Rows and columns of this matrix are summed in two ways. The number of zeros is counted (and recorded as "compatibilities") and the numbers of extra steps are summed.

The compatibility matrix provides information of two kinds. It shows which characters provide "good" patterns and thus are relatively close to the presumed correct cladogram. Such characters would have high column compatibilities, *i.e.*, have a large number of characters compatible with their pattern and consequently few extra steps in their column. The compatibility matrix also supplies us with information about "poor" characters which are those whose patterns have relatively few characters compatible with them and call for a large number of extra steps, and which also fit poorly to most other patterns, showing few compatibilities and requiring many extra steps. Thus in Table 3 the pattern of character 4 has only 2 compatibilities and requires 11 extra steps by the other characters in order to conform to it. In addition, character 4 is incompatible with all other patterns and requires 18 extra steps to fit it to the patterns of the other characters. Such characters may be poor because of miscoding of their character states. The latter can arise from errors in transcription of data (such a case occurred in the analysis of the horses discussed below), or by an incorrect interpretation of the evolutionary trends in character states. Coding a character 0, 1, 2, 3 implies that evolution has proceeded in steps from 0 to 3. If, in fact, evolution proceeded from state 3 to 0, the compatibility matrix would show the miscoded character to be poor as a pattern and in fitting other patterns. The assumption of irreversibility of evolutionary steps may not be true in a specific case. When state 2 of a character arose from state 1 as well as by reversion from state 3, the character will show up as miscoded, if we consider all OTU's exhibiting the operationally homol-

TABLE 3. *Compatibility matrix.*

		Patterns						Compatibilities	Extra steps
		1	2	3	4	5	6	7		
Characters	1.	X	2	2	2	2	1	1	0	10
	2.	1	X	1	2	0	1	0	2	5
	3.	2	2	X	4	1	2	1	0	12
	4.	2	3	4	X	3	3	3	0	18
	5.	1	1	1	3	X	1	1	0	8
	6.	0	0	0	0	0	X	0	6	0
	7.	0	0	0	0	0	0	X	6	0
Compatibilities:		2	2	2	2	3	1	2	14	–
Extra steps:		6	8	8	11	6	8	6	–	53

ogous character state 2 as identical. Thus the method provides a check on some of its assumptions. If we knew which instances of apparent state 2 were really reversions from state 3, we could recode these as state 4 to preserve the linear sequence, which would improve the pattern and fit of this character in the compatibility matrix. It has therefore been our practice to exclude characters which show few compatibilities and large numbers of extra steps in their columns and rows on the assumption that these characters are miscoded. The reconstruction of the cladogram is then carried out without considering these characters which are later fitted separately to the reconstructed cladogram. Frequently reexamination of poor characters and fitting them to the final reconstruction in their own most parsimonious sequence will reveal the source of miscoding and permit their use in subsequent studies.

A number of different approaches to the reconstruction of the cladogram have been developed. None work perfectly so that they directly provide the most parsimonious solution. All methods provide a *procladogram*, which represents a state of considerable parsimony but must be adjusted by inspection or preferably a systematic program of trial and error to change it to the final most parsimonious arrangement. Our first approach fitted all characters to the cladogram of a good pattern as defined above, making adjustments as necessary. The steps necessary to fit

TABLE 4. *The monothetic method for reconstructing cladograms.*

A. Data matrix for group A (Table 1) with characters 6 and 7 recoded and character 4 omitted.

Characters	OTU's 7	8	13	14	15	25	28	Cycle 1, Step 1
1	1	0	1	0	1	1	1	OTU's 14 and 15 have 6 zeros each. Removal of OTU 14 leaves no "non-zero" rows. Removal of OTU 15 leaves rows 3 and 5 non-zero and 7– all-zero. Therefore, remove OTU 15.
2	1	1	0	0	0	2	0	
3	1	1	1	2	0	2	2	
5	1	1	2	1	0	1	3	
6+	1	0	0	0	0	0	0	
6–	0	0	0	0	0	1	0	
7+	0	0	0	0	0	0	1	
7–	0	0	0	0	1	0	0	
Number of zeros	3	5	5	6	6	3	4	

B. Data matrix A with OTU 15 removed.

Characters	OTU's 7	8	13	14	25	28	Cycle 1, Step 2
1	1	0	1	0	1	1	
2	1	1	0	0	2	0	
3	1	1	1	2	2	2	Subtract unity from rows 3 and 5; delete row 7–.
5	1	1	2	1	1	3	
6+	1	0	0	0	0	0	
6–	0	0	0	0	1	0	
7+	0	0	0	0	0	1	
7–	0	0	0	0	0	0	

C. Data matrix B with unity subtracted from rows 3 and 5 and row 7– deleted.

Characters	OTU's 7	8	13	14	25	28	Cycle 2, Step 1
1	1	0	1	0	1	1	Recompute number of zeros for remaining OTU's. OTU's 8 and 14 have 6 zeros each. Removal of either 8 or 14 leaves no non-zero rows. Removal of both 8 and 14 leaves row 1 nonzero. Therefore, remove OTU's 8 and 14 together.
2	1	1	0	0	2	0	
3	0	0	0	1	1	1	
5	0	0	1	0	0	2	
6+	1	0	0	0	0	0	
6–	0	0	0	0	1	0	
7+	0	0	0	0	0	1	
Number of zeros	4	6	5	6	3	3	

D. Data matrix C with OTU's 8 and 14 removed.

Characters	OTU's 7	13	25	28	Cycle 2, Step 2
1	1	1	1	1	
2	1	0	2	0	Subtract unity from row 1. Row 1 becomes all-zero, so delete.
3	0	0	1	1	
5	0	1	0	2	
6+	1	0	0	0	
6–	0	0	1	0	
7+	0	0	0	1	

E. Data matrix D with row 1 deleted.

Characters	OTU's 7	13	25	28	Cycle 3, Step 1
2	1	0	2	0	Recompute number of zeros for remaining OTU's. OTU 13 has 5 zeros. Removal of OTU 13 leaves no non-zero rows. Similarly for OTU 7 with 4 zeros. Removal of OTU's 7 and 13 leaves row 3 non-zero. Therefore, remove OTU's 7 and 13.
3	0	0	1	1	
5	0	1	0	2	
6+	1	0	0	0	
6–	0	0	1	0	
7+	0	0	0	1	
Number of zeros	4	5	3	3	

F. Data matrix E with OTU's 7 and 13 removed

Characters	OTU's 25	28	Cycle 3, Step 2
2	2	0	
3	1	1	Subtract unity from row 3. Row 3 becomes all-zero, so delete. OTU's 25 and 28 are a terminal bifurcation.
5	0	2	
6+	0	0	
6–	1	0	
7+	0	1	

all characters are marked on the chosen pattern cladogram (Fig. 1). Branches of this basic cladogram are rearranged if this achieves greater parsimony of evolutionary steps. The basic outline of the tree is retained but the provisionally shared steps in cluster (7–25) must be resolved. In small studies, such as this one, it is simplest to make a frequency distribution of steps shared by OTU pairs, triplets, etc., which are subsets of a cluster. Not counting character 4, as explained in the results for Group A, we find that OTU's 8 or 14 have only one shared step with OTU 25 which, however, has 2 steps in common with OTU 7. Since OTU 8 shares a step with both OTU's 7 and 25, OTU 14 is least related to the cluster and becomes its basal branch. The origin of the stem for OTU 14 is at the same point as that for the stem leading to OTU's 13 and 28 because there are no steps common to all of cluster (7–25) that do not also affect OTU's 13 and 28. The joint step of OTU's 14 and 25 (character 3) cannot be made compatible with such an arrangement and, therefore, must be an example of parallelism. Next to arise from the stem bearing the rest of the cluster is OTU 8 because it shares one step with OTU's 7 and 25. The latter share two steps and represent the terminal branches of the former cluster.

The completed reconstruction can be inspected in Fig. 2. In this cladogram evolutionary steps for characters 8 through 12 are also added. These were not included in the data matrix or the computation, because they occurred in the derived state in only one OTU each. The steps for these characters are compatible with the general cladogram and emphasize that OTU 25 is very specialized.

In larger clusters, frequency distributions of shared steps are more tedious without a digital computer. Another approach consists of a single-linkage cluster analysis (Sokal and Sneath, 1963) of the OTU's based on the number of common evolutionary steps.

The third approach is the monothetic method which works directly from the original data matrix. However, all characters having both positive and negative states must be recoded as two characters as shown in Table 4A for the data matrix of Table 1. Although the method is monothetic in operation, its results are polythetic (Sokal and Sneath, 1963, p. 13). The steps are as follows.

1. Zero states are counted for each OTU (column). This provides some measure of "primitiveness," as OTU's with greater numbers of characters in state "0" should branch off the main trunk of the cladogram near its base.
2. The OTU with the greatest number of zeros is removed and the remaining data matrix is checked for rows (characters) without zeros. (If there are ties for largest number of zeros, the first of the tied OTU's is removed.)
3. If no "non-zero" rows appear, the OTU with the next largest number of zeros is removed (or the second of the tied OTU's is removed) and the previously removed OTU column is placed back into the matrix.
4. If there are still no non-zero rows, then

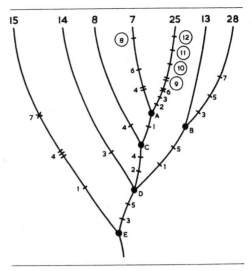

FIG. 2. Reconstruction of cladogram of Group A of the hypothetical animals, based on the pattern cladogram of Fig. 1. The OTU's and evolutionary steps are indicated as in Fig. 1. Characters 8 through 12 (circled) were not included in the data matrix (Table 1) or Fig. 1 because they occurred in the derived state in only one OTU each. The total number of evolutionary steps, not counting characters 8 through 12, is 23 compared with the minimum number of 15 steps from Table 1. The ancestors of the OTU's at the tips are indicated by black circles and identified by capital letters. Their probable character states can be easily obtained from the cladogram by going from the base to the ancestor. Thus for ancestor C, characters 1 through 7 will be in states 0, 1, 1, 1, 1, 0, 0, respectively.

the first and second OTU's are removed simultaneously. If this still leaves no non-zero rows, the OTU columns are placed back into the matrix and the OTU with the next largest number of zeros is removed. If no non-zero rows appear, the third and first, third and second, or all three OTU's are removed in that order. This process is systematically continued until at least one non-zero row appears (Table 4A).
5. When a non-zero row appears after the removal of one or more OTU's, the OTU or OTU's whose removal from the matrix produced a non-zero row is drawn as a branch from the base of the procladogram (Fig. 3A).
6. Unity (1) is then subtracted from each character state code in each non-zero row (Table 4B). This is repeated, if the row remains non-zero after the subtraction of one. Another branch, the main trunk of the procladogram, is drawn adjacent to the branch bearing the removed OTU or OTU's and evolutionary steps

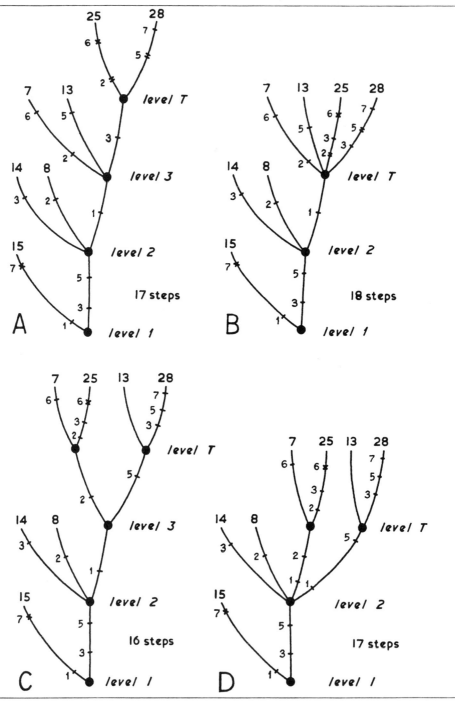

FIG. 3. Steps in the reconstruction of the cladogram of Group A (Fig. 2) by the monothetic method. Symbolism as in earlier figures. Level numbers and T refer to levels of furcation from 1 to terminal level. A. Procladogram resulting from monothetic technique illustrated in Table 4. Total number of evolutionary steps is 17. B. OTU 25 moved down one branching point; 18 evolutionary steps result. C. OTU's 25 and 7 grouped as are OTU's 13 and 28. Achieved parsimony of 16 steps. This is equally

representing the subtraction of one are drawn on that trunk with a number beside it for identification representing the character affected (Fig. 3A). At this time, any rows which are left all-zero may be removed from the matrix and dropped from subsequent consideration. Steps for such characters are placed on the branch leading to the removed OTU's (Fig. 3A).

7. The number of column zeros for the remaining OTU's is recomputed and the process is repeated through a second cycle (Tables 4C and 4D) and continued until the reconstruction of the procladogram is completed (Tables 4E and 4F).

8. When the procladogram (an initial approximation of the most parsimonious solution) is completed, all character states (evolutionary steps) are added to the cladogram and final readjustments for greater parsimony are made (Fig. 3A).

Because all of the methods devised so far yield only good approximations of the most parsimonious solution, it is necessary to test for parsimony and to make adjustments when these are indicated. This can best be practiced with a systematic procedure. First remove all internodes, *i.e.*, segments of stems between furcations, which do not bear any evolutionary steps because there is no reason to assume separate branching points in the absence of intervening evolutionary steps. Next move all common evolutionary steps found on adjacent branches to the base stem of these branches. This practice is parsimonious by making one evolutionary step do the work of two. Finally, and of most consequence, is the trial and error moving of branches which we shall illustrate in Fig. 3. From the procladogram resulting from the monothetic method (Fig. 3A) we try moving one of the terminal branches, the branch carrying OTU 25 or OTU 28, down one furcation to the furcation level 3 indicated in Fig. 3A. This now becomes furcation level T (the terminal level) and OTU's 25, 28, 7, and 13 all emerge from this point

(Fig. 3B). While our original procladogram required 17 evolutionary steps, the new adjustment requires 18 steps, and is thus moving away from the intended direction. However, we can now check members of the cluster 25, 28, 7, and 13, for common steps which can be removed to base stems. We find in Fig. 3C that OTU's 25 and 7 can be placed together with character step 2 made common, and OTU's 28 and 13 can be joined with character step 5 in common. This necessitates parallel steps in OTU's 25 and 28 for character 3, which previously was a common step, but we have now reduced the number of steps for the cladogram to 16. This is the same number required by the true cladogram but Fig. 3C is not the correct solution. Moving the branch which bears OTU's 7 and 25 in Fig. 3C from furcation level 3 to level 2 (Fig. 3D) will result in an equally parsimonious cladogram which is the same as Fig. 2, the correct solution. This illustrates an important point. Different, but equally parsimonious solutions may occur and in order to distinguish between them one must have added information from further characters. When character 4 was laid on the cladogram of Fig. 3C, without correcting it for apparent miscoding, this took an additional eight steps. Further rearrangement following the principles outlined above reduced the added steps for character 4 to only seven and yielded the correct solution. However, when character 4 is fitted and recoded in its most parsimonious sequence to either the cladogram of Fig. 3C or of Fig. 2, it results in only four additional steps and the two solutions remain equally parsimonious. Therefore, in such cases, additional characters must be sought in order to find the most probable solution to

←

parsimonious but not identical to cladogram in Fig. 2. Addition of character 4 to this cladogram and further adjustments for parsimony result in a cladogram identical to Fig. 2. D. Branch bearing OTU's 7 and 25 move down one branching point; 17 evolutionary steps result. If OTU's 8, 7, and 25 are now rearranged so that they share their common step for character 2, the cladogram of Fig. 2 is obtained.

the cladogram and to the recoding of character 4.

We should point out that the small studies reported in this paper are based on very few characters and decisions on alternative cladograms are frequently taken on the saving of a single step. When more characters are studied, such decisions generally are more soundly based. However, even in larger studies equally parsimonious solutions may occur in certain portions of the tree, where the determination of structure depends on very few evolutionary steps.

Three computer programs have been developed by Ronald Bartcher for carrying out the above methods of numerical cladistics. The first program calculates a compatibility matrix, finds the optimal pattern, and then fits the characters to this pattern. A second program carries out the monothetic method of finding a procladogram. The third program parctices parsimony on a cladogram for any given data matrix. The output is shown as an actual cladogram with the evolutionary steps marked in. These programs, called CLADON I, II, and III, respectively, were prepared in FORTRAN IV, for the IBM 7040 with 16K memory, at The University of Kansas Computation Center. As currently written they can handle 30 OTU's and 50 characters. Persons interested in obtaining copies of CLADON for adaptation to their computational equipment are invited to write to the authors. To provide some idea of running time, four of the reconstructions reported in this paper (including one with 14 characters) took approximately one-half minute at a cost of $1.50 by CLADON I, or two minutes ($6.00) by CLADON II including drawing of cladograms. Attempts to improve parsimony by CLADON III (unsuccessful—apparently correct solutions were obtained by either CLADON I or II) took five minutes.

RESULTS

The technique was applied experimentally to data from several hypothetical cases, including Group A, seven OTU's from Camin's imaginary animals, which furnished the illustrative example of the technique section. Data from several groups of real organisms were also analyzed.

Group A

The cladogram of Fig. 2, which was obtained directly from the compatibility matrix (Table 3), proved to be entirely correct. Besides leading to the reconstruction, the compatibility matrix provides additional information of interest. Characters which provide poor patterns usually fit well to many other patterns and are, therefore, usually compatible with the optimal pattern (e.g., characters 2, 6, and 7). Some characters which provide poor patterns may fit only moderately well to other patterns. This usually indicates some lack of parsimony, i.e., parallel evolution for that character (e.g., characters 1 and 3 in Table 3 and Fig. 2). We have already noted character 4 which provides a poor pattern and also fits quite poorly to most other patterns. In view of these considerations, we excluded character 4 from procedures for finding the cladogram. However, as we have seen in the monothetic method, it was necessary to employ character 4 in order to obtain the correct solution. That it is not unduly discordant can be seen from the moderate number of extra steps in rows and columns. Nevertheless, when equally parsimonious solutions occur, it is probably preferable to seek additional new characters in order to resolve such solutions. Examination of the true phylogeny of the OTU's in Group A revealed an error in tracing the forms, which unintentionally produced reversibility in character 4. When character 4 is recoded to fit the cladogram of Fig. 2 in its most parsimonious sequence, reversibility plus parallelism is revealed.

Another use of a reconstructed cladogram is to predict the character states of the ancestral forms at the branching points (see Fig. 2). Because no evolutionary changes have taken place since OTU 28 branched off their common ancestral stem, OTU 13 is identical with the ancestral form B, for the characters under consideration. This relationship presents a method for introducing fossil forms into a study along with recent forms. All OTU's which show no evolutionary steps subsequent to their last point of branching can be considered ancestral to all OTU's derived subsequent to the branch. An analysis of

Group A, including fossils *A* through *E*, again resulted in the correct cladogram.

Two other studies of simulated phylogenies showed that cladograms with little or moderate amounts of parallelism were reconstructed without error. Further work with hypothetical phylogenies led to the following tentative conclusions.

1. If reversible characters greatly outnumber irreversible ones, the most parsimonious tree will probably be incorrect. However, if the reversible characters are not highly correlated with each other, the compatibility matrix will provide criteria for removing them prior to the reconstruction and, if the remaining irreversible characters are numerous enough to give the correct cladogram, the most parsimonious tree will be correct.

2. Parallel or miscoded convergent characters (Sokal and Camin, 1965) will not be detected as such when they outnumber divergent characters and will show as recent divergences in the most parsimonious tree. If divergent characters are more numerous than others, the most parsimonious tree will show all evolutionary steps correctly.

Fossil Horses

W. A. Clemens of the Zoology Department, The University of Kansas, kindly provided us with data on lineages of fossil horses. These lineages are reputedly among the best known in the animal kingdom. He chose species within those genera believed to represent some of the major lines of horse evolution (see Fig. 4), although the actual species are not necessarily in the direct cladistic lines. Characters chosen (Table 5) were among those considered significant by authorities in the field and for which data on all species used in the analysis were available. Table 6 shows the data matrix and the compatibility matrix from which the reconstruction shown in Fig. 4 can be obtained. Presumed ancestral forms are shown circled and by dashed stems.

The reconstruction of equid cladistics is correct according to the studies of Stirton

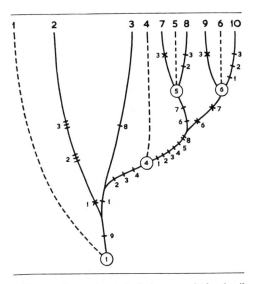

Fig. 4. Reconstructed cladogram of the fossil horses based on data in Table 6. The OTU code numbers and evolutionary steps are indicated as in Figs. 1 and 2. Character 3 has been recoded as explained in the text. Dashed branches represent OTU's ancestral to others in the study. Their code numbers are also shown in circles at points of branching. There are 31 evolutionary steps in this cladogram, compared with a minimum number of 20 (Table 6). OTU numbers represent the following fossil horse species: 1. *Mesohippus barbouri* Schlaikjer, Oligocene, data from Schlaikjer (1932, 1935). 2. *Hypohippus osborni* Gidley, Miocene, data from Gidley (1907) and Osborn (1918). 3. *Archaeohippus blackbergi* (Hay), Miocene; following White (1942), *A. nanus* is regarded as a synonymous species. Generic reference follows Stirton (1940). Data from Simpson (1932), White (1942), and Bader (1956). 4. *Parahippus pristinus* Osborn, Miocene, data from Osborn (1918). 5. *Merychippus* (*Merychippus*) *seversus* (Cope), Miocene, data from Downs (1956, 1961). 6. *Merychippus* (*Protohippus*) *secundus* Osborn, Miocene, data from Osborn (1918). 7. *Nannipus* cf. *minor* (Sellards), Pliocene, data from Lance (1950). 8. *Neohipparion occidentale* (Leidy), Pliocene, data from Gregory (1942). 9. *Calippus placidus* (Leidy), Pliocene, data from Gidley (1906, 1907) and Osborn (1918). 10. *Pliohippus mexicanus* Lance, Pliocene, data from Lance (1950).

(1940) and of Simpson (1951). During the analysis two characters appeared to be miscoded. One of these (character 7) was discovered to have been erroneously transcribed and is shown corrected in Table

TABLE 5. *Characters and character states of fossil horses.*

	Character State Codes* A	B		Character State Codes* A	B
Character 1. FEET:			4. CROCHET:		
3 toes, all supporting, phalanges broad	−1	0	Absent or present in only a few individuals	0	0
3 toes, lateral digits reduced in size and slightly shorter than middle digit, phalanges broad	0	−	Present, but prefossette closure not complete	1	−
			Prefossette closure complete	2	1
3 toes, lateral digits reduced in size and slightly shorter than middle digit, phalanges narrow	1	1	5. CEMENT:		
			Absent or irregularly developed	0	0
			Present	1	1
3 toes, lateral digits distinctly shorter	2	2	6. FOSSETTES:		
1 toe, laterals reduced to splints	3	3	With simple borders	−1	−1
2. LENGTH OF METATARSAL III:			Fossette closure not complete	0	0
Less than 140 mm	0	0	With complex borders	1	1
140 to 169 mm	1	−	7. PROTOCONE:		
170 to 200 mm	2	1	Linked to protoselene	−1	−1
Greater than 200 mm	3	2	Fossette closure not complete	0	0
3. AVERAGE CROWN LENGTHS (anteroposterior) of 4th upper premolar and 2nd upper molar:			Separated from protoselene	1	1
			8. CROWN HEIGHT:		
Less than 14.5 mm	0	0	Brachydont	0	0
14.5 to 16.9 mm	1	1	Hypso-brachydont	1	1
17.0 to 20.9 mm	2	−	Hypsodont	2	2
Greater than 20.9 mm	3	2	9. METALOPH AND ECTOLOPH:		
			Not connected	0	−
			Connected	1	0

* Column A. Character states for entire study. Column B. Character states for restricted study of species from genera represented in late Miocene or early Pliocene faunas (species 2, 3, 7, 8, 9, and 10), recoded when necessary.

6A. The other was character 3, average crown length of two cheek teeth, which Clemens, unknown to us, deliberately had coded from the point of view of operational homology, *i.e.,* from smallest to largest, in order to test the method. It is generally assumed from fossil evidence that each of the subgenera of *Merychippus,* here represented by OTU's 5 and 6, gave rise to at least one lineage of horses of larger size and another lineage in which there was a reduction in size. Interestingly this interpretation is sustained by character 3, reflecting changes in dimensions of the dentition, but not character 2, length of metatarsal III. This might be an artifact reflecting the choice of OTU's.

If the characters are operationally coded, reversals of evolutionary trends tend to confound the true cladogenesis of the data. When the compatibility matrix indicated character 3 to be miscoded, it was laid aside until after the cladogram had been reconstructed from the other characters. However, within clusters (5, 7, 8) and (6, 9, 10) we were unable to differentiate species 5 and 7 or 6 and 9. The generally accepted phylogenetic interpretation of the evolutionary changes for character 3 emerged automatically when the assumption of irreversibility was relaxed and the character was fitted most parsimoniously to the cladogram. Thus by a consistent application of the technique the originally miscoded character was recognized and contributed to the final reconstruction.

For a subsequent analysis of "recent" forms we selected six species from genera represented in late Miocene or early Pliocene faunas (species 2, 3, 7, 8, 9, and 10). Although some of the ancestral genera (species 1, 4, 5, and 6) survived into this period, their characteristics were assumed to be unknown. However, the predicted characteristics of the ancestral genera are correct and the cladogram emerges as before. Character 9 was not used since it is invariant in this analysis. Thus, no prediction could be made about the nature of character 9 for OTU 1. Also, since species 5 and 6 were not included, it was not possible to infer the evolutionary changes of character 3 correctly.

TABLE 6. *Fossil horses.*

A. Data matrix.

Characters	OTU's 1	2	3	4	5	6	7	8	9	10	Character states (c)	Minimum steps (c − 1)
1.	0	−1	1	1	2	2	2	2	2	3	5	4
2.	0	3	0	1	2	2	2	3	2	3	4	3
3.	0	3	0	1	2	2	1	3	1	3	4	3
4.	0	0	0	1	2	2	2	2	2	2	3	2
5.	0	0	0	0	1	1	1	1	1	1	2	1
6.	0	0	0	0	1	−1	1	1	−1	−1	3	2
7.	0	0	0	0	1	−1	1	1	−1	−1	3	2
8.	0	0	1	0	2	2	2	2	2	2	3	2
9.	0	1	1	1	1	1	1	1	1	1	2	1
												20

B. Compatibility matrix.

Characters	Patterns 1	2	3	4	5	6	7	8	9	Compatibilities	Extra steps
1.	X	4	5	1	1	1	1	1	0	1	14
2.	5	X	1	3	3	4	4	3	0	1	23
3.	6	1	X	3	3	5	5	3	0	1	26
4.	1	3	4	X	1	1	1	1	0	1	12
5.	0	1	2	0	X	0	0	0	0	6	3
6.	1	2	4	0	0	X	0	0	0	5	7
7.	1	2	4	0	0	0	X	0	0	5	7
8.	1	3	5	1	1	1	1	X	0	1	13
9.	0	1	1	1	1	1	1	1	X	1	7
Compatibilities:	2	0	0	3	2	2	2	3	8	22	–
Extra steps:	15	17	26	9	10	13	13	9	0	–	112

Other Organisms

The Fusulinidae, a group of paleozoic protozoa were analyzed using data obtained from Dunbar (1963), Dunbar and Henbest (1942), and Dunbar and Skinner (1937) by Roger Kaesler of the Geology Department at The University of Kansas. Cladograms of genera as well as species corresponded well with ideas on cladistic relationships expressed by Dunbar. Studies of 25 species of bees of the *Hoplitis* complex and of 24 genera of Mecoptera (scorpionflies) were also carried out using recent material. In both instances the cladograms obtained by our method corresponded well with ideas on cladistic relationships expressed by authorities in the field. Separate publications on all of these studies are in preparation. Such studies are continuing and are suggesting methods for analyzing OTU's of supraspecific rank.

DISCUSSION

General Considerations

From the findings reported above it would appear to be possible to deduce cladistic sequences from the characteristics of recent organisms. It may be argued that the cladistic solutions obtained from recent organisms merely reflect the thinking of the taxonomists who furnished us with the data. This is true in a general sense. However, the method proposed here tests the assumptions behind character coding. Repeatedly, through our methodology, we have been able to point out to our colleagues errors in their reasoning about evolutionary trends in characters.

While evolutionists probably have a relatively thorough understanding of modes of evolutionary change, assumptions about the relative frequencies of these phenomena may be in error. We therefore do not know how frequently assumptions 3, 4, and 5 about character coding will be valid in any given study.

The correctness of our approach depends

on the assumption that nature is indeed parsimonious. Alternative, equally parsimonious solutions may appear and the choice between them may not be evident from the data at hand. While the addition of a single new character may permit a decision, we should be on our guard against relying too firmly on the cladograms so obtained. Far-reaching decisions about stems are sometimes taken on the weight of a single evolutionary step. Partly this has been due to our choice of few characters for the analyses which initially were carried out by hand. While it is remarkable that even with few characters we obtained results consistent with the known facts, it is obvious that more characters would make decisions on junctions less likely to be dependent on the presence or absence of single evolutionary steps. The larger studies with more extensive suites of characters currently being processed by computer should lead to firmer cladograms. The probability of correctness of any portion of the tree varies with the relative reliability of our interpretation of any one character and with the number of characters and character states on which it is based.

The method as described above assumes equal probability of all evolutionary steps[5] after the characters have been coded. The method of coding characters and our initial assumptions about the evolutionary trends do reflect judgments based on biological knowledge of the material. The criteria by which this may be done have been outlined by several authors (Hennig, 1957; Maslin, 1952). Thus in actuality all evolutionary steps are not assumed to be equally probable.

The method proposed here is not substantially different from the conventional cladistic approaches of phylogenists. It simply quantifies and systematizes these procedures, making them objective in the process and permitting them to be put on

a computer. Thus, they have the same relation to conventional (cladistic) phylogeny that numerical taxonomy has to conventional phenetic taxonomy. Just as the study of numerous characters and the preparation of dendrograms in numerical taxonomy enhance knowledge and understanding of systematic relationships, so an analysis of cladogenesis along the lines proposed here leads systematists to critical tests of their ideas and assumptions about a phylogeny.

The proposed method does not weight characters equally in the construction of the cladogram, since compatible characters are preferred over those that are incompatible. Characters with few states tend to be more compatible than those with many. Since evolutionary steps are equally weighted, those with more states will be more heavily weighted. However, the weighting procedure agrees with the principles of numerical taxonomy (Sokal and Sneath, 1963); it is automatic and *a posteriori*, based on the entire available evidence rather than on *a priori* or character-by-character weighting as employed in conventional phylogenetic procedures.

Technical Points

The method illustrated here and several variations currently being investigated are empirical approaches to finding the most parsimonious cladogram. A possible pattern cladogram might be a phenogram, if phenetics is closely related to cladistics. However, cladograms and phenograms will be similar only when similarities are due to recent divergence. It is to be expected that the two types of dendrograms will not be entirely alike because they measure different aspects of phyletic relationship. Locating the cladogram requiring the minimum number of evolutionary steps by trial and error is a stupendous computational task, but might be made manageable by a Monte Carlo method. We have therefore attempted to reach a near parsimonious solution by one of the methods reported above, before applying trial and

---

[5] This point has been called to our attention by E. C. Minkoff, Harvard University.

error improvements. An analytical mathematical solution which would give the single most parsimonious cladogram is a difficult mathematical problem.

We do not yet know how to evaluate all of the information in the compatibility matrix. It indicates when some of our assumptions, such as irreversibility in evolution, are wrong. The matrix may also point out cases in which the basic assumption of evolutionary parsimony is invalid.

A feature for changing the primitive state of a character can be built into the computer program to try a variety of assumptions adopted by the operator to reveal the most parsimonious, internally consistent evolutionary pattern. In this way we could investigate hypotheses about evolutionary trends in the character. There seems to be no fundamental obstacle to assuming more complicated evolutionary trends than the V-shaped ones discussed above. Thus a character coded

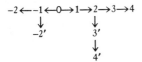

could be included in the computations.

We have as yet no technique for dealing with cases of hybridization. Similarly, we have not yet explored the consequences of missing data for character states of given OTU's. These could be handled by adding the characters to the cladogram after the construction is complete, in the same way in which we now add the characters which differ in a single state for one OTU only. The cladogram might then provide a means of predicting the missing character states as it does for the character states of ancestral forms.

Implications for Systematics

The development of a technique for deducing cladistic relationships among organisms appears to furnish a base for classification alternative to the phenetic system espoused by numerical taxonomy. Since the present method appears to be the "phyletic" one professed by orthodox taxonomists, it might appear that classifications should be established on the basis of it or a similar method. However, we have pointed out elsewhere (Sokal and Camin, 1965) that phyletic relationships are always a composite of phenetic, cladistic, and chronistic relationships not always clearly separated in the minds or writings of systematists and that systematics as a whole must be based on all of these considerations. The degree to which phenetics and cladistics coincide is not yet known, although we may assume it to be considerable. Since no operational system for combining phenetic and cladistic relationships is available we must choose between a phenetic or cladistic basis for classification. We distinguish here between "systematics" and "classification," the former including not only the study of the order of living things but also the causes and processes bringing this about, while classification is simply the arrangement of organized nature into categories for the convenience of biologists. For a variety of reasons, detailed elsewhere (Sokal and Camin, 1965), it would seem that a phenetic basis is preferable for classification in this narrow sense until an operational system, combining cladistics and phenetics can be established.

Comparison of phenograms with cladograms may lead to the resolution of phenetic resemblance into its components discussed by us in detail in Sokal and Camin (1965). If by a comparative cladistic and phenetic study of a group of organisms it has been shown that an apparently similar character in two organisms could not possibly be due to primitive patristic similarity, it must therefore be a case of parallelism (derived patristic similarity) or of classical convergence. This will stimulate biologists to a study of the underlying structural and physiological phenomena which lead to an apparently identical result. Such studies may permit the separation of parallelisms from divergence and may facilitate the recognition of characters which have been miscoded as convergent. The joint consideration of chronistics, cladistics, and phenetics will also enable objective measurement of evolutionary rates.

SUMMARY

A method is described for reconstructing presumed cladistic evolutionary sequences of recent organisms and its implications are discussed. Characters of the organisms to be studied are presented in a data matrix of the type employed in numerical taxonomy with the character states arrayed according to a presumed evolutionary sequence. The reconstruction proceeds on the hypothesis that the minimum number of evolutionary steps yields the correct cladogram. The method has been programmed for computer processing.

LITERATURE CITED

ARNETT, R. H., JR. 1963. The phenogram, a method of description for studies on *Oxacis* (Coleoptera, Oedemeridae). Coleopt. Bull., **17**: 6–18.

BADER, R. S. 1956. A quantitative study of the Equidae of the Thomas Farm Miocene. Bull. Mus. Comp. Zool., **115**: 47–78.

DOOLITTLE, R. F., AND B. BLOMBÄCK. 1964. Amino-acid sequence investigations of fibrinopeptides from various mammals: evolutionary implications. Nature, **202**: 147–152.

DOWNS, THEODORE. 1956. The Mascall fauna from the Miocene of Oregon. Univ. California Publ. Geol. Sci., **31**: 199–354.

———. 1961. A study of variation and evolution in Miocene *Merychippus*. Contrib. Sci., Los Angeles County Mus., **45**: 1–75.

DUNBAR, C. O. 1963. Trends of evolution in American fusulines. *In* Von Koenigswald, G. H. R., *et al.* (eds.), Evolutionary trends in Foraminifera. Elsevier Publ. Co., New York, pp. 25–44.

DUNBAR, C. O., AND L. G. HENBEST. 1942. Pennsylvanian Fusulinidae of Illinois. Bull. Illinois State Geol. Surv., **67**: 1–218.

DUNBAR, C. O., AND J. W. SKINNER. 1937. Upper Paleozoic ammonites and fusulinids. 2. Pernian Fusulinidae of Texas. Univ. Texas Bull., **3701**: 517–825.

EDWARDS, A. W. F., AND L. L. CAVALLI-SFORZA. 1964. Reconstruction of evolutionary trees. *In* Phenetic and phylogenetic classification. Systematics Assoc. Publ., **6**: 67–76.

GIDLEY, J. W. 1906. New or little known mammals from the Miocene of South Dakota. Bull. Amer. Mus. Nat. Hist., **22**: 135–153.

———. 1907. Revision of the Miocene and Pliocene Equidae of North America. Bull. Amer. Mus. Nat. Hist., **23**: 865–934.

GREGORY, J. T. 1942. Pliocene vertebrates from Big Spring Canyon, South Dakota. Univ. California Publ. Geol. Sci., **26**: 307–446.

HENNIG, W. 1957. Systematik und Phylogenese. Ber. Hundertjahrfeier Deutsch. Entomol. Ges., pp. 50–70, Berlin.

LANCE, J. F. 1950. Paleontología y estratigrafía del Plioceno de Yepomera, Estado de Chihuahua. Iª parte: Equidos, excepto Neohipparion. Bol. Inst. Geol. Mex., **54**: 1–83.

MASLIN, T. P. 1952. Morphological criteria of phylogenetic relationships. Syst. Zool., **1**: 49–70.

MAYR, E. 1965. Numerical phenetics and taxonomic theory. Syst. Zool., **14**: 73–97.

OSBORN, H. F. 1918. Equidae of the Oligocene, Miocene, and Pliocene of North America; iconographic type revision. Mem. Amer. Mus. Nat. Hist. (n.s.), **2**: 1–330.

SCHLAIKJER, E. M. 1932. The osteology of *Mesohippus barbouri*. Bull. Mus. Comp. Zool., **72**: 391–410.

———. 1935. Contributions to the stratigraphy and palaeontology of the Goshen Hole area, Wyoming. IV. New vertebrates and the stratigraphy of the Oligocene and early Miocene. Bull. Mus. Comp. Zool., **76**: 97–189.

SIMPSON, G. G. 1932. Miocene land mammals from Florida. Bull. Florida Geol. Surv., **10**: 7–10.

———. 1951. Horses. The story of the horse family in the modern world and through sixty million years of history. Oxford Univ. Press, New York, 247 pp.

———. 1963. The meaning of taxonomic statements. *In* S. L. Washburn [ed.], Classification and human evolution. Aldine, Chicago, pp. 1–31.

SOKAL, R. R., AND J. H. CAMIN. 1965. The two taxonomies: areas of agreement and of conflict. Syst. Zool., in press.

SOKAL, R. R., AND P. H. A. SNEATH. 1963. Principles of numerical taxonomy. W. H. Freeman and Co., San Francisco and London, 359 pp.

STIRTON, R. A. 1940. Phylogeny of North American Equidae. Univ. California Publ. Geol. Sci., **25**: 165–198.

THROCKMORTON, L. H. 1965. Similarity versus relationship in *Drosophila*. Syst. Zool., in press.

WHITE, T. E. 1942. The lower Miocene mammal fauna of Florida. Bull. Mus. Comp. Zool., **92**: 1–49.

WILSON, E. O. 1965. A consistency test for phylogenies based on contemporaneous species. Syst. Zool., in press.

QUANTITATIVE PHYLETICS AND
THE EVOLUTION OF ANURANS

Arnold G. Kluge and James S. Farris

Classical evolutionary taxonomy has been widely criticized for the lack of precision in its methods, while the far more precise numerical phenetic taxonomy has been even more widely censured for its failure to take into account the evolutionary basis of relationships among organisms. We believe it is worth while to develop still another taxonomic methodology, incorporating the precision of numerical techniques and the power of evolutionary inference. We refer to this hybrid methodology as quantitative phyletic taxonomy.

In the present paper we combine an exposition of techniques of quantitative phyletics with some rationale for them, and with examples of their application in the form of a study of the relationships between families of anuran amphibians. We have chosen anurans both because of their intrinsic interest and because of the long history of controversy surrounding frog classification. Since the relationships of frogs have been much debated, several discussions of taxonomic principles are included in the literature. These discussions provide a convenient set of reference points through which we can readily discuss the philosophy underlying the techniques of quantitative phyletics.

To achieve a convenient framework for discussing principles, we have sacrificed in this paper some detail on the frogs themselves. We have used only those relatively few characters that have been commonly

referred to in the literature on anurans. We have deleted some families of frogs from the study in order to produce evolutionary trees directly comparable to those given by authors who have studied only a few families. As a least common denominator of sets of families to be included, we chose those treated by Inger (1967) in his most recent work on anuran phylogeny. We realize that the use of a restricted number of family names may cause some confusion as to just what our assertions on frog affinities are. To alleviate this confusion partially, we have included in Appendix I a list, according to family, of those genera that we have referred to in the text.

CHARACTER WEIGHTING

Quantitative phyletic analysis differs from other taxonomic methods in that it attempts to employ biological information in selecting optimal coding of characters and in weighting of characters; its objective is to discover the evolutionary relationships of organisms, rather than simply the phenetic relationships. The weighting of characters is done in the interest of efficiency in discovering the relationships; if convergence is a real phenomenon, then not all characters are equally correlated with the evolutionary history of organisms. If valid means of character weighting can be found, they will tend to improve our chances of inferring the correct phylogeny. In order to achieve an accurate estimate of the real relationships of organisms, however, we must carefully select our procedures in such a way that personal bias has little chance of influencing the outcome of the analysis and that the methods are not implicitly circular in the logic of inferring the evolutionary pathways.

Several methods for prior selection of "good" taxonomic characters have been suggested, but not all are equally well suited to quantitative analysis. Among the criteria that are the most subjective are those that depend upon weighting characters according to the individual taxonomist's opinions (presumably formed before any taxonomic analysis has been performed) concerning the functional importance of characters, the significance of the biological roles of characters, the implications of functional relationships between characters, and the "most logical" direction of evolution for characters. We regard evidence of this type as too conjectural to be of any importance in taxonomic procedure. That such methods depend on an individual's understanding of a phenomenon immediately opens the possibility of endless argument between different taxonomists with different understandings of the same situation. Where such evidence for weighting has been used in a qualitative manner, it cannot be regarded as well founded. Serious taxonomic errors may be produced directly by the taxonomist's speculation on biological phenomena. An example of the perils of using subjective weighting is the work of Ghiselin (1966), who produced quite plausible-sounding "biological" arguments to the effect that torsion must have preceded coiling in the evolution of gastropods, since otherwise some intermediate form would pass through an adaptively impossible stage. Batten, Rollins and Gould (1967), however, noted that the subclass Cyclomya consists entirely of proto-gastropods with coiling but no torsion. We believe that "biological" and "functional" evidence for inferring importance and direction of evolution of characters should be excluded from

objective taxonomic studies, at least until such evidence can be interpreted in a more rigorous way than is generally possible.

A form of subjective taxonomic judgment, which we believe represents a type of weighting, is the interpretation of some characters being more "fundamental" than others. Usually, the worker who uses this type of reasoning decides which characters are most fundamental, and then he creates the major branches of a phylogenetic tree in such a way that the principal groups are each characterized by a single state of one of the fundamental characters. Finer branches of the tree are created later on the basis of other evidence.

Often this type of weighting leads to phylogenetic conclusions that seem indefensible from any other standpoint. Hecht (1963), for example, concluded that Orton's (1953) tadpole types were the most reliable indicators of anuran relationships, and then he went on to conclude that the aberrant tadpoles of the Microhylidae implied an ancient origin for that group. Hence, the Microhylidae occupied alone one of the branches of one of the earliest furcations of Hecht's tree. We cannot condone such a conclusion. That the Microhylidae differ in some set of features from other anurans could be taken as evidence for an ancient origin for that group only if it were supposed that change in those features were more probable in the distant past than in the more recent past. We know of no way in which evidence on the truth of that supposition can be obtained from available data.

Griffiths (1963) also seemed to rely on fundamental characters. He argued that the number of presacral vertebrae is a highly significant character and concluded that the Anura are diphyletic, one phyletic line characterized by nine presacrals, the other by eight or less. Quite aside from the issue of whether number of presacrals is a "good" character, it is clear that Griffiths overemphasized it. We could conclude from the distribution of presacral counts in frogs that the Anura are diphyletic only if we assumed that it is evolutionarily impossible for a frog with nine (or eight) presacrals to be the ancestor of one with eight (or nine). We do not see how such an assumption could be defended. Griffiths never attempted to do so.

Inger (1967) introduced a new form of character weighting in his modification of Wilson's (1965) concept of "uniqueness." A character state is unique if the set of organisms that possess it form a monophyletic group. That is, it has originated just once during phylogeny. According to Wilson, a state is *unique and unreversed* if it appears just once in phylogeny, and is never lost in any of the lines. possessing it. Inger used "unique" to mean "unique and unreversed." Using the term in Inger's sense, there is a complete correspondence between the set of organisms that have a unique character state and the set of members of a monophyletic group. Inger pointed out quite correctly that unique character states, if they can be identified, can serve as excellent indicators of phyletic relationships. The difficulty lies in the identification.

Inger (1967:381) offered four criteria by which unique character states can be recognized: (1) there is no obvious selective difference between the states of the character; (2) the state occurs in many taxa of the group being studied; (3) the character has low variability within taxa; (4) the unique state has an unusual developmental pattern. We do not believe the first and last of these to be realistic criteria. To establish that there is no selective difference between states of a character, it would be necessary to understand completely the selective forces shaping the species under study. Such knowledge is not available at the present level of development of evolutionary theory, and to try to fill the gaps in knowledge with speculation would only lead to the same kind of difficulties discussed above in connection with "understanding." Use of "un-

usual developmental pattern" is equally unsatisfactory for want of a stable criterion of "unusual"-ness. Criteria 2 and 3 are much more useful. They are both functions of variation within OTUs (Operational Taxonomic Units, see Sokal and Sneath, 1963) of some rank, and therefore capable of being objectively measured. Note that if criteria of uniqueness are restricted to measures of within-OTU variability, then uniqueness becomes operationally equivalent to conservatism as estimated through the within-OTU variability. This is not a surprising connection. We would expect that a highly conservative character would be more likely to have a state characteristic of a monophyletic group than would a less conservative character. Further, since conservative characters by definition evolve slowly, large reversals would be less likely than in less conservative characters, so that the states of a highly conservative character would have a higher probability of being unreversed than would states of a less conservative character. While the concept of uniqueness is an important one in evolutionary taxonomy, uniqueness is simply another facet of conservatism, when viewed from the standpoint of prior weighting of characters.

Inger (1967) has found reason to dismiss certain sets of characters altogether before the taxonomic analysis is initiated. For example, he generalized (p. 370) that "fossils cannot contribute much to our understanding of the phylogeny within the Order Salientia." And, as evidence for this generalization he stated (p. 369) that "The simplification of the skeleton [of anurans] . . . makes parallelism within the order or convergence between families likely. . . ." As an example of the simplicity of the skeletal system, he considered the shape of the vertebral centrum, and from that discussion he concluded that the character does not provide "a sound clue to phylogeny" (p. 370). His argument for this thesis is of two kinds, only one of which appears to be valid. Inger referred to a "relatively simple change in developmental pattern involving only one structure, one kind of tissue, and one process" (p. 370) as the cause of the interfamilial and intrafamilial variability in the shape of the centrum. On the contrary, we would argue that the phenotypic simplicity, and even genotypic simplicity if it were known, of the skeletal system of anurans need not imply evolutionary simplicity. We are not aware of any work that has demonstrated the genetic basis of the formation of the anuran centrum or that the genetic basis is any simpler than that, for example, underlying cytochrome c. Cytochrome c is both genotypically and phenotypically "simple" and, yet, it appears to be of considerable value in reconstructing the phylogeny of life (Fitch and Margoliash, 1967). In summary, we contend that there need not be a relationship (1) between "simplicity" of genotype and "simplicity" of phenotype, (2) between simplicity of genotype and/or phenotype and taxonomic variability, and (3) between simplicity of genotype and/or phenotype and their use in the reconstruction of phylogeny.

One method of weighting characters that does seem to be objective is the use of variation within OTUs as an index to the relative evolutionary rates of characters. Farris (1966) pointed out that the variability of a character within a species (or a higher taxonomic category) would be expected to be inversely related to the conservatism of the character. He suggested weighting characters by dividing each unit character difference by the standard deviation of the character within biological populations. The relationship between intra-OTU variability, evolutionary rate, and character weighting is easily expressed. The rate at which a character can change in evolution is necessarily limited by the variability of the character within populations. Selection, no matter how intense, cannot change the average character state of a population in some unit time by a greater amount than the range of values

available in the population in that unit time. If a character has high variability within OTUs, then a large difference between two OTUs does not imply lack of close relationship, since the variable character could have changed rapidly. If on the other hand, a character has low variability within OTUs, then a large difference between OTUs is probably indicative of lack of close relationship, since the highly stable character probably could not evolve rapidly. Hence, in drawing taxonomic conclusions, we place greater weight on characters with low variability within OTUs.

If the rate of evolution of characters is directly related to the amount of their within-population variation, multiplying each character by the reciprocal of the within-population standard deviation would be expected to transform the characters onto new scales in which the rates of evolution would be approximately equal over characters. If the characters in a study are so scaled that they have approximately equal average rates of evolution, we can use another type of weighting; for if the rates of evolution are approximately equal, the characters should all display about the same range of variation *between* OTUs, unless some characters are markedly more prone to convergence than are others. Characters prone to convergence would tend to show less variation between OTUs than other characters. Thus, the between-OTU variation (on the *transformed* character scales) can be used directly to weight characters. If we want to use weighting of this type without first transforming the characters, we would do so by multiplying each of them by the ratio of the between-OTU variation divided by the within-OTU variation.

In this paper OTUs are families and the characters have all been given a binary coding. We have weighted characters inversely according to their variation within OTUs. Our rationale for this kind of weighting is simply that a character known to vary within a group of closely related species is apparently capable of changing quite rapidly in evolution, hence seeming to be less conservative than a character known to vary only between distantly related forms. For a character that in reality has only a few discrete states, the conservatism corresponds inversely to the probability of a change from one state to another in a unit time. Most of the characters that we have coded as discrete binary variables, on the other hand, probably pass through a number of morphologically transitional stages in the process of changing from one of the coded states to another. In such characters, the conservatism corresponds inversely to the rate at which such a transition can take place. It is realistic to speak of the "rate" of even a binary character precisely, because the actual evolution is more continuous than the discrete coding would suggest. Since the rate at which a transition between two morphological states might be accomplished no doubt depends on the complexity of the genetic control of the character, indeed, on a wide variety of biological parameters, our weighting procedure is to some extent equivalent to weighting based more directly on detailed biological information. However, we do not assume this connection between ours and other conceivable weighting methods. Our position is simply that the rate of evolution itself is an accurate guide to the reliability of characters for phylogenetic inference. The outstanding advantage of using the variability of a character within OTUs—an indirect index to the rate of evolution—to carry out character weighting is that the weighting is objectively fixed by the data itself; it can be performed even when the detailed biological information necessary for valid application of other weighting schemes is not available.

ESTIMATION OF PRIMITIVE STATES

We use the term "primitive state of a character" to refer to that state of a character which we infer to have been present in the common ancestor of the set of OTUs in the study. We do not intend by "primitive" any of the other attributes of characters often associated with that term. In particular, we do not consider a primitive state to be unique or unreversed necessarily, nor do we imply that a character is conservative or irreversible merely by asserting that it has a primitive state.

Subjective taxonomic studies often make use of the idea that a character has a particular direction of evolution, which cannot be reversed. Ideas of this type are used to infer the primitive condition of a character, as well as to infer membership of monophyletic groups through the assumption that character states are both unique and unreversed. This is a different sort of reasoning from that involved in asserting that a particular character state is likely to be unreversed because the character is highly conservative; irreversibility is often assumed even for characters that do not appear to be very conservative. While we have no doubt that many characters have indeed evolved primarily in one direction, we doubt that any character is completely irreversible. Further, we are aware of no objective criteria by which the relative reversibilities of characters might be established prior to a taxonomic analysis. In this study we exclude prior speculation on irreversibility and assume that all characters are at least potentially capable of undergoing reversal. To infer the primitive states of characters, we rely on available fossil material and on the criteria for primitiveness established by Wagner (see Wagner, 1961 and references therein, and Kluge, 1967). In order of reliability these criteria are:

(1) The primitive state of a character for a particular group is likely to be present in many of the representatives of closely related groups.
(2) A primitive state is more likely to be widespread within a group than is any one advanced state.
(3) The primitive state is likely to be associated with states of other characters known from other evidence to be primitive.

In applying our criteria 1 and 2, we note that "closely related groups" can be selected through estimates of overall similarity that make no assumptions about primitive conditions. In evolutionary studies, the overall similarity would be computed through a weighted average unit character difference, where the weighting is done on the basis of conservatism as estimated by intra-OTU variability. In deciding on the "widespread" of criterion 2, we do not simply count numbers of taxa showing a particular state. A character state is widespread if it occurs in several taxa that otherwise have little in common. Thus, if a particular phyletic line were much more successful than others and produced many more species, genera, or families, we would not erroneously consider the state of a character in that line to be primitive merely because many taxa showed that condition.

CONSTRUCTION OF TREES

In constructing a phyletic tree, we employ the method of Wagner (1961), which is a simplified procedure for producing a most parsimonious tree in the sense of Camin and Sokal (1965). We choose this algorithm for its ease of calculation and because it makes no assumptions about reversibility of characters. Further, it is capable of being applied to continuously coded data and, hence, to data in which the characters are rescaled in order to weight for conservatism. If c is the relative conservatism of a character with numerical states x, y, and z, the weighting is accomplished simply by multiplying each state to obtain new values cx, cy, cz. When weighting of this type is used, the most parsimonious tree based on the weighted data is influenced more by the most conservative characters. In these capabilities the method of Wagner differs from the procedure of Camin and Sokal, and is, we believe, superior to that procedure.

We provide here a short description of the Wagner method for constructing trees. For further discussion and derivations of results, the reader is referred to Farris (in press). We use the following conventions: X(A,i) denotes the state of character i for OTU A, and the *difference*, D(A,B), between OTU A and OTU B is defined to be

$$D(A,B) = \sum_{i} |X(A,i) - X(B,i)| . \quad (1)$$

The objective of the Wagner method is to form a network, or tree, by connecting all original OTUs and realize in the process a minimum number of changes ("steps" in the sense of Camin and Sokal, 1965) on the tree. This is a network of minimum *length* in the space in which "length" is defined a certain way. To define the length of a tree, we note first that on a tree, each OTU is connected directly to one of the branching points on the tree (ie., the most recent depicted ancestor of that OTU). For example, in Fig. 1B, Y is the most recent ancestor of A. In that figure, ANC(Y) is an ancestor of A, but it is not the most recent ancestor of A. The connection between OTU A and its most recent ancestor we shall call *interval* A, using the OTU name to index the interval. The difference, as defined in equation (1), between OTU A and its most recent ancestor, will be called the *length of interval* A. The *length of the tree* is the sum of the lengths of all intervals of the tree. The tree of minimum length is defined to be *most parsimonious*. We assume throughout that the X(A,i) and the D's computed from them are weighted values. It should be clear that the choice of weighting coefficients can usually affect which tree is most parsimonious.

A most parsimonious tree usually has incorporated into it one or more hypothetical intermediates. These are artificial OTUs used as branching points on the tree. Their purpose is to minimize the length of the tree, and their character states are chosen to achieve that end.

The Wagner method itself proceeds as follows:

(1) Choose an ancestor OTU. Go to 2.

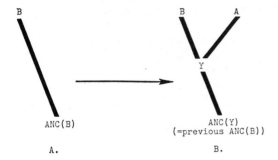

FIG. 1.—Inserting a new OTU and intermediate onto a tree. **A.** Before insertion. **B.** After insertion. In **A** there is one interval, B. In **B** there are three intervals, B, A, and Y.

(2) Find the OTU that has the smallest difference, as defined in equation (1). Connect it to the ancestor to form an interval. Go to 3.
(3) Find the unplaced OTU, A, that differs least from the ancestor. Go to 4.
(4) Find the interval from which the OTU identified in 3 differs least. The difference, $D(A,INT(B))$, between OTU A and interval B, is computed as follows: $INT(B)$ is a connection between OTU B and OTU B's most recent ancestor on the existing tree. Let this most recent ancestor be denoted $ANC(B)$. Then $D(A,INT(B)) = (D(A,B) + D(A, ANC(B)) - D(B,ANC(B)))/2$. Go to 5.
(5) Attach OTU A to the interval found in 4, denoted B. To do this construct an intermediate, Y, and insert it into the tree. The insertion is shown in Fig. 1. For each character, i, $X(Y,i)$ is computed as the median of $X(A, i)$, $X(B,i)$ and $X(ANC(B),i)$. Go to 6.
(6) If any OTUs remain unplaced, go to 3. Otherwise, stop.

Readers who so desire may obtain a FORTRAN IV program to perform the above algorithm from the junior author.

THE PARSIMONY CRITERION

The use of most parsimonious trees has been attacked by Inger (1967) and by Rogers, Fleming, and Estabrook (1967). Inger stated (p. 369) that "[Parsimony] does not adequately take into account the numerous parallelisms that may occur within a taxon. Neither does it allow for alteration in the conditions of selection that may lead a population to head first in one

genetic direction and then in another. . . ." and (p. 381) "I can adduce no biological reasons for using this criterion as a basis for choosing [among?] many alternatives." We believe that these criticisms lack force. The parsimony criterion, like any other criterion used in evolutionary study is intended ultimately to detect parallelism. Parsimony does this by erecting an evolutionary pattern that is most consistent with available data. Parallelisms can then be detected once the evolutionary pattern is established. Certainly one could not objectively detect parallelism by assuming that it existed prior to the analysis! In using parsimony, we take the possibility of reversals into account by assuming no irreversibility of characters and by using the Wagner method. Parsimony operates by finding a pattern of relationships that is most consistent with the data. This may not be a "biological" reason for choosing between alternative trees; but the principle of tailoring theories to fit known facts is an irreplaceable part of science in general.

Rogers et al. (1967) contend that the use of the parsimony criterion assumes that evolution itself is parsimonious, and that while the most parsimonious tree may indeed be the tree most likely to be true, it still may have very low probability of being correct—so low, in fact, as to invalidate the principle of choosing the most parsimonious tree. That the first claim is untrue is shown by the fact that a most parsimonious tree may show a large number of convergent and parallel changes, demonstrating that evolution is not parsimonious. The second criticism is implicitly a criticism of the use of any kind of maximum likelihood estimation procedure. Considering the general and successful use of maximum likelihood estimators in statistics, this argument cannot be accorded much weight.

A MEASURE OF CONSISTENCY

As we have indicated above, one of our objectives is to produce a tree that is most consistent with the original data. In comparing trees, we may wish to measure the degree to which they are consistent with data. Camin and Sokal (1965) use "number of extra steps" to measure the deviation of the tree from a perfect fit to data. We will use a somewhat more general index of consistency for this purpose.

We now define the *index of consistency*, c. The *range*, r, of character i, r(i), is defined as the difference between the numerically largest and numerically smallest states of the character. The *size, R,* of the data is defined as

$$R = \sum_i r(i).$$

Letting L stand for the length of the tree, we define the index of consistency of a tree to a set of data as $c = \dfrac{R}{L}$, where R, L, and the tree have been computed on the set of data for which c is specified. The value of c lies between 0 and 1. It is 1 if there is no convergence on the tree, and tends to 0 as the amount of convergence on the tree increases. Since c is monotone decreasing on L, c is maximal over trees for a set of data on the most parsimonious tree. The index c is influenced by weighting coefficients, it being assumed that L and the r(i) are computed on weighted data.

We use c instead of "number of extra steps," because it varies between fixed limits, and can be used on weighted and continuously coded data and to compare the fits of trees to different data sets. We note that $c = \dfrac{R}{R + S}$, where S is "extra length," is some weighted continuous analog of "number of extra steps."

PRELIMINARY WAGNER TREES

One of the first steps in our taxonomic analysis was the computation of Wagner trees using an initial suite of characters taken from the literature. We followed Inger's (1967) coding of these characters.

We present the results of this analysis here, both as an example of the capabilities of most parsimonious trees to summarize data and because the preliminary trees bear on our later conclusions.

We formed a Wagner tree for the data presented in Inger's Table 1. The cladogram is depicted in our Fig. 2. We have made one modification in Inger's coding, that of arciferal versus firmisternal pectoral girdle condition. Inger's state g, "transitional," is a description of the variation in the character in a family and does not correspond to any condition of an organism. Accordingly, we have eliminated that state from the analysis by breaking each family having that state into two OTUs, one coded for arciferal, G, and one for firmisternal, g. How well our tree fits the data depends partly on exactly how one numerically encodes two of Inger's characters. The tadpole spiracle and the type of centrum each have three states, and for each Inger gave one state as being primitive and the other two as derived. He did not stipulate, however, whether one derived state is supposed to be a precursor to the other, or whether both derived states are supposed to arise directly from the primitive state. If we make no assumptions about obligatory sequential relations between character states, we are effectively using a *trivial metric* on that character (viz., a unit character difference between two OTUs is zero if the two OTUs have the same state in the multistate character, and is one otherwise). With such a metric, the number of evolutionary changes necessary to fit the data is 12, and the tree has 20 steps, so that there are eight extra steps. Some of the loss of parsimony in the tree of Fig. 2 is owing to Inger's opinions as to primitive states. If the state in the Ascaphidae is taken to be primitive for each character, a tree with 19 steps can be obtained with the same cladistic topology (Fig. 3).

While we do not prefer the trivial metric, its use is necessary in order to compare the fit of our Fig. 2 to the data of Inger's three

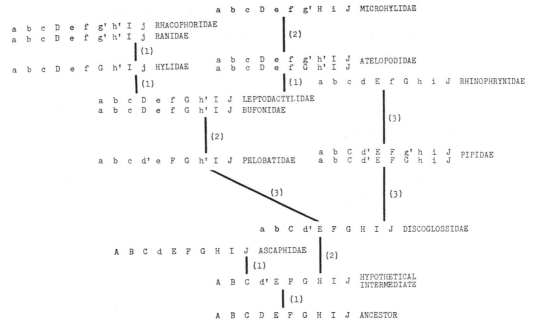

Fig. 2.—A maximum parsimony phylogeny of anuran families constructed according to the procedure of Wagner (1961). The 10 characters are those used by Inger (1967); see page 8 for further discussion. The phylogeny has a total length of 20, and it includes four cases of simple homoplasy, c, d, f and i, one example of multiple homoplasy, g', and two evolutionary reversals, D and H (reversals are only another form of homoplasy).

trees as summarized in his Table 2. We are forced to compare fits through the table because Inger did not list the character states of the intermediates on his trees. (We presume throughout that Inger's listings, "Fig. a," "Fig. b," and "Fig. c" in his Table 2 refer to his Figs. 4, 5, and 6, respectively). His Table 2 gives the number of "convergent changes" for his three trees. These values can be converted into the number of extra steps, the usual measure of degree of parsimony of a cladogram (Camin and Sokal, 1965), only if it is assumed that the metric on individual characters is trivial. Under that assumption, the number of extra steps for a tree is equal to the sum over characters of extra steps for each character. The number of extra steps for a character is the sum over character states of a function of number of convergent changes (nc) for the state. The function is

$$f(nc) = 0, \text{ if } nc = 0$$
$$f(nc) = nc - 1, \text{ if } nc \geq 1.$$

The necessary computation on Inger's Table 2 reveals that his Figs. 4, 5, and 6 have 15, 11, and 10 extra steps, respectively.

Thus, for the most consistent assumptions about primitive states, the Wagner tree for Inger's data has c = .63, while the trees that Inger proposed have c between .445 and .545; these c values were all computed with all characters assigned the same weight. This illustrates how the most parsimonious tree is the most consistent one for a set of data.

CHARACTER ANALYSIS AND RECORDING

That the highest c value, .63, for Inger's data is substantially less than unity indicates disagreement between the characters, as coded, on the relationships of frog families. It is possible that all this

197

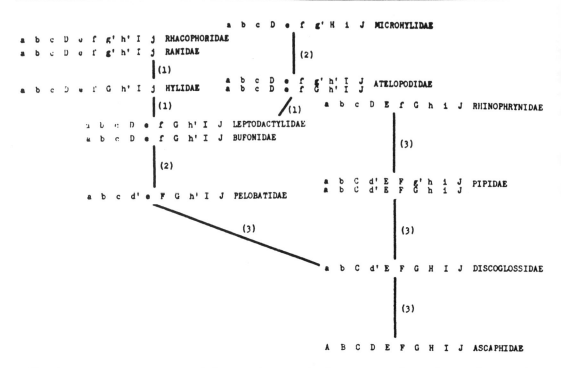

Fɪɢ. 3.—A maximum parsimony phylogeny of anuran families constructed according to the procedure of Wagner (1961). The 10 characters are those used by Inger (1967), except that all character states exhibited by the Ascaphidae are treated as primitive; see page 8 for further description of data modification. Compared to Fig. 1, this phylogeny is shorter by one evolutionary step, 19, and it includes one less homoplasy, d, and one extra reversal, D.

disagreement is owing to true convergence within frogs, (viz., that some structure in line A is identical to a structure in line B, though they have evolved independently). Such convergence can be detected only after one succeeds in forming a fairly accurate tree of frog relationships.

Some "convergence" however, may be owing to the miscoding of data. That is, two structures that actually differ and are nonhomologous may have been coded erroneously as the same state of some character. This pseudoconvergence would also be detectable on an accurate tree. Unlike true convergence, however, pseudoconvergence is capable of being detected simply by scrutinizing the organisms in the study, without reference to any tree-forming procedure.

We have undertaken a morphological re-analysis of the characters in the study.

The aims of this analysis were to establish the relative variation of the characters within OTUs, to rectify as many cases of miscoding as possible, and to achieve a numerical coding of the data that most accurately represents our state of knowledge of the data. Detailed discussion of individual characters has been relegated to Appendix II. We will summarize here the main features of our procedures.

Since we expect the reliability of characters for evolutionary inference to be inversely related to within-OTU variability, we have sought to define character states so as to minimize their within-OTU variation. For example, in defining the states "arciferal" and "firmisternal" of the character "pectoral girdle type" we have used the definitions of those terms proposed by Griffiths (1963), rather than the classical definitions as used, for example, by Inger

TABLE 1. NUMBER OF PRESACRAL VERTEBRAE.

Ascaphidae	9 (8[a])
Discoglossidae	9 or 8 (7[a])
Pipidae	8 or 7 (6 and 5[a])
Rhinophrynidae	8
Pelobatidae	9 or 8 (7[a])
Bufonidae	8 or 7
Atelopodidae	8, 7, 6 or 5[b]
Leptodactylidae	8 or 7
Hylidae	8 or 7
Ranidae	8 or 7
Rhacophoridae	8 or 7
Microhylidae	8 or 7

[a] Apparently owing to a synostosis of two or more vertebrae; not coded for in Table 2.
[b] The presence of five vertebrae in *Oreophrynella* corresponds more closely to the extreme reduction in atelopodids (*sensu stricto*) than bufonids. This suggests that *Oreophrynella* is an atelopodid and not a bufonid.

(1967), because the newer definitions lead to lower within-OTU variation for this character.

Some characters have been deleted from the character set because of extremely high within-OTU variability. This is done because highly variable characters are difficult to code meaningfully for OTUs of family rank. Further, for some such characters we suspect much of the variation to be purely phenotypic and so not readily subject to evolutionary interpretation.

Characters which contained more than one piece of information have been subdivided into a series of characters. For example, "tadpole type" has been resolved into "mouth armed or not," "spiracle median or sinistral," "spiracle anterior or posterior" and "operculum origin." Degree of subdivision of characters is a function of redundancy of information content.

After deletions, we retained 6 of the 10 "characters" used by Inger (1967). Following subdivision, these 6 "characters" are represented as 11 unit characters. Each unit character has been resolved into two states, so that each is conveniently numerically coded as a binary variable (viz., each character takes on numerical values 0 and 1). Steps of the analysis and coding procedure are summarized in Fig. 6 and Tables 1–4.

Most of the 11 binary characters are assigned weight 1. Characters IIIa and IIIb (centrum type), IV (pectoral girdle type), and Vb (median versus sinistral spiracle) were assigned lower weight because of within-OTU variation. The weights are given in Table 3.

CLADISTIC CONCLUSIONS

A most parsimonious tree for the recoded, weighted data is presented in Figs. 4 and 5. We emphasize that we do not regard these diagrams as the "Ultimate Truth" on frog relationships. We do feel, however, that they are somewhat more reliable than the conclusions of, say, Hecht (1963), Griffiths (1963), or Inger (1967), by virtue of the attention we have paid to the way in which our trees were constructed and the character analysis. Fig. 4 is a ground-plan diagram in the sense of Wagner (1961). Fig. 5 is a more conventional phyletic tree diagram with the same cladistic relationships of OTUs shown in Fig. 4.

Although in constructing our tree we have attempted merely to find the phylogeny most consistent with the 11 characters coded, our results are very similar to the classical notions of anuran phylogeny held by many workers in the past (see the numerous dendrograms reproduced by Hecht, 1963, and Inger, 1967). It seems that the opinions of those workers were based in part on data other than what we coded (the actual data used is not explicitly stated by many). If this is true, it would appear that our tree has considerable predictive value. By a similar argument, we would expect that other trees would have less predictive value, to the extent that they differ from ours. Our phylogeny is relatively robust in the sense that major changes in topology can be achieved only by adding many discordant characters or by extreme weighting of characters. The detailed topology of the Wagner diagram (Fig. 4) would probably be somewhat labile under the addition of new characters. For example, a new intermediate form con-

TABLE 2. THE VARIABILITY (A) AND TAXONOMIC DISTRIBUTION (B) OF SIX CHARACTERS.[a]

TABLE 2A

	States and Polarity	
Characters	(primitive) ————————→	(derived)
I Tail muscle (*M. caudalipubo- ischiotibialis*) (present or absent)	present (A) ————————→ absent (a)	
II Ribs (free, fused or lost)	free in both subadults and adults (B) ————→	"fused" in both sub- adults, and adults (b′) ↗ free in subadults, fused in adults (b) ————————→	lost in both subadults and adults (b*)
III Vertebral ossification (modes of)	ectochordal (C) ←	stegochordal (c′) holochordal (c)
IV Pectoral girdle (epicoracoidal carti- lages free—arciferal, or fused—firmister- nal)	arciferal (D) ————————→ firmisternal (d)	
V Spiracle	coded according to Fig. 6	
VI Scapula and clavical (overlap or juxtapose)	overlap (F) ————————→ juxtapose (f)	

[a] Table 2A above corresponds to Table 1 of Inger with the following exceptions: the number of presacral vertebrae, *M. adductor longus, M. sartorius*, cornified beaks and denticles, and vent characters are not included (see text for reasons). The scapula-clavicle character is not the same as that used by Inger (see Appendix II).

TABLE 2B

					Character states	
Families	Characters I	II	III	IV	V	VI
Ascaphidae	A	B	C	D	E, e, e*	F
Discoglossidae	a	B	c′	D	e	F
Pipidae	a	b	c′	D, d	E*	F
Rhinophrynidae	a	b*	C	D	E*	F
Pelobatidae	a	b′	c′	D	e′	f
Bufonidae	a	b′	c	D	e′	f
Atelopodidae	a	b′	c	D	e′	f
Leptodactylidae	a	b′	c	D	e′, e*′, e**	f
Hylidae	a	b′	c	D	e′, e*″, e***, e⁻	f
Ranidae	a	b′	c	d	e′, e**′	f
Rhacophoridae	a	b′	c	d	e′	f
Microhylidae	a	b′	c	d	e″, e**″	f

necting the Pipidae and the Rhinophrynidae might replace the current direct connection between those two taxa. Such altera- tions of the ground-plan diagram would not, however, affect the topology of the phyletic tree (Fig. 5). It is in this sense

TABLE 3. CHARACTERS OF TABLE 2 CODED IN A BINARY FORM.[a]

	Binary form of characters										
	I	II			III		IV	V			VI
Families		a	b	c	a	b		a	b	c	
Ascaphidae	0	0	0	0	0	0	0	0	0	0	0
Discoglossidae	1	0	0	0	0	1	0	0	0	0	0
Pipidae	1	0	1	0	0	1	0	1	–	–	0
Rhinophrynidae	1	–	–	1	0	0	0	1	–	–	0
Pelobatidae	1	1	1	0	0	1	0	0	1	1	1
Bufonidae	1	1	1	0	1	0	0	0	1	1	1
Atelopodidae	1	1	1	0	1	0	0	0	1	1	1
Leptodactylidae	1	1	1	0	1	0	0	0	1	1	1
Hylidae	1	1	1	0	1	0	0	0	1	1	1
Ranidae	1	1	1	0	1	0	1	0	1	1	1
Rhacophoridae	1	1	1	0	1	0	1	0	1	1	1
Microhylidae	1	1	1	0	1	0	1	0	0	1	1
Weight of character	1	1	1	1	¼	¼	½	1	½	1	1

[a] See Appendix II for further explanation.

that we consider our conclusions to be robust. For example, to disengage the Microhylidae from the line leading to the Ranidae and Rhacophoridae, and to shift its origin to the Ascaphidae-Rhinophrynidae-Pelobatidae lines as Inger (1967) proposed, would require that the pseudo-operculum (character Va) be considered primitive or that spiracle position (Vb) be weighted about five times more than any other character.

The greatest single point of difference between Inger's conclusions and ours is the placement of the Microhylidae. His opinions on that point seem to be predicated on the variability of the "firmisternal" condition (p. 369) and the uniqueness of the sinistral spiracle (p. 378). We have removed the variability of the "firmisternal" condition by a more judicious coding of the character. As we pointed out above, the irreversibility of sinistral spiracle does not follow from its uniqueness. Further, in the Hylidae we have pointed out that *Phyllomedusa* has a nearly median spiracle, *Nototheca fissilis* a median spiracle, and that *Hyla goeldii* is polymorphic (including the extremes of median and sinistral open-

TABLE 4. OCCURRENCE OF THE CHARACTER STATES OF BEAKS AND DENTICLES IN ANURAN FAMILIES.[a]

	Beak		Denticles (rows)			
	present	absent	triple	double	single	absent
Ascaphidae	×	×	×	×	×	×
Discoglossidae	×		×	×	×	
Pipidae		×			?	×
Rhinophrynidae		×				×
Pelobatidae	×	×			×	×
Bufonidae	×	×			×	×
Atelopodidae	×				×	
Leptodactylidae	×	×		?	×	×
Hylidae	×	×		?	×	×
Ranidae	×	×		?	×	×
Rhacophoridae	×	×			×	×
Microhylidae		×				×

[a] The questionable occurrence of states (?) are discussed in Appendix II.

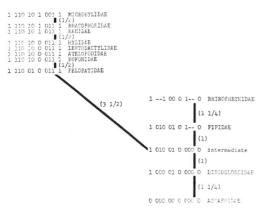

Fig. 4.—A phylogeny of anuran families constructed according to the maximum parsimony procedure of Wagner (1961). The six sets of binary coded characters are discussed on pages 19–32; see Tables 2 and 3 for summary. The phylogeny has a total length of 9½. Character IIIb is both convergent and reversed, and Vb is reversed.

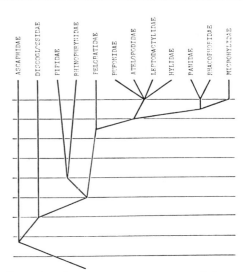

Fig. 5.—The maximum parsimony phylogeny of anuran families (of Fig. 4) translated into the more classically formed dendrogram. The horizontal parallel lines represent equal degrees of evolutionary divergence.

ings). The most obvious interpretation of these facts is that there has been at least one case of reversal from sinistral to a median condition in the hylids. If the reversal has occurred in the hylids, why could it not have also occurred in the microhylids? The median spiracle of almost all microhylids is not only median, but considerably distant from the branchial chamber; in ascaphids and discoglossids the spiracle is on the edge of the chamber. Of our 11 characters, the microhylids differ from the ranid-rhacophorid group in only one. Therefore, we believe that available evidence quite strongly supports our placement of the microhylids with the ranid-rhacophorid line.

Our Fig. 4 is so drawn as to imply that some of the modern families of frogs are derived from other modern families, and this requires clarification. The Wagner diagram is not intended to imply, for example, that the common ancestor of ranids and microhylids was one of the known ranids—or even any ranid. The ground-plan diagram merely implies that the common ancestor of ranids and microhylids was like modern ranids with respect to the 11 characters coded on the diagram. On the other hand, we cannot concur with Inger's criticism (p. 372) of postulating the derivation of one modern taxon from another. The "modern" families of frogs certainly had fossil members. We see no reason why, for instance, all species currently classified in the Microhylidae could not have a common ancestor that would fit within the Ranidae, as that family is usually conceived. Our position would be that while modern frog families may not be derived from each other, there is no biological law that would prevent this from being the case. To take a well-known example, we believe it is quite reasonable to state, "Mammals are derived from reptiles," implying merely that all mammals had a common ancestor that was a reptile, quite independently of the fact that Mammalia and Reptilia are both "modern" classes.

DISCUSSION

If the weighting criterion that we have used is a biologically realistic one, we would expect that characters with high variability within-OTUs would be more prone to

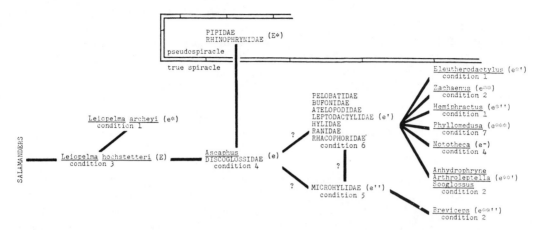

FIG. 6.—The probable course of evolution of the major operculum and spiracle conditions found in anurans. Each condition is defined and other taxonomic examples are presented on pages 26–30.

convergence than would other characters. Inspection of Fig. 4 seems to indicate that this is indeed the case. That this agreement of our tree with our expectations is not an artifact of our weighting procedure is indicated by two points. As we noted above, the Wagner tree for these data is robust and could be changed only by extremely different weighting coefficients from those used. In fact, *the* most parsimonious tree for these data would have the same form as our Fig. 4, if all the characters were weighted equally, or if the weighting coefficients were varied in any manner from equal weighting within factors of five! Certainly, the Wagner tree for our data does not depend strongly on the choice of weighting factors. The second consideration is that much the same picture of amounts of convergence in different characters is obtained from the Wagner tree for Inger's (1967) data (our Fig. 2). These data were not selected or weighted by us in any way. When analyzed by a most parsimonious tree, the data on frogs, both before and after our recoding, seem to show a stable set of relationships. It is not trivial that the concordance between single characters and the pattern of the data as a whole are largely predictable from the within-OTU variabilities of char-

acters. We believe that this predictability provides support for the practice of weighting by variation within OTUs.

The selection, weighting, and recoding of characters that we performed was done entirely on the basis of morphological data that had not been taken fully into account by earlier workers. It is interesting that the character reevaluation resulted in a data set with a high consistency index (c = .84). Related to this point is the fact that the characters showing the most convergence on the preliminary Wagner tree (Fig. 2) turned out to be the ones which seemed morphologically to be the most in need of recoding. That this is the case is again not trivial; it seems to suggest that most parsimonious trees have some power to detect miscoded characters. Trees produced by other means, for example, Fig. 4 of Inger (1967), seem to have less power in this respect. We would expect that a reasonably reliable method for producing evolutionary trees would indeed have some power of this kind. We would predict, in fact, that detection of miscodings in "rough" sets of data would be one of the most valuable applications of most parsimonious trees. Since Wagner trees seem to have some ability to predict which characters are miscoded or over-weighted, even when

other sources of weighting and coding information are unavailable, it should be
possible to construct algorithms to find a
series of trees, all but the first, based in
part on weighting information supplied by
the previous trees. Such algorithms would
provide weighting coefficients generated
entirely by a computer. We are currently
investigating that type of program.

SUMMARY

A new taxonomic methodology, termed
quantitative phyletic taxonomy, is proposed.
It is founded on the precision of numerical
techniques and the power of evolutionary
inference. Quantitative phyletic analysis
employs biological information in selecting
optimal coding and weighting of characters.
The objective of quantitative phyletics is to
discover the evolutionary relationships of
organisms.

Characters are weighted according to
the concept of conservatism as estimated
through within- and between-OTU variability. Multiplying each character by the
reciprocal of the within-population standard deviation transforms characters on to
scales in which rate of evolution would be
approximately equal over characters. With
equal rate of evolution, the range of variation between OTUs should be approximately equal; characters prone to convergence would tend to exhibit less between-
OTU variability and they are weighted
accordingly. The advantage of weighting
characters by their OTU variability is that
the weighting is fixed objectively by the
data itself.

The primitive state of a character, that
which occurred in the common ancestor of
a set of OTUs, is inferred from evidence
derived from fossils and on the degree and
kind of distribution of the character states
over the OTUs. Preliminary estimates of
overall similarity, that make no assumptions
about primitiveness, can be used to infer
the character state distribution. We assume that all characters are, at least potentially capable of undergoing reversal.

The Wagner method for producing most
parsimonious trees is described and an
algorithm presented. The method makes
no assumptions about reversibility of characters, and it can be applied to continuous
variables and to those that are weighted.
OTU linkage is determined by a difference
equation, and the interval length between
pairs of OTUs is equal to the sum of their
character state differences. The length of
the tree is the sum of the lengths of all
intervals of the tree, and the tree of minimum length is the *most* parsimonious.
Hypothetical OTUs are formed usually at
branching points to minimize the length
of the tree. The parsimony criterion is
likened to a kind of maximum likelihood
estimation procedure, and it is intended
ultimately to detect homoplasy. A general
index of consistency is defined; it measures
the deviation of a tree from a perfect fit to
data.

The Wagner method for producing most
parsimonious trees was applied to a suite
of characters from anuran families which
were taken directly from the literature.
These data had been used previously to
form trees by different numerical taxonomic methods. The most parsimonious
tree gave the best fit to the data. That the
highest consistency value is substantially
less than unity indicates considerable disagreement between characters, and the
discordance is owing to either true or
pseudoconvergence, or both. Pseudoconvergence is produced by miscoding data
and it is capable of being detected simply
by scrutinizing the organisms without reference to any tree forming procedure. In
our reanalysis and recoding of the original
suite of characters, we defined character
states so as to minimize their within-OTU
variation. This is in accord with our thesis
that the most reliable characters for evolutionary inference are the most conservative. Some characters were deleted because of their extremely high within-OTU
variability, while others were subdivided
into a set of independent variables. After

reanalysis and recoding, the remaining characters were each resolved as a binary variable and weighted according to their within-family variability. The final most parsimonious tree, using these data, appears to have considerable predictive power and to be relatively robust. The close relationship of the Microhylidae to the Ranidae is a major difference between our tree and the conclusions of other recent investigators.

REFERENCES

ALCALA, A. C. 1962. Breeding behavior and early development of frogs of Negros, Philippine Islands. Copeia, 1962:679–726.

ARCHEY, G. 1922. The habitat and life history of *Liopelma hochstetteri*. Rec. Canterbury Mus., 2:59–71.

BAIRD, D. 1965. Paleozoic lepospondyl amphibians. Amer. Zool., 5:287–294.

BATTEN, R. L., H. B. ROLLINS, AND S. J. GOULD. 1967. Comments on "The adaptive significance of gastropod torsion." Evolution, 21:405–406.

BOULENGER, G. A. 1891. A synopsis of the tadpoles of the european batrachians. Proc. Zool. Soc. London, 1891:593–627.

BOULENGER, G. A. 1897. The tailless batrachians of Europe. Ray Soc., London, 210 p.

BRAUER, A. 1898. Ein neuer Fall von Brutpflege bei Fröschen. Zool. Jahrb. Abt. Syst., 12: 89–94.

CAMIN, J. H., AND R. R. SOKAL. 1965. A method for deducing branching sequences in phylogeny. Evolution, 19:311–326.

CEI, J. M. 1968. Notes on the tadpoles and breeding ecology of *Lepidobatrachus* (Amphibia: Ceratophryidae). Herpetologica, 24:141–146.

COCHRAN, D. M. 1955. Frogs of southeastern Brazil. Bull. U.S. Nat. Mus., 206:1–423.

DUGÈS, A. 1834. Recherches su l'ostéologie et la myologie des batraciens a leurs différens ages. J. B. Baillière, Paris. 216 p.

DUNLAP, D. G. 1960. The comparative myology of the pelvic appendage in the Salientia. J. Morph., 106:1–76.

DUNLAP, D. G. 1966. The development of the musculature of the hindlimb in the frog, *Rana pipiens*. J. Morph., 119:241–258.

EMELIANOFF, S. V. 1925. Zur Frage uber die Entwickelung der Wirbelsäule bei den Amphibien. Rev. Zool. Russe, Moscow, 5:53–72.

ESTES, R. 1965. Fossil salamanders and salamander origins. Amer. Zool., 5:319–334.

FARRIS, J. S. 1966. Estimation of conservatism of characters by constancy within biological populations. Evolution, 20:587–591.

FARRIS, J. S. 1969. Maximum likelihood and maximum parsimony in numerical cladistics. Evolution, 23:(in press).

FITCH, W. M., AND E. MARGOLIASH. 1967. Construction of phylogenetic trees. Science, 155: 279–284.

GALLIEN, L., AND CH. HOUILLON. 1951. Table chronologique du development chez *Discoglossus pictus*. Bull. Biol. France Belgique, 85:373–375.

GHISELIN, M. T. 1966. The adaptive significance of gastropod torsion. Evolution, 20:337–348.

GOIN, C. J., AND O. B. GOIN. 1962. Introduction to herpetology. W. H. Freeman and Co., San Francisco. 341 p.

GRIFFITHS, I. 1956. The status of *Protobatrachus massinoti*. Nature, 177:342–343.

GRIFFITHS, I. 1959. The phylogeny of *Sminthillus limbatus* and the status of the Brachycephalidae (Amphibia Salientia). Proc. Zool. Soc. London, 132:457–487.

GRIFFITHS, I. 1960. The phylogenetic status of the Sooglossinae. Ann. Mag. Nat. Hist., 13: 626–640.

GRIFFITHS, I. 1963. The phylogeny of the Salientia. Biol. Rev., 38:241–292.

GRIFFITHS, I., AND A. LEITÃO DE CARVALHO. 1965. On the validity of employing larval characters as major phyletic indices in Amphibia, Salientia. Rev. Brasiliera Biol., 25:113–121.

HECHT, M. K. 1962. A reevaluation of the early history of the frogs. Part I. Syst. Zool., 11:39–44.

HECHT, M. K. 1963. A reevaluation of the early history of the frogs. Part II. Syst. Zool., 12: 20–35.

HEWITT, J. 1919. *Anhydrophryne rattrayi*, a remarkable new frog from Cape Colony. Rec. Albany Mus., 3:182–189.

INGER, R. F. 1954. Systematics and zoogeography of Philippine Amphibia. Fieldiana: Zool., 33:183–531.

INGER, R. F. 1958. Comments on the definition of genera. Evolution, 12:370–384.

INGER, R. F. 1966. The systematics and zoogeography of the Amphibia of Borneo. Fieldiana: Zool., 52:1–402.

INGER, R. F. 1967. The development of a phylogeny of frogs. Evolution, 21:369–384.

JAMESON, D. L. 1957. Life history and phylogeny in the salientians. Syst. Zool., 6:75–78.

KAMPEN, P. N. VAN. 1923. The Amphibia of the Indo-Australian Archipelago. E. J. Brill, Leiden. 304 p.

KEY, K. H. L. 1967. Operational homology. Syst. Zool., 16:275–276.

KLUGE, A. G. 1966. A new pelobatine frog from the Lower Miocene of South Dakota with a discussion of the evolution of the *Scaphiopus-Spea* complex. Contr. Sci. Los Angeles Co. Mus. 113, 26 p.

KLUGE, A. G. 1967. Higher taxonomic categories of gekkonid lizards and their evolution. Bull. Amer. Mus. Nat. Hist., 135:1–60.

LAMOTTE, M. 1963. Contribution à l'étude des batraciens de l'ouest africain. XVII. Le développement larvaire de *Hymenochirus* (*Pseudhymenochirus*) *merlini* Chabanaub. Bull. Inst. fond. d'Afrique noire, 25:944–953.

LEE, A. K. 1967. Studies in Australian Amphibia. II. Taxonomy, ecology, and evolution of the genus *Heleioporus* Gray (Anura: Leptodactylidae). Australian J. Zool., 15:367–439.

LIMBAUGH, B. A., AND E. P. VOLPE. 1957. Early development of the Gulf Coast toad, *Bufo valliceps* Wiegmann. Amer. Mus. Novitates, no. 1842, 32 p.

LIU, C. 1950. Amphibians of western China. Fieldiana. Zool. Mem, 2:1–400.

LUTZ, B. 1944. Biologia e taxonomia de *Zachaenus parvulus*. Bol. Mus. Nac., Zool., 17:1–66.

LUTZ, B. 1948. Ontogenetic evolution in frogs. Evolution, 11:29–39.

MOOKERJEE, H. K. 1931. On the development of the vertebral column of Anura. Philos. Trans. Roy. Soc. London (B), 219:165–196.

MOOKERJEE, H. K. 1936. The development of the vertebral column and its bearing on the study of organic evolution. Proc. Indian Sci. Congr. Zool., 23:307–343.

MOOKERJEE, H. K., AND S. K. DAS. 1939. Further investigation on the development of the vertebral column in Salientia (Anura). J. Morph., 64:167–209.

NEWTH, D. R. 1949. A contribution to the study of fore-limb eruption in metamorphosing Anura. Proc. Zool. Soc. London, 119:643–659.

NIEUWKOOP, P. D., AND J. FABER. 1956. Normal table of *Xenopus laevis* (Daudin). North-Holland Publ. Co., Amsterdam. 243 p.

NOBLE, G. K. 1922. The phylogeny of the Salientia. I. The osteology and the thigh musculature; their bearing on classification and phylogeny. Bull. Amer. Mus. Nat. Hist., 46:1–87.

NOBLE, G. K. 1927. The value of life history data in the study of the evolution of the Amphibia. Ann. New York Acad. Sci., 30:31–128.

NOBLE, G. K. 1929. The adaptive modifications of the arboreal tadpoles of *Hoplophryne* and the torrent tadpoles of *Staurois*. Bull. Amer. Mus. Nat. Hist., 58:291–334.

NOBLE, G. K. 1931. The biology of the Amphibia. McGraw-Hill Book Co., New York. 577 p.

ORTON, G. L. 1946. Larval development of the eastern narrow-mouthed frog, *Microhyla carolinensis* (Holbrook), in Louisiana. Ann. Carnegie Mus., 30:241–248.

ORTON, G. L. 1949. Larval development of *Nectophrynoides tornieri* (Roux), with comments on direct development in frogs. Ann. Carnegie Mus., 31:257–277.

ORTON, G. L. 1953. The systematics of vertebrate larvae. Syst. Zool., 2:63–75.

ORTON, G. L. 1957. The bearing of larval evolution on some problems in frog classification. Syst. Zool., 6:79–86.

PARSONS, T. S., AND E. E. WILLIAMS. 1963. The relationships of the modern Amphibia: A Reexamination. Quart. Rev. Biol., 38:26–53.

PETERS, J. A. 1964. Dictionary of herpetology. Hafner Publ. Co., New York. 392 p.

PIVETEAU, J. [ed.]. 1955. Traité de Paléontologie. Masson et Cⁱᵉ, Paris. 1113 p.

POWER, J. H., AND W. ROSE. 1929. Notes on the habits and life histories of some Cape Peninsula Anura. Trans. Roy. Soc. South Africa, 17:109–115.

PROCTOR, J. B. 1921. On the variation of the scapula in the batrachian groups Aglossa and Arcifera. Proc. Zool. Soc. London, 1921:197–214.

PYBURN, W. F. 1967. Breeding and larval development of the hylid frog *Phrynohyas spilomma* in southern Veracruz, México. Herpetologica, 23:184–194.

RIDEWOOD, W. G. 1897. On the development of the vertebral column in *Pipa* and *Xenopus*. Anat. Anz., 13:359–376.

RITLAND, R. M. 1955a. Studies on the postcranial morphology of *Ascaphus truei*. I. Skeleton and spinal nerves. J. Morph., 97:119–178.

RITLAND, R. M. 1955b. Studies on the postcranial morphology of *Ascaphus truei*. II. Myology. J. Morph., 97:215–282.

ROGERS, D. J., H. S. FLEMING, AND G. ESTABROOK. 1967. Use of computers in studies of taxonomy and evolution. *In* T. Dobzhansky, M. K. Hecht, and W. C. Steere [eds.], Evol. Biol. (Appleton-Century-Crofts, New York), 1:169–196.

ROMER, A. S. 1966. Vertebrate paleontology. Univ. Chicago Press, Chicago. 468 p.

SOKAL, R. R., AND P. H. A. SNEATH. 1963. Principles of numerical taxonomy. W. H. Freeman and Co., San Francisco. 359 p.

STARRETT, P. 1960. Descriptions of tadpoles of Middle American frogs. Misc. Publ. Mus. Zool. Univ. Michigan, 110, p. 1–38.

STARRETT, P. 1967. Observations on the life history of frogs of the family Atelopodidae. Herpetologica, 23:195–204.

STEEN, J. C. VAN DER. 1930. De ontwikkeling van het occipito-vertebrale streek van *Microhyla*. Leiden, Thesis.

STEPHENSON, E. M. 1960. The skeletal characters of *Leiopelma hamiltoni* McCulloch, with particular reference to the effects of heterochrony on the genus. Trans. Roy. Soc. New Zealand, 88:473–488.

STEPHENSON, N. G. 1955. On the development of the frog, *Leiopelma hochstetteri* Fitzinger. Proc. Zool. Soc. London, 124:785–795.

TIHEN, J. A. 1960a. Two new genera of African bufonids, with remarks on the phylogeny of related genera. Copeia, 1960:225–233.

TIHEN, J. A. 1960b. On *Neoscaphiopus* and

other Pliocene pelobatid frogs. Copeia, 1960:
89–94.

TIHEN, J. A. 1965. Evolutionary trends in frogs.
Amer. Zool., 5:309–318.

TURNER, F. B. 1952. The mouth parts of tadpoles of the spadefoot toad, Scaphiopus ham-
mondi. Copeia, 1952:172–175.

VILLIERS, C. G. S., DE. 1929a. Some features
in early development of Breviceps. Ann.
Transvaal Mus., 13:142–151.

VILLIERS, C. G. S., DE. 1929b. The development
of a species of Arthroleptella from Jonkershoek,
Stellenbosch. South African J. Sci., 26:481–
510.

WAGNER, W. H., JR. 1961. Problems in the
classification of ferns, p. 841–844. In Recent
advances in botany. Univ. Toronto Press,
Toronto.

WALKER, C. F. 1938. The structure and systematic relationships of the genus Rhinophrynus.
Occ. Paps. Mus. Zool. Univ. Michigan 372,
p. 1–11.

WATSON, D. M. S. 1940. The origin of frogs.
Trans. Roy. Soc. Edinburgh, 60:195–231.

WILLIAMS, E. E. 1959. Gadow's arcualia and
the development of tetrapod vertebrae. Quart.
Rev. Biol., 34:1–32.

WILSON, E. O. 1965. A consistency test for
phylogenies based on contemporaneous species.
Syst. Zool., 14:214–220.

APPENDIX I

The following is a list of the extant anuran
genera that we cite in the text. They have been
arranged according to those families recognized
by Inger (1967). In that the families were not
defined, by either characters or taxa-included, we
have been forced to make our placements by inference (both from Inger, 1967, and Griffiths,
1959, 1960, and 1963).

Ascaphidae: *Ascaphus, Leiopelma.*
Discoglossidae: *Alytes, Barbourula, Bombina,
Discoglossus.*
Pipidae: *Hymenochirus, Pipa, Pseudhymenochirus,
Xenopus.*
Rhinophrynidae: *Rhinophrynus.*
Pelobatidae: *Aelurophryne, Leptobrachium, Mego-
phrys, Pelobates, Pelodytes, Scaphiopus, Scutiger.*
Bufonidae: *Bufo, Cacophryne, Nectophrynoides.*
Atelopodidae: *Atelopus, Brachycephalus, Melano-
phryniscus, Oreophrynella.*
Leptodactylidae: *Ceratophrys, Eleutherodactylus,
Engystomops, Heleioporus, Heleophryne, Lepi-
dobatrachus, Leptodactylus, Limnodynastes,
Mixophyes, Notaden, Odontophrynus, Physaelae-
mus, Pleurodema, Rhinoderma, Sminthillus,
Telmatobius, Zachaenus.*
Hylidae: *Centrolene, Cochranella, Hemiphractus,
Hyla, Nototheca, Phrynohyas, Phyllomedusa,
Pseudis, Ptychohyla, Teratohyla.*
Ranidae: *Anhydrophryne, Arthroleptella, Caco-*

*sternum, Gigantorana, Hyperolius, Oxyglossus,
Ptychadena, Rana, Sooglossus, Staurois, Tricho-
batrachus.*
Rhacophoridae: *Afrixalus, Chiromantis, Kassina,
Rhacophorus.*
Microhylidae: *Breviceps, Chaperina, Elachistocleis,
Hoplophryne, Kalophrynus.*

APPENDIX II

Tail muscle.—Inger (1967:371) noted that the
"tail muscles," presumably referring to both the
M. pyriformis and the *M. caudalipuboischiotibialis*
(see Griffiths, 1963:262), are primitive "according to everyone's opinions," and he coded them
together as two character states, namely present
in the Ascaphidae and absent in all other modern
families (p. 375, Table 1). He reasoned (p.
370) that "As frogs are one of the very few
groups of vertebrates that have lost their tails and
the only amphibians known to have done so,
taillessness is almost certainly one of the derived
conditions of the order." We agree with the kind
of reasoning employed by Inger. In this example,
however, it must be pointed out that, as recently
as 1955, Ritland (1955a; also see 1955b:259,
272–9) discussed the likelihood that the *M.
caudalipuboischiotibialis* is merely a coccygeal
head of the *M. semimembranosus*, not homologous
with one of the three tail wagging muscles of
salamanders and, therefore, not a primitive feature
of the tailed pro-Salientia. In addition, it must be
further noted that the *M. pyriformis* appears to be
uniformly present in all families of anurans with
the exception of *Pelobates, Pipa* and *Hymenochirus*
(Noble, 1922:33; Dunlap, 1960:17). In the
absence of other critical investigations into this
problem, we have chosen tentatively to accept
the classical thesis that the *M. caudalipuboischiotibialis* of anurans is homologous with one of the
tail wagging muscles in salamanders. We have
coded the character in two states in anurans,
present or absent (Tables 2 and 3). Based on our
criterion 1, we have coded the former state as
primitive and the latter as derived. In Table 3
the primitive state is coded as 0, the derived as 1.
The *M. pyriformis* has not been encoded because of the kind and degree of variability that
it exhibits.

Presacral vertebrae.—Inger (p. 372) referred to
the number of presacral vertebrae as two character states, 9 and < 9 (p. 375, Table 1). He
accepted the presacral number of nine as diagnostic
of only the ascaphids among modern anurans,
thereby not considering the variability previously
described by Tihen (1960b; also see Kluge, 1966:7,
8) for pelobatids, which have eight or nine, and
discoglossids, which also have eight or nine
(Boulenger, 1897:39, reported on specimens with
as many as 11 vertebrae, but this probably included the sacrum and the urostyle). The references to eight vertebrae in *Leiopelma* and *Ascaphus*

(Ritland, 1955a:142; Stephenson, 1960:480), seven in pelobatids and discoglossids, and six and five in pipids all appear to involve a synostosis of two or more of the normally free nine, eight or seven presacral vertebrae in their respective families. It seems reasonable to us that only this specific category of variability may be logically ignored (see Table 1) in the discussion of most of anuran phylogeny, since it probably represents a developmental anomaly (*sensu stricto*).[1] On the contrary, however, it seems that the presence of nine vertebrae cannot be ignored in pelobatids and discoglossids on this basis. In these two families the individual nine free presacral vertebrae very often all appear to be normal. The larger number seems to be a reflection of the more posterior position of the sacral vertebra, relative to those species which have eight presacral vertebrae, and not some subdivision of a trunk segment. The number and position of the spinal nerves appear to be the best indication of the normalcy of the vertebral column (Ritland, 1955a).

As shown in Table 1 we have attempted to recognize the "normal" variability in vertebral number in anuran families, and it seems reasonable to code the states according to the progressive trend of decrease in number of vertebrae (9 → 8 → 7 → 6 → 5). Our interpretation of primitive and derived states rests on the assumption that pro-anurans, e.g., *Triadobatrachus* (*Protobatrachus*), salamanders and caecilians, and all other vertebrates for that matter, have larger numbers than modern anurans (*fide* fossils and criterion 1).

It appears likely that Inger's reason for considering all vertebrae variability of less than nine as only a single character state is best illustrated by his statement that (p. 372) "the lower number eight or less could very well have developed several times in independent lines and could have been derived from a stock (ascaphoid?) which had nine." In this example he has lumped character states on the basis of their variability and potential independent loss, while in the same example he did not consider the condition of nine vertebrae in pelobatids and discoglossids.

We argue that when character states are so similar as to appear to be homologues (as at least a first approximation, see Key, 1967), then convergence can only be discerned after a phylogeny has been constructed using more than one character. This procedure is particularly critical for homonomous series of structures, such as vertebrae, where some total meristic of the individual parts is the character.

We have not used the number of presacral vertebrae in our phyletic reconstruction because of the lack of critical study of normal variability, particularly among the more primitive frogs.

---

[1] Exceptional in *Pelodytes* where the actual fusion of the first and second vertebrae has been fixed evolutionarily.

Ribs.—Inger (1967:371, 375, Table 1) treated ribs as two character states, either present in some stage of their life history, or absent. He did not specify the absence as owing to actual loss of the center of ossification, or that the center had fused to some part of the vertebra. In our opinion, to more accurately describe the known variation in anurans, this character must be coded in at least the following three states: (1) ribs present in both subadults (unfortunately precise aging, in the form of normal tables, is available for only a few species) and adults, (2) ribs present in subadults, but absent in adults, or (3) ribs absent in both subadults and adults. Noble (1931: 233) pointed out that ribs are present in pipid larvae and that they fuse to the vertebral diapophyses in adults. As a more precise example, in the pipid *Xenopus laevis* ribs appear relatively late, stage 52 of its normal table, and only become fused to the diapophyses in postmetamorphic individuals, following stage 66 (Nieuwkoop and Faber, 1956: 107). It is this specific process of rib fusion during ontogeny that most strongly supports our contention that character state (2), as described above, must be encoded.

The cladogram derived from Inger's data (Fig. 1) indicates that ribs are absent (c) in the Rhinophrynidae and in the Pelobatidae and all other derived families. Superficially then, the absence of ribs in these two phyletic lines must be considered an example of independent loss. The following data seem to suggest, however, that independent evolution is not involved. The cleared and stained tadpoles of *Rhinophrynus dorsalis* that we prepared, at stages approximately equivalent to 47 to 58 of *Xenopus laevis* (Nieuwkoop and Faber, 1956), do not appear to possess ribs, either as separate centers of ossification, or as presumptive synostotic areas at the ends of the vertebral diapophyses. In addition, the similar proportions of diapophysis length to vertebra width, including intra- and intersegmental comparisons of vertebrae two through four, in tadpoles of these same stages with adults, further suggests that ribs have not developed after stage 58 and then fused to the ends of the diapophyses. Lastly, the ends of the diapophyses of postmetamorphic *Rhinophrynus* do not indicate, even vaguely, the presence of fused ribs. In contrast to these findings, our examination of numerous cleared and stained tadpoles and recently transformed individuals of the pelobatid genus *Scaphiopus* (*bombifrons, couchii, hammondii, intermontanus*) seems to suggest that in this phyletic line the absence of ribs is owing to rib-diapophysis fusion, similar to that so readily seen in pipids. For example, in a recently transformed *Scaphiopus holbrookii* (20 mm snout to vent length), one can clearly see a large separate center of ossification at the distal end of both diapophyses of the fourth, postcranially located, vertebra. If this condition can be confirmed in additional material of S. *holbrookii*, and demonstrated for other vertebrae, as well as in other pelobatids and for other families derived from that stock, then a case of independent evolution in the cladogram can be explained away. Ridewood (1897) indicated the likelihood of rib-diapophysis fusion in *Pelobates*, and in one series of *Microhyla* van der Steen (1930) found what appeared to be a separate anlagen, which later fused with short processes on the third pair of arches. And, rib-diapophysis fusion appears to be an established fact in at least two other families, Bufonidae and Ranidae, since Mookerjee (1931:191) described in detail a condition similar to S. *holbrookii* in *Bufo* and *Rana*. Mookerjee's (p. 191) point that "the mesenchymatous rib is from the beginning formed close to the distal end of the rib-bearer [= diapophysis]", particularly in the case of *Rana* and *Bufo*, might form some basis for the generally accepted thesis that ribs are absent ("lost") in the more advanced frogs. The data on *Scaphiopus, Pelobates, Microhyla, Rana* and *Bufo*, although admittedly very meager, tentatively suggest that the character of ribs should be coded as (1) free ribs present in both subadults and adults, (2) free ribs present in subadults, fused to diapophyses in adults, (3) ribs "fused" to diapophyses in subadults and adults, and (4) ribs lost in both subadults and adults (see Table 2). This form of coding removes the case of independent loss of ribs in anurans if state 4 is derived from 2, independent of 3. The primitive state (1) is inferred from fossil evidence and criterion 1.

In Table 3 the rib character (II) has been coded as (a) rib free in subadult = 0, rib fused in subadult = 1, (b) rib free in adult = 0, rib fused in adult = 1, and (c) rib present = 0, rib absent = 1. This form of binary coding seems to contain all of the relevant information conveyed in Table 2.

Vertebral ossification.—Inger (1967:372–3) used the three developmental modes of ossification of the centrum of vertebrae described in detail by Griffiths (1963:256–61)—ectochordy, holochordy and stegochordy—as another major character in his analysis of anuran phylogeny. He discussed Griffiths' contention that the adult husk-like ectochordal vertebra is primitive, and the partial husk-type stegochordal and the solid holochordal are derived, and concluded that there is another interpretation possible, which he followed (see p. 375, Table 1). His reasoning for this action is as follows (p. 373): "As most extinct lepospondylous amphibians had holochordal vertebrae . . . , holochordy is probably the primitive state and ectochordy and stegochordy represent derived conditions." He concluded that (p. 373) "the only evidence for the primitive nature of ectochordy is its appearance as an early ontogenetic state in many frogs."

Our discussion of this character is presented in two parts. In the first section, we will accept without qualification, as did Inger, Griffiths' concept of the kinds of vertebral ossification and their taxonomic distribution. In that context we will focus on the following points: (1) that the Lepospondyli are ancestral to the Lissamphibia is questionable, (2) and even if the Lepospondyli are ancestral to modern amphibians, the predominance of an ectochordal-like vertebra, not holochordal, in adults of that extinct group suggests that the former condition is more likely to be primitive, and (3) there are other, and more relevant, data than the time of appearance in ontogeny which suggest that ectochordy is primitive. It should be emphasized that the strict use of this character, as defined by Griffiths (p. 258), involves some knowledge of the "morphogenetic [ontogenetic] pattern," which of course is available for few, if any, fossil forms. The second part of our discussion is devoted to a reinterpretation of the kinds of vertebral ossification and their taxonomic distribution in anurans. We believe the reinterpretations to be correct over those of Griffiths'. However, we have purposely ignored ours in the final reconstruction of anuran phylogeny (Tables 2 and 3; Figs. 4 and 5). The reason for this procedure is given in the section on character analysis and recoding (page 9–11).

In a relatively recent paper, Parsons and Williams (1963) reexamined the relationships of the three modern amphibian orders, as the monophyletic Lissamphibia, and concluded that it is impossible to put forward even a tentative theory as to which Paleozoic amphibians are ancestral to the living group(s) (p. 48). Further, Baird (1965) and Estes (1965) recently presented evidence to support the derivation of the Lissamphibia from the Labyrinthodontia rather than from the Lepospondyli. To avoid confusion we note that the meaning in Estes' statement (p. 33), that it "is as possible to derive the vertebrae of modern amphibians from a primitive rhachitomous type as it is from a lepospondylous type," almost certainly follows Williams' (1959) use of vertebral terminology and not that of Griffiths (see Peters, 1964:187, 304) and it can therefore be disregarded in the immediate discussion.

It appears that the lepospondyls (aistopods, nectrideans and microsaurs) are characterized by vertebral centra which ossify as cylinders around the notochord, and it was by virtue of these "husk vertebrae" that they were named Lepospondyli (see Baird, 1965:287). In addition, apparently most of the other groups of primitive amphibians, the Labyrinthodontia, have lepospondylous-like vertebrae, either completely or incompletely encircling the notochord (terminology sensu Baird; see Romer, 1966:94, Fig. 128, and Piveteau, 1955). Watson's (1940:224) statement that "The vertebrae of *Miobatrachus* [now considered a synonym of the rhachitome labyrinthodont *Amphibamus*] so much resemble those which must have existed in the ancestors of the Anura [*Protobatrachus*]" emphasizes the great degree of similarity between these major taxonomic assemblages (also see Hecht, 1962:41). We can see no major gross morphological difference between the lepospondylous kind of vertebrae and the ectochordal type found in adults of the anuran families Ascaphidae and Rhinophrynidae; we are, however, making this comparison without knowledge of the pattern of development in the Lepospondyli and other primitive amphibians.

Evidence that ectochordy is primitive and that stegochordy and holochordy are derived obtains from the following points: (1) that the other members of the Lissamphibia, the salamanders and the caecilians, have ectochordal-like vertebrae, (2) the earliest (Triassic) and most primitive frog-like form known *Triadobatrachus* [= *Protobatrachus*] (Piveteau, 1955; Romer, 1966: 100; Watson, 1940), which may be represented by a larva or late metamorphic individual (Griffiths, 1956; also see Hecht, 1962:43), has ectochordal-like vertebrae, and (3) the predominance of ectochordal-like vertebrae in both Labyrinthodontia and Lepospondyli (*fide* fossil evidence and criterion 1).

To assume that ectochordy is a derived condition almost certainly creates numerous cases of convergence within the Anura and in the other major groups of the Amphibia, particularly the Labyrinthodontia; a more parsimonious interpretation would be to consider that state primitive as we have done (Table 2). In recoding the states of this character we have denoted both stegochordy and holochordy as derived. It appears that the change from ectochordy to either of these states involves at least a subtle increase in the deposition of calcium salts and a change in the site of calcium deposition relative to the notochord. The actual existence of the notochord in the more advanced families of frogs depends (basically) on whether one looks at the gross or microscopic anatomy and from which level in the ontogenetic continuum the study material is taken. In consideration of this point, it is not surprising that the cladogram given in Fig. 4 indicates states of homoplasy. We have used this character, even in its recoded form, primarily because of its historical popularity among anuran taxonomists.

The fact that heterochrony is more than just a possibility in some frogs (Stephenson, 1960), and that the development of the centrum is likely to be relatively strongly influenced by such a phenomenon, also should be taken into consideration when using this character in more detailed phyletic studies. Inger considered the possibility of heterochrony (p. 373), but he used it as the only evidence in support of ectochordy as the

primitive state (see our statement number 3, p. 21).

In Table 3 it has been necessary to code this character (III) in the binary forms (a) notochord retained in adult = 0, notochord not retained in adult = 1, and (b) centrum ossification ventral in position = 0, ossification not ventral in position = 1. This manner of coding does not appear to sacrifice any information conveyed in Table 2.

Inger seemingly followed Griffiths (1960, 1963), who in turn almost certainly obtained his basic interpretation of the different developmental patterns of the anuran centrum and intervertebral body from Mookerjee (1931). Our brief review of Mookerjee's major works on this subject (1931, 1936, and Mookerjee and Das, 1939), coupled with our own observations on whole and thick sectioned and stained vertebrae of a few anuran taxa, has led us to conclude that reinterpretation is in order.

Mookerjee recognized two different modes of vertebra development, perichordal and epichordal, neither of which necessarily involves any reference to the Gadovian concept of arcualia (see Mookerjee, 1936:317). According to Mookerjee's interpretation, the perichordal type of centrum formation relates to the chondrification and eventual ossification of sclerotomic cells around the notochordal sheaths, while the epichordal type involves the chondrification and ossification only of the dorsal, or the dorsal and lateral, parts of the perichordal tube of sclerotomic cells. In the latter mode, some ventral portion of the tube remains membranous and ultimately degenerates along with the notochord. The credit for the actual discovery of the two modes should probably go to Dugès (1834). Griffiths (1963) referred to the former type as the ectochordal developmental pattern, and to the latter as stegochordal; he coined the new terms only because of the historical association of perichordal and epichordal with Gadow's concept of vertebrae formation from arcualia. To these two developmental modes, Griffiths added his third category (that of holochordy) and, furthermore, he recognized two kinds of stegochordy on the basis of their supposedly different developmental pathways.

According to Griffiths (1963:260), holochordal vertebrae are those where "the notochord is completely replaced by osteoid tissue." By this definition, holochordy is only a developmental continuation of the ectochordal (perichordal) mode. Griffiths' distinction between the two kinds of stegochordy (his Fig. 5; b3 and d3) is based on whether the ventral and lateral walls of the perichordal tube become cartilaginous before they degenerate. We see no major reason for recognizing Griffiths' terms ectochordal and stegochordal over perichordal and epichordal, respectively, *sensu* Mookerjee; in the remainder of this part of the discussion we will use the latter terms.

We have retained Griffiths' "holochordal," but not as a state of the perichoral-epichordal character. This new usage obtains from the fact that it denotes how much of the area, once occupied by the notochord, is replaced by bone; it does not denote where in the perichordal tube the centrum develops. Lastly, our observations, and apparently those of Mookerjee, do not support the contention that there are two stegochordal modes of development, at least in those families to which Griffiths applied the term.

Mookerjee and Das (1939) discussed a number of taxonomically important morphologic features in the pattern of centrum development that appear to have been overlooked by Griffiths and Inger. Mookerjee and Das recognized two kinds of perichordal and epichordal modes. These are based on whether or not the ventral hyaline cartilage (sensu Mookerjee and Das) develops in the sclerotomic tube of cells sometime during ontogeny. The ventral cartilage does not take part in the formation of the epichordal centrum because it is that ventral part of the perichordal tube which always degenerates. The cartilage seems to degenerate in all of the perichordal centra as well. They pointed out that in some frogs the hypochord in the region of the urostyle fuses to the epichordal portion of the perichordal tube following the degeneration of the notochord and the ventral hyaline cartilage. The presence or absence of the ventral hyaline cartilage, and the loss or fusion of the hypochord to the epichordal centrum, seem to show considerable intrataxonomic and regional (in the same vertebral column) variability, which suggests that they may not be particularly useful taxonomic characters. Thirdly, and of obvious importance to any further study of modes of vertebral development is their observation that all of the frogs that they examined had a perichordal atlas (except *Rhacophorus maculatus*). Finally, and of considerable importance, is their discovery of a form of epichordy in *R. maculatus*, which in some characteristics is intermediate between the epichordal and perichordal extremes described by Griffiths. In general, the intermediate characteristics in the mid-trunk centra in *R. maculatus* are (1) only a relatively small part of the perichordal tube degenerates, and therefore the centrum is formed of a large part of the perichordal tube (in the adult, the centrum is intermediate in depth), and (2) the notochord does not appear to completely degenerate and therefore it seems to give rise to the large spaces in the adult centrum (see Mookerjee and Das, 1939, Pl. 2, Fig. 30). Our observations (as discussed below) add support to the idea that there is a continuum of subtle change between the two developmental extremes. The numerous levels of intermediacy suggest that the two categories cannot continue to be recognized unless one adopts the dichotomous conditional of "perichordal tube continuous, or discontinuous." We believe this conditional to be unrealistic, since it seems that almost all epichordal frogs have a perichordal atlas.

The following summary of the taxonomic distribution of perichordal and epichordal vertebrae is based on the observations of Mookerjee (1931) and Mookerjee and Das (1939), and our own: Ascaphidae, Rhinophrynidae, Bufonidae, Ranidae, and Microhylidae—perichordal; Discoglossidae, Pelobatidae, Atelopodidae, Hylidae, Leptodactylidae, and Rhacophoridae—both perichordal and epichordal; Pipidae—epichordal.

Our study of histological preparations of nearly continuous developmental series (*sensu* neurula to post-metamorphosis) of *Ascaphus truei*, *Rhinophrynus dorsalis*, *Bufo marinus*, *Rana sylvatica*, *Cacosternum capense*, and *Breviceps mossambicus* indicates that they are perichordal and that *Xenopus gilli* is epichordal. Our series of *Discoglossus pictus* is also nearly continuous, and the species is perichordal (*sensu stricto*), unlike the epichordal condition reported for the only other discoglossid studied, *Bombina bombina* (Mookerjee, 1931). Our interpretation of perichordy in *Discoglossus* is based on the fact that the ventral part of the perichordal tube does not disintegrate, although it becomes fibrous and shrinks, thereby restricting the notochordal cavity to a very small oval space.

Of the pelobatids that we have examined, *Scaphiopus bombifrons* and *S. intermontanus* are perichordal. Although our developmental series of *S. couchii*, *S. h. holbrookii* and *S. holbrookii hurteri* are not continuous we believe the former two taxa are perichordal. In contrast to *S. bombifrons* and *S. intermontanus*, the ventral part of the perichordal tube of *S. couchii* and *S. h. hurteri* remains fibrous until well after metamorphosis, at which time it finally ossifies. The mid-trunk centra of *S. h. holbrookii* appear to be epichordal, like that condition reported for *Pelobates fuscus* (Emelianoff, 1925). Our examples of megophryine pelobatids, *Leptobrachium hasseltii* and *Megophrys monticola*, are perichordal.

In the Atelopodidae and the Rhacophoridae we have not been able to study continuous developmental series. Our observations indicate that in the former family, *Atelopus minutus* is very likely epichordal, while *A. varius cruciger* and *Brachycephalus ephippium* are almost certainly perichordal. In the latter family, *Afrixalus weidholzi* and *Kassina senegalensis* are perichordal, in contrast to Mookerjee and Das' conclusion that *Rhacophorus maculatus* is epichordal.

Our histological preparations of the Hylidae and Leptodactylidae, in particular, support our contention that there are conditions between the extremes of perichordal and epichordal which cannot be placed without serious question in either category. In the Hylidae, *Hyla arborea japonica*, *H. cadaverina*, *H. cinerea*, *H. eximia*, *H.*

septentrionalis and *Pseudis limellum* are classified as epichordal; the epichordal condition in *Pseudis* has been reported previously by Mookerjee (1931). In all of the *Hyla*, the ventral part of the perichordal tube and the notochord appear to degenerate. The mid-trunk centra of adult *H. septentrionalis* are vacuolated like those of the epichordal *Rhacophorus maculatus* (Mookerjee and Das, 1939: Pl. 2, Fig. 30). These spaces may be equated to that small part of the notochordal canal that is left within the epichordal arc. In *Ptychohyla* there are two distinct degrees of perichordy, one of which is very close to the epichordal type found in *Hyla*. In *P. schmidtorum* the centrum is very shallow like that of *Hyla*. Unlike those species, however, the ventral part of the perichordal tube becomes fibrous and shrinks (it does not degenerate) and forms a horizontal band across the bottom of the shallow epichordal arc of cartilage, wherein a very small part of the notochord remains. In contrast to *P. schmidtorum*, *P. spinipollex* is typically perichordal; the centrum is relatively deep and its ossification resembles that of *Rana* as described by Mookerjee and Das.

In the Leptodactylidae we have found typical perichordal centra in *Ceratophrys ornata*, *Heleophryne purcelli*, *Leptodactylus ocellatus*, *Notaden nicholsi*, *Pleurodema diplolistris* and *Telmatobius marmoratus*. Epichordal centra are present in *Eleutherodactylus biporcatus*, *E. rugulosus*, and *Leptodactylus podicipinus*. Our developmental series of the following leptodactylids lack some critical stages and our conclusions are, therefore, tentative: *Eleutherodactylus rhodopis*, *Leptodactylus labialis*, *L. pentadactylus*, *L. fuscus*—perichordal, and *Physaelaemus cuvieri*—epichordal. The latter group of species have centrum depths intermediate between the perichordal and epichordal size extremes of Griffiths.

Although the number and kinds of taxa that have been studied by us are not extensive (further work is in progress), we believe that they are sufficient to indicate that considerable intrafamilial variability exists among anurans, and that there is a continuum of change between the states perichordal and epichordal. In the future studies that are obviously required to better document this variability and to assess the states of the continuum, we believe the critical developmental stages to be studied most intensively are those during and shortly after metamorphosis. It is during these stages that the notochord and the ventral and lateral walls of the perichordal tube disintegrate. Because of the obvious cases of regional and ontogenetic variability and the numerous states of intermediacy between perichordal and epichordal, we believe that all future studies should uniformly focus on a specific vertebra, or set of vertebrae, such as the third, fourth or fifth. It is only in this way that the homologic equivalence throughout all studies will be maximized and

the taxonomic value of the character more accurately evaluated. We have noted during our study that all adult frogs with epichordal centra seem to exhibit relatively very shallow centra, with the shallowness being directly proportional to the degree of epichordy. In addition to these characteristics, most of the kinds of epichordal centra stained with Alizarin Red-S exhibit a granular appearance in metamorphosing and post-metamorphic individuals, unlike those with the perichordal pattern. These kinds of evaluations may provide a reasonably accurate assessment of the type of developmental pattern in the absence of continuous series.

Griffiths and Inger appear to have greatly overgeneralized the taxonomic distribution of holochordy. By definition, it cannot be present in the extreme epichordal Hylidae, Atelopodidae, Leptodactylidae, and Rhacophoridae. In addition to these exceptions, we have found numerous examples of species with perichordal vertebrae in adult ranid and rhacophorid genera alone (e.g., *Rana*, *Gigantorana*, *Hyperolius*, and *Kassina* and *Afrixalus*). It must be emphasized that this kind of perichordal centrum differs from that in ascaphids, because (1) the notochordal canal does not completely pass through the entire length of the centrum, since the ossified intervertebral body is usually fused to one end of it, and (2) the persistent notochord may be relatively hard, but nevertheless it is fatty in texture and it does not contain any detectable amount of calcium salts. The variability that we have found in this character also indicates that it too possesses relatively little information on interfamilial relationships.

The terms procoelous and opisthocoelous are commonly used to describe to which end of the centrum the intervertebral body attaches. When the intervertebral body does not attach to the centrum, the term amphicoelous is often applied. Unlike most interpretations, however, we believe the most judicious coding of the characters to be as follows (1) intervertebral body free or fused to the centrum, and (2) intervertebral body fused to the anterior end of the centrum (opisthocoelous) or to the posterior end (procoelous). The other frequently used term, diplasiocoelous, has been interpreted in at least two very different ways by anuran systematists. It has been defined as the biconcave eighth (or last) postcranial vertebra, and as that vertebral column wherein the eighth vertebra is biconcave and the remainder are procoelous (Noble, 1931; Goin and Goin, 1962: 218). If one chooses the first definition, then diplasiocoelous seems to be equivalent to saying "the eighth presacral vertebra is amphicoelous." If the second definition is followed, which is the usual case, then diplasiocoelous obviously is not comparable to the character states procoelous and opisthocoelous. If one restricts his studies to what he believes to be homologous vertebrae, say the third, fourth or fifth, then the condition of the homonomous series of elements as a whole (diplasiocoely) is not relevant.

We believe that the concept of vertebral articulation in anurans must be changed based on the general similarity of centrum and intervertebral body formation in any one body segment, and the numerous levels of intermediacy between perichordy and epichordy. It seems only reasonable that for every state of intervertebral body development there should correspond a state of procoely, or opisthocoely. To use the two extremes of perichordy and epichordy as an example, it would follow that the procoelous or opisthocoelous perichordal vertebrae are not equivalent to the procoelous, or opisthocoelous, epichordal vertebrae. The degree to which they are homologous depends on how much of the perichordal tube is included in the intervertebral body.

M. adductor longus and M. sartorius.—Inger (1967:371, 375, Table 1) used the condition of the *M. adductor longus* and the *M. sartorius* as 2 of his 10 characters (see his footnote, p. 380). And, he referred to the overall evolutionary trend of progressive fragmentation and specialization of musculature—from fishes to amphibians to reptiles to mammals—as his evidence for determining primitive and derived states of these specific muscles. He coded the *M. adductor longus* as absent (primitive) or present (derived) and the *M. sartorius* as combined with the *M. semitendinosus* (primitive) or split off from it (derived). He explicitly ruled out the likelihood that the absence of these muscles in primitive anurans could be attributed to a secondary loss by reason of the general increase in locomotor efficiency resulting from division and specialization of the thigh musculature. The relationship between locomotor efficiency and muscle complexity has not yet been documented, and, in any event, it would not appear to apply in view of the high degree of similarity between primitive salamanders and frogs such as *Ascaphus* (Noble, 1922:45–46, 55–57). The following data indicate that there is considerably greater variation within the presently recognized families than Inger recorded and that there is more direct evidence available for inferring primitive and derived states.

Inger (1967:375, Table 1) recorded the *M. adductor longus* as absent in the Ascaphidae, Discoglossidae, Pipidae and Rhinophrynidae, and as present in the Pelobatidae, Bufonidae, Atelopodidae, Leptodactylidae, Hylidae, Ranidae, Rhacophoridae, and Microhylidae. No literature is cited for the source of these data and those on the *M. sartorius*, and no intrafamilial variation was noted. However, according to Noble's (1922: 23–57) classic study of anuran thigh musculature and more recently Dunlap's (1960:1–76) detailed work, there are the following exceptions. The

M. adductor longus is absent in only two discoglossid genera, *Alytes* and *Bombina*; it is present in the genus *Discoglossus*. In the family Pelobatidae the muscle is present in *Pelobates* and absent in *Scaphiopus* and both interspecifically and intraspecifically variable in *Megophrys*. Moreover, in the Leptodactylidae the muscle is absent in *Mixophyes* and one of two specimens of *Eleutherodactylus tubulus* [?], the atelopodid genera *Atelopus* and *Oreophrynella*, and the two rhacophorid genera, *Chiromantis* and *Rhacophorus*, that have been examined. In addition, Tihen's (1960a:227, Table 1, 232) study of the Bufonidae indicated that of the 10 genera recognized, the *M. adductor longus* is present only in *Bufo* and that it has been lost independently in the family at least three times (see Tihen's Fig. 1, p. 229).

Inger (1967:375, Table 1) recorded the *M. sartorius* as combined with the *M. semitendinosus* (which is equivalent to saying "adductor longus absent" in that it too is combined with another muscle, the *M. pectineus*) in the Ascaphidae, Discoglossidae, Pipidae and Pelobatidae, and as separate (which is equivalent to saying "adductor longus present" in that it has separated from the *M. pectineus*) in the Rhinophrynidae, Bufonidae, Atelopodidae, Leptodactylidae, Hylidae, Ranidae, Rhacophoridae, and Microhylidae. Again, no intrafamilial variation was noted, but according to Noble (1922:29–31) and Dunlap (1960:6–9) there appear to be the following exceptions. The *M. sartorius* is only partially separated from the *M. semitendinosus* in the pipid genus *Xenopus* (possibly interspecifically variable—compare Noble to Dunlap), in the discoglossid genus *Discoglossus*, and in *Rhinophrynus* and *Limnodynastes* of the Rhinophrynidae and Leptodactylidae, respectively. The variability in these morphologically intermediate genera, in terms of the degree of separation of the *M. sartorius* from the *M. semitendinosus* and the considerably varied origins and insertions of the heads of the two muscles (Dunlap, 1960:8, 1966), suggests that homoplasy may be involved in all cases.

The two-headed nature of the Mm. *sartoriosemitendinosus* in most of the more primitive frogs (Dunlap, 1960:8), the only partial separation of the *M. sartorius* from the *M. semitendinosus* in *Xenopus* (e.g., *X. mulleri*, Noble, 1922:30–31, Pl. X, Fig. 2), and the intraspecific variability in *Megophrys montana* and the other morphologically intermediate genera (Dunlap, 1960) points to more direct evidence for the form in which Inger has coded the character states.

It would appear that the intraspecific and interspecific variability described for various anuran species, genera, and families, and in particular that morphogenetic sequence described in numerous *Megophrys* species (Noble, 1922:26), and the physical relationship of the *M. adductor longus* to the *M. pectineus* (Dunlap, 1960:9–11) provides better evidence for ascertaining the primitive and derived states of that character than does the very general evolutionary trend cited by Inger. That the *M. adductor longus* may have evolved and been lost independently in the order Salientia alone is almost certain. Arguments similar to these can also be made for the *M. sartorius*. We cannot justify the use of these muscle characters. They appear to contain little information on relationships at higher taxonomic levels in anurans and nothing is gained by considering them further.

Pectoral girdle.—Like many others before him, Inger (1967:369, 375, Table 1) used the condition of the epicoracoid cartilages of the pectoral girdle as a taxonomic character. In general he appears to have followed the more classical definition of character states (*sensu* Cope-Boulenger) delimited by Noble (1931). These states are (1) epicoracoidal cartilages overlapping and free—arciferal, (2) epicoracoidal cartilages fused on the midline—firmisternal, and (3) a "transitional" condition. The actual source of Inger's taxonomic data are difficult to ascertain for the following reasons: (1) he defines "transitional" as "In families that are sufficiently variable to be considered intermediate"; this does not correspond to Noble's (1922) intermediate "arcifero-firmisternal" condition defined on the basis of morphology, and (2) Inger did not mention the variation in both the arciferal and firmisternal categories (*sensu* Noble) which were explicitly noted by Noble (1922, 1931) and Griffiths (1959, 1960, 1963).

Inger (p. 369) cited the exceptional intermediate conditions found in the leptodactylid *Sminthillus* (first discussed in detail by Noble, 1922) and the bufonid *Cacophryne* (*sensu* Griffiths, 1959:480) as evidence that incomplete firmisterny evolved independently. He then used this example to "raise the possibility" that the major firmisternal families, Microhylidae and Ranidae, share this resemblance as the result of independent evolution. From Inger's criticism of Griffiths' thesis that the ranids and microhylids are closely related (p. 373), we must assume that he was aware of the fact that Griffiths had redefined the epicoracoid character of the pectoral girdle. As Griffiths summarized (1963:264, 1959:472), criteria based only on degree of fusion, freedom and overlapping of the epicoracoid cartilages are "incapable of exact taxonomic application," and for this reason he redefined the character in terms of "whether or not they possess posteriorly-directed epicoracoid horns." Griffiths contended (p. 265) that the two character states defined on this basis "agree broadly with, respectively, the Arcifera and Firmisternia (*sensu* Boulenger) except that such forms as *Rana rugulosa*, *R. tigrina* and *Atelopus*, etc. are correctly designated." He also pointed out that this scheme avoids Noble's (1922) enigmatic "arcifero-firmisternal" condition.

If we accept Griffiths' criteria for defining the states of the epicoracoid character, then the following taxonomic distribution obtains: All anuran families considered herein are uniformly arciferal except the Ranidae, Rhacophoridae and Microhylidae (Griffiths, p. 273), which are uniformly firmisternal, and except the Pipidae which according to Griffiths' study (p. 271) has two arciferal genera (*Xenopus* and *Pipa*) and two firmisternal genera (*Hymenochirus* and *Pseudhymenochirus*). We have recorded the states of this character according to Griffiths' definition and his taxonomic survey in the absence of data to the contrary.

We must come to some decision as to which of the two states, arcifery or firmisterny, is primitive. It is clear that one cannot rely on the fossil record for an interpretation, since epicoracoid cartilages have yet to be found preserved. Nor does a comparison, in the sense of Griffiths' definition, with salamanders appear to be logical. Therefore, we are forced to rely on the estimate that arcifery is primitive (our criterion 2); this results solely from the fact that the greater proportion of anuran taxa exhibit that condition (Table 2). In Table 3, the character (IV) is translated into a binary form (arciferal $= 0$, firmisternal $= 1$).

Spiracle.—Inger discussed at length (1967: 373–383) the usefulness of "two" morphological characters of larvae proposed by Orton (1953, and later expanded upon by her in 1957) and the taxonomic distribution of the states of those characters. The two characters are (1) spiracle paired or single, emerging from the ventral midline or the left side of the body, and (2) cornified beaks and denticles present or absent (denticles in one or two rows per ridge, or absent, were also conditions described by Orton, 1953). In attempting to substantiate the taxonomic usefulness of these characters, Inger criticized Griffiths (1963), who had concluded that the intrataxonomic variability (= "polymorphism" of Griffiths) of the characters was too great for them to be of any real value in delimiting major groups of anuran taxa (also see Griffiths and Carvalho, 1965). The data of most of the other workers (Boulenger, 1891; Walker, 1938; Turner, 1952) that Griffiths cited in support of his conclusion, do not appear to document his thesis (see below), nor do his references to the "medio-ventral" condition of the spiracle in the hylid *Phyllomedusa trinitatus* and the presence of denticles in *Pseudhymenochirus*. Our discussion of the three larval characters attempts to give a better estimate of the degree of variability than that presented by Griffiths.

Inger's statement (p. 375) that "Each of . . . the four types of tadpoles . . . is characteristic of one or several families of adults, and each family of adults has only one of these major types of free swimming tadpole" might be attributed to Orton's knowledge of variability in 1953. This relationship results from her statement (1957:84) that "The apparent integrity of the four larval types [I–IV] and their apparent reliability as indicators of basic evolutionary lines are inferred from their restriction to, and constancy within, the particular taxonomic group which each larval type characterizes." Such character constancy appears to have been responsible for her postulate that the tadpoles of the Pipidae and Rhinophrynidae (Type I) and the Microhylidae (Type II) are the most primitive. Her awareness of the importance of understanding variability is clearly recognized, however, when she noted that future discoveries would probably necessitate changing her conclusions (p. 85).

The fact that the "operculum" and, therefore, the spiracle of pipids is not homologous to the operculum of fish, salamanders and most other amphibians is such a discovery that radically changes Orton's thesis. Nieuwkoop and Faber (1956:141) described the development of the "operculum," and thereby the "spiracle," in the pipid *Xenopus laevis* as follows: "At stage 40 the third visceral arch has formed an ectodermal fold, the operculum, which grows caudalwards. At stage 44 it overarches the sinus cervicalis, leading to the formation of the cavity of the filter apparatus or the gill chamber. After the external gills have been completely covered over, only a narrow opening is left at stage 46. This opening forms a ventral oblique slit at stage 48, while the operculum has formed a thin caudal border. At stage 50 a fold has grown out from the hyoid arch, forming the basal portion of the operculum while the original outgrowth of the third arch forms its apical portion." We have found a similar developmental pattern in our whole material of *X. laevis*, and in *X. gilli* and *Rhinophrynus dorsalis*. These observations suggest that the entire pipid-rhinophrynid phyletic line has evolved a unique operculum and spiracle. It therefore follows that these data must be coded separately from the usual condition, namely (1) true operculum and spiracle formed mostly of an ectodermal outgrowth of visceral arch II (hyoid), and (2) pseudo-operculum and spiracle formed mostly of an ectodermal outgrowth of visceral arch III; an outgrowth from arch II occurs late and forms only the base of the pseudo-operculum. The relatively late ontogenetic appearance of the ectodermal fold from arch II in pipids and rhinophrynids, and the fact that the operculum is a composite (not in the sense of Orton, 1949:263–264) formed of folds from arches II and III, suggests that this character state is unique among fishes, salamanders and anurans. It follows that the pipid-rhinophrynid state is almost certainly derived. The apparent restriction of the composite pseudo-operculum and pseudo-spiracle to the phyletic line appears to correspond with the evolution of a very different type of gill chamber. In addition to this correspondence, the Pipidae exhibit a further unique

modification of the gill chamber, namely the foreleg develops in a separate enclosed space whose walls appear to be derived from the presumptive opercular epithelium just before the opercular folds begin to form (Newth, 1949; Nieuwkoop and Faber, 1956). Among anurans, this chamber appears to be restricted to pipids, and it would therefore be judged the derived state.

It is convenient for us to note at this time that we have been able to confirm Griffiths' statement (1963:254) that the leptodactylid *Lepidobatrachus* has paired spiracles (Cei, 1968; Cei, pers. comm.) and that these paired openings are not the result of the eruption of the forelegs (UMMZ 128836). On the other hand, Griffiths (1963; Griffiths and Carvalho, 1965) has repeatedly cited Turner's (1952) work on *Scaphiopus* as evidence of polymorphism in the oral characters of tadpoles. And, that his work can be cited as evidence is predicated on Turner's contention that S. *bombifrons* and S. *intermontanus* are conspecific with S. *hammondii*, which does not appear to be the case (Kluge, 1966). Owing to the lack of critical early developmental stages of *Lepidobatrachus*, we have not been able to determine whether its operculum is derived from visceral arch II or III. Therefore, we have not attempted in this paper to encode the paired spiracle condition in the Leptodactylidae.

If we can temporarily ignore the distinction between the categories "free-living larvae" and "direct development" in frogs with a type "1" operculum (see above), and consider all embryos at a stage close to the onset of metamorphosis (*sensu* tail reabsorption), then at least seven major known modifications of the spiracle (*sensu* Orton, and Inger) can be easily discerned. These modifications, as characterized and exemplified taxonomically below, are recognized on the basis of the degree of development of the true operculum and the dermal fold (Orton, 1949:263) and the extent and place of their fusion (*viz.*, in the formation of the spiracle). Life history terminology follows Jameson (1957:76, Table 1).

Condition 1. Spiracle absent (e.g., direct developing Leptodactylidae, such as *Eleutherodactylus*, and Ascaphidae, *Leiopelma archeyi*, Stephenson, 1955:787, and Archey, 1922; and the hylid genus *Hemiphractus* where the eggs and developing young are carried on the back of the parent, Orton, 1949:264). The development of the operculum appears to be completely suppressed and the dermal fold is only poorly developed, thereby leaving a wide gap through which the branchial region and the foreleg buds are exposed.

Condition 2. Spiracle absent (e.g., in such nonaquatic embryos as the ranid genera *Anhydrophryne*, Hewitt, 1919; *Arthroleptella*, Power and Rose, 1929, and de Villiers, 1929b; and *Sooglossus*, Brauer, 1898; the microhylid genus *Breviceps*, de Villiers, 1929a; and tentatively suggested for the terrestrial nest building leptodactylid genus

Zachaenus, Lutz, 1944). In contrast to condition
1, where the operculum is absent, and conditions
4–7, where the operculum is present and partially
fused, the operculum-dermal fold fusion is complete, and therefore the branchial region and the
foreleg buds are entirely covered.

Condition 3. Spiracle absent (e.g., the ascaphid
Leiopelma hochstetteri, Stephenson, 1955:787–
788). Like condition 2, with the exception that
the operculum does not fuse with the dermal
fold, the forelegs are usually covered by the
posterior edge of the well-developed operculum.

Condition 4. Spiracle single and restricted to
the ventral mid-line very near the posterior rim
of the branchial chamber (e.g., the ascaphid
genus *Ascaphus*, the Discoglossidae, and the hylids
Hyla goeldii [intraspecifically variable] and
Nototheca fissilis; Griffiths and Carvalho, 1965).
The position of the single spiracle and its slit-like
external opening are owing to the absence of
operculum-dermal fold fusion on the midline very
close to the rim of the branchial chamber.

Condition 5. Spiracle single and restricted to
the ventral midline, with the external aperture,
in contrast to condition 4, usually located immediately below the vent (e.g., most Microhylidae).
The arboreal east African *Hoplophryne rogersi*
(Noble, 1929:301, Fig. 6) exemplifies the variant
where the external aperture is not located immediately below the vent, but is closer to the
middle of the coiled gut. The South American
Elachistocleis ovalis exemplifies an additional
variant where the spiracular opening is sinistral
and directed upwards along the caudal fin (Griffiths and Carvalho, 1965). The usual position of
the spiracle and its relatively complex tube-like
canal and external aperture are owing to the
absence of operculum-dermal fold fusion on the
midline. Early in development the external aperture of the spiracle is slit-like and is located very
close to the rim of the branchial chamber and
only later grows over the coiled gut on the ventral
midline towards the vent in the form of a long
tube. This ontogenetic pattern is clearly shown
by Orton (1946:243, Figs. 1a–d). There appears
to be a weak positive relationship in conditions
4 to 7 between the length of the free tubular
portion of the spiracular canal and the shape of
its external aperture, slit-like or round, relative
to the distance the external aperture is from the
rim of the branchial chamber (*viz.*, the longer
the free part of the spiracular tube and the more
oval the form of the external aperture, the longer
the spiracular canal).

Condition 6. Spiracle single and restricted to the
left side of the body, usually much above the
ventrolateral margin of the trunk (e.g., the majority of the species in the Pelobatidae, Ranidae,
Rhacophoridae, Bufonidae, Atelopodidae, Leptodactylidae, and Hylidae, *fide* Inger, 1967). The
sinistral position of the spiracle and its relatively

complex tube-like free extension and usually
round external aperture are owing to the absence
of operculum-dermal fold fusion near the left
posterolateral corner of the branchial chamber.
Early in development the external aperture of the
spiracle is usually slit-like and located very close
to the rim of the branchial chamber, and it only
later grows around the left side of the body in
the form of a tube. This state is not restricted to
larvae with aquatic development (e.g., it occurs
in the larvae of the bufonid *Nectophrynoides
tornieri* and the leptodactylid *Rhinoderma* which
have direct development and where the embryo
is carried by the parent until birth). Defined only
by the position of the spiracular opening, without
reference to the body, the microhylid *Elachistocleis
ovalis* would be referred to condition 6.

Condition 7. Spiracle single and restricted to
slightly left of the ventral midline (e.g., the
arboreal nest building hylid genus *Phyllomedusa*,
Starrett, 1960; and *Hyla goeldii* [intraspecifically
variable], Griffiths and Carvalho, 1965). The
nearly midline position of the sinistrally located
spiracle and its slit-like external aperture are owing
to the absence of operculum-dermal fold fusion
slightly to the left of the midline near the posterior
rim of the branchial chamber. There is variation
in the position of the external aperture of the
spiracle in both sinistral conditions 6 and 7,
particularly in the latter, but of the two conditions,
Phyllomedusa and *Hyla goeldii* appear to be exceptional in the degree to which the spiracle is
located close to the midline. We cannot agree
with Griffiths' (1963) statement that the spiracle
in *Phyllomedusa trinitatis* is located exactly on the
midline and, therefore, in some sense equivalent
to condition 4 (see Griffiths and Carvalho, 1965:
115).

To evaluate the polarity of all seven character
states, it is most convenient to consider them in
the form of a binary set of comparisons. First,
we can ask, is the absence (condition 1) or the
presence of a true operculum (all other conditions) primitive? The great similarity of anlage,
developmental pattern, and final form of the
operculum of fishes, salamanders and frogs, ex-
cepting that condition found in pipids and
rhinophrynids as described above, clearly suggests
that it is primitive (criterion 1). We have acknowledged this dichotomy by recognizing the character as true operculum present (primitive) or
absent (derived); see Table 2 and Fig. 6. Next
we can evaluate the relative primitiveness of the
operculum in terms of complete fusion with the
dermal fold (condition 2) versus only partial
fusion (conditions 4 to 7) or no fusion at all
(condition 3). Again, similar to the previous
argument, our conclusion that condition 2 is probably derived rests on the fact that the anuran taxa
that exhibit that modification appear to represent
relatively minor exceptions (criteria 2 and 3)

compared to the majority operculum-spiracle condition in the major taxonomic categories to which they are usually assigned on the basis of other characteristics. For example, the genera *Anhydrophryne*, *Arthroleptella*, and *Sooglossus*, and *Breviceps*, and *Zachaenus* are almost certainly related to three natural groups of genera, namely the Ranidae, Microhylidae and Leptodactylidae, respectively; the vast majority of these related taxa have aquatic larvae with a single sinistral spiracle (operculum-dermal fold fusion incomplete). Our argument is the most parsimonious one, for to assume that all of the examples of condition 2 are homologues, and the taxa possessing this modification are primitive, dictates that the spiracle has evolved independently many times and is not homologous (see Fig. 6).

The explanation of condition 3 in *Leiopelma hochstetteri* must be considered differently from the above, because a simple parsimonious taxonomic approach cannot be applied. This obtains if we accept the Ascaphidae as a natural group of only two genera, *Leiopelma* and *Ascaphus*. *Leiopelma* has direct development (operculum absent or present and completely free), while *Ascaphus* has a typical aquatic larva with a spiracle (the operculum and dermal fold are fused in part). The fact that the operculum of *L. hochstetteri* is similar to the probable ancestral form, namely that seen in most salamander larvae where the spiracle has not formed owing to the absence of operculum-dermal fold fusion, suggests that its state is primitive (criterion 1). Additional support for this relationship obtains from Stephenson's (1955) point of view (which counters Noble's, 1927:63, conclusion) that the mode of development of *Leiopelma* is more like urodeles than frogs. In the absence of relevant data to the contrary, we tentatively recognize the condition 3 "spiracle" (found in *L. hochstetteri*) as primitive. Accepting this thesis requires that the condition of *L. archeyi* be derived from condition 3 of *L. hochstetteri* and that conditions 4 to 7 are also derived from 3. If data are found to contradict our contention that one of the states is primitive, condition 3 will still be considered primitive relative to 1, but in itself derived from some anuran where the spiracle (*sensu stricto*) is present (e.g., like conditions 4 to 7).

Our binary comparison of the remaining conditions 4 to 7 involves the position of the external aperture of the spiracle relative to the rim of the branchial chamber. In condition 4 the spiracle is slit-like and located on or very close to the posterior rim of the chamber. The ontogenetic pattern leading up to this final position has been investigated in some detail by Gallien and Houillon (1951) in *Discoglossus pictus* (also see Boulenger, 1897:98, for data on *Bombina variegata pachypus*). In contrast to this final ontogenetic position, the external aperture of the spiracle in all remaining

conditions 5 to 7, following an extended developmental sequence, are located some relatively great distance from the rim of the chamber. It seems most likely that the primitive condition is 4 (*sensu Ascaphus* and Discoglossidae), wherein the external aperture is not carried any relatively great distance, if indeed it is moved at all, from the rim of the chamber owing to the development of only a short spiracular tube. Our argument here is based on the similarity of the earlier phases of the ontogenetic pattern of all anurans and, that to realize the final position of conditions 5 to 7, the pattern always passes through a stage like the position of condition 4 (criteria 2 and 3).

The median spiracular opening in *Nototheca fissilis* is almost certainly derived from condition 6 (typical sinistral spiracle). Here again, as we discussed in greater detail above, our argument rests on criterion 2; our stand results from our accepting *Nototheca fissilis* as a hylid and recognizing that almost all other hylids have a sinistral spiracle and that a similar sinistral spiracle is present in most other families (e.g., Pelobatidae, Ranidae, Rhacophoridae, Bufonidae, Atelopodidae, and Leptodactylidae). We fully recognize that this implies an *a priori* decision as to the limits of the higher taxonomic categories. This problem is not encountered when the study is initiated at the individual or infraspecific populational level, and moreover, closely related groups can be recognized prior to analysis of the group under study through estimates of overall similarity that make no assumptions about primitive conditions. The developmental sequence of spiracle formation in *Nototheca fissilis* should be examined closely for evidence of similarity to condition 6.

Within the remaining conditions, 7 (nearly median spiracle, as exemplified by the hylid genus *Phyllomedusa* and 80 per cent of the *Hyla goeldii* examined, Griffiths and Carvalho, 1965) is almost certain to have been derived from condition 6. Our argument for this conclusion is based on criterion 2. That *Hyla goeldii* is polymorphic, namely it exhibits conditions 4, 7 and probably 6, clearly indicates the probable pattern of evolutionary reversal (*viz.*, from condition 6 to 7 to 4). It is interesting that *Hyla goeldii* and *Nototheca fissilis* are probably very closely related (Cochran, 1955:191).

We know of no biologically realistic *a priori* way to relate the remaining conditions 5 and 6, in terms of polarity. We cannot accept Inger's (1967:377) reasoning that the sinistral spiracle is derived from one in a median position because "In organisms that are essentially bilaterally symmetrical, a bilateral developmental pattern seems more primitive than an asymmetrical one." Because of this difficulty we have coded condition 5 (e.g., Microhylidae) as derived from both condition 4, independent of 6 and from 6, and similarly we have coded condition 6 as derived from both condition 4, independent of 5, and from 6 (see Fig. 6). There can be little doubt that the "sinistral" spiracle of *Elachistocleis ovalis* is derived from the typical median position of the Microhylidae (criterion 2).

From the preceding discussion it seems quite clear that the "spiracle character" embodies relatively little information that relates to the interpretation of phylogeny. Like the other characters, we have considered the different modifications in a binary form (Table 3, V). We believe the following three codings contain most of the phyletically useful information: (a) true operculum = 0, pseudo-operculum = 1; (b) true spiracle not strongly sinistral = 0, true spiracle strongly sinistral = 1; and (c) true spiracle located on or near margin of branchial chamber = 0, true spiracle not located on or near margin of branchial chamber = 1.

Cornified beaks and denticles.—The second larval "character" that Inger (1967:375, Table 1) used, the presence or absence of cornified beaks and denticles, deserves reinterpretation and must be reevaluated as a meaningful estimate of phyletic affinity. To code both beaks and denticles as a single character is inconsistent with the known variability that each of these two sets of structures exhibits, namely they are to a large degree independently variable (see Table 4). One can only accept them as a single character at the family level if the intrataxonomic variability is not coded for (the procedure followed by Inger). Moreover, Orton (1953:69) recognized not only the presence or absence of rows of denticles as discrete states, but also referred to double versus single rows of denticles per ridge (apparently she overlooked the "triple" rows in some discoglossids and *Ascaphus*). The denticles and the beak should be treated as separate characters and the form of the rows of denticles (triple, double or single per ridge) included as states of that character (Table 4).

As seen in Table 4, both beak and denticle characters are highly variable. It is because of this degree of variability that we believe there is little, if any, diagnostic value derived from using either of them in the phyletic analysis of the classically delimited anuran families considered herein. We acknowledge that in most of these families a beak and a single row of denticles are present in the great majority of the species. In addition, it is clear that in some species where the beak and the rows of denticles, or the denticles alone, are absent their loss is positively correlated with the occurrence of some form of direct development; however, this correlation does not hold for all examples of beak-denticle loss. Even if all of the relatively few exceptions of beak and denticle loss in each family could somehow be rationally explained on the basis of cause and effect related to direct development, the char-

acters would still not be admissible for the following reasons. There are many kinds and degrees of direct development involved (Lutz, 1948; Jameson, 1957). Direct development is known in two of the three families, Pipidae and Microhylidae, which primitively are almost certain to have had a free swimming kind of tadpole without beaks and denticles. Also, there is the purely practical problem of assessing the considerable degree of ontogenetic change (Limbaugh and Volpe, 1957). For meaningful interspecific comparisons, the times of appearance and disappearance of beaks and denticles in the premetamorphic phase of ontogeny must be determined. These necessary data are given in the form of normal tables, of which not more than 20 species of anurans have been described in sufficient detail (Nieuwkoop and Faber, 1956). An additional problem will be to include the condition where the larval mouth parts are typical, except that they are not cornified (e.g., *Rhinoderma*).

Of all of the states of the beak and denticle characters delimited in Table 4, the presence or absence of double or triple rows of denticles might be considered taxonomically useful. However, even here we do not have confidence in the condition of the character, because of the extreme difficulty in distinguishing between species with double rows per ridge (sensu *Ascaphus*) and those that have two "single" rows very close together (e.g., ranids *Trichobatrachus robustus*, *Rana rugulosa* and *R. crassipes*, hylid *Hyla claresignata*, and the leptodactylid genus *Heleophryne*). Equally important, these taxonomic examples almost certainly indicate that tooth rows have secondarily evolved *de novo* in anurans, and in more than one phyletic line. Future studies that consider denticle characteristics must attempt to homologize the individual rows and treat them as separate characters.

Griffiths' (1963:254) statement that small denticles are present in the pipid *Pseudhymenochirus* (presumably *P. merlini*) must be reinvestigated, because Lamotte (1963:946) described them as absent in his study of *P. merlini*. If Griffiths' observation can be confirmed, then another major parameter of denticle variation must be taken into consideration.

Vent (larvae).—The position of the external aperture of the vent, either median or dextral, has been noted in most of the descriptions of anuran larvae from the mid-Eighteenth Century to the present day. Noble (1927:74) pointed out that, although the position of the anal opening had been considered as both a "good" generic and familial character for many years, its variability was too great for it to be of any real significance. Apparently for similar reasons, Orton (1953, 1957) altogether omitted the position of the vent in her survey of taxonomic characters of anuran larvae.

Inger (1967:375, Table 1; also see 1958:382) did not discuss the conclusions of Noble, Orton and others, and he used the position of the anus in his analysis of anuran phylogeny. He stated (p. 379) that "The position of the vent is remarkably constant within families" and reasoned that "As the anus is median in adult frogs and most vertebrates, including other amphibians, I am treating the dextral vent of larvae as the derived state." We do not agree that the character is "remarkably constant within families."

To assess the variability of the position of the vent in aquatic larvae, we cursorily surveyed the relevant literature at hand. Our sources were Alcala (1962), Griffiths and Carvalho (1965), van Kampen (1923), Lee (1967), Liu (1950), Noble (1927), Pyburn (1967), Starrett (1960, 1967), numerous papers by Lamotte and his colleagues, 1954 to 1965, and Inger (1954, 1966). The exceptions to Inger's familial definitions of the state of the vent are listed below in parentheses. Pelobatidae—median (*Aelurophryne brevipes*, *A. glandulata, Leptobrachium, Megophrys oshanensis, Scutiger pingii, S. popei, S. schmidti*); Atelopodidae—median (*Melanophryniscus moireirae*); Leptodactylidae—median (*Ceratophrys, Engystomops pustulosus, Heleioporus, Odontophrynus americanus*); Hylidae—dextral (*Centrolenella fleischmanni, C. granulosa, C. prosoblepon, C. reticulata, C. spinosa, Hyla annectans, H. montana, Phrynohyas venulosa*); Ranidae—dextral (*Oxyglossus laevis, O. lima, Ptychadena maccarthyensis, P. mascareniensis, P. submascareniensis, P. tournieri, Rana jerboa, Staurois chunganensis, S. richetti*); Microhylidae—median (*Chaperina fusca, Elachistocleis ovalis, Kalophrynus* sp.). Many are both intrafamilial and intrageneric exceptions, and the degree of variability within both taxonomic categories ranges from single exceptions to nearly 40 percent of those species surveyed. It should be pointed out that a polymorphic condition might be inferred if age is not taken into account in those larvae that have dextral vents. A truly polymorphic situation has been reported by Lee (1967, pers. comm.) in the leptodactylid genus *Heleioporus*. Both the median and dextral vents have been found in the same egg mass in each of the following species: *H. albopunctatus, H. eyrei, H. inornatus*, and *H. barycragus*. From these findings, we cannot accept the position of the vent as a meaningful character in delimiting anuran relationships. Our cursory survey of the literature strongly suggests that both states of that character have evolved numerous times in parallel.

Scapula and clavicle.—Inger (1967) gave some consideration to the use of the scapula to clavicle ratio as a character in his analysis of anuran phylogeny. He noted that a short ratio (greater than three, *fide* Griffiths, 1963) "may be" primitive, and as evidence for this possibility he used the argument of (p. 374) "the occurrence of short

scapulae in the families of frogs generally held to be primitive." He (p. 380) used the character in the phylogeny generated by his computer program, coding the character as ≥ 3 or < 3 with the primitive state not specified. The same character was also tabulated in his other phylogenies (Figs. 4 and 5, Table 2).

Griffiths (1963:265) almost certainly served as Inger's primary source of data on the scapula to clavicle ratio; he in turn explicitly (at least in part) obtained some information from Proctor (1921). Griffiths' contention (p. 265) that the Ascaphidae, Discoglossidae and Pipidae have a scapula to clavicle [= precoracoid of Proctor] ratio greater than three is not supported by Proctor's raw data. We recorded the following ratios from skeletal material in our possession, which add further evidence that the "primitive" families have a ratio greater than three, while the more advanced families (including the Rhinophrynidae) have a ratio of less than two.

Ascaphidae: *Ascaphus truei* (n = 4; orv = 1.4–1.9; x̄ = 1.58[1]; also see Ritland, 1955a:146, Fig. 6). Also see Stephenson (1960:481, Fig. 4) for contradictory data on *Leiopelma*. Discoglossidae: *Alytes obstetricans* (n = 3; orv = 2.1–2.3; x̄ = 2.17); *Barbourula busuangensis* (n = 1; 2.8); *Bombina bombina* (n = 4; orv = 2.9–3.3; x̄ = 3.1); *B. orientalis* (n = 2; orv = 1.9–2.2; x̄ = 2.05); *B. variegatus* (n = 1; 3.1); *Discoglossus pictus* (n = 5; orv = 2.2–2.7; x̄ = 2.36). Pipidae: *Hymenochirus boettgeri* (n = 1; 1.4); *Pipa pipa* (n = 2; orv = 2.9–3.3; x̄ = 3.1); *Xenopus gilli* (n = 1; 2.8); *X. laevis* (n = 3; orv = 2.8–3.6; x̄ = 3.07); *X. mulleri* (n = 5; orv = 2.6–3.2; x̄ = 2.88).

We believe that there is a very general trend in the scapulae of frogs towards increased size and complexity (see Tihen, 1965:311), but the trend is not a simple dichotomy as Inger and Griffiths accepted; the ratio exhibits considerable variation within genera, and between families of primitive frogs as described above. Equally important, before the character can be considered again, the degree of ontogenetic change (which we believe to be significant) must be described, and a more precise method of measuring the elements must be defined and followed by all investigators. The acute angle of the clavicle of some species of anurans makes that measurement extremely difficult to take and certainly contributes to the greater variance in those species.

Griffiths (1963:265) first discussed the scapula to clavicle ratio in the context of two other characters: (1) proximal end of scapula, uncleft or cleft, and (2) scapula overlain anteriorly by the clavicle or not. We have attempted to substantiate Griffiths' contention that the Ascaphidae and the Pipidae have a cleft scapula. The degree

---

[1] n = number of specimens examined; orv = observed range of variation; x̄ = mean.

of variation that we found intrataxonomically at all levels in the Pipidae, and the conspicuous effects of increased ossification of the glenoid cartilage with age, suggests that a much more detailed study of the character is in order before it can be considered further. The second scapula character of Griffiths shows considerably greater promise in elucidating anuran phylogeny. We have examined representatives of all genera of the Ascaphidae (except *Leiopelma*), Discoglossidae, Pipidae, and Rhinophrynidae and find that the distal end of the clavicle overlaps much of the scapula. All of the material of the other families that we have examined shows little or no anterior overlap of the clavicle (the plane of contact is nearly vertical). Of the forms examined in the latter group of families, *Pelodytes punctatus* most closely approaches the overlapped state of the former group of families.

We have used the scapula to clavicle overlap character instead of the scapula to clavicle ratio character of Griffiths. We cannot accept the scapula to clavicle ratio as a very meaningful taxonomic character, because of its extreme degree of variability as described above. We believe the scapula to clavicle overlap character is consider-

ably less variable than Griffiths' other scapula associated characters within the classically accepted families used herein, and the simple dichotomy of overlap or no overlap is certainly much easier to record. We have coded the overlapped state as primitive (0 in Table 3) on the basis of similar conditions occurring in almost all other amphibians where the clavicle is present and where there is no suggestion of loss of limbs (*fide* fossil evidence and criteria 1 and 2). Before we can consider the dichotomy of clavicle to scapula overlap with greater confidence, all species of *Leiopelma* must be studied and preadult stages of *Xenopus* species must be followed through metamorphosis. In the case of adult *Xenopus*, the scapula and clavicle are fused together and, thereby, make the overlap of the two elements difficult to interpret.

Museum of Zoology and Department of Zoology, University of Michigan, Ann Arbor, Michigan 48104. Present address of junior author is Department of Biological Sciences, State University of New York at Stony Brook, Stony Brook, New York 11790.

12

METHODS FOR COMPUTING WAGNER TREES

James S. Farris

The Wagner Ground Plan Analysis method for estimating evolutionary trees has been widely employed in botanical studies (see references in Wagner, 1961) and has more recently been employed in zoological evolutionary taxonomy (Kluge, 1966; Kluge and Farris, 1969). Wagner Trees are one possible generalization of the most parsimonious trees of Camin and Sokal (1965). The Wagner technique is of considerable interest for quantitative evolutionary taxonomists because it is readily programmable and because the type of tree produced can tractably be extended to applications in a variety of novel quantitative phyletic techniques.

In this paper I shall formalize the concept of a Wagner Network and discuss a number of algorithms for calculating such networks. The rationale for the methods described will not be treated extensively here, as it is published elsewhere (Kluge and Farris, 1969).

REPRESENTATION OF TREES

An evolutionary tree, T, can be represented as an ordered pair, $T = (N, f)$, where N is a collection of nodes of the tree and f is a function that assigns to a member, n, of N a unique node, $f(n)$, such that n is directly derived from $f(n)$ according to the tree, T. We shall call $f(n)$ the *immediate ancestor* of n. Fig. 1(i) depicts a tree with nodes A, B, C, D, E, F, G, and *ancestor function*:

$$f(A) = B$$
$$f(C) = B$$
$$f(B) = D$$
$$f(F) = E$$
$$f(G) = E$$
$$f(E) = D$$

Note that $f(.)$ is not defined for D, the ancestor of the whole tree. While D is the ancestor of the whole tree, it is not true, for example that $f(A) = D$, because D is not the *immediate* ancestor of A.

We shall consider only pairs, $T = (N, f)$, that have a unique node, P, in N such that if n is any element of N, then there exists a non-negative integer, K, for which $f^K(n) = P$. The function $f^K(.)$ is the K-fold composition of f on itself; i.e. $f^4(x) = f(f(f(f(x))))$. These pairs are just the trees that have a unique "root" and on which every member of N is connected to the tree so that it is a descendent of the "root." The trees we consider are, in short, those that fit the usual taxonomic notion of "tree."

NETWORKS

Trees are directed entities in which the root is presumed to represent a point chronologically prior to any descendent point. We can think of a tree as being specified in two components, the first being the relative position of the nodes in the branching pattern, and the second being the location of the root. If the root is not specified, we have an "undirected tree," or a *network*. A network with a certain set of nodes may correspond to a wide class of trees with the same nodes, each tree differing from the others in the class only in the position of its root. Figs. 1(i) and 1(ii) show two trees that differ only in the location of their roots. Fig. 1(iii) depicts the generalization of both trees: a network with the "same" branching pattern, but with no root specified at all.

The central problem of evolutionary taxonomy is to construct an evolutionary tree.

221

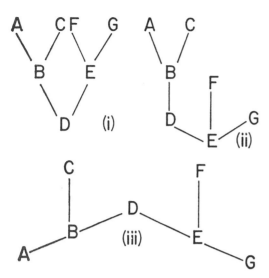

FIG. 1.—Three equivalent Wagner Networks, two of which, ((i) and (ii)) are also Wagner Trees.

One possible way to approach the problem is to first construct a network corresponding to the class of trees that contains the desired tree and then to select a tree from the class by specifying the root node. One possible advantage of this approach is that the form of the network may be used in inferring the proper position of the root.

There are many ways of specifying networks. The most convenient method for our purposes is to subvert the notation for trees. A network, W, will be defined as an ordered pair, $W = (N, g)$, consisting of a class, N, of nodes, and a function, g, that assigns to a member, n, of N a unique node $g(n)$ in N such that n is directly connected to $g(n)$. The g function differs conceptually from the f function only in that no directionality of the linkage between $g(n)$ and n is implied. We will treat only networks, $W = (N, g)$, for which N contains a unique *base element*, Q, such that for any element, n, in N there is a non-negative integer, K, for which $g^K(n) = Q$. These are just the simply connected networks, and they are the only networks that can have the same form as a tree of the type we consider. If $T = (N, f)$ is *any* member of

the class of trees that have the same form as a network, $W = (N, g)$, then W can be represented as (N, g^*), where $g^* = f$. We have listed above the ancestor function for Fig. 1(i). The ancestor function for Fig. 1(ii) is

$$f(A) = B$$
$$f(C) = B$$
$$f(B) = D$$
$$f(D) = E$$
$$f(F) = E$$
$$f(G) = E.$$

It should be clear that the ancestor function of either Fig. 1(i) or Fig. 1(ii) can serve as the *connection function* of Fig. 1(iii).

LENGTH

In constructing an evolutionary tree it is necessary to choose among a large number of possible alternative trees. One way of making the choice is to select the tree that implies the minimum amount of evolutionary change between the OTUs. Camin and Sokal (1965) took such an approach in introducing the idea of *most parsimonious* trees as trees with the smallest number of "steps" (changes in integer-valued characters). The notion of *length* of a tree is a generalization of "number of steps."

A data matrix, X, assigns a state, $X(A, i)$, to node A for character i. The *difference* between two nodes, A and B, is defined to be

$$d(A, B) = \sum_i | X(A, i) - X(B, i) |. \quad (1)$$

Two nodes, A and $f(A)$, are the end points of a unique internode of a tree, $T = (N, f)$, for A in N. The length of the internode between A and $f(A)$ is defined to be $d(A, f(A))$.

The tree $T = (N, f)$ has just K-1 internodes if K is the number of nodes in N. The *length* of a tree is defined to be

$$L(N, f) = \sum_{n \neq P} d(n, f(n)), \quad (2)$$

where P is the ancestor of the tree. The length, $L(N, f)$, of a tree, $T = (N, f)$, is the sum of the lengths of the internodes of the tree.

If the data matrix, X, takes on only integer values, (1) clearly gives the "number of steps" between A and f(A). Length as defined here is thus related to the concept of parsimony as used by Camin and Sokal. Equation (1), however, is defined for any real-valued data matrix, X, so that we can speak of the length of a tree in terms of continuously coded characters.

The length of a network is defined analogously to the length of a tree. For a network, $W = (N, g)$, the length is

$$L(N, g) = \sum_{n \neq Q} d(n, g(n)), \qquad (3)$$

where Q is the base element of N.

WAGNER TREES AND NETWORKS

The most parsimonious trees of Camin and Sokal (1965) are trees with a minimum number of steps, provided they have no evolutionary reversals. Analogously, we will be interested in trees of minimum length. The assumption of irreversibility of evolution will not be made, however.

We define a *Wagner Tree* for a set, S, of OTUs as a tree, $T = (N, f)$ such that

    1) S is a subset of N

    2) if $T' = (N', f')$ is any other tree satisfying

(1), then $L(N, f) \leq L(N', f')$.

A *Wagner Network* for a set, S, is analogously defined as a network, $W = (N, g)$ such that:

    1) S is a subset of N

    2) if $W' = (N', g')$ satisfies (1), then $L(N', g') \geq L(N, g)$.

Any of the nodes of a network can be used as the root of a corresponding tree without altering its length (since (2) and (3) are the same equation). Since irreversibility of evolution is not assumed, any tree generated from a network in this way is a legitimate candidate as a Wagner Tree. Hence a Wagner Network on S can be converted into a Wagner Tree for S with no change in length. It is thus possible to find a Wagner Tree by finding a Wagner Net-

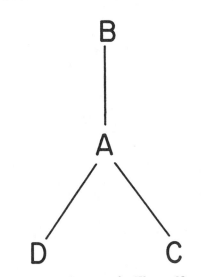

FIG. 2.—A local region of a Wagner Network.

work. This simplification of the task of finding a tree is possible only if evolutionary reversals are permitted.

HYPOTHETICAL TAXONOMIC UNITS

The nodes of a Wagner Tree are not required all to be OTUs, and in fact it is often necessary to use nodes not in S in order to achieve a shortest tree for S. The nodes of a Wagner Tree, $T = (N, f)$, that are not OTUs are purely hypothetical objects chosen simply to allow the length of the tree to be minimized. I shall refer to hypothetical nodes as Hypothetical Taxonomic Units (HTUs).

The definition of Wagner Trees implies a useful property of HTUs: the character state values of an HTU are related to the character state values of the OTUs and the other HTUs in a simple way. Consider the four-node network of Fig. 2 and suppose that it is a Wagner Network. Whether Fig. 2 is interpreted as a tree or a network is immaterial. Fig. 2 can be considered as a section of a more extensive network.

The length of the network of Fig. 2 is $d(A, D) + d(A, C) + d(A, B)$. Recalling (1), this can be rewritten as

$$\sum_i | X(A, i) - X(D, i) | +$$

$$\sum_i | X(A, i) - X(C, i) | +$$

$$\sum_i | X(A, i) - X(B, i) | =$$

$$\sum_i [| X(A, i) - X(D, i) | +$$
$$| X(A, i) - X(C, i) | +$$
$$| X(A, i) - X(B, i) |]. \qquad (4)$$

Since (4) is completely additive over characters we can seek values $X(A, i)$ to minimize (4) one character at a time. For character i, let

$$a = X(A, i)$$

$p =$ the largest of $X(B, i)$, $X(C, i)$, $X(D, i)$

$q =$ the median of $X(B, i)$, $X(C, i)$, $X(D, i)$

$r =$ the smallest of $X(B, i)$, $X(C, i)$, $X(D, i)$.

The ith component of (4) is equal to

$$| a - p | + | a - q | + | a - r |. \qquad (5)$$

The optimal value of $a = X(A, i)$ minimizes (5). We can see that if a is chosen outside the interval (r, p), the value of (5) must exceed $p - r$. If a is chosen inside (r, p), the first and last terms of (5) sum to $p - r$, and if $a = q$, the total value of (5) is $p - r$. The optimal value of a is thus q, and in general, the optimal value of $X(A, i)$ is the median of $X(B, i)$, $X(C, i)$, $X(D, i)$ if A is connected just to the nodes B, C, D, of a Wagner Network.

The *median-state property* of HTUs is used in the algorithms described later to specify optimal HTUs by constructing them one character state at a time.

INTERVAL DIFFERENCES

Some calculations for constructing Wagner Trees can be simplified by a relation on the differences between HTUs and OTUs. The relation is a corollary of the median-state property of HTUs.

Suppose that a node, B, is connected to two other nodes, C and D, through an optimal HTU, A (Fig. 2). We know from the argument above that for each character the state of A is the median of the states of B, C, and D. Then for any character, i,

$X(A, i)$ lies between $X(B, i)$ and $X(C, i)$
$X(A, i)$ lies between $X(B, i)$ and $X(D, i)$
$X(A, i)$ lies between $X(C, i)$ and $X(D, i)$.

Consequently,

$$| X(B, i) - X(C, i) |$$
$$= | X(A, i) - X(B, i) | + | X(A, i) - X(C, i) |;$$

$$| X(B, i) - X(D, i) |$$
$$= | X(A, i) - X(B, i) | + | X(A, i) - X(D, i) |;$$

$$| X(C, i) - X(D, i) |$$
$$= | X(A, i) - X(C, i) | + | X(A, i) - X(D, i) |. (6)$$

Summing over characters and using (1), (6) yields

$$d(B, C) = d(A, B) + d(A, C); \qquad (7)$$
$$d(B, D) = d(A, B) + d(A, D); \qquad (8)$$
$$d(C, D) = d(A, C) + d(A, D). \qquad (9)$$

Adding (7) and (8), we obtain

$$d(B, C) + d(B, D)$$
$$= 2d(A, B) + d(A, C) + d(A, D). \quad (10)$$

By (9), (10) becomes

$$d(B, C) + d(B, D) = 2d(A, B) + d(C, D). (11)$$

Therefore,

$$d(A, B) = (d(B, C) + d(B, D) -$$
$$d(C, D))(\tfrac{1}{2}). \qquad (12)$$

Equation (12) gives the difference between a node, B, and the optimal HTU through which B may be connected to two other nodes, C and D. It will be useful to think of the connection between C and D as an object called *interval*, $INT(C, D)$, whose difference from a node, B, may be calculated as

$$d(B, INT(C, D)) = (d(B, C) + d(B, D) -$$
$$d(C, D))(\tfrac{1}{2}). \qquad (13)$$

PRIM NETWORKS

We now consider a number of algorithms for calculating approximations to Wagner Networks. It is important to realize that the current level of development of algorithms for finding Wagner Networks is comparable to the state of methods of finding Camin-Sokal Trees prior to the work of Estabrook (1968): several reliable heuristic programs exist, but no existing algorithm is certain to produce a Wagner Network for an arbitrary set of data.

We shall first treat Prim Networks. A Prim Network, for our purposes, is a Wagner Network, subject to the constraint that the set of nodes, N, is identical to the set of OTUs, S. Thus, no HTUs are constructed. Prim Networks are so called because they were introduced into evolutionary taxonomy by Edwards and Cavalli-Sforza (1963), who named them with reference to the work of Prim (1957).

Prim Networks are quite crude approximations to Wagner Networks, but have the advantages that they can be computed exactly and very efficiently. Prim Networks may be most useful in evolutionary studies as tools for preliminary analyses of data, the more elaborate programs described below being reserved for refining final conclusions. A Prim Network can provide a fairly accurate picture of a Wagner Network. Fig. 3(i) depicts a Wagner Tree for the anuran data of Kluge and Farris (1969), while Fig. 3(ii) shows a tree produced from a Prim Network for the same data. In this instance, the Prim and Wagner Networks indicate virtually identical relationships and differ in length by only 1 unit.

A Prim Network may be computed as follows:

1) Pick an OTU, say Q, as a starting point. It does not matter which OTU is used. Go to 2.

2) Find the OTU in S that is closest to Q. Connect it to Q to form a network with one linkage. Go to 3.

3) Compute the difference between each unplaced OTU and the network. The difference between an OTU, A, and a network, $W = (N, g)$, with nodes in N is defined to be $\min_{n \text{ in } N}(d(A, n))$. Go to 4.

4) Find the OTU that is closest to the network. Add it to the network by connect-

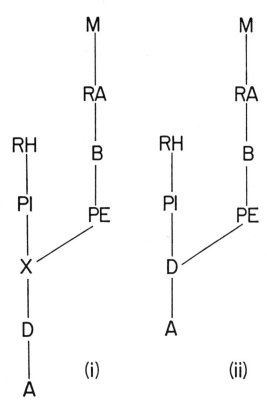

FIG. 3.—Evolutionary trees for families of anurans, produced by the Wagner Method (i), and from a Prim Network (ii). Legend: A: Ascaphidae, D: Discoglossidae, X: a hypothetical intermediate, PI: Pipidae, RH: Rhinophryhidae, PE: Pelobatidae, B: Bufonoid complex, RA: Ranoid complex, M: Microhylidae. (cf. Kluge and Farris, 1969).

ing it to the node from which it differs least. Go to 5.

5) If any OTUs remain unplaced, go to 3. Otherwise, stop.

A FORTRAN IV program to perform this algorithm is short enough to be profitably described here.

```
      DO 101 I = 2, IT
      LIN(I) = I
      JB(I) = 1
      DB(I) = 0.
      DO 101 J = 1, N
101   DB(I) = DB(I) + ABS(X(J, I) - X(J, 1))
      ITO = IT - 1
      DO 102 IP = 2, ITO
      D = DB(LIN(IP))
```

```
        IB = IP
        K = IP + 1
        DO 103 I = K, IT
        IF(DB(LIN(I)). GE. D) GOTO 103
        D = DB(LIN(I))
        IB = I
103     CONTINUE
        JEXT = LIN(IB)
        LIN(IB) = LIN(IP)
        LIN(IP) = JEXT
        DO 104 I = K, IT
        D = 0.
        L = LIN(I)
        DO 105 J = 1, N
105     D = D + ABS(X(J, L) – X(J, JEXT))
        IF(D. GE. DB(L)) GOTO 104
        DB(L) = D
        JB(L) = JEXT
104     CONTINUE
102     CONTINUE
```

The program operates as follows. N is the number of characters; IT is the number of OTUs; and X(J, I) is the data matrix. The *first* subscript of X indexes characters, the second, OTUs. OTU 1 is chosen as the starting point and the list, LIN, is loaded with the numbers of the unplaced OTUs (all the OTUs except 1). Each unplaced OTU is assigned a difference, DB(I), from the network, and DB(I) is initialized to the difference between OTU I and OTU 1. Each OTU, I, is assigned a closest OTU, JB(I), on the network. JB(I) is initialized to 1 for each OTU other than 1. These initializations are all performed by the 101 loop.

In the 103 loop the unplaced OTU that is closest to the network (has the smallest DB value) is located. The list LIN is then updated so that the numbers of the unplaced OTUs are stored in positions K, . . . , IT of LIN. The OTU selected in loop 103 is called JEXT, and the number of JEXT is saved in position IP of LIN, JEXT being the IPth OTU added to the network. The reason for saving the number of JEXT in LIN is that it is convenient to have the connection function printed out in the order in which the OTUs are placed on the network.

In the 104 loop the difference between each unplaced OTU, L, and JEXT is computed as D. If D exceeds DB(L) no action

is taken. If D is smaller than DB(L), DB (L) and JB(L) are updated.

At each stage of the computation, for any unplaced OTU, I, DB(I) is the difference between I and the network, while JB(I) is the number of the network node closest to I. At the end of execution, JB(I) is the connection function of the Prim Network, and DB(I) is the length of the internode between I and JB(I).

The algorithm presented here has some advantages over other ways of computing Prim Networks. The program above computes the difference between OTUs I and J, I < J, just once and either saves the value in DB or discards it. An OTU-by-OTU matrix of differences is not stored, so that Prim Networks for quite large sets of data can be found on relatively small computers. Further, inspection of the program will reveal that it produces a network for IT OTUs by performing (IT-3) (IT-2) comparisons of differences. We may contrast this amount of work with that required by a Prim Network forming procedure described by Edwards and Cavalli-Sforza (1963). "[The Prim Network] can be found by listing all the distances between points in increasing order, and successively allocating segments to these distances, omitting any segment which completes a loop." The minimum number of comparisons needed to order the distances between IT OTUs is set by information theory at $(\frac{1}{2})(IT)$ $(IT-1)(\log_2((\frac{1}{2})(IT)(IT-1)))$. The program presented here can thus be made more computationally efficient than other programs for computing Prim Networks, because it forms the networks by doing less work.

A working version of the FORTRAN IV Prim Network Program is available from the author, as is an IBM 360/67 Assembler Language routine using a similar algorithm.

THE WAGNER METHOD

The "Wagner Method" presented here is a computerized version of the original Wagner procedure for producing trees (Wagner, 1961, and references therein). This method has been employed by Kluge and Farris (1969).

In the Wagner Method, the tree is formed by adding OTUs one at a time to a tree that initially consists of a single node—the ancestor. The ancestor may be "hypothetical" in that it is not an existing OTU. The "hypothetical" ancestor is, however, treated as an OTU rather than an HTU in that the character states of the ancestor are fixed, not computed by the algorithm.

The order in which OTUs are added to the tree is determined by the rank order of the *advancement index*. For OTU I, the advancement index is defined to be $d(I, A)$, where A is the ancestor. OTUs with small advancement indices are added to the tree first. At each stage, the placement of the next OTU to be added is determined through the interval distance formula, and a new HTU to connect an OTU to the network is formed using the median-state property.

The algorithm of the Wagner Method is then:

1) Select an ancestor, A. Go to 2.

2) Compute the advancement index, AD $(I) = d(I, A)$ for each OTU, I. Go to 3.

3) Find the OTU with smallest advancement index. Connect it to the ancestor to form a tree with one linkage (one interval). Go to 4.

4) Find the unplaced OTU, B, with smallest advancement index. Go to 5.

5) Find the interval (linkage), INT(C, $f(C)$), for C a node of the tree such that $d(B, INT(C, f(C)))$ is minimal. Go to 6.

6) Construct an HTU, Y, as the median of B, C, and $f(C)$. Go to 7.

7) Update the ancestor function:
$$f(Y) = f(C)$$
$$f(B) = Y$$
$$f(C) = Y.$$
Go to 8.

8) If any OTUs remain unplaced, go to 4. Otherwise stop.

The simple Wagner algorithm can be modified by changing the criterion for the order in which the OTUs are added to the

tree. I have experimented with programs in which the addition sequence is given by

1) scanning the tree by the interval-distance formula at each step to determine which unplaced OTU is closest to the tree (as opposed to being closest to the ancestor),

2) adding OTUs to a Wagner Tree with ancestor A in the same order as they would be added to a Prim Network with base element A, and

3) adding OTUs with large advancement indices first.

None of these modifications has shown particular superiority to the original algorithm in forming ordinary Wagner Trees. Methods (2) and (3) have proved to be superior to the original algorithm in applications where Wagner programs are used in successive weighting procedures (see Farris, 1969).

The ancestor in the simple Wagner algorithm can be used to impart a direction to the tree, but it need not be interpreted in this way. If a real OTU is used as the "ancestor" in the Wagner algorithm, the output can be used as a Wagner Network.

ROOTLESS WAGNER METHODS

The simple Wagner algorithm does not impose irreversibility on the trees it produces, and from this standpoint, the choice of an ancestor may not be crucial. It has been found, however, that the form of the tree produced by a simple Wagner algorithm can be changed by altering the ancestor. The Rootless Wagner algorithms have been developed as an approach to reducing the dependency of the tree on the inferred ancestor.

A Rootless Wagner algorithm is produced from the simple algorithm described above by modifying the first 4 steps:

1) Select a pair of OTUs, A and D such that $d(A, D) \geqslant d(I, J)$ for any OTUs, I and J, in the study. Link them to form an interval. Go to 2.

2) Compute the "advancement index" of each OTU, I, as $d(I, INT(A, D))$. Go to 3.

3) Find the OTU, $C \neq A, D$, with the

largest "advancement index." Form an HTU, Y, as the median of A, B, and C. Construct an ancestor function:

$$f(A) = Y$$
$$f(D) = Y$$
$$f(C) = Y.$$

Go to 4.

4) Find the unplaced OTU, B, with the largest "advancement index." Go to 5.

As with rooted Wagner procedures, a number of Rootless Wagner algorithms can be generated by changing the criterion for the addition sequence of OTUs. Several addition criteria have been tried, but none seem to be appreciably more effective than the one given above. The criterion described can operate effectively in successive weighting applications.

The Rootless Wagner algorithm has the advantage that it can produce a Wagner Network with no reference to an ancestor at all. The algorithm can be used, however, to produce a directed Wagner Tree, simply by including a postulated ancestor among the OTUs to be analyzed. Hypothetical ancestor "OTUs" are given no special treatment by a Rootless Wagner procedure. It is still possible for the form of the tree to be influenced by the inclusion of an ancestor.

OPTIMIZATION OF TREES

A drawback of the algorithms described above is that the HTUs produced during the stepwise addition of OTUs to the tree may not be the optimal ones for the complete tree. The algorithms in which intervals are created by choosing the most disparate available OTUs are particularly subject to this difficulty. Suppose, for example, that the initial pair, A and D, of OTUs selected by the Rootless Wagner procedure have quite similar states, $X(A, i)$ and $X(D, i)$ for the ith character; i.e., $| X(A, i) - X(D, i) |$ is a relatively small number. There cannot, of course, be many such characters, since A and D were selected for the large value of $d(A, D)$ (cf. Eq. (1)). The presence of a few such characters is nonetheless possible. Now any HTU,

Y, generated by the existing Rootless Wagner algorithms as a connector of some OTU to $INT(A, D)$ will have for character i a state, $X(Y, i)$, that lies numerically between $X(A, i)$ and $X(D, i)$, inclusive. It is quite probable that several OTUs, C, E, and F, say, will be independently connected to INT(A, D), say through HTUs, R, Y, Z, respectively. Then $X(R, i)$, $X(Y, i)$ and $X(Z, i)$ will all lie numerically between $X(A, i)$ and $X(D, i)$. Now suppose that C, E, and F are described by a state, $x = X(C, i)$, that does not lie between $X(A, i)$ and $X(D, i)$. Then the output of the Rootless Wagner procedure will effectively assert that state x is convergently present in OTUs C, E, and F, while the similarity of $X(A, i)$ and $X(D, i)$ is homologous. A more parsimonious interpretation is that state x is homologously present in C, E, and F, while the similarity of $X(A, i)$ and $X(D, i)$ is convergent. The latter interpretation can be imposed without changing the form of the Wagner Network. It is necessary only to alter the character state values of HTUs R, Y, Z.

A tree with a fixed branching form, that is, a fixed set of cladistic relationships, can be optimized for the parsimony criterion by computing for it an optimizing set of HTUs. I shall describe a procedure for doing so below. In the description, I shall utilize the concept of the *state set*, $S(Y, i)$ of an HTU, Y, for a character, i. The state set is a closed interval describing a range of character state values applicable to Y and i. The term *cladistic difference* is used in the sense of Farris (1967). In particular, two nodes are said to have *cladistic difference unity* if they share an immediate common ancestor.

The procedure is most conveniently described in sections. The first is the clustering procedure, as follows. Consider a set K of clusters to have initially as its elements just all the OTUs in the study, each OTU being considered as a "cluster" of one OTU. Then:

1. Select any two clusters, A and B, that have cladistic difference unity according to the tree being optimized.

2. Remove A and B from K.

3. Place $Y = f(A) = f(B)$ in K.

4. For every character, i, compute $S(Y, i)$ as described below.

5. If K has more than one element, return to (1).

The rules for computing new state sets (step (4)) are:

If clusters A and B in K are united to form $Y = f(A) = f(B)$ then $S(Y, i)$ is

R-1: the intersection of $S(A, i)$ and $S(B, i)$, *provided* that intersection is not empty; otherwise, $S(Y, i)$ is

R-2: the smallest closed interval of one of the forms $[a_i, b_i]$ or $[b_i, a_i]$, where a_i is an element of $S(A, i)$, and b_i is an element of $S(B, i)$.

The state sets, $S(Z, i)$, of HTUs, Z, will generally not be singleton, so that the character states, $X(Z, i)$ of Z, will not generally be unique. The ambiguity of the states assigned to HTUs is reduced to a minimum by

R-3: if an HTU, F, with non-singleton state set, $S(F, i)$, has $f(F) = H$, replace $S(F, i)$ with the intersection of $S(F, i)$ and $S(H, i)$.

R-3 is most easily applied through a second pass through the tree after the clustering cycle that computes state sets of HTUs has been completed.

An example of the optimizing process may prove helpful. We use the hypothetical data:

OTU	Character 1	2	3	4	5
A	2	1	0	0	1
B	1	2	0	0	0
C	0	0	1	2	0
D	0	0	2	1	0
E	0	0	0	0	1
F	0	0	0	0	0

We shall optimize the tree of Fig. 4(i) for these data. For the example, I shall indicate state sets by expressions of the form $[x, y]$, where x and y are the bounds of the range of state values in a set $S(A, i)$.

The initial cluster set, K, contains A, B, C, D, and E. Clustering according to Fig. 4, we first unite A and B to form Y. By R-2

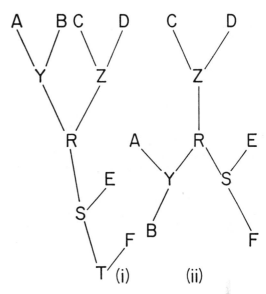

FIG. 4.—A tree (i) and a network (ii) to be optimized.

we assign to the state sets of Y the ranges of variation between A and B. Then Y has the state sets $[1, 2]$, $[1, 2]$, $[0, 0]$, $[0, 0]$, and $[0, 1]$ for the five characters respectively. Similarly, Z has state sets $[0, 0]$, $[0, 0]$, $[1, 2]$, $[1, 2]$, and $[0, 0]$. For the first four characters, the state sets of Y and Z have empty intersections, so that the corresponding state sets of R are computed according to R-2. For character five, the state sets of Y and Z have a non-empty intersection, so that R-1 applies. R then has state sets $[0, 1]$, $[0, 1]$, $[0, 1]$, $[0, 1]$, $[0, 0]$. Similarly, the first four state sets of S can be computed as the intersections of the state sets of R and E, while the last state sets of S is the range between the "1" state of E and the "0" state of R. S has state sets $[0, 0]$, $[0, 0]$, $[0, 0]$, $[0, 0]$, and $[0, 1]$. The intersection of the state sets of S and F is non-empty for all characters. Thus, T has state sets $[0, 0]$, $[0, 0]$, $[0, 0]$, $[0, 0]$, and $[0, 0]$.

Making the second pass through the tree, we replace the state sets of Y with their intersections with the state sets of R, obtaining new state sets $[1, 1]$, $[1, 1]$, $[0, 0]$, $[0, 0]$, and $[0, 0]$ for Y, and proceed similarly for the other HTUs. The character states of the HTUs are in the case uniquely determined and are

	Character				
HTU	1	2	3	4	5
Y	1	1	0	0	0
Z	0	0	1	1	0
R	0	0	0	0	0
S	0	0	0	0	0
T	0	0	0	0	0

Note that the optimized lengths of INT (R, S), INT(S, T), and INT(T, F) are zero.

The same procedure as described above can be used also to optimize the HTUs of approximated Wagner Networks simply by imposing an arbitrary direction on the connection function of the network. The tree example just performed is equivalent to optimizing the network of Fig. 4(ii).

DISCUSSION

The HTU optimizing procedure can be used to increase the parsimony of trees and networks generated by Wagner programs. It can also be used, however, to assign an optimal set of HTUs to the dendrogram generated by any sort of numerical taxonomic clustering procedure.

Since any dendrogram can now be assigned a parsimony-optimal set of HTUs, any dendrogram can also be assigned a length, and, consequently, a measure of parsimony. It is therefore practical to use parsimony as an optimality criterion on any sort of dendrogram, including strictly phenetic ones. The phenetic desirability of doing so is, of course, open to doubt and will have to be extensively investigated. It does seem possible, however, that parsimony will provide a general optimality criterion that lacks some of the less appealing qualities of the cophenetic correlation coefficient, now widely used as an optimality measure on phenograms (see Farris, 1969a).

The possibility of calculating an optimal set of HTUs for a tree of any branching form also allows us to compare the relative parsimony of intuitively derived and numer-

ically derived hypotheses on phylogeny. The relative degree of the likelihood of alternative phylogenetic theories can thus be assessed, as can the relative efficiency of intuitive and numerical techniques.

Finally, since optimal HTUs can be calculated after a tree has been formed, we are freed of the necessity to compute trees by the sort of stepwise procedure used in the Wagner programs. We may be able to reduce the amount of computer time necessary to obtain taxonomic conclusions by using conventional clustering methods to obtain the form of the tree. HTUs could then be computed separately. The type of clustering scheme used will need to be carefully selected, however. Most phenetic clustering methods generate branching forms with an unacceptably low degree of parsimony. I am currently investigating clustering criteria to allow clustering programs to generate branching forms optimizable to acceptably small length.

REFERENCES

CAMIN, J. H., AND R. R. SOKAL. 1965. A method for deducing branching sequences in phylogeny. Evolution, 19:311–327.

EDWARDS, A. W. F., AND L. L. CAVALLI-SFORZA. 1963. Reconstruction of evolutionary trees. In Phenetic and phylogenetic classification. Systematics Assoc. Publ., 6:67–76.

ESTABROOK, G. F. 1968. A general solution in partial orders for the Camin-Sokal model in phylogeny. J. Theoret. Biol., 21:421–438.

FARRIS, J. S. 1969. A successive approximations approach to character weighting. Syst. Zool., 18:374–385.

FARRIS, J. S. 1969a. On the cophenetic correlation coefficient. Syst. Zool., 18:279–285.

KLUGE, A. G., AND J. S. FARRIS. 1969. Quantitative phyletics and the evolution of anurans. Syst. Zool., 18:1–32.

PRIM, R. C. 1957. Shortest connection networks and some generalizations. Bell Syst. Tech. J., 36:1389–1401.

WAGNER, W. H. 1961. Problems in the classification of ferns, p. 841–844. In Recent advances in botany. Univ. Toronto Press, Toronto.

Biological Sciences, State University of New York, Stony Brook, New York 11790.

Contribution No. 12 from the program in Ecology and Evolution, State University of New York at Stony Brook, Stony Brook, New York 11790.

13

THE PHYLOGENY OF GUTIERREZIA: AN ECLECTIC APPROACH

OTTO T. SOLBRIG

Solbrig, O. T. (Gray Herbarium, Harvard University, Cambridge, Mass.). The phylogeny of *Gutierrezia*: an eclectic approach. Brittonia **22**: 217–229. 1970.—One approach to the problem of deducing the genealogy of a set of organisms is to propose several hypotheses using different procedures, based on different evolutionary inferences. Such an approach was followed here, and four different phylogenies were constructed, three of them computer built. Consistency with independently obtained phenetic, cytological, and phytogeographical data was used to select the most probable phylogenetic tree among the four. It is shown that the most probable tree is one constructed under the assumption that character states found close to the mean and/or modal values for the genus are primitive. It is also shown that certain phylogenetic conclusions obtain in all four phylogenies.

A phylogeny[1] is an hypothesis that states the probable evolutionary history of a group of organisms. Such schemes are often developed as two-dimensional representations of the many dimensions of evolution. In constructing a phylogeny, one finds that certain constraints are therefore imposed. In the past, phylogenies have been established often on purely subjective criteria, born from the investigator's intimate knowledge of the plants and without a clear enunciation of the premises and methods employed. The large degree of subjectivity and personal bias involved in such a procedure has been criticized repeatedly (cf. Kluge, 1969). Attention to the procedures by which phylogenies are constructed and employment of more rigorous numerical methods allow us, however, to avoid some of the criticisms that have been leveled against phylogenies and against taxonomic procedures in general (Sokal & Sneath, 1963). "Proof" of a phylogeny is still its consistency with paleobotanical findings, phytogeographical evidence, genetic affinities, and phenetic similarities of the species. These criteria, although not truly unrelated, can be established more or less independently of each other. All these techniques and approaches promise more objective and realistic phylogenies than were possible in the past.

The object of the present paper is to formulate several possible phylogenies of *Gutierrezia,* a genus of Compositae-Astereae. To produce these various schemes, slightly different techniques and different initial assumptions were used. The various phylogenies were then compared with cytological, morphological, and phytogeographical evidence to determine which scheme seems the most probable. This paper is intended as a contribution to the understanding of the evolution of *Gutierrezia* and not as a contribution to numerical techniques, which have been borrowed from others and simply applied here. Detailed discussion of the techniques used by me, and the evolutionary inferences behind them have been provided by Kluge & Farris (1969).

---

[1] The term "phylogeny" lends itself to confusion since it is used to represent both the actual evolutionary history of a taxon (that can never be truly and completely ascertained) and the inferred history from available data (paleontological, phytogeographical, morphological, etc.). It is used in this second sense in this paper.

TABLE I

MAJOR CHARACTERISTICS OF *Gutierrezia* SPECIES

Species	Plant Height (cm)	Habit	Leaf			Involucre		Flower			Ligule length	Pappus length	No. Chr.
			Length	Width	Shape	H	Ratio	No. L.	No. T.	Color			
sarothrae (0)	25(0)	P(0) Gl(0)	40(0)	2.0(0)	linear (0)	6.0(0)	0.5(0)	5(0)	4(0)	Y (0)	2.5(0)	1.0(0)	4, 8
serotina (1)	25(0)	P(0) Gl(0)	20(0)	1.0(0)	" (0)	3.5(1)	0.9(0)	7(0)	8(0)	Y (0)	2.5(0)	1.4(0)	4
bracteata (1)	60(0)	P(0) SG(½)	40(0)	1.0(0)	" (0)	6.0(0)	0.4(½)	6(0)	8(0)	Y (0)	3.0(0)	1.5(0)	8, 12
californica (2)	40(0)	P(0) Op(1)	20(0)	2.0(0)	" (0)	8.0(1)	0.5(0)	7(0)	12(0)	Y (0)	3.0(0)	1.0(0)	12
microcephala (4)	60(0)	P(0) Gl(0)	35(0)	3.0(0)	" (0)	3.5(1)	0.3(1)	1(1)	1(1)	Y (0)	2.5(0)	2.0(0)	8, 16
grandis (4)	75(½)	P(0) SG(½)	50(1)	4.0(1)	" (0)	7.0(1)	0.6(0)	7(0)	5(0)	Y (0)	2.5(0)	1.0(0)	—
texana (3½)	50(0)	A(1) Gl(0)	40(0)	3.0(0)	linear to lin. lanceo. (0)	3.5(1)	1.1(1)	12(½)	10(0)	Y (0)	2.5(0)	1.0(0)	4
glutinosa (4½)	50(0)	A(1) Gl(0)	25(0)	3.0(0)	" (0)	4.0(1)	1.5(1)	12(½)	30(1)	Y (0)	2.5(0)	1.0(0)	4
mandonii ssp. *mandonii* (2½)	15(1)	P(0) Gl(0)	15(½)	1.5(0)	linear (0)	4.5(1)	0.7(0)	9(0)	10(0)	Y (0)	3.0(0)	1.0(0)	12
mandonii ssp. *iserni* (½)	50(0)	P(0) Gl(0)	25(0)	2.0(0)	" (0)	5.0(0)	0.7(0)	9(0)	10(0)	YW(½)	2.5(0)	1.0(0)	12
mandonii ssp. *gillesii* (1)	50(0)	P(0) Gl(0)	20(0)	1.5(0)	" (0)	5.0(0)	0.7(0)	9(0)	10(0)	W(1)	3.0(0)	1.0(0)	20
spathulata (1½)	25(0)	P(0) Gl(0)	20(0)	5.5(1)	linear- lanceo. (0)	5.0(0)	0.8(0)	7(0)	10(0)	YW(½)	3.0(0)	2.0(0)	—
ameghinoi (3)	25(0)	P(0) SG(½)	15(½)	2.0(0)	linear (0)	6.0(0)	1.0(1)	8(0)	13(0)	Y (0)	3.0(0)	2.5(1)	—
espinosae (3½)	50(0)	P(0) SG(½)	10(1)	3.5(1)	linear- lanceo. (0)	5.5(0)	0.7(0)	8(0)	13(0)	Y (0)	4.0(1)	1.5(0)	—
tallalensis (4)	80(1)	P(0) Op(1)	30(0)	3.5(1)	" (0)	5.5(0)	0.8(0)	8(0)	13(0)	YW(0)	3.5(1)	1.5(0)	—
resinosa (5)	100(1)	P(0) Op(1)	30(0)	1.5(0)	linear (0)	7.0(1)	0.5(0)	7(0)	9(0)	YW(½)	2.0(½)	2.5(1)	28
baccharoides (7)	5(1)	P(0) Cu(1)	10(1)	3.5(1)	lanceo. (1)	5.5(0)	1.0(1)	8(0)	10(0)	Y (0)	3.5(1)	1.0(0)	—
neaeana (4½)	35(0)	P(0) SC(½)	17(½)	2.5(½)	lanceo. spath. (1)	6.0(0)	1.0(1)	10(0)	12(0)	Y (0)	5.0(1)	1.5(0)	—
gayana (7½)	75(½)	P(0) Gl(0)	40(0)	1.5(0)	linear (0)	10 (1)	1.5(1)	15(1)	45(1)	W(1)	6.0(1)	.5(1)	16
repens (8)	10(1)	P(0) Cr(1)	25(0)	4.0(1)	lanceo. (1)	6.0(0)	1.1(1)	14(1)	22(1)	Y (0)	5.0(0)	1.5(0)	—
ruiz-lealii (7½)	15(1)	P(0) Cu(1)	25(0)	3.5(1)	" (1)	6.0(0)	1.0(1)	12(½)	16(0)	Y (0)	6.0(1)	2.5(1)	16

P = perennial; A = annual; Gl = globose; SG = Semi-globose; Op = Open; Cu = cushion-plant; Cr = creeping; Y = yellow; W = white; YW = yellow and white pop. Numbers in parentheses are values used for the Ground-plan divergence index. Ratio = W/H.

TABLE II

MATRIX OF PHENETIC DIFFERENCES BETWEEN SPECIES IN THE PRIM NETWORK

	1	2	3	4	5	6	7	8	9	10	11	12	13	14	15	16	17	18	19	20	21
ameghinoi 1	x	14.4	16.9	22.5	19.9	13.1	24.5	21.1	14.9	11.2	30.7	9.1	9.0	30.3	4.3	26.4	16.3	5.7	24.4	21.0	22.4
baccharoides 2		x	31.8	37.4	5.0	28.0	39.4	36.0	29.3	26.1	45.6	5.8	15.3	15.4	10.6	41.3	31.2	20.6	9.5	35.8	31.2
bracteata 3			x	5.6	36.0	7.6	17.0	4.2	9.4	5.7	13.8	26.0	25.9	46.4	21.2	9.5	10.8	11.2	40.5	13.5	14.9
californica 4				x	42.4	13.2	24.6	9.8	15.0	11.3	19.4	31.6	31.5	52.8	26.8	15.1	16.4	16.8	46.9	21.1	22.5
espinosae 5					x	33.0	44.4	41.0	34.8	31.1	50.6	10.8	20.3	10.4	15.6	46.3	36.2	25.6	4.5	40.9	42.3
gillesii 6						x	15.2	11.8	5.6	1.9	21.4	22.2	22.1	43.4	17.4	17.1	7.0	7.4	37.5	11.7	13.1
glutinosa 7							x	23.2	9.6	13.3	32.8	33.6	33.5	54.8	28.8	28.5	18.4	18.8	48.9	3.5	17.1
grandis 8								x	13.6	9.9	9.6	30.2	30.1	51.4	25.4	5.3	15.0	15.4	45.5	19.7	21.1
iserni 9									x	3.7	23.2	24.0	23.9	45.2	19.2	18.9	8.8	9.2	39.3	6.1	7.5
mandonii 10										x	19.5	23.0	20.2	41.5	15.5	15.2	5.1	5.5	35.6	9.8	11.2
microcephala 11											x	39.8	39.7	61.0	35.0	4.3	24.6	25.0	55.1	29.3	30.7
neaeana 12												x	9.5	21.2	4.8	35.5	25.4	14.8	15.3	30.1	31.5
repens 13													x	30.7	4.6	35.4	25.3	14.7	24.8	30.0	31.4
resinosa 14														x	26.0	56.7	46.6	36.0	5.9	51.3	52.7
ruiz-lealii 15															x	30.7	20.6	10.0	20.1	25.3	26.7
sarothrae 16																x	20.3	20.7	50.8	25.0	26.4
serotina 17																	x	10.6	40.7	14.9	16.3
spathulata 18																		x	30.1	15.3	16.7
taltalensis 19																			x	45.4	46.8
texana 20																				x	13.6
gayana 21																					x

For the last ten years the genus *Gutierrezia* has been studied in the field and laboratory by means of a multidimensional approach, involving chromosome investigations, biometrical analysis of wild populations, hybridization attempts, morphological, anatomical, and genecological studies, transplant experiments, and taximetric analysis. Details of many of these studies, as well as keys and descriptions, have been published elsewhere (Solbrig, 1960, 1964, 1965, 1966; Rüdenberg & Solbrig, 1963) and will not be elaborated here. Attention will be focused instead on the construction of a phylogeny of *Gutierrezia*.

METHODS

The general procedure followed in the investigation of *Gutierrezia* has been as follows. On the basis of field observations, analysis of the external morphology, and biosystematic considerations, 19 species and three subspecies have been recognized (Solbrig, 1960, 1966). These constitute the OTU's (Operational Taxonomic Units, fide Sokal & Sneath, 1963), the relationships of which are being investigated. Each of these is assumed to be monophyletic for the purpose of the present study. In reality, each species is formed by a number of populations that are not necessarily related in a linear fashion (Solbrig, 1964, 1965). The characters used are modal values for the species, obtained after measuring over 1000 herbarium specimens. The set of characters comprises all the independent features that were found to vary in this group of very similar species. The mean value for numerical characters and the modal value for meristic traits were used in each case. This procedure does not allow one to evaluate the validity of the OTU's. However, for the purpose of the present analysis, their delineation is assumed to be correct since we are trying to assess the historical relationships of the various species within the genus rather than the validity of each species. Also, the question of the status of the genus *Gutierrezia*, as understood by me, is not posed.

Any numerical taxonomic analysis consists of two parts: (1) obtaining a similarity coefficient, e.g., a generalized distance, a coefficient of association or a coefficient of correlation; and (2) the linking together of the OTU's in a two-dimensional graph on the basis of the similarities coefficient obtained in (1). In the present study the so-called Prim Network technique has been used in solving both aspects of the numerical taxonomical problem. The Prim Network technique was developed in 1957 by R. C. Prim in order to determine the shortest possible network of direct links between a given set of telephone terminals. His technique can be applied as well to problems in systematic biology.

As stated above, two major steps are involved in the computational procedure. The first consists of setting up a "distance table" (an $n \times n$ taxa matrix) which gives the distance between all the taxa (Table II). The distance measure (D) used here is the sum over all characters of the absolute value of the difference between the character states in taxon A and B.

$$D(A, B) = \sum_i |X(A, i) - X(B, i)| \qquad (1)$$

where $X(A, i)$ denotes the state of character "i" for OTU A.

The second step is to connect the various taxa in such a way as to produce the shortest total distance. The detailed instructions for doing this are given by Prim (1957). Basically, the procedure consists of choosing the shortest links connecting any two species, and then by a process of elimination, adding more links until all species are connected in a network. This network can have any form, from completely linear—that is, all species can be arranged in a straight line—to perfectly radial—that is, all species are connected to a central one. The Prim network was used as an aid in assessing phenetic relationships.

Phylogenetic relationships between OTU's have been assessed using the Groundplan-Divergence Index of Wagner (1961, 1966); see also Kluge & Farris, 1969). To use the Groundplan method, one must find the most primitive state of each character. Accordingly, characters that occur commonly in the genus are considered primitive. The situation in related genera is also taken into consideration in assessing a character, whether advanced or primitive. Wagner's method is a simplified procedure for providing the most parsimonious tree in the sense of Camin & Sokal (1965). The objective is to form a network, or tree, by connecting all original OTU's and realize in the process a minimum number of changes ("steps," fide Camin & Sokal, 1965) on the tree. This is a network of minimum length in the space in which length is defined as the sum of the lengths of all intervals of the tree, an interval being the connection between a given OTU and its most recent tentative ancestor. The difference, as defined by a given equation, between a given OTU and its most recent tentative ancestor, is the length of the interval.

The most parsimonious tree usually incorporates one or more hypothetical intermediates. These are artificial OTU's used as branching points in a tree. Their purpose is to minimize the length of the tree, and their character states are chosen to achieve that end (Kluge & Farris, 1969).

The Wagner method can be computed using two procedures. One is a manual, qualitative procedure in which characters that are considered primitive (using the criteria enunciated above) are assigned a value of 0, and every advanced character is assigned a value of 1 (and intermediate values for intermediate character states). In this way, an index is obtained for each species (Table I). Species are then arranged in a "target" with the distance from the center given by their respective indices. Connections between species, as well as hypothetical ancestors, are determined by the Groundplan, so that each taxon (real or hypothetical) is separated by one character state difference.

The second procedure is the entirely numerical and computerized method of J. Farris. In this method an ancestor is given, and the OTU A that has the smallest interval length as defined by Equation (1) is chosen. From among the unplaced OTU's, the one that differs least from the ancestor (that is, has the smallest interval length), OTU B is chosen. An intermediate is then produced that connects OTU's A and B with the ancestor in such a way that the resulting tree is the most parsimonious one (for details of the algorithm and the computational procedure, cf. Kluge & Farris, 1969). Both procedures were used in this study.

RESULTS

The procedure used is as follows: the phenetic, cytological, and phytogeographical relationship of the various species were assessed independently. All of the data were then combined in constructing several phylogenies. The concordance of the phylogeny with the relationships arrived at on the basis of the independent analysis of the phenetic, cytological, and phytogeographical criteria was used to determine which phylogeny was most probably correct.

ASSESSMENT OF PHENETIC AFFINITIES

Phenetic affinities can be assessed in different ways with varying degrees of subjectivity. One method is the subjective grouping of the species of a genus into sections, and within each section the informal grouping of species on the basis of certain "key" characters. The assumption is, of course, that species within a group are more similar phenetically to each other than to species in other groups, species in a section being phenetically more similar to each other than to species in other sections, and so on. It may be argued that assessing phenetic relations on the basis of one or a very few characters is poor practice. Because taxonomists usually base their classifications on more data than they explicitly express in keys, a method such as the hierarchical taxonomic system is valid as a first approximation to the evaluation of phenetic affinities. I believe, however, that every effort should be made to quantify phenetic relations. The problem so far seems to be that the various methods and coefficients available (Sokal & Sneath, 1963) do not necessarily produce the same values for the degree of affinity between taxa when given the same data. In the present study we have used the so-called Prim Network (Kluge, 1969) (cf. Methods section).

Several observations can be made from Fig. 3, which represents the Prim Network. The first and most interesting is that all the South American species are phenetically more similar among themselves than they are to any North American species (shown in the figure by hatching); conversely the North American species are more similar to each other than to any of the South American species. Second, *Gutierrezia texana* and *G. glutinosa*, the two annual species, are more similar phenetically to each other than they are to any of the perennial species. Another observation that emerges is that although the Prim program is set up to draw up non-linear as well as linear relations, most of the similarities appear to be linear and additive with very little branching in the resulting diagram other than in the North-South American "interphase."

CYTOLOGICAL DATA

Two kinds of chromosome data can be considered to assess relationships: number and morphology.

Eleven of the 19 species have been studied cytologically. This group includes seven of the eight North American and four of the South American species. Altogether over 100 populations were investigated, varying from 59 populations of the North American species *G. sarothrae* to only one of the South American *G. gayana* and *G. ruiz-lealii*. All species investigated had a chromosome number of $n = 4$ or a multiple of 4. The highest number found was $n = 28$ (14-ploid); all multiples between 4 and 28 were found with the exception of the dodecaploid level ($n = 24$) (Table I).

A highly interesting observation is the presence of more than one level of ploidy in the same species: diploids ($n = 4$) and tetraploids ($n = 8$) in *G. sarothrae*; tetraploids ($n = 8$), hexaploids ($n = 12$), and octoploids ($n = 16$), in *G. microcephala*; hexaploids ($n = 12$), and decaploids ($n = 20$) in *G. mandonii*. When investigated in detail, these species showed that there was no absolute difference other than chromosome number by which different levels of ploidy within a species could be identified (Solbrig 1964, 1965). It also was discovered that the morphological variability both within and between populations, regardless of ploidal level, is very great.

The North American species of *Gutierrezia* have chromosome numbers that vary from $n = 4$ to $n = 16$, while the South American species, on the other hand, vary from $n = 12$ to $n = 28$. The South American species appear therefore to be higher polyploids in general than the North American species.

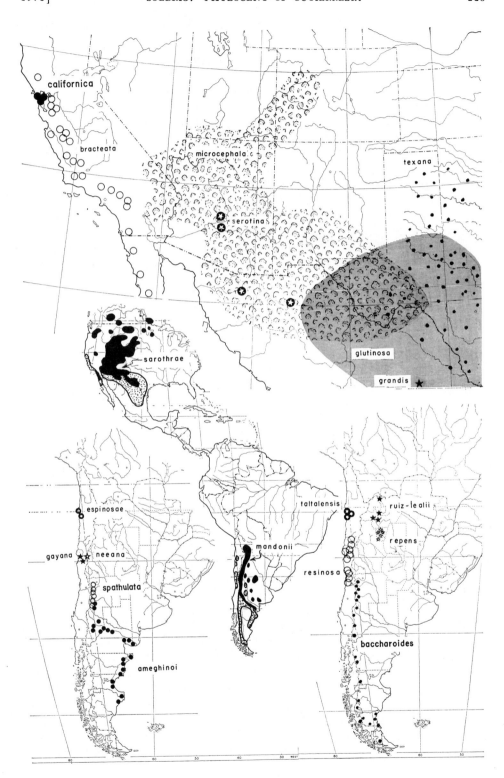

Fig. 1. Distribution of species of *Gutierrezia* in North and South America.

237

All perennial species analyzed, including those of both North and South America, have an identical basic karyotype (Rüdenberg & Solbrig, 1963), and the polyploids repeat this basic karyotype in their genome. This is evidence for an increase rather than a reduction in chromosome number, since otherwise it cannot be comprehended why a species with a high chromosome number would repeat a basic karyotype of four chromosomes.

The annual species show a karyotype that is clearly distinct from that of the perennials. The annuals are diploid, and the differences between the two basic karyotypes can be best interpreted as the result of inversions and translocations. It cannot be stated on cytological grounds whether the karyotype of the annual or that of the perennial species is primitive.

PHYTOGEOGRAPHICAL CRITERIA

The genus occupies a disjunct distribution with one set of species restricted to the dry and semi-dry areas of northern Mexico and the western United States, and the other set found in the lower Andean ranges from Bolivia to the Magellan Straits and surrounding areas (Map, Fig. 1). As pointed out earlier, the species within each of the sets are more similar to each other in their morphology than they are to species of the other group. The cytological data do not contradict this statement. Therefore, it appears likely that the species in one area have evolved from a single immigration from species from the other region. I believe that the direction of the interchange was from North America to South America. This statement is based on the following evidence: (1) the South American species are high polyploids, while the North American ones contain the diploids, and no species of the latter has more than 16 pairs of chromosomes. From the conclusions reached in the section on cytology, this would indicate that the North American species are more primitive; (2) the Astereae are very rich in genera and species in North America and relatively scarce in genera in South America. There are no genera closely related to *Gutierrezia* in South America, but several in North America. This observation reinforces the conclusion arrived at in point (1).

THE RECONSTRUCTION OF PHYLOGENY

The material presented so far seems to indicate that: (1) the North American and South American species form two independent evolutionary stocks; (2) that the annual species form an evolutionary line independent of the perennial ones; and (3) that no species is ancestral to one with a lower chromosome number than itself. The data by themselves (Solbrig, 1965) also allow certain other hypotheses. For example, *G. bracteata* and *G. californica* are closely related on the basis of their phenetic similarity, their same chromosome number, and their close geographical position. However, it is the object of this study to avoid such subjective judgments and to try to express the relationships in the form of a numerical value.

Four different phylogenetic trees were constructed. The first one, which is depicted in Fig. 2, was produced according to the manual procedure for the Groundplan-Divergence Index method of Wagner. In this first tree, South and North American species were arbitrarily considered as separate evolutionary lines. The North American species are further subdivided into two lineages: 1) that of the perennial species, with *G. sarothrae* as the ancestral form; and 2) that which includes the two annual species. The South American species form three phylogenetic branches: the first comprises the four Chilean species; the second, the two species

238

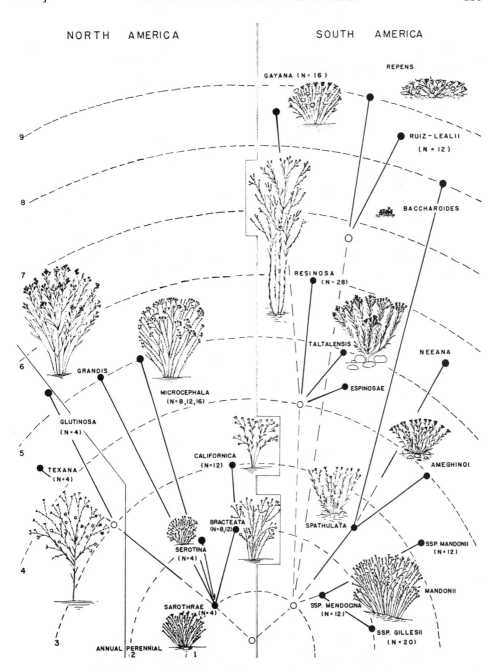

Fig. 2. Wagner "target" diagram indicating a probable phylogeny for *Gutierrezia* according to the non-computerized method (for further details see text). Habit of plants drawn to scale; broken lines indicate less well established relationships. Ssp. *mendocina* = ssp. *iserni*.

of northern Argentina, *G. repens* and *G. ruiz-lealii*; and the third, the line formed by *G. ameghinoi*, *G. neaeana*, and *G. baccharoides*. The three subspecies of *G. mandonii* are close to the ancestral form that gave rise to the South American species according to this first scheme.

239

TABLE III

CHROMOSOME NUMBERS IN *Gutierrezia*

	North America	South America
$n = 28$		*G. resinosa*
$n = 24$		
$n = 20$		*G. mandonii*
$n = 16$	*G. microcephala*	*G. gayana*
$n = 12$	*G. californica, G. bracteata, G. microcephala*	*G. mandonii, G. ruiz-lealii*
$n = 8$	*G. sarothrae G. bracteata, G. microcephala*	
$n = 4$	*G. sarothrae, G. serotina, G. texana, G. glutinosa*	

Three other phylogenies were produced using the method of Farris. In each one, different criteria of "primitiveness" were used. In the first computer run, the following set of characters were considered primitive: perennial, globose habit; lanceolate leaves; large heads with numerous flowers and a campanulate involucre; yellow ligules, 5 or more mm. long; a pappus of bristles. These postulates are a distillation of those of Bentham (1873), Cronquist (1955), and others, as to what the "primitive" characters are in the tribe Astereae of Compositae. No species of *Gutierrezia* fits all these characteristics, but *G. gayana* approximates them most closely, and was therefore designated as the ancestral species for the first run. The resulting phylogeny is depicted in Fig. 4. It can be seen that it does not accord with the phytogeographical criteria that were established (in this phylogeny a line of South American species gives rise to South and North American species), nor with the cytological assumptions (hexaploids and decaploids are more primitive than diploids). Furthermore, it splits the subspecies of *G. mandonii* into independent evolutionary lines. Finally, it seems to indicate the evolution of great diversity between *G. gayana* and *G. mandonii* with little speciation.

A second computer run was then made using entirely different postulates of primitiveness. These were arrived at by the author subjectively and are: perennial, globose habit; linear leaves; small heads (1–2 mm.) with few flowers (4 or 5) and a narrow, turbinate involucre; yellow, short ligules (1–2 mm.), and a pappus of awns. These characters are the prevailing ones among species of *Gutierrezia*, but are not very common among related genera. *Gutierrezia sarothrae* is the species that most closely approximates these characteristics, and consequently was designated as the ancestral species for the second computer run. The resulting phylogeny is depicted in Fig. 5. This does not agree very well with the cytological criteria either, and as with the first run, it indicates evolution of great diversity between certain species with little speciation. Figure 5 is similar to Fig. 2, the phylogeny having been arrived at by using Wagner's Groundplan Index, and in both cases similar, but not identical, concepts of primitiveness were used.

A third run was made using no extant species as primitive, but a hypothetical species with the following characters: medium sized heads (3–4 mm.) with 8–10 flowers and a medium sized involucre; yellow, short ligules (1–2 mm.), a pappus of awns; globose, perennial habit; four chromosomes, and native to North America. These characteristics of the hypothetical species were arrived at by choosing the modal character states for each of the characters studied. They are therefore a kind

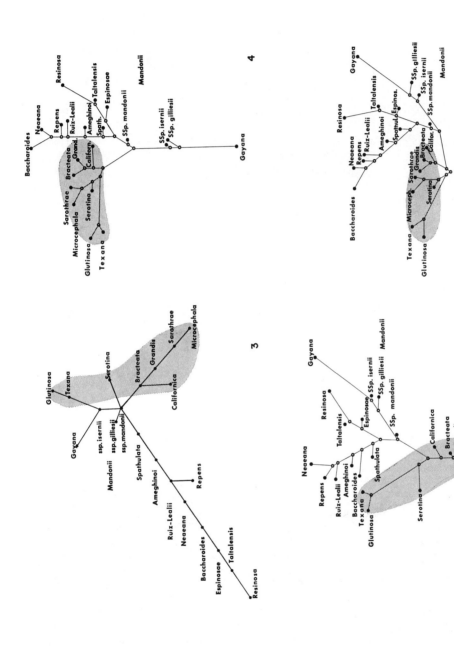

Fig. 3–6. Fig. 3. PRIM diagram indicating phenetic affinities between all species. Fig. 4, 5, 6. Three hypothetical phylogenies considering *G. gayana*, *G. sarothrae*, and a hypothetical ancestor as primitive (for further details see text). North American species in stippled area; South American species outside of stippled area.

of central tendency measure. The resulting phylogeny is shown in Fig. 6. This meets all the phytogeographical and cytological criteria that were set up, as well as the experimental and field observations.

DISCUSSION

Prior to the establishment of numerical methods, phylogenies were constructed using mostly subjective criteria, born from an intimate knowledge of the group by the investigator. Criticisms were directed at the resulting phylogeny and not at the way in which it was reconstructed. The establishment of objective criteria to construct phylogenies such as in Wagner's Groundplan-Divergence Method, permits the evaluation of both the criteria that the investigator is following and the actual phylogeny which is reconstructed. In turn, it allows the investigator to be more autocritical and more consistent. If we assume that the original data (paleobotanical, morphological, genetical, etc.) are correct, the phylogeny constructed which uses them should represent the most plausible one (probabilistically speaking).

However, since there is as yet no absolute way of establishing "primitiveness," there is no way of determining what phylogenetic tree is most likely to be correct. This holds true even if a procedure, such as parsimony, is available to construct the most probabilistic tree. One answer to this dilemma is to use more than one criterion of primitiveness, each incorporating a different set of subjective criteria, and to construct different trees. The resulting phylogenies can then be compared with other criteria such as phytogeography or genetics, and the one showing the best fit is selected as the most plausible. This is the procedure that was used in the present study. Furthermore, the various trees can be compared in search of common features. If such common features are found, their validity is enhanced since they are truly independent from the criteria of primitiveness.

In the present case five such features that were found are:

1) The two annual species are very closely related to each other, more than either is to any perennial species.

2) *Gutierrezia gayana* is the most dissimiliar and might possibly belong to another genus.

3) The North American and South American species form two groups of closely related taxa. This lends further support to the idea that there was only one dispersal event between the two areas.

4) *Gutierrezia sarothrae* is not an ancestral species as had been previously postulated on the basis of purely subjective criteria (Solbrig, 1964).

5) The three Chilean species, *G. resinosa*, *G. taltalensis*, and *G. espinosae*, are closely related, and so are the three mountain species from northern Argentina, *G. repens*, *G. neaeana*, and *G. ruiz-lealii*. This is particularly interesting in the case of *G. neaeana*, a species of uncertain affinities until now.

Points 1–3 had been inferred before; points 4 and 5 are new insights derived from this analysis.

In summary, any phylogeny will approximate the true evolutionary history of a group insofar as the criteria used in its construction and the assumptions made are correct. However, since there is no absolute way to test these criteria, particularly assumptions regarding "primitiveness," there is no way of ascertaining which phylogeny is correct. By constructing several phylogenetic trees using alternative assumptions of primitiveness, the investigator can test their agreement with independently obtained evidence. It also permits the isolation of those parts of the trees for which there is agreement independent of the criteria used, and those relations that are affected by conflicting criteria. In this way problems that require more work can be isolated.

The use of this methodology in the present paper has reinforced some previously proposed hypothesis (e.g., South American species more advanced than North American) and negated others (e.g., *Gutierrezia sarothrae* as a primitive species). In this respect it has produced some valuable new insights. The method is advanced as of general validity.

ACKNOWLEDGMENTS

I am grateful to Drs. Arnold Kluge and James Farris for technical assistance and to Drs. Kluge and Warren H. Wagner, Jr., for reading the manuscript. A grant from the National Science Foundation is gratefully acknowledged.

LITERATURE CITED

Bentham, G. 1873. Notes on the classification, history, and geographical distribution of Compositae. Jour. Linn. Soc. Bot. (London) 13: 335–577.

Camin, J. H. & R. R. Sokal 1965. A method for deducing branching sequences in phylogeny. Evolution 19: 311–326.

Cronquist, A. 1955. Phylogeny and taxonomy of the Compositae. Amer. Midl. Nat. 53: 478–511.

Farris, J. S. 1966. Estimation of conservatism of characters by constancy within biological populations. Evolution 20: 587–591.

————— 1970. Maximum likelihood and maximum parsimony in numerical cladistics. Evolution 24: (in press).

Kluge, A. G. 1969. The evolution and geographical origin of the New World *Hemidactylus mabouiabrookii* complex (*Gekkonidae, Sauria*). Misc. Publ. Museum Zoology, Univ. Michigan 138: 1–78.

————— & J. S. Farris 1969. Quantitative phyletics and the evolution of Anurans. Syst. Zool. 18: 1–32.

Prim, R. C. 1957. Shortest connection networks and some generalizations. Bell System Tech. Jour. 36: 1389–1401.

Raven, P. H. & H. J. Thompson 1964. Haploidy and angiosperm evolution. Amer. Nat. 98: 251, 252. [Letter to the Editor.]

Rüdenberg, Lily & O. T. Solbrig 1963. Chromosome number and morphology in the genus *Gutierrezia* (Compositae-Astereae). Phyton (Argentina) 20: 199–204.

Solbrig, O. T. 1960. Cytotaxonomic and evolutionary studies in the North American species of *Gutierrezia* (Compositae). Contr. Gray Herb. (Harvard Univ.) 188: 1–63.

————— 1964. Infraspecific variation in the *Gutierrezia sarothrae* complex (Compositae-Astereae). Contr. Gray Herb. (Harvard Univ.) 193: 67–115.

————— 1965. The California species of *Gutierrezia* (Compositae-Astereae). Madroño 18: 75–84.

————— 1966. The South American species of *Gutierrezia*. Contr. Gray Herb. (Harvard Univ.) 197: 3–42.

Sokal, R. & P. H. A. Sneath 1963. Principles of Numerical Taxonomy. San Francisco, California: W. H. Freeman. xvi + 359 pp.

Wagner, W. H., Jr. 1961. Problems in the classification of ferns. Recent Advances in Botany 1: 841–844.

————— 1966. Modern research on evolution in ferns. pp. 164–184. *In*: Jensen, W. A. & L. G. Kavaljian (Editors), Plant Biology Today: Advances and Challenges. Ed. 2. Belmont, California: Wadsworth Publishing Co., Inc. x + 208 pp.

Part V

CHARACTER COMPATIBILITY ANALYSIS

Editors' Comments
on Papers 14, 15, and 16

A second general approach to the estimation of the branching patterns of evolution and the sequence of character state change is character compatibility analysis. This method permits the identification of characters for which advanced states are likely to have been derived uniquely during the course of evolution of the members of the group being studied. These unique changes could have occurred at the time of speciation events (cladogenesis) or without the occurrence of speciation events (anagenesis). In this method, the patterns of agreement between pairs of cladistic characters are examined. When character polarities have been assessed, this method is equivalent to the Hennigian argumentation method (Estabrook, 1978; Duncan, 1984).

Agreement between characters is addressed in the three papers in this section, with refinements in this concept proposed by each author. Wilson (Paper 14) proposes a method to test the subsets defined by the derived character states of various characters providing a basis for a test of the consistency or correspondence between characters and the phylogenetic relationships that are proposed by each character. LeQuesne (Paper 15) examines the concept of uniquely derived character states. He notes that two characters each with two states must conflict if all four combinations of character states occur in the organisms under study. At least one of the characters will not have been uniquely derived during the course of evolution. The idea of addition of subsets and comparison of distribution and occurrence of character states has provided the basis for an important series of papers by Estabrook and coworkers. One of these (Paper 16) is reproduced here. It outlines the concepts central to character compatibility analysis, including the definition of cladistic characters, convex character states, and their relationship to the consistency test of Wilson, an explicit statement of compatibility and its relationship to uniquely derived character states, and the use of compatibility analysis at the primary and secondary levels in the analysis of the relationships of a group of species. One of the most significant points made in this paper is that the identification of compatible characters provides a basis for identifying characters that could be reflective of the historically true evolutionary history of a group of organisms.

REFERENCES

Duncan, T., 1984, Willi Hennig, Character Compatibility, Wagner Parsimony, and the "Dendrogrammaceae" Revisited, *Taxon* **33:**785–791.

Estabrook, G. F., 1978, Some Concepts for the Estimation of Evolutionary Relationships in Systematic Botany, *Syst. Bot.* **3:**146–158.

14

Copyright © 1965 by the Society of Systematic Zoology
Reprinted from *Syst. Zool.* **14**:214–220 (1965)

A Consistency Test for Phylogenies Based on Contemporaneous Species

EDWARD O. WILSON

It is commonly stated that phylogenies deduced from data about contemporaneous species cannot be "proved" because, obviously, evolution is a past event recoverable only from fossils. This is true to the extent that no proof exists which has the decisiveness of a witnessed event or the consistency of a physical measurement. Yet scientific proof is rarely direct and is always relative in degree. Evolutionary hypotheses might never be definitive by the standards of experimental biology, but they are valid if they are both falsifiable and heuristic. That is, to be valid they should make concrete predictions that are capable of being negated if the hypothesis is false; and they should point the way to deeper, more meaningful investigations if they are momentarily upheld. Phylogenetic taxonomy has been open to criticism not so much for indirection as for its lack of techniques of formal analysis that render its hypotheses falsifiable and heuristic.

One such procedure that might be employed involves the "weighting" of characters with reference to their phylogenetic significance. Taxonomists intuitively select character states which they postulate to define monophyletic sets of species. The ideal character contains some state that both uniquely defines a set of species and has not been reversed in evolution, so that all existing species which possess this state can be said to have descended from one species in the past that evolved the state. For every such character state that can be identified, a branch in the phylogenetic tree can be added. This extreme form of phylogenetic hypothesis, then, is initiated as a hypothesis about unique, unreversed characters. The formulation perhaps cannot be decisively

proven on the basis of contemporaneous species. But can it be disproved? And is anything of biological significance to be gained by the procedure? The following test hopefully gives an affirmative answer to both questions. It is not original in the sense that it offers something very new to taxonomic thinking. Instead, its purpose is to express one common intuitive taxonomic procedure in a new, more rigorous form.

Definitions. Consider a series of m unique, unreversed character states a_1, a_2, a_3, . . . , a_m *each representing a different character*, as yellow dewlap and flattened tail can be said to be states of two separate characters (dewlap color and tail shape) in lizard species. These particular states are interpreted to have appeared during a speciation episode that has resulted in a monophyletic taxon of n contemporaneous species. They now exist in any combination in various of the n species. At one extreme, they may be totally lacking in a given species; at the other extreme, all m character states may occur together in a given species. By *unique* is meant that a given character state a_i appeared in the past only once and in one species. It now exists in one or more descendant species. By *unreversed* is meant that the state has never been lost, i.e., has never reverted to a prior state, in any of the species giving rise to the contemporaneous taxon. The character state itself might have arisen *de novo* as a new structure, it might have appeared as a new state in a series of discrete character states, or it might be arbitrarily recognized as some point and beyond in a continuous morphocline. The taxon therefore is to be treated as a sample space whose points are contemporaneous species; and the character states are events that can occur on the points in the sample space. It is desirable that the character states in the hypothesis be chosen initially for considerations not having to do with the way they jointly define sets of species in the taxon. The properties that can be expected to induce the choice include uniqueness with reference to other taxa, structural complexity, and absence of other states that are clearly annectant or derivative and degenerate in nature.

The hypothesis. The m character states are unique and unreversed.

Testing the hypothesis.[1] Let us label the states such that the possession of a_1 completely defines a set of contemporaneous species A_1, a_2 defines a smaller or equal set A_2, a_3 defines a still smaller or equal set A_3, and so on. Three possible alternative outcomes can now be simply stated (Fig. 1).

I. If the sets of species defined by the character states are non-overlapping, i.e., $A_1 \cap A_2 \cap A_3 \cap \ldots \cap A_m = 0$, the hypothesis cannot be tested.

II. If the sets are overlapping but do not enclose each other (form a series of proper subsets), in the order a_1, a_2, a_3, \ldots, a_m, the hypothesis is rejected.

III. If the sets are overlapping and A_2 is wholly enclosed in (is a proper subset of) A_1 (and A_3 is enclosed in A_2, and so on) the arrangement is consistent with the hypothesis but does not definitely prove it. If Situation III holds, the following phylogenetic hypothesis is also consistent: $A_o \cap \tilde{A}_1$, $A_1 \cap \tilde{A}_2$, \ldots, $A_{m-1} \cap \tilde{A}_m$, A_m are the contemporaneous branches of a phylogenetic tree of the kind illustrated in Figure 2.

Examples of reasonably long sequences of character states that pass the consistency test are probably familiar to most taxonomists. Two such sequences from the ants

[1] The following conventions from set theory are used: A_1 symbolizes the set of all species that possess character state a_1. $\tilde{A}_1$ symbolizes the set of all species that do *not* possess character state a_1. $A_1 \cap A_2$ indicates those species that are in both A_1 and A_2, i.e., possess both character states a_1 and a_2. $A_1 \supset A_2$ means that A_2 is contained wholly within A_1, i.e., all species that possess a_2 also possess a_1.

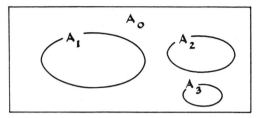

I. Test not applicable

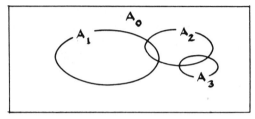

II. Test failed, hypothesis rejected

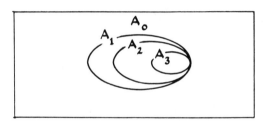

III. Test passed, hypothesis not rejected

FIG. 1. The consistency test. The rectangle encloses the taxon under consideration. It must be reasonably discrete in many characters from all other taxa. Each ellipse encloses a set of species A_i characterized by a character state a_i hypothesized to be unique and unreversed. Only one state per character is considered. A_o is the set of species not bearing any character state hypothecated to be unique and unreversed.

(family Formicidae) are given in Table 1.

Suppose that the m character states are mutually consistent, as in Situation III. Although it is not possible from this fact alone to prove the phylogenetic hypothesis of Figure 2, we might still be able to narrow the permissible alternative explanations somewhat. First, consider the model which is the opposite of the one under consideration, namely that the m states have appeared and disappeared in a random manner with reference to each other during the evolution of the taxon. This hypothesis can be tested in the following manner. Imagine the circumstance, among all possible circumstances, in which there would exist the highest probability of the nested pattern arising by chance combinations alone. This is the simple case illustrated in Figure 3. There are $m + 1$ species evolving separately during the time that the characters are fixed (at random with reference to each other) to produce the nested pattern. Given that at the time m of the species acquired a_1, $m - 1$ acquired a_2, $m - 2$ acquired a_3, *et seq.*, with a single species acquiring a_m, the probability that the resulting sets A_i could be nested by chance alone is

$$P(A_1 \supset A_2 \supset \ldots \supset A_m)$$
$$= \frac{[m!2!][(m-1)!3!]\ldots[3!(m-1)!][2!m!]}{[(m+1)!]^{m-1}}.$$

All other situations in which the character states are fixed independently to give a nest of sets are equally or less probable. In other words, the equation above, based on the random model illustrated in Figure 3, gives an upper limit for the probability that the m character states were evolved randomly with reference to each other. Applying it for various values of m we find that $P(A_1 \supset A_2 \supset \ldots \supset A_m)$ is 6/125 for four character states, 1/225 for five character states, 16/84,035 for six character states, and 9/153,664 for seven character states. In order for this explicit formulation to be valid it is necessary that the character states be chosen initially without reference to the kind of classification they would engender in the taxon. In practice such selection would come about in the first study of a group of species, before the distribution of various character states with reference to each other are considered.

In sum, if four or more character states hypothecated to be unique and unreversed

Sets of Contemporaneous Species

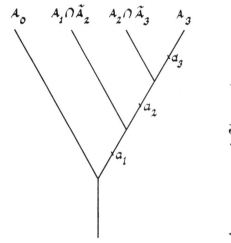

FIG. 2. The phylogenetic hypothesis (cladogram) that is permissible if the character states a_1, a_2, and a_3 pass the consistency test. Each character state represents a *different* character. A_o, $A_1 \cap \tilde{A}_2$, $A_2 \cap \tilde{A}_3$, and A_3 represent sets of contemporaneous species. The nodes labeled a_1, a_2, and a_3 mark the appearance of these character states in time. The ends of the branches are arbitrarily arranged along equal intervals because the consistency test by itself gives no information about over-all similarity of the sets of species.

then pass the consistency test, we are reasonably justified in considering them correlated in some historical manner, regardless of the pathways of speciation taken by the taxon in the past. With much more confidence, this rule can be based on five or more characters.

Suppose the consistency test is thus passed with reasonable confidence. There are five alternative ways in which the character states could be correlated:

1) The m character states were fixed at random. Later, there was differential survival among the species according to their respective combinations of the m states, resulting in the modern consistent pattern. Unless we also postulate genetic drift, this explanation subsumes that the ways that the m states are combined are at one time selectively neutral and later selectively significant. The probability certainly exists

TABLE 1—SEQUENCES OF INTERCONSISTENT CHARACTER STATES IN THE FORMICIDAE.

	Character State	Group Defined within the Taxon
Series No. 1 (Taxon = Aculeate Hymenoptera)	Metapleural gland	Family Formicidae
	Pulvinate poison gland	Subfamily Formicinae
	Sepalous proventriculus	"Section Euformicinae"
	Dense, appressed pilosity in discrete soldier caste	Subgenus *Machaeromyrma* of *Cataglyphis*
Series No. 2 (Taxon = Genus *Lasius*)	"*Niger*-type" male mandible	*Lasius* exclusive of Subgenus *Chthonolasius*
	Metapleural guard hairs lost in female castes	Subgenus *Dendrolasius*
	β-form queen	*L. teranishii* and *L. spathepus*
	Appendages covered with long, silvery pilosity	*L. spathepus*

but seems intuitively relatively small. *Or,*

2) A superordinate character state, e.g., a_1 with reference to a_2, always or with very high frequency appears soon after the subordinate appears; but it also originates in a certain fraction of the species without the subordinate character state. This possibility seems even more remote than (1). *Or,*

3) A subordinate state, e.g., a_2 with reference to a_1, occurs only after the superordinate state is present. But it can still be nonunique and reversible within the species bearing the superordinate state. For example, a_2 could still appear and disappear many times over in species bearing a_1. This is perhaps more likely than (1) and (2); however, if the subordinate character states really could appear and disappear in multiple fashion within the set of species bearing superordinate character states, it would be necessary for each of the subordinate states to have changed in concert to preserve the consistency observed in the contemporaneous taxon. *Or,*

4) The character states could first have appeared together and then been lost in concert to produce the precise pattern. *Or,*

5) The states are unique and unreversed within the taxon. This seems the most likely hypothesis. It is certainly the simplest.

Heuristic value. Consistent phylogenetic schemes, even when based entirely on con-temporaneous species, are useful for two reasons: they serve to confirm the identity of the most unusual and stable character states, and they make exact predictions about state combinations in the species yet to be discovered. While remaining explicitly vulnerable, they are a valuable scientific procedure, comprising that part of taxonomic research which has the greatest general interest. This positive aspect of phylogenetic analysis holds whether or not enough characters pass the consistency test to allow the random hypothesis to be rejected. It also holds whether or not the phylogeny deduced is correct in detail and regardless of its effect on formal classification.

An example illustrating the heuristic value of cladistic analysis can be taken from my recent revision of the ant genus *Aenictus* (Wilson, 1964). Two character states, the "Typhlatta spots" of the head and presence of teeth on the anterior margin of the clypeus, were among those initially guessed to be unique and unreversed, but they proved not to be interconsistent. In particular, *Aenictus currax* and *A. huonicus* possessed Typhlatta spots but appeared to lack clypeal teeth. Since these two species are very similar in all characters studied, they were inferred to be closely related. Also, since they are both endemics of New Guinea,

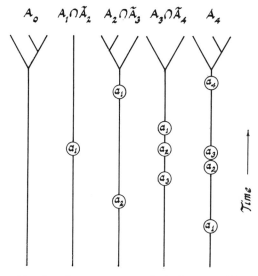

A_0 $A_1 \cap \tilde{A}_2$ $A_2 \cap \tilde{A}_3$ $A_3 \cap \tilde{A}_4$ A_4

FIG. 3. The situation in which there would be the highest probability of m character states (in this case $m = 4$) passing the consistency test while evolving in a random manner. There are $m + 1$ species during the appearance of the states. The placement of the numbers indicates the appearance of the states in time. They are scattered arbitrarily in this diagram to suggest the condition of randomness. All other situations would give equal or higher probabilities. After the states appear the $m + 1$ species may or may not speciate further to produce the contemporaneous taxon, as exemplified by the irregular branching near the ends of the phyletic lines.

which lies on the periphery of the range of the species group to which they belong, it was guessed that any deviant character states shown by them would be more likely to be derived than original. This second deduction was based on a rule shown by Indo-Australian ant species generally. A closer, second examination of the *Aenictus* species resulted in support for the hypothesis: workers of *A. currax* were found to have hidden, rudimentary teeth. As a consequence, it was concluded that clypeal teeth have been lost secondarily in *A. huonicus*. It is my impression that similar logical sequences are often, even routinely followed in taxonomic revisions. Taxonomists seldom spell their procedures out, however, as I have done in the *Aenictus* revision.

Relation to classification. Formal classifications need not be isomorphic with phylogenetic schemes that are simply cladistic in nature. The sets of species $A_i \cap \tilde{A}_{i+1}$ defined by characters that continue to pass the consistency test may or may not be recognized as taxa. It is conceivable, for example, that a species in A_2 differs from one in $A_1 \cap \tilde{A}_2$ only by the character state a_2 but is different from other species in A_2 by many other characters. In this case it would be valid taxonomic procedure either to lump $A_1 \cap \tilde{A}_2$ and A_2 or to combine the one species from A_2 with $A_1 \cap \tilde{A}_2$; and it would be dubious procedure to split $A_1 \cap \tilde{A}_2$ and A_2 as taxa. This conclusion has been reached by members of both the phylogenetic (Simpson, 1961) and numerical schools (Sokal and Sneath, 1963).

Even so, the present study together with independent and parallel attempts to formalize cladistic analysis (e.g., the articles by Sokal and Camin and by Throckmorton in this issue of *Systematic Zoology*) indicate that we can hope to distinguish with confidence between "constant characters" and "fickle characters." In the aggregate, constant characters reflect phylogenies more accurately than fickle characters, and it would appear that insofar as we wish to transmit evolutionary information in our classifications constant characters should be given greater weight. Such classifications may not be as stable and reproducible as those based on the averaged similarity of unweighted characters, but they have more biological interest. Taxonomy should be more than the blind clustering of taxa according to over-all similarity, as suggested by the "numerical taxonomists." In spite of the attractive simplicity of the latter technique and its undoubted usefulness in special cases, it seems to be of dubious value as a broad philosophy of classification. The main objection is that numerical taxonomy has up to the present offered little hope of yielding new biological information, precisely because it has not been constructed with reference to any real biological questions. Put in another way, taxonomy is a language that can be designed according to any one of many sets of rules. The rules selected should be of maximum heuristic value; beyond that, it is only necessary that they be stated very plainly. While not intending to disparage multivariate statistics or the considerable technical achievements of the numerical taxonomists, I would regard a taxonomy based automatically and *a priori* on unweighted characters as a desirable measure only in cases where phylogenetic hypotheses cannot in any way be tested. Even at its best this procedure should never be accepted as a doctrine. In fact, it seems more likely than ever before that taxonomists will eventually develop standard methods for the combination of phenetic measures and cladistic inferences into truly phylogenetic classifications. To do so would be one of the great achievements of modern evolutionary biology.

Summary

Cladograms of contemporaneous species are most rigorously constructed from a hypothesis which postulates unique, unreversed character states (Fig. 2). Many such phylogenetic schemes, if false, can be quickly discarded by a simple consistency test illustrated in Figure 1. If a set of four or more character states $a_1, a_2, a_3, \ldots, a_m$ found in m different characters in a taxon are selected without reference to the grouping of species within the taxon and then are found to characterize successively smaller sets of species in such a way as to pass the consistency test, it is reasonable to conclude that the character states evolved in the taxon in a non-random manner with respect to each other. The relation of this inference to phylogeny and the heuristic value of phylogenies based solely on contemporaneous species are discussed.

Acknowledgments

I am very grateful to Eli Minkoff and Angelo Serra for critical readings of the manuscript. Several other persons, including A. F. Bartholomay, W. H. Bossert, W. L. Brown, H. E. Evans, R. Inger, E. MacLeod, E. Mayr, G. G. Simpson, R. R. Sokal, and R. W. Taylor, have discussed various aspects of the problem and provided help and encouragement. The consistency test was developed in conjunction with a recent systematic study (Wilson, 1964) of the Indo-Australian doryline ants supported by a grant from the National Science Foundation.

REFERENCES

SIMPSON, G. G. 1961. Principles of animal taxonomy. Columbia University Press, New York, 247 p.

SOKAL, R. R., and P. H. A. SNEATH. 1963. Principles of numerical taxonomy. W. H. Freeman, San Francisco, 359 p.

WILSON, E. O. 1964. The true army ants of the Indo-Australian area (Hymenoptera: Formicidae: Dorylinae). Pacific Insects 6:427–483.

Appendix

The following is a proof of the proposition that the model in which $m \geqslant 3$ character states are fixed in $m + 1$ species gives the highest probability, among all possible models, that the m character states could have evolved at random with respect to each other and still have been fixed interconsistently. Let the array of numbers of species in each group A_i vary and any given array be labeled with a number j; in the extreme case $j = 1$, there exists the extreme model just cited in which the number of species in A_0 is $m + 1$, the number in A_1 is m, the number in A_2 is $m - 1$, and so on. In short, the array $j = 1$ contains the smallest number of species possible. In a second array $j = 2$, A_0 might contain $m + 2$ species, A_1 m species, A_2 $m - 1$ species, and so on. The probability that a given array occurred as the character states were fixed is p_j and $\sum_j p_j = 1$.

The probability that in an array j the m character states would be fixed in a given pattern with respect to one another can be designated q_{jk}. In particular let us label as $k = \alpha$ the condition in which the character states turn out to be interconsistent. What is desired is the maximum value of $\sum_j p_j q_{j\alpha}$ for all possible arrays j. Now it is intuitively apparent and has been borne out by inspection of many cases (but not formally proved for all possible cases) that where $m \geqslant 3$ the maximum value of $q_{j\alpha}$ is obtained when the array j contains the smallest number of elements, i.e., in the case $j = 1$. The maximum value for $q_{j\alpha}$ is $q_{1\alpha}$, which is a constant when m is chosen. Consider any array j that occurred in evolution. Then given some value of m, $\sum_j p_j q_{j\alpha} \leqslant \sum_j p_j q_{1\alpha}$ for all j and, since $\sum_j p_j q_{1\alpha} = q_{1\alpha} \sum_j p_j$ for the special "alpha case" and $\sum_j p_j = 1$, it follows that $\sum_j p_j q_{j\alpha} \leqslant q_{1\alpha}$.

EDWARD O. WILSON is in the Biological Laboratories, Harvard University, Cambridge, Massachusetts 02138.

15

A METHOD OF SELECTION OF CHARACTERS IN NUMERICAL TAXONOMY

WALTER J. LE QUESNE

The basis of conventional taxonomy has been the selection of characters by a subjective process in such a way as to produce the most self-consistent scheme of relationships between species. In numerical taxonomy, the normal approach is to use as wide a range of characters as possible and to try to choose these in an objective way with the purpose of removing the subjective element as far as practicable. The method proposed below was evolved with the plan of eliminating some out of the range of characters used, in a purely objective manner, in the hope of leading to relationships which were somewhat more likely to be phylogenetically valid. It rests, however, on somewhat different assumptions from the method proposed by Camin and Sokal (1965). Application to observed data can also lead to some interesting deductions concerning the speciation process.

The 'uniquely derived character' and logical consequences.—If one is studying the taxonomy of a group, a character that has evolved only in one direction on a single occasion in its history is more likely to give an unambiguous indication of its phylogeny: this concept will be defined as that of the *uniquely derived character*, using the word 'uniquely' in the same sense of occurring only once. Let us first suppose that all the characters have been reduced to a purely dichotomous form (with alternatives A and B). Now, if character 1 had the ancestral character-state 1_A which once during the development of the group altered to 1_B and if the independent character 2 originally represented by character-state 2_A also once altered to 2_B, not more than three out of the four possible combinations of these two character-states (*i.e.*, 1_A2_A, 1_A2_B, 1_B2_A and 1_B2_B) will be found. If, therefore, for any pair of dichotomous characters, all four combinations are found within the group, either character 1 or character 2 is not a uniquely derived character, or alternatively neither of them is. In practice, this will apply whether A or B is the ancestral character in each case.

It must be pointed out, however, that if three or less of the possible combinations are found, it does not necessarily prove that characters 1 and 2 are both uniquely derived characters, but only that they may possibly be. If, for example, character 2 changed from 2_A to 2_B on two or more occasions on each of which character 1 was in the same state, only three combinations will be found. Moreover, if the four combinations have been evolved during the history of the group, one of these may have died out again or alternatively not be represented in the material studied.

It may also be noted that if we consider two characters, both of which are expressed as both A and B character-states in the group of species considered, and find that only two out of the above-mentioned four possible combinations are found, these two characters are, in the group concerned, completely correlated and can be applied together in the formation of any key. All such relationships are thus well worth noting.

If we allow characters also to be expressed in a form where three possibilities are in a series (*e.g.*, a measurement which could be above, within or below the median range, represented by A, X and B respectively), it too can be seen that if all four possible combinations 1_A2_A, 1_A2_B, 1_B2_A and 1_B2_B of the extreme form occur, character 1 and character 2 cannot both be uniquely derived characters.

Application to a data matrix.—To apply this principle to a specific group, a data matrix is first prepared in the usual way, giving the character-state (expressed as A, B or X, as defined above) for each species and character (*i.e.*, 'test' in the sense of Sokal and Sneath) considered. Then each of the possible pairs of characters is considered to see how many of the possible combinations (excluding any terms including X) AA, AB, BA and BB occur among all the species. A matrix is drawn up with a space for the relationship between each pair of characters and a mark put on this in each case where all four possibilities have been found: this will be called the *character-pair matrix*.

At this stage, the number of marks corresponding to each character (in its combination with each of the other characters) is counted. Although it is not possible to say that any particular character cannot be a uniquely derived character, that with the greatest number of marks is the least likely to be so. Vertical and horizontal lines are drawn to eliminate the marks concerned and the number of remaining marks again counted in the case of all the other characters and the process repeated until all the marks have been eliminated.

Ideally, the characters left can be used to work out the phylogeny, at least in part. In cases where the selection of characters for elimination can be done in a number of ways that do not differ very widely in their probability, any phylogenetic deductions will be uncertain, but at least a figure can be given for the maximum number of characters which could be uniquely derived.

The coefficient of character-state randomness.—If none of the characters were uniquely derived, one would not expect to find all four combinations of character-states for every pair of characters. In fact, utilizing the number of species found in each character-state for each character, it is possible to calculate the probability of finding the four character-state combinations for each pair of characters, assuming purely random distribution of the character-states among the species. If the number of pairs of characters for which the four combinations of character-states are actually observed (in part or the whole of the character-pair matrix) is divided by the sum of the calculated probabilities on the above assumptions, one gets a ratio which can be called the *coefficient of character-state randomness*. This can most conveniently be expressed as a percentage, 0% representing entirely apparently uniquely derived characters and 100% representing no apparent correlation in the distribution of the character-states.

Application to data on Argodrepana.—The above principles can be simply illustrated by application to the data on *Argodrepana* (Lepidoptera, Drepanidae) quoted by Wilkinson (1967). This is a simple case, applied only to seven species, and more complex examples, such as that of *Teldenia*, will be discussed in separate publications. In the latter case, at least, the coefficient of character-state randomness is much higher than in *Argodrepana*, and consequently it is not easy to deduce the most probable phylogeny.

Only characters where both character-states A and B were represented by at least two species were used, since otherwise all four combinations cannot possibly be obtained. In order to simplify some of the subsequent calculations, it is convenient to define the less frequent of the character-states used in the analysis for each character as A and the more (or equally) so as B. To comply with this, the following substitutions were made in the symbolic representation in Wilkinson's paper:

Character 1; 1 becomes A and 2 becomes B

TABLE 1. DATA MATRIX FOR *Argodrepana* SPP., MODIFIED FROM DATA OF WILKINSON (1967).

Character	n_A	n_B	n_x	Species						
				1	2	3	4	5	6	7
9	2	5	0	B	B	B	B	B	A	A
12				B	B	B	B	B	A	A
13				A	A	B	B	B	B	B
17				B	B	B	B	B	A	A
30				A	A	B	B	B	B	B
34				B	B	B	B	B	A	A
36				A	A	B	B	B	B	B
44				B	B	B	A	A	B	B
45				B	B	B	A	A	B	B
46				B	B	B	B	B	A	A
58				B	B	B	B	A	B	A
71				A	A	B	B	B	B	B
80				A	A	B	B	B	B	B
83				B	B	B	A	A	B	B
11	2	4	1	B	B	X	B	B	A	A
68				B	B	B	A	B	A	X
76				A	A	X	B	B	B	B
2	2	3	2	A	A	X	X	B	B	B
3				A	A	X	X	B	B	B
25	2	2	3	B	X	A	X	A	B	X
1	3	4	0	B	A	A	B	A	B	B
47				B	B	A	B	A	A	B
75				B	B	A	A	A	B	B

Characters 2, 3; C becomes B

Characters 9, 12, 17, 46; C becomes A and A becomes B

Character 11; C becomes A and D becomes B

Characters 13, 47, 75, 80; 1 becomes A and O becomes B

Character 25; 150 becomes X (A = below, B = above)

Characters 30, 36, 44; B becomes A and O becomes B

Character 34; E becomes A

Character 45; C becomes A and O becomes B

Character 58; B becomes A and E becomes B

Character 68; 3, 2, 1 become A, X, B respectively

Character 71; 2 becomes A and 1 becomes B

Character 76; 4, 1, 2 become A, X, B respectively

Character 83; 2 becomes A and O or 1 becomes B

If n_A, n_B, n_x represent respectively the number of species for which the character-state is A, B or X (the latter including those for which it is undetermined), the characters are grouped according to values as shown in Table 1.

On comparing each pair of characters and marking with a cross where all four combinations of character-states occur, we obtain the character-pair matrix shown in Table 2. The number of marks for each character in combination with each of the others is given at the foot of the column, designated as N_x.

Characters 1, 47, 58 and 68 are successively deleted from this matrix by crossing out the vertical and horizontal lines corresponding to these four characters. This eliminates all marks. Thus, 19 out of the 23 characters suitable for this analysis behave as uniquely derived characters.

The following patterns of character-states are found among the remaining 19 characters, allowing an X to be replaced by either an A or a B to fit into the scheme.

Species	Characters
1 2 3 4 5 6 7	
A A B B B B B	2, 3, 13, 30, 35, 71, 76, 80
B B B B B A A	9, 11, 12, 17, 34, 46
B B B A A B B	44, 45, 83
B B A A A B B	25, 75

In the first three of these groups, we have examples of completely correlated characters. In the last case we have correlated characters 25 and 75 by regarding one species with wing-span 150mm as A and two as B, so that this relationship is of doubtful validity.

In every case of a uniquely derived character either all species with character-state A or all species with character-state B must be directly derived from a common ancestor. If all the above characters are actually uniquely derived, one can put forward several possible cladograms. The most probable of these can be derived on the assumption that all species with character-state A (a smaller group in each case than all with character-state B) are directly derived from

TABLE 2. CHARACTER-PAIR MATRIX BASED ON DATA FOR *Argodrepana*.

Character	1	2	3	9	11	12	13	17	25	30	34	36	44	45	46	47	58	68	71	75	76	80	83
1	–																						
2	x	–																					
3	x		–																				
9				–																			
11					–																		
12						–																	
13	x						–																
17								–															
25									–														
30	x									–													
34											–												
36	x											–											
44	x												–										
45	x													–									
46															–								
47	x			x	x	x		x			x		x	x	x	–							
58	x			x	x	x		x			x		x	x	x	x	–						
68													x	x		x		–					
71	x																		–				
75	x																x	x		–			
76	x																		x		–		
80	x																					–	
83	x																x	x	x				–
Nx	14	1	1	2	2	2	1	2	0	1	2	1	4	4	2	13	12	5	1	3	2	1	4

a common ancestor, from which one can deduce the following:

species 1 and 2 are directly derived from a common ancestor;

species 6 and 7 are directly derived from a common ancestor;

species 4 and 5 are directly derived from a common ancestor;

species 3 and the common ancestor of 4 and 5 are directly derived from a common ancestor.

We thus derive the following cladogram to show the phylogeny:

This diagram is not intended to be a phenogram or a phylogram.

Alternatively, the data left in Table 2 after elimination of the apparently unsuitable characters can be used to work out similarity coefficients between the species in the normal way and these converted into a "phenogram." In this case, I believe that the latter would necessarily display one of the cladistic relationships derived on logical grounds.

Coefficient of character-state randomness in Argodrepana.—For the calculation of the coefficient of character-state randomness (as defined above), we have to find the probability of all four character-state combinations being present in each of the fourteen possible combinations from pairs of values of n_A, n_B and n_x. The symbol N_1 represents the number of characters with the first pattern of values of n_A n_B and n_x and N_2 that with the second, N_{12} is the number of combinations of pairs consisting of one of each of these patterns. By multiplying the probability in each case by N_{12} and then adding the products, the expected value corresponding to a completely random

TABLE 3. CALCULATION OF PROBABILITIES OF ALL FOUR CHARACTER-STATE
COMBINATIONS BEING PRESENT IN THE DIFFERENT CASES.

1) n_A	n_B	n_x	N_1	2) n_A	n_B	n_x	N_2	Prob.	N_{12}	Prob. $\times N_{12}$
2	5	0	14	2	5	0	14	0.4762	91	43.33
				2	4	1	3	0.3810	42	16.00
				2	3	2	2	0.2857	28	8.00
				2	2	3	1	0.1905	14	2.67
				3	4	0	3	0.5714	42	24.00
2	4	1	3	2	4	1	3	0.3048	3	0.91
				2	3	2	2	0.2286	6	1.37
				2	2	3	1	0.1524	3	0.46
				3	4	0	3	0.5714	9	5.14
2	3	2	2	2	3	2	2	0.1429	1	0.14
				2	2	3	1	0.0558	2	0.11
				3	4	0	3	0.5143	6	3.09
2	2	3	1	3	4	0	3	0.3429	3	1.03
3	4	0	3	3	4	0	3	0.8571	3	2.57
							Totals		253	108.8

distribution of character-states is obtained, as shown in Table 3.

Using the whole of the above data in conjunction with the fact that the number of marked positions on the character-pair matrix is 40, the coefficient of character-state randomness is $40/108.8 \times 100\% = 36.8\%$.

As pointed out earlier in this paper, this figure indicates the extent to which the pattern of character-states is random rather than that to be expected were the characters uniquely derived. Here its relatively low value suggests that the pattern is considerably ordered.

Conclusions.—This method will enable an objective assessment of the probable phylogeny of certain groups and in all cases will give a numerical measure of the extent to which speciation can be regarded as being due to uniquely derived characters (or at least to characters which appear to be unique). Before we can make any generalizations about the results obtained, the method will have to be applied to a number of cases, preferably not all in closely related groups.

If a data matrix has been assembled for purposes of numerical taxonomy, this approach should not be difficult to apply. In cases where the numbers of species or characters is not too large (*e.g.*, 32 species and 30 to 40 characters or 13 species and 80 characters), the character-pair matrix can be worked out in a few hours without mechanical aids, but in large groups the task could be computerized.

ACKNOWLEDGMENTS

I wish to thank Dr. R. E. Blackith, Mr. J. C. Gower and Dr. C. Wilkinson for their valuable comments (*in litt.*) on the above-mentioned ideas.

REFERENCES

CAMIN, J. H. AND R. R. SOKAL. 1965. A method for deducing branching sequences in phylogeny. Evolution, 19:311–326.
WILKINSON, C. 1967. A taxonomic revision of the genus *Teldenia* Moore (Lepidoptera: Drepanidae, Drepaninae). Trans. R. Ent. Soc. Lond., 119:303–362.

16

AN APPLICATION OF COMPATIBILITY ANALYSIS TO THE BLACKITHS' DATA ON ORTHOPTEROID INSECTS

George F. Estabrook, Joseph G. Strauch, Jr., and Kent L. Fiala

Consideration of the agreement among characters has long been an important component in systematics. It has probably been tacitly practiced by some systematists since the time of Haeckel (1866), who apparently based descriptions of types on collections of character states that tended to co-occur. Agreement between a character and a classification is central to the ideas of Hennig (1966). The concept of agreement between characters was explicitly discussed for plant systematics by Sporne and in this context has become known as the Doctrine of Correlations (Sporne, 1975, and references therein). Recently, Wilson (1965), Camin & Sokal (1965), and Le Quesne (1969), among others, have discussed related ideas. The aim of compatibility analysis is to reveal the pattern of agreement among characters, to guide the monographer's choice of those that may be most useful in reconstructing phylogenetic history.

CHARACTERS

A (non hierarchical) classification of the group under study is a collection of subsets of that group, with the property that each member of the study group belongs to exactly one subset. The subsets are called the classes of the classification. A (non hierarchical) classification of the group under study, derived from a single basis for comparison is called a (qualitative) character. Each class in this classification is called a character state. A character with one state classifies the study group into one class, i.e., it considers all members of the study group to be the same. A character with two states classifies the study group into two classes and thus is a basis for distinguishing members of one class (character state) from members of the other class (character state). A character may have as many states as there are entities in the study group, but generally qualitative characters have from one to few states. A description of the distinguishing property of the members of a character state is usually associated with the character state. Often this description is used as a name for the character state. A description of the basis for comparison is often used as a name for the character.

A qualitative character becomes a cladistic character when an estimate of the evolutionary relationshps among its states is made. This estimate takes the form of a tree diagram with (real or hypothetical) character states at the branch points and branch ends. Such a character state tree is an hypothesis asserting which states are primitive or derived.

An estimate of relationships for a study group is defined as a tree diagram with members of the study group at the ends

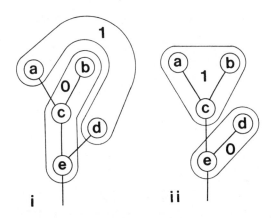

Fig. 1.—Convex Character States. This tree is an estimate of the phylogenetic history for evolutionary units a–e. The character shown in i does not support this estimate because state 1 is not convex on the tree shown (because the path between a and d includes c and e, which are not members of the group defined by state 1). The character shown in ii supports this estimate because both of its states are convex on the tree shown.

cussed in more detail by Estabrook et al. (1975). Ideas basic to this concept were presented by Wilson (1965) and Le Quesne (1969). It is discussed in a more general context by Estabrook (1972) and McMorris (1975).

Two characters are called *compatible* if there is at least one estimate of relationships for the group under study that they both support. If there exists *no* estimate of relationships for the group under study that two characters can simultaneously support, then that pair of characters will be called incompatible.

The concept of compatibility defined here differs from that of Camin and Sokal (1965). Their concept defines the compatibility of one character with the "pattern" determined by another character. Unlike the symmetric concept presented here, their scheme allows character K to be compatible with the "pattern" of character L, even if character L is incompatible with the "pattern" of character K.

The concept that characters as classifications can be more, or less, alike leads naturally to the suggestion that one approach to classification might be to seek that classification that agrees most with all the characters. Estabrook (1971) discussed this concept in detail. It is clear, however, that most of the progress in our understanding of the relationships among creatures has come, not through the construction of classifications that get better support from the extant characters as presently conceived, but through new understandings of what should be the bases for comparison, and of how these bases for comparison should be used to structure data into characters (Huxley 1867; Stebbins 1974). The problem of character construction is closely related to the problem of homology, discussed in general by Jardine (1969) and in the present context by Estabrook (1972). Crovello (1970) reviewed the analysis of characters, and Bisby (1970) discussed the problem of character weighting. Giles (1963) proposed that differences among suggested

of the branches and ancestors (usually hypothetical but possibly members) at the branch points (see Fig. 1). A collection of entities from the study group enlarged by the inclusion of hypothetical ancestors, if any, is convex for an estimate of relationships if any real or hypothetical entities that lie on the unique path along phyletic lines between two members of this collection are also members of this collection (see Fig. 1). Note that both monophyletic and paraphyletic groups are convex. A group that contains the most recent common ancestor of any two of its members (Farris, 1974) is not necessarily convex.

A qualitative character supports an estimate of relationships if the hypothetical ancestors (if any) can be assigned to states of the character so that each character state is convex for that estimate. A cladistic character supports an estimate of relationships if, in addition, the specified directions of evolutionary changes in the character state tree are consistent with the estimate. This concept is defined and dis-

classifications of Dermaptera (earwigs) are mainly due to differences among their proponents on the relative importance of various characters. Thus, at least some of the characters considered important by these proponents are incompatible with each other.

COMPATIBILITY ANALYSIS

The basic idea of compatibility analysis is to use the patterns of agreement and disagreement among characters to choose characters that might best reflect true relationships. If we assume that *some* estimate of relationships is historically correct, then clearly some characters support this estimate and others do not. If we could classify the characters that we assemble for a systematic study into those that support historical truth (i.e., true characters) and those that do not (i.e., false characters), then we could choose the true ones as the basis for reconstructing true relationships. Obviously this is impossible.

By definition, however, any pair of *true* characters must be compatible, while false characters are less likely to be compatible. Thus, all the true characters will belong to the same collection of mutually compatible characters, and the *prima facie* suggestion is that the largest (or the least likely by some reasonable stochastic model) such collection contains them. The characters in this collection will be called primary characters. Since they all support the same estimate of relationships, it is merely a logical exercise (i.e., no "quantitative" method is required) to construct this estimate.

To conduct a compatibility analysis, a collection of lower taxa (usually species, although higher taxa may be used) representing the variation in a higher taxon is chosen. The problem of the choice of the membership of the higher taxon to be studied was discussed by Strauch (1976). This collection is called the study collection, and the lower taxa comprising it are called evolutionary units (EU's).

Data must be in the form of a study collection for which cladistic characters have been defined. Data can be concisely presented as a "basic data matrix" in which each row represents a character and each column represents an EU; at the intersection of each row and column appears the name of the state of character "row" to which belongs EU "column." It is especially important to cite or include a published description of characters and a basic data matrix in any revisionary taxonomic work.

Every possible pair of characters is analyzed to see if there is an estimate of relationships that is supported by both members of the pair. An algorithm to perform this analysis was shown to be mathematically sound by Estabrook et al. (1976a, 1976b). These authors also showed that there exists an estimate of evolutionary relationships supported by all the characters in any collection of cladistic characters for which every pair is compatible. Thus, all the characters in any pairwise compatible collection can simultaneously support an estimate of evolutionary relationships. A pairwise compatible collection is termed a clique, from the equivalent graph theoretical concept (Kemeny et al., 1966).

All cliques for a study collection can be found. Usually the largest clique is chosen to be the clique of primary characters, but biological factors may support another choice. If the characters in one clique *all* involve the making of sound, for example, while those in another clique involve many different functions, then the second collection might be selected even though it contains fewer characters. Characters that divide the study collection into states more nearly equal in size may be favored, since such characters are statistically less likely to have many random compatibilities, and hence less likely to be included in a large clique. Another alternative might be to choose that clique whose characters distinguish the most EU's. Often large cliques have many characters in common. When no other criterion seems available for choosing among two or more large cliques,

these common characters provide an estimate of relationships that all characters in
all these cliques support.

Since all primary characters are mutually compatible, an efficient algorithm
reveals the simplest evolutionary tree that
they all support.

ALGORITHM TO REVEAL THE SIMPLEST ESTIMATE SUPPORTED BY COMPATIBLE CHARACTERS

If there are n EU's, including a real or
hypothetical root, the simplest tree diagram supported by all the characters in a
given clique can be represented in an $n \times n$
matrix in which $M(i,j) = 1$ if and only if
EU i is immediately ancestral (in all
characters of the clique) to EU j. To construct the matrix:

1. Fill the matrix with zeros.
2. Make all possible pairwise comparisons
   between EU's. Each such comparison
   can have one of three outcomes:
   a. EU i and EU j belong to the same
      state in each character. Make no
      entry in the matrix, and identify EU's
      i and j as members of an equivalence
      class. In the future, skip any com-
      parisons involving EU j.
   b. EU i and EU j differ in some char-
      acters, and EU i (or j) always belongs
      to the more derived state in these
      characters. Enter a 1 in $M(j,i)$ (or
      $M(i,j)$).
   c. EU i and EU j differ in some char-
      acters, but sometimes EU i belongs
      to the more derived state and some-
      times EU j does. Make no matrix
      entry because the EU's are on dif-
      ferent branches.
3. Clear the matrix of any entries for EU's
   that were subsequently eliminated in
   step 2a, so that each equivalence class
   has a single representative in the matrix.
4. For each j, examine all entries $M(i,j)$
   for $i = 1$ through n. If for any j there
   are multiple ones, perform the following steps:

   a. If, say, $M(i,j) = 1$ and $M(k,j) = 1$,
      examine the value of $M(i,k)$.
   b. If $M(i,k) = 0$, let $M(k,j) = 2$. Other-
      wise, let $M(i,j) = 2$.
   c. Continue until a single entry of 1
      remains.
5. The matrix is complete. For each i,
   examine all entries $M(i,j)$ for $j = 1$
   through n. Identify all entries of 1. If
   any EU i is the immediate ancestor of
   multiple EU's, perform these steps:
   a. Calculate the character states of the
      most recent common ancestor of each
      pair of derived EU's.
   b. If these calculated states differ from
      the character states of EU i, identify
      them as the states of a hypothetical
      EU.

SECONDARY CHARACTERS

Typically, primary characters suggest
major lines of descent but leave unresolved
questions of relationship within them.
Where more resolution of detail is desired,
the characters can often be augmented by
secondary characters which make new
distinctions. This is done by choosing a
subcollection of EU's that is convex on the
estimate provided by the primary characters and that contains the unresolved area.
In practice groups that are monophyletic
for this estimate are usually chosen. Characters not formerly compatible with all the
primary characters may now become compatible with them in the context of this
smaller collection of EU's. Each subordinate clique should contain as a subset the
clique used to define the phyletic line to
which the subordinate clique applies. This
could necessitate rejecting the largest subordinate clique in favor of a smaller one.

The idea of using secondary characters
for estimating relationships on restricted
parts of a phylogenetic tree is consistent
with the belief of many systematists that
some characters which vary within a large
group are only important for certain subgroups (Stebbins 1974). This process of
finding cliques for smaller convex subsets
of the study can be repeated as often as

appropriate. It is probably best to discontinue the analyses when the number of EU's, or characters which vary among them, becomes small relative to the entire data set.

COMPATIBILITY ANALYSES OF THE DATA OF BLACKITH AND BLACKITH

Blackith and Blackith (1968) presented a study of 12 Orthopteran taxa described by 92 cladistic characters. In our analyses of these data, we omitted their EU Proscopiidae because states for several of its characters were not available. Omitting the Proscopiidae necessitated omitting their character number 47 as well since only the Proscopiidae had the derived state of that character.

We used their decisions that the states scored 0 in their Table 2 represented the primitive state for each character and that the states scored 1 in their Table 2 represented the derived state for each character. The compatibility analysis of the 91 characters which varied among the 12 EU's for which all characters were scored yielded two largest cliques of 23 characters each:

Clique I: 1, 2, 12, 13, 15, 17, 20, 24, 29, 34, 35, 38, 42, 46, 59, 63, 68, 69, 70, 71, 72, 73, 74, 77, 79, 80, 81, 82, 84, 86, 87, 89, 91.

Clique II: 1, 2, 12, 13, 15, 17, 21, 24, 29, 34, 35, 38, 42, 46, 59, 63, 68, 69, 70, 71, 72, 73, 77, 78, 79, 80, 81, 82, 84, 86, 87, 89, 91.

These cliques differ by only two characters; characters 20 and 74 in Clique I are replaced by characters 21 and 78 in Clique II. These characters are involved only in determining the relationships among groups D, E, and F (code names of Le Quesne [1972] for each group are listed in the legend of Fig. 2). The 31 characters which both cliques have in common (the intersection of the cliques) were chosen as the set of primary characters. The tree defined by the primary characters is shown in Fig. 2. Open circles represent the hypothetical

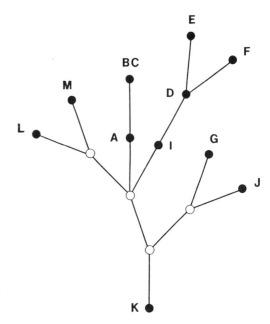

FIG. 2.—Tree determined by primary characters. In this figure and throughout, the associations between taxa and letters are: A, Gryllacrididae; E, Cyrtacanthacridinae; F, Pyrgomorphidae; G, Eumastacidae; I, Tetrigoidea; J, Phasmatodea; K, Dermaptera; L, Blattodea; M, Mantodea.

EU's generated by application of the algorithm described above; solid circles represent real EU's. When a real EU is connected directly to another real EU on this tree, such as F connected to D, it does not imply that one EU is the ancestor of the other, but only that the characters defining the tree do not distinguish some EU's from their ancestors. Thus, in Fig. 2, D, E, and F share a common ancestor, but none of the characters used to devise the tree separates D from the ancestor which it shares with E and F. This relationship is shown with dashed lines in Fig. 4.

To resolve further the relationships among groups A, B, C, D, E, F, I, L, and M, found to be a monophyletic group on the tree determined by the primary characters (Fig. 2), a second analysis was made of the 73 characters which varied

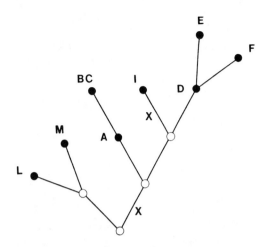

FIG. 3.—Tree determined by reanalysis of a convex group.

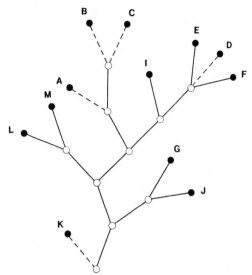

FIG. 4—Final estimate by compatibility analysis, with modern forms moved from ancestral positions with dashed lines.

among them. Two largest cliques of 35 characters were found:

Clique III: 3, 12, 14, 15, 19, 20, 26, 28, 29, 30 32, 33, 35, 36, 37, 38, 42, 49, 50, 51, 52, 59, 61, 62, 69, 72, 73, 74, 77, 82, 84, 86, 87, 88, 91;

Clique IV: 3, 12, 14, 15, 19, 21, 26, 28, 29, 30, 32, 33, 35, 36, 37, 38, 42, 49, 50, 51, 52, 59, 61, 62, 69, 72, 73, 77, 78, 82, 84, 86, 87, 88, 91.

Within the groups being analyzed, all of the characters in these cliques are compatible with all of the primary characters. As in the previous analysis, these cliques differ by only two characters; characters 20 and 74 in Clique III are replaced by characters 21 and 78 in Clique IV. The intersection of these cliques was taken as the best set of characters for this analysis. The characters which appear in this set but not in the list of primary characters (found in the first analysis) are the secondary characters for this study; they are important only in the smaller phyletic line analyzed in the second analysis. The tree defined by the second analysis (Fig. 3) shows that the secondary characters define two steps (marked with X's) unresolved by the primary characters. The results of the two

analyses are combined in Fig. 4 to give an overall estimate of the evolutionary relationships of these insects, as indicated by the Blackiths' data. In Fig. 4 any group not distinguished from its ancestor by the

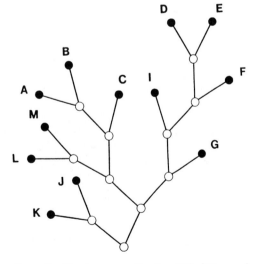

FIG. 5.—Parsimony estimate (Blackith and Blackith, 1968).

characters used is set off from its ancestor by a dashed line.

DISCUSSION

The Blackiths' (1968) data were chosen for this illustration because they are published and because they were used by Le Quesne (1972, 1974) to illustrate his methods. Le Quesne's (1969) original method for finding the "maximum number of characters which could be uniquely derived" is ancestral to our method, but there are important differences. Le Quesne's method is restricted to two-state characters and does not require characters to be cladistic as here defined. Only among cladistic characters does mutual compatibility imply simultaneous support of an estimate of phylogeny; but see McMorris (in press) for more discussion of this point. In any case, Le Quesne's (1972) methods of character selection do not entail mutual compatibility. For these reasons, none of the sets of characters he selected from the Blackiths' data are in fact cliques. Not surprisingly, therefore, none of his cladograms exactly agrees with the tree that we found (Fig. 4) from the same data.

Our tree also differs from that found by the Blackiths (Fig. 5); they used the method of Camin and Sokal (1965), which constructs a most parsimonious tree using all the data with the only restriction being that no character may have reversals of states. Fig. 5 suggests that the suborders Ensifera (Groups A–C) and Caelifera (Groups D–I) are not closely related. Our reanalysis of the Blackiths' data agrees with conventional classifications (Borror & Delong, 1971; Blackith & Blackith, 1968; Key, 1970) of the Orthoptera, that these two groups (with the exception of the Eumastacidae, Group G) are sister groups. A close relationship between the Blattodea (Group L) and the Mantodea (Group M) is indicated by all analyses discussed here. Our results indicate a close relationship between the Eumastacidae (G) and the Phasmatidae (J), while the method of Camin and Sokal indicates a close rela-

tionship between the Phasmatidae (J) and the Dermaptera (K). Neither of these results agrees with the conventional views of the relationships of these groups (Key, 1970).

As the Blackiths recognized, their data set is inadequate in several ways. Among these are underrepresentation of the diversity of the groups studied and the use of "composite specimens" comprised of parts taken from different species and in some cases from different genera. Furthermore, the data heavily represent characters not previously used to estimate relationships in the Orthoptera while omitting many of the characters traditionally used for this purpose. Thus, whether or not the relationships that have been estimated from the Blackiths' data seriously challenge currently held views of relationships within the Orthoptera awaits further study.

The theoretical and mathematical basis for the method of character compatibility has been discussed elsewhere (Estabrook, 1972; Estabrook et al., 1975, 1976a, 1976b) and the method has been successfully applied to studies of relationships of cichlid fishes (Cichocki 1976) and of charadriiform birds (Strauch 1976). In this paper we have discussed its relationship to some systematic practices of long standing and exemplified it with the data of Blackith and Blackith (1968). Source code, instructions, and a worked example for a computer program which finds the compatibilities in a data set, finds cliques from the resulting compatibility matrix, and constructs tree diagrams from the resulting cliques, are available from the authors.

ACKNOWLEDGMENTS

We would like to thank The University of Michigan Computing Center, Dr. R. C. F. Bartels, Director, and Division of Biological Sciences, Dr. W. R. Dawson, Chairman, for the computer services that supported this study. KLF was supported by a NSF Graduate Fellowship during the completion of this work.

REFERENCES

BISBY, F. A. 1970. The evaluation and selection of characters in Angiosperm taxonomy: an example from *Crotalaria*. New Phytologist 69: 1149–1160.

BLACKITH, R. E., AND R. M. BLACKITH. 1968. A numerical taxonomy of Orthopteroid insects. Austr. J. Zool. 16:111–131.

BORROR, D. J., AND D. M. DeLONG. 1971. An introduction to the study of insects, 3rd Ed., Holt, Rinehart and Winston, New York.

CAMIN, J. H., AND R. R. SOKAL. 1965. A method for deducing branching sequences in phylogeny. Evolution 19:311–326.

CICHOCKI, F. G. 1976. Cladistic history of cyclid fishes and reproductive strategies of the American genera *Acarichthys*, *Biotodoma*, and *Geophagus*. Vol. I. Ph.D. Thesis, University of Michigan, Ann Arbor, Michigan.

CROVELLO, T. J. 1970. Analysis of character variation in ecology and systematics. Ann. Rev. Ecol. Syst. 1:5–98.

ESTABROOK, G. F. 1971. Some information theoretic optimality criteria for general classification. Math Geol. 3:203–207.

ESTABROOK, G. F. 1972. Cladistic methodology: A discussion of the theoretical basis for the induction of evolutionary history. Ann. Rev. Ecol. Syst. 3:427–456.

ESTABROOK, G. F., C. S. JOHNSON, JR., AND F. R. McMORRIS. 1975. An idealized concept of the true cladistic character. Math BioSci. 23: 263–272.

ESTABROOK, G. F., C. S. JOHNSON, JR., AND F. R. McMORRIS. 1976a. An algebraic analysis of cladistic characters. Discrete Math. 16:141–147.

ESTABROOK, G. F., C. S. JOHNSON, JR., AND F. R. McMORRIS. 1976b. A mathematical foundation for the analysis of cladistic character compatibility. Math BioSci. 29:181–187.

FARRIS, J. S. 1974. Formal definitions of paraphyly and polyphyly. Syst. Zool. 23:548–554.

GILES, E. T. 1963. The comparative external morphology and affinities of the Dermaptera. Trans. R. Ent. Soc. London 115:95–164.

HAECKEL, E. 1866. Generelle Morphologie der Organismen. Berlin.

HENNIG, W. 1966. Phylogenetic Systematics (translated by D. D. Davis and R. Zangerl). Univ. of Illinois Press, Urbana, Illinois.

HUXLEY, T. H. 1867. On the classification of birds; and on the taxonomic value of the modifications of certain cranial bones observable in that class. Zool. Soc. Lond., Proc. 1967:415–472.

JARDINE, N. 1969. The observational and theoretical components of homology: a study based on the morphology of the dermal skull-roofs of rhipidistian fishes. Biol. J. Linn. Soc. 1:327–361.

KEY, K. H. L. 1970. Orthoptera. *In* Insects of Australia, Melbourne U. Press, Carlton, Victoria, pp. 323–347.

LE QUESNE, W. J. 1969. A method of selection of characters in numerical taxonomy. Syst. Zool. 18:201–205.

LE QUESNE, W. J. 1972. Further studies based on the uniquely derived character concept. Syst. Zool. 21:281–288.

LE QUESNE, W. J. 1974. The uniquely evolved character concept and its cladistic application. Syst. Zool. 23:513–517.

McMORRIS, F. R. 1975. Compatibility criteria for cladistic and qualitative taxonomic characters. *In* Estabrook, G. F. (ed.), Proceedings of the Eighth International Conference on Numerical Taxonomy. W. H. Freeman and Company, San Francisco.

McMORRIS, F. R. (in press). On the compatibility of binary qualitative taxonomic characters. Accepted for publication by Bull. Math. Biol.

STRAUCH, J. G., JR. 1976. The cladistic relationships of the Charadriiformes. Ph.D. Thesis, University of Michigan, Ann Arbor, Michigan.

SPORNE, R. K. 1975. The Morphology of Angiospermes. Cambridge University Press, Cambridge, England.

STEBBINS, G. L. 1974. Flowering plants: Evolution above the species level. Belknap Press, Cambridge, Massachusetts.

WILSON, E. O. 1965. A consistency test for phylogenies based on contemporaneous species. Syst. Zool. 14:214–220.

Manuscript received June, 1976
Revised April, 1977

Part VI

STATISTICAL APPROACHES TO
PHYLOGENETIC INFERENCE

Editors' Comments
on Papers 17, 18, and 19

Edwards and Cavalli-Sforza (Paper 17) introduced two important concepts in cladistics. The idea of methodological parsimony (minimum evolution) was first introduced in an abstract for a talk in 1964 and later described in more detail in the paper included here. Their second major contribution was to view the estimation of branching patterns of evolution in a statistical sense. In their model, such estimation is analogous to a random-walk phenomenon with a constant probability of branching and a constant rate of walking. They present undirected cladograms for human blood groups superimposed on the geographical distribution of these groups to illustrate the use of this method.

Fitch and Margoliash (Paper 18) also made an early and very significant contribution to the use of statistical inference to estimate evolutionary history. Their method has been most influential in molecular approaches to cladistics. Theirs is a parsimony method based on unweighted pair-group clustering. In this method, the percentage standard deviations of the difference between the tree distances and the observed distances are minimized.

During the last ten years, the most ambitious effort to view the entirety of cladistic methodology in a statistical framework has been undertaken by Felsenstein. Paper 19 outlines the basis for his maximum-likelihood model of cladistic methods and describes how existing methods perform within such a model and when they are likely to fail. Finally, the relationships between parsimony and compatibility methods are outlined, demonstrating that within such a statistical framework, each class of methods has advantages and limitations.

17

Reconstruction of Evolutionary Trees

A. W. F. EDWARDS & L. L. CAVALLI-SFORZA

*(International Laboratory of Genetics and Biophysics, Naples (Pavia Section),
Istituto di Genetica, Università di Pavia, Italy)*

CONTENTS

1. INTRODUCTION

THE extent to which the classification of a group of organisms may be used to make inferences about phylogenetic relationships has long been a controversial point in biology. Today, particularly in view of the development of 'numerical taxonomy' it is important to consider, more deeply than hitherto, the whole question of evolutionary inferences and their relation to taxonomy. This paper contains some suggestions for studying the problem, which lead to practical methods of estimation. First, however, it is convenient to consider the logical nature of taxonomic studies in general.

2. TAXONOMY

We suspect that the present confusion of ideas about the logical basis of taxonomy is partly due to a failure to distinguish between different types of taxonomic enquiry, and that this failure may have been induced by the fact that all taxonomic studies use much the same type of data. Every kind of taxonomic enquiry must have its own particular purpose and logical basis; indeed, the logical structure must arise naturally from the particular purpose, which therefore needs to be specified in detail. A classification created with no particular purpose in mind is likely to be logically indefensible and partially useless; for when it is put to a specific* use, for which it was not, of course, designed, it will

---

* We use *specific* and its derivatives in their everyday senses.

most probably be inefficient. In the same way, a motor vehicle built without a specification is unlikely to turn out to be a fast sports car: it is more likely to be a heterogeneous collection of ill-suited mechanical bits and pieces.

Though the need to specify the purpose of a taxonomic study has, from time to time, been admitted, some confusion has been generated by the concept of the 'general purpose' classification. To say that the purpose of a classification is "general" is, in our view, too vague to be of use in its construction. If we may return to the motor vehicle, a 'general purpose' car can be specified only because the particular purposes to which it may be put, and their relative importance, can themselves be specified rather accurately. Thus the designer may anticipate that the proposed vehicle will spend 50% of its time carrying only the driver, 20% carrying two people, and so forth, and that it will be required to carry so much in the way of goods so often, and go up hills of such-and-such a gradient once a year on average. On the basis of such information an acceptable compromise design will be achieved. There are three important things to note in this example. First, that 'general purpose' is only meaningful if it is regarded as a collection of weighted specific purposes: secondly, that the outcome of a general purpose specification is unlikely to fulfil any of the specific purposes as well as a specially designed vehicle could; and, thirdly, that the justification of the general purpose vehicle is an economic one, because for optimum service all the time one would need as many vehicles as there are uses, but this would be very expensive: only rich farmers can afford a Jaguar *and* a Land-Rover.

In taxonomy the situation is much the same; the only justification for a general purpose classification is that we cannot afford the time or the money to create the separate specific classifications that we really need. It seems to us to have no purely scientific justification, and it is surely an illusion to imagine that a 'general purpose' classification approaches an unattainable 'ideal' one in the same way that a circle drawn on a piece of paper approaches a perfect circle. We thus hold that the most essential prerequisite of a classification is its purpose, clearly defined, and that this is no less true in the case of a 'general purpose' classification, which should only seek to be a compromise solution to the problem of satisfying specific purposes economically.

Given the purpose to which a classification is to be put, its design, the choice and weighting of the characters to be used, and indeed its logical structure, follow, just as the design of a motor car follows from its specification. Naturally, there is likely to be room for subjective judgements in the choice and weighting of characters, and we must not be surprised to find two taxonomists disagreeing even when the specification of the classification seems quite rigid, just as there is room for motor car designers to interpret a specification subjectively. But the scope for subjective decisions is directly related to the comprehensiveness of the specification, which should be complete if objectivity is very important. Conversely, when there is virtually no specification at all, as in the 'general purpose' case as used hitherto, there is immense scope for subjective decisions, and attempts to find a logical justification where none exists are doomed to death by drowning in a sea of indefensible statements.

Rather as Fowler apologizes for taking all his examples of bad English from *The Times*, and insists that this is not because *The Times* contains more bad English than most newspapers, but because it is the newspaper he prefers, so we apologize for taking some statements about general purpose classification from the admirable summary of Numerical Taxonomy by Sneath and Sokal (1962) and quoting them:

"Such [natural] classifications may be termed 'general' and are for general purposes, in distinction to classifications made with a special purpose in mind, which may be called 'special' classifications. This implies among other things that in constructing general classifications characters have equal weight...".

"The taxonomic equivalence of all characters can be seen most clearly if we attempt to construct an objective criterion for weighting characters. If we cannot decide how to weight the features we must give them equal weight, unless we propose to allocate weight on irrational grounds".

"We hold the view that a 'natural' or orthodox taxonomy is a general arrangement intended for general use by all kinds of scientists; it cannot therefore give greater weight to features of one sort, or it ceases to be a general arrangement".

The logical frailty of these attempts to justify a certain procedure when the purpose of the intended classification has not been carefully specified should be a warning to us all.

There seem to us to be three major types of taxonomic study, each of entirely different logical structure, and it is of the utmost importance to differentiate them. The first, which we may call Pure Taxonomy, is solely concerned with the processing of information. It is descriptive. Each organism can only be described in terms of the characters it exhibits, but since some of these characters will be common to two or more of the organisms, it is clearly unnecessary to describe each organism separately, and some form of classification will present all the information in a simpler form. This approach has been developed by Maccacaro (1958) and Rescigno and Maccacaro (1961), and has not, in our view, received the attention it deserves. It should be noted that the problem of choosing and weighting characters is irrelevant in *pure* taxonomy: *given* the characters, the problem is to classify the organisms according to them in such a way that the maximum amount of information is retained. Thus, given a set of snooker balls, it is possible to classify the balls by colour without loss of any information about this character.

The second type of taxonomy may be called "applied", for it seeks to classify the organisms in such a way that the resulting classification shall be the best obtainable for the specified purpose at hand. Thus tyre manufacturers are primarily interested in classifying motor vehicles by the size and number of their wheels, ferry operators by their lengths, and customs officials by their contents. Once the purpose of the classification has been clearly specified, the taxonomist can set about choosing and weighting the characters he considers appropriate, though, as we mentioned above, if the specification is not sufficiently comprehensive, different taxonomists may have different views about the characters.

The third type of taxonomy involves what Sneath and Sokal (1962) call "phylogenetic speculation". We may refer to it as phylogenetic taxonomy, and note that it does not primarily involve classification, since its avowed purpose is the estimation of the evolutionary tree which supposedly unites the organisms being studied. Characters must be chosen and weighted according to their evolutionary significance, and judgement of this is likely to be subjective. However, the more that is known about the genetics of the organisms, the easier it will be to agree about the relevance of characters. In Pavia over the past two years we have studied in considerable detail the problem of the statistical estimation of the form and dimensions of evolutionary trees, and we describe our current approach below. The logical basis of such studies seems to be rather simple, probably because their purpose can be so simply and clearly defined, but it is important to realise that the procedure is not primarily classificatory, although, once an estimate of the form of an evolutionary tree has been produced, organisms can of course be grouped according to relative evolutionary divergence.

Our three-taxon classification of taxonomy, and our belief that it holds the key to a satisfactory logical structure of taxonomic studies, will, no doubt, be criticized. It may well be asked, for example, whether the classification of sheep as sheep and goats as goats * is pure, applied, or phylogenetic in type, and, if applied, what is the specific purpose of the classification? Is it not really a general purpose classification? Our answer to this is that the classification *was* indeed applied, but that the original division into sheep and goats was for a purpose which is lost in antiquity. However, it was found that the division was useful in other ways, or, to look from a different point of view, classification of the animals for other purposes frequently led to the same division. The division was evidently rather stable, so much so that it was appropriate to name the animals in the two classes as sheep and goats. Such a classification is acceptable to us for two main reasons: first, because sheep and goats are really rather different, and intermediate forms do not exist, so that there is never any doubt as to whether an animal is a sheep or a goat. The subject matter of the classification is 'robust' with respect to the different purposes to which the classification is to be put. Only robust classes of objects acquire names. But secondly, and more important, the classification is acceptable because we understand its limitations. We know that classification into sheep and goats is not much use if we are interested in the proportion of the combined flock that is male, or the proportion that is under a certain age. In fact

* The terms "sheep" and "goat" refer here to the West European domesticated races of *Ovis aries* L. and *Capra hircus* L., respectively.

our classification is not really general purpose at all; though it arose out of an applied taxonomy, because of the stability of the classes, it has become a pure taxonomy, descriptive in intent. When we say "sheep" we mean, according to the Oxford Dictionary, a "timid gregarious beast kept in flocks for the mutton or lamb and wool and leather it yields". A goat, on the other hand, is "a lively wanton strong-smelling usually horned and bearded ruminant quadruped".

Our present problem, however, is not that of considering how classifications arose historically, because we are trying to set up a classification which is logically acceptable, and does not rely on age-old decisions. It is as though we were confronted with a field full of sheep and goats for the first time, and were asked to classify the animals. Our reaction must be to ask to what purpose the classification will be put, or, alternatively, to ask what are the characters that we should consider. Thus we can embark on an applied or a pure taxonomy, whichever is appropriate. But if we are asked to classify the animals "for general purposes" we must insist on knowing *what* purposes, though, of course, there is nothing wrong in admitting that the animals seem at first sight to be divisible into two homogeneous classes, and that this classification will no doubt serve the majority of purposes included under the title "general". *But* this fact derives from the particular data we are using, and cannot be used as a justification for a taxonomic procedure whose logical structure must be independent of the organisms being considered. The fact that different characters and different taxonomic methods lead to the same segregation of the animals into sheep and goats tells us something about sheep and goats, but nothing about the logical basis of taxonomy.

3. Reconstruction of Evolutionary Trees

Having detached phylogenetic taxonomy from purely classificatory studies, we may now proceed to examine it more closely. The characters which are fundamental to differentiation are the base-sequences of DNA. These are at present unavailable directly, although methods have recently been suggested to measure the extent to which two DNA's match each other for base sequence. In the absence of such data, use must be made of genotypic, or even phenotypic, differences. At the genotypic level the relative importance of various gene substitutions in evolution is a matter which only detailed study can reveal; it will depend upon the past and present selective forces, mutation rates and population structures and sizes. At the phenotypic level such information is even less readily available.

As has been found with classification, it is convenient to represent the variation amongst the organisms geometrically, the chosen characters being represented by axes in a multidimensional space, and the organisms, specified by these characters, being represented as points in this hyperspace. If a time dimension, everywhere normal to the character space, is added, the course of evolution — were it but known — could be seen as a 'tree', sometimes branching as groups of organisms diverged, sometimes growing together as hybridization took place, and with many branches ending in extinction before the remainder intercepted the 'now' character space to indicate the current disposition of the organisms. This approach presupposes that the character-space is continuous, which will normally be an acceptable approximation. In cases in which discreteness is apparent the space will take the form of a lattice of points.

The proper basis for a study of evolutionary divergence will be provided by means of that transformation of the space-time which, as far as is known, makes a unit vector, in whatsoever direction (normal to time) and in whatsoever part of the space-time, correspond to a fixed amount of evolutionary divergence — divergence being properly defined for the problem under consideration. Such a transformed space-time would thus be everywhere isotropic with respect to evolutionary progress. Transformation is, in this context, equivalent to character-weighting.

Put thus formally, it will be obvious that the knowledge necessary for the reconstruction of evolution is rarely available, but, as in other problems of statistical estimation, deductions can always be made provided the assumptions on which they rest are remembered.

The present problem is to estimate the form of the evolutionary tree given only the information contained in the 'now' character-space. Fossils, which would appear at other time levels in space-time, will not be treated at present, although we would like to stress particularly that it is clear from this approach that there is no substantial logical difference about estimating the course of evolution with or without fossil evidence.

In our isotropic space we may think of evolution as a branching random walk, with a constant probability of branching and a constant rate o walking: that is, after the elapse of a certain time interval, the probability distribution in space of the position of a population will be normal, with mean at its original position and variance proportional to the time elapsed. The constant of proportionality will depend on many parameters, such as the population size and the type and intensity of selection, and will usually be unknown. Since an evolutionary tree uniting n points (without loops) is bound to contain $n-1$ branching points, the probability of branching is not a parameter which has to be estimated, although it is worth noting that the theory of birth and death processes may have some interesting things to say about the expected form of an evolutionary tree.

Now the probability density at a distance d spatially and t temporally from a point in the space-time of p spatial dimensions is given by

$$\sigma^{-p} (2\pi t)^{-\frac{1}{2}p} \exp \{-d^2/(2t\sigma^2)\}$$

where σ^2 is the constant of proportionality mentioned above. The log-likelihood is therefore given by

$$-\{d^2/(2t\sigma^2) + \tfrac{1}{2}p \log (2t\sigma^2) + \tfrac{1}{2}p \log \pi\}$$

Writing T for $2t\sigma^2$, omitting the constant, and changing the sign, the expression becomes

$$d^2/T + \tfrac{1}{2}p \log T$$

Each arm of a postulated evolutionary tree will have an expression of this type associated with it, d being its spatial length and T proportional to its temporal length. Thus the likelihood of the tree will be maximized if the sum of these expressions, over all the arms, is minimized (since we have changed the sign, and the branching points, being constant in number, are irrelevant).

We have found that the mathematics of the estimation procedure can be made quite compact using a matrix notation, and we do not anticipate any major difficulties in the corresponding computer programmes, although these are not yet working. However, as will be mentioned below in connexion with our example, other methods have been used which may be expected to give similar results.

Thus maximum-likelihood estimation is not difficult once the topological form of the tree has been specified, but unfortunately there will generally be too many forms for the maximum likelihood of each to be evaluated: with n points to unite there are $(2n-5)!/(n-3)!\ 2^{n-3}$ trees, or more than two million with ten points. We therefore have to resort to some prior method of clustering. Initially we developed our own method, based on the analysis of variance (Edwards and Cavalli-Sforza, 1963b), but we have finally adopted a method which is particularly simple and rapid. Prim (1957) has shown that the network of minimum length uniting n points, but with the segments constrained to meet only at these given points, can be found by listing all the ditances between points in increasing order, and successively allocating segments to these distances, omitting any segment which completes a loop. The resulting net gives some indication of the types of tree it will be worthwhile to use in the maximum-likelihood estimation procedure. The relationship is somewhat intuitive at the moment, but once one is "on the track" of a good topology, experience shows what sort of changes will be likely to increase the overall likelihood. In particular, if the length of an arm is estimated as zero, a change in the topology of that part of the tree is indicated.

We are now left with the problem of finding the correct transformation of the character space. The transformation is, of course, a reflection of the genetic assumptions which are being made, and these will be peculiar to each case. For the moment it will be sufficient to indicate how we have proceeded in a particular example.

4. The Recent Evolution of Man

(a) *Materials*

As our example, we decided to study the phylogeny of fifteen samples of human populations, using as characters the frequencies of various blood-group alleles. This choice was made because of the availability of the data, the intrinsic interest of the problem, and the amount of knowledge about the genetic situation. Data on the blood-group systems A_1A_2BO, Rh (four sera), MNSs, Fy and Di were obtained for fifteen populations, three from each continent, with the invaluable help of Dr. Mourant and Mrs. Sobczak, of the Medical Research Council Blood Group Reference Library, London. In all, twenty 'genes' were distinguished. Nearly every sample consisted of more than 70 individuals, and was tested for each of the five groups; although there is a certain amount of heterogeneity with respect to the sera used, and the methods of calculating the gene frequencies, these data were considered adequate for a trial of the methods.

The five blood groups were assumed to be independent, because of the absence of evidence for linkage between them. Each allele within a system, and each of the systems, was given equal weight. In view of the apparent absence or weakness of selective forces in these blood groups, and, in cases where differential mortality has been demonstrated, of the impossibility of setting up selective models of sufficient validity, it has been assumed for our preliminary work that selective forces are absent, and that the observed divergences are, or can be treated as being, due to drift. Boyd's comment (Boyd, 1963), that "unless the blood groups are adaptive, they are not going to be very useful in racial classification", is not acceptable: any variable, even a random one, which shows a persistently high correlation between successive generations, may be useful. If differential selective forces are operating, the simultaneous use of several independent loci should render their effects on the analysis minimal. All the populations have been treated as though they were, and always have been, of the same size and structure.

(b) Methods

It is now necessary to set up a character-space on the basis of these assumptions, following the methods outlined above. Since the normal scale of gene frequencies is anisotropic in the sense that the effects of drift are differently measured in different parts of the scale, a transformation is necessary. The 'square-root' transformation given by

$$\sin^2\theta = p, \qquad \cos^2\theta = q,$$

is appropriate for two alleles at a locus (Fisher, 1954), since it stabilizes the binomial variance throughout the range, and Fisher (personal communication) has pointed out that it may be extended to the case of many alleles. With n alleles, populations will be represented as points on $(1/2^n)$th of the surface of the unit hypersphere in n dimensions, the angular distance between populations with gene frequencies $(p_1, q_1, r_1,...)$ and $(p_2, q_2, r_2,...)$ being given by

$$\text{Cos}\,\alpha = \sqrt{p_1p_2} + \sqrt{q_1q_2} + \sqrt{r_1r_2}+...$$

Since the character-space is curved, it will be necessary to transform it into a Euclidean space before the available maximum-likelihood treatment can be applied. As in map projections, some desirable qualities of the attribute-space will be lost in the transformation. The most appropriate transformation is worth a separate study, but the simplest approach is to use, as the measure of distance between two points, the chord — the Euclidean straight line — rather than the arc of the great circle on the surface of the hypersphere. This is equal to $\sqrt{(2 - 2\cos a)}$. The maximum possible error thus incurred is given by the ratio of the length of the arc of a quadrant of a circle to that of the corresponding chord, or $\sqrt{2} : \frac{1}{2}\pi$, amounting to a maximum underestimate of distance of 10%.

Given the Euclidean distance between two populations for each locus, the total distance over all loci is found by taking the square root of the sum of the squared distances for individual loci, by Pythagoras' theorem, since the character-spaces for the loci are mutually orthogonal. The resulting array of pairwise distances between populations is ready for finding the Prim network. In order to use the maximum-likelihood computer program it is necessary to set up a system of Cartesian coordinates for the points representing the populations, and this may be done on a computer from the pairwise distances. The space thus created is the final version of the character-space, but with an arbitrary frame of reference. The unit distance, however, is not arbitrary, and corresponds to an amount of evolution equal to that incurred in a single gene substitution at a single locus.

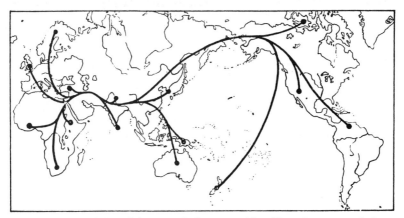

FIG. 1. Topology of the minimum-evolution tree uniting fifteen human populations; constructed on the basis of the frequency of blood-group alleles.

Unfortunately, as has been mentioned, our maximum-likelihood programmes are not yet working, but, using this character-space, our earlier "method of minimum evolution" (Edwards and Cavalli-Sorza, 1963a) gave a 'best' tree of the topological form shown in the figure (Fig. 1). It is probable that this method gives a tree which is approximately the same as the projection of the maximum-likelihood tree onto the 'now' character space. Whilst we would be the first to admit that we have been fortunate to get such a reasonable result from preliminary data, we feel that it is at least encouraging, and justifies our intention to develop our methods further, and apply them to more extensive bodies of information.

5. ACKNOWLEDGEMENTS

This work has been supported by a grant from the U. S. Atomic Energy Commission and by Euratom-CNR-CNEN contract no.012-61-12 BIAI.

REFERENCES

BOYD, W. C., 1963. Genetics and the human race. *Science*, **140**: 1057-1064.

EDWARDS, A. W. F., and CAVALLI-SFORZA, L. L., 1963a. The reconstruction of evolution. Unpublished paper read at 142nd meeting of the *Genetical Society of Great Britain*, London, July 1963. (Abstract in *Ann. hum. Genet.*, **27**: 104-105, and in *Heredity, Lond.*, **18**: 553.)

—— —— 1963b. A method for cluster analysis. Unpublished paper read at *5th Int. Biometric Conf.*, Cambridge, 1963.

FISHER, R. A., 1954. *Statistical Methods for Research Workers*, ed. 12. Oliver & Boyd, Edinburgh. 356 pp.

MACCACARO, G. A., 1958. La misura della informazione contenuta nei criteri di classificazione. *Annali Microbiol.*, **8**: 231-239.

PRIM, R. C., 1957. Shortest connection networks and some generalizations. *Bell Syst. tech. J.*, **36**: 1389-1401.

RESCIGNO, A., and MACCACARO, G. A., 1961. The information content of biological classifications. In: C. Cherry (ed.), *Information Theory*, 437-446. Butterworth and Company, London.

SNEATH, P. H. A., and SOKAL, R. R., 1962. Numerical taxonomy. *Nature, Lond.*, **193**: 855-860.

18

Copyright © 1967 by the American Association for the Advancement of Science
Reprinted from *Science* **155:**279–284 (1967)

Construction of Phylogenetic Trees

A method based on mutation distances as estimated from cytochrome *c* sequences is of general applicability.

Walter M. Fitch and Emanuel Margoliash

Biochemists have attempted to use quantitative estimates of variance between substances obtained from different species to construct phylogenetic trees. Examples of this approach include studies of the degree of interspecific hybridization of DNA (*1*), the degree of cross reactivity of antisera to purified proteins (*2*), the number of differences in the peptides from enzymic digests of purified homologous proteins, both as estimated by paper electrophoresis-chromatography or column chromatography and as estimated from the amino acid compositions of the proteins (*3*), and the number of amino acid replacements between homologous proteins whose complete primary structures had been determined (*4*). These methods have not been completely satisfactory because (i) the portion of the genome examined was often very restricted, (ii) the variable measured did not reflect with sufficient accuracy the mutation distance between the genes examined, and (iii) no adequate mathematical treatment for data from large numbers of species was available. In this paper we suggest several improvements under categories (ii) and (iii) and, using cytochrome *c*, for which much precise information on amino acid sequences is available, construct a tree which, despite our examining but a single gene, is remarkably like the classical phylogenetic tree that has been obtained from purely biological data (*5*). We also show that the analytical method employed has general applicability, as exemplified by the derivation of appropriate relationships among ethnic groups from data on their physical characteristics (*6, 7*).

Dr. Fitch is an assistant professor of physiological chemistry at the University of Wisconsin Medical School in Madison. Dr. Margoliash is head of the Protein Section in the Department of Molecular Biology, Abbott Laboratories, North Chicago, Illinois.

Determining the Mutation Distance

The *mutation distance* between two cytochromes is defined here as the minimal number of nucleotides that would need to be altered in order for the gene for one cytochrome to code for the other. This distance is determined by a computer making a pairwise comparison of homologous amino acids (8). For each pair a *mutation value* is taken from Table 1 which gives the minimum number of nucleotide changes required to convert the coding from one amino acid to the other. The table is derived from Fig. 2 of Fitch (9) except that, as a result of the work of Weigert and Garen (10) and Brenner, Stretton, and Kaplan (11), the uridyl-adenosylpurine trinucleotide is now treated as a chain-terminating codon. This change of codon meaning, although it does not affect the method of calculation, does cause the mutation values for amino acid pairs involving glutamine with cysteine, phenylalanine, tyrosine, serine, and tryptophan to become 1 greater than in the table previously published (12). Also, misprints involving the leucine-glycine and valine-cysteine pairs have been corrected. To maintain homology, deletions, all of which occur near the ends of the chains, are represented by X's. The amino- and carboxyl-terminal sequences in which deletions occur are shown in Table 2. Thus all cytochromes are regarded as being 110 amino acids long. If the homologous pairing includes an X, no mutation value is assigned.

For each possible pairing of cytochromes, the 110 mutation values found are summed to obtain the minimal mutation distance. For purposes of calculation, these mutation distances are proportionally adjusted to compensate for variable numbers of pairs of residue positions in which at least one member contains an X. For example, the number of X-containing amino acid pairs occurring between the *Saccharomyces* and *Candida* cytochromes c is 1, whereas that between two mammalian cytochromes c is 6. Thus the known mutation distance of the former pairing is multiplied by 110/109 whereas that of the latter is multiplied by 110/104. The results for 20 known cytochromes c, rounded off to the nearest whole number, are shown in the lower left half of Table 3.

The basic approach to the construction of the tree is illustrated in Fig. 1, which shows three hypothetical proteins, A, B, and C, and their mutation distances. There are two fundamental problems: (i) Which pair does one join together first? (ii) What are the lengths of legs a, b, and c?

As a first approximation, one solves problem (i) simply by choosing the pair with the smallest mutation distance, which in this case is A and B, with a distance of 24. Hence A and B are shown connected at the lower apex in Fig. 1. To solve the second problem, one notes that the distance from A to C, 28, is 4 less than the distance from B to C. Hence there must have been at least 4 more countable mutations in the descent of B from the lower apex than in the descent of

Table 2. Areas of cytochromes c involving deletions. The first seven and the last four amino acids of the cytochromes c for the 20 species studied are shown. Deletions are represented by X's. Sequences are reported in the single-letter code of Keil et al. (21), a key to which is provided in Table 1.

Amino terminal positions 1–7	Organism	Carboxyl terminal positions 107–110
PLPFGQY	*Candida*	LXSI
XEGFILY	*Saccharomyces*	LXCG
XXYFSLY	*Neurospora*	LELX
XXYVPLY	Moth	SEXI
XXYVPLY	Screwworm fly	LSEI
XXXXXXY	Tuna	LESX
XXXXXXY	All other vertebrates	No deletions

A. Thus if $a + b = 24$ and $b - a = 4$, then $a = 10$, $b = 14$, and therefore $c = 18$. Note that an exact solution is obtained from which a reconstruction of the mutation distances precisely matches the input data.

When information from more than three proteins is utilized, the basic procedure is the same, except that initially each protein is assigned to its own subset. One then simply joins two subsets to create a single, more comprehensive, subset. This process is repeated according to the rules set forth below until all proteins are members of a single subset. A phylogenetic tree is but a graphical representation of the order in which the subsets were joined.

In the present case, we start with 20 subsets, each subset consisting of a single cytochrome c amino acid sequence. To determine which two subsets should be joined, all possible pairwise combinations of subsets are in turn assigned to sets A and B, with all remaining subsets in each case assigned to set C. In each alternative test all proteins are thus a part of one of the three sets. The three sets are treated exactly as in the preceding example, except that now the mutation distances used are averages determined from every possible pairing of proteins, one from each of the two sets whose average mutation distance is being calculated.

One arbitrarily accepts, from among all the possible pairings examined, that assignment of protein subsets to sets A, B, and C which provides the lowest average mutation distance from A to B. The leg lengths are then calculated and recorded. Henceforth the proteins of A and B so joined are treated as a single subset, and the entire procedure described in the preceding paragraph is repeated. Thus the number of subsets, originally equal to the number of pro-

Table 1. Mutation values for amino acid pairs. Each value is the minimum number of nucleotides that would need to be changed in order to convert a codon for one amino acid into a codon for another. The table is symmetrical about the diagonal of zeros. Letters across the top represent the amino acids in the same order as in the first column and conform to the single-letter code of Keil, Prusik, and Sörm (21).

	A	C	E	F	G	H	I	L	M	N	O	P	Q	R	S	T	U	V	W	Y
Aspartic acid	0	2	2	2	1	1	2	1	3	1	1	2	2	2	3	2	1	2		1
Cysteine	2	0	2	1	3	2	3	2	3	2	1	2	3	1	1	1	2	2	2	1
Threonine	2	2	0	2	2	2	1	1	1	1	2	1	2	1	1	2	2	2	1	2
Phenylalanine	2	1	2	0	3	2	3	2	2	2	1	3	2	1	2	1	1	1		2
Glutamic acid	1	3	2	3	0	2	1	1	2	2	2	2	1	2	2	2	2	1	3	1
Histidine	1	2	2	2	2	0	2	2	3	1	1	1	1	1	2	3	1	2	2	2
Lysine	2	3	1	3	1	2	0	2	1	1	2	2	1	1	2	2	2	2	2	2
Alanine	1	2	1	2	1	2	2	0	2	2	2	1	2	2	1	2	2	1	2	1
Methionine	3	3	1	2	2	3	1	2	0	3	2	2	1	2	2	1	1	1		2
Asparagine	1	2	1	2	2	1	1	2	2	0	1	2	2	2	1	3	2	2	1	2
Tyrosine	1	1	2	1	2	1	2	2	3	1	0	2	2	2	1	2	2	2	2	2
Proline	2	2	1	2	2	1	2	1	2	2	0	1	1	1	2	1	1	2		2
Glutamine	2	3	2	3	1	1	1	2	2	2	2	1	0	1	2	2	1	2	3	2
Arginine	2	1	1	2	2	1	1	2	1	2	2	1	1	0	1	1	1	2	2	1
Serine	2	1	1	1	2	2	2	1	2	1	1	1	2	1	0	1	1	2	1	1
Tryptophan	3	1	2	2	3	3	2	2	3	3	2	2	1	1	1	0	1	2	3	1
Leucine	2	2	2	1	2	1	1	1	1	2	2	1	2	1	1	1	0	1	1	2
Valine	1	2	2	1	1	2	1	1	1	2	2	2	2	2	2	2	1	0	1	1
Isoleucine	2	2	1	1	3	2	2	1	1	2	2	3	2	3	1	3	1	1	0	2
Glycine	1	1	2	2	1	2	2	1	2	2	2	2	1	1	1	2	1	2		0

teins (N), is reduced by 1 with each cycle. In this fashion, after $N-1$ joinings of subsets, the initial phylogenetic tree will have been produced. Because average mutation distances are now being used, the solutions obtained are very unlikely to permit an exact reconstruction of the input data.

Testing Alternative Trees

Because of the arbitrary nature of the rule by which proteins are assigned to sets A and B, the initial tree will not necessarily represent the best use of the information. To examine reasonable alternatives, one simply constructs another tree by assigning an alternative pair of protein subsets to sets A and B whenever the mutation distance between the two subsets is not greater by some arbitrary amount than that between the members of the initial pair used in constructing the initial phylogenetic tree (13). The tree that is less satisfactory on the basis of criteria set forth below is discarded, and other alternatives are tested.

The best of 40 phylogenetic trees so far examined is presented in Fig. 2. Each juncture is located on the ordinate at a point representing the average of all distances between the juncture and the species descendant from it. The mutation distance to any one descendant may be more or less than the ordinate value.

By summing distances over the tree, it is possible to reconstruct values (upper right half of Table 3) comparable to the original input mutation distances (lower left half of Table 3).

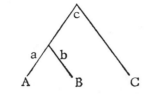

Mutation Distances

	B	C
A	24	28
B		32

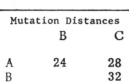

Fig. 1. Calculation of observed mutation distances. The upper apex represents a hypothetical ancestral organism that divided into two descending lines, one of which subsequently also divided. Thus we have three present-day species, A, B, and C. The number of observable mutations that have occurred in a particular gene since the A and B lines of descent diverged are represented respectively by a and b. The number of mutations that separate the lower apex and C is represented by c. The sums of $a + b$, $a + c$, and $b + c$, then, are the mutation distances of the three species as currently observed.

The 20 species are indicated in the last column; the identifying numbers in the first column and the top row of the table may be used as coordinates. Thus the tabulated values interrelating the human and horse cytochromes at coordinates (1,4) and (4,1) are mutation distances of 17 and 15 respectively, the former being the input datum, the latter having been obtained from the tree by reconstruction. If the absolute difference between two such mutation dis-

tances $| (i,j) - (j,i) |$ is multiplied by 100 and divided by (i,j), the result is the percentage of change from the input data. If such values are squared and the squares are summed over all values of $i < j$, the resultant sum (Σ) may be used to obtain the percent "standard deviation" (4) of the reconstructed values from the input mutation distances. The number of mutation distances summed is $N(N-1)/2$, or 190 for our case. If this number is reduced by 1, divided into the sum Σ, and the square root taken, the result is the percent "standard deviation." Since the standard deviation is a larger number than the standard error, the probable error, or the average deviation, the percent "standard deviation" is used here, it being less likely to create overconfidence in the significance of a result (4).

The Statistically Optimal Tree

In testing phylogenetic alternatives, one is seeking to minimize the percent "standard deviation." The scheme shown in Fig. 2 has a percent "standard deviation" of 8.7, the lowest of the 40 alternatives so far tested. The percent "standard deviation" for the initial tree was 12.3.

In addition to using a gene product to discover evolutionary relationships among several species, one can similarly delineate evolutionary relationships among different genes. Our procedure constructs, from the amino acid sequences of human alpha, beta, gamma, and delta hemoglobin chains and whale myoglobin (15), the gene phylogeny

Table 3. Minimum numbers of mutations required to interrelate pairs of cytochromes c. Values in the lower left half of the table are mutation distances as determined from the amino acid sequences and, prior to rounding off, were used to derive Fig. 2. Values in the upper right half of the table are reconstructed distances found by summing the leg lengths in Fig. 2. The references cited in the last column are to studies of the amino acid sequences of the cytochromes c of the indicated species.

Protein	1	2	3	4	5	6	7	8	9	10	11	12	13	14	15	16	17	18	19	20	
1		1	13	15	15	13	11	14	15	15	16	16	17	29	29	30	33	64	62	68	Man (22)
2	1		12	15	14	12	11	13	15	14	15	15	16	28	29	29	32	63	61	67	Monkey (*Macacus mulatta*) (23)
3	13	12		9	8	6	7	8	13	13	13	14	15	26	27	27	30	61	59	65	Dog (24)
4	17	16	10		1	5	10	11	15	15	16	16	17	29	29	30	33	64	62	68	Horse (25)
5	16	15	8	1		4	9	10	14	14	15	15	16	28	28	29	32	63	61	67	Donkey (26)
6	13	12	4	5	4		7	8	13	12	13	13	14	26	27	27	30	61	59	65	Pig (27)
7	12	11	6	11	10	6		7	11	11	12	12	13	24	25	25	29	60	57	63	Rabbit (30)
8	12	13	7	11	12	7	7		13	13	14	14	15	27	27	28	31	62	60	66	Kangaroo (*Canopus canguru*) (28)
9	17	16	12	16	15	13	10	14		3	3	3	8	26	27	27	30	61	59	65	Chicken (17)
10	16	15	12	16	15	13	8	14	3		4	4	8	26	27	27	30	61	59	65	Pekin duck (29)
11	18	17	14	16	15	13	11	15	3	4		2	9	27	27	28	31	62	60	66	Pigeon (29)
12	18	17	14	17	16	14	11	13	3	4	2		9	27	27	28	31	62	60	66	King penguin (*Aptenodytes patagonica*) (29)
13	19	18	13	16	15	13	11	14	7	8	8	8		28	29	29	32	63	61	67	Snapping turtle (*Chelydra serpentina*) (31)
14	20	21	30	32	31	30	25	30	24	24	28	28	30		33	34	37	68	66	72	Rattlesnake (*Crotalus adamanteus*) (32)
15	31	32	29	27	26	25	26	27	26	27	26	27	27	38		35	38	69	67	73	Tuna (33)
16	33	32	24	24	25	26	23	26	25	26	26	28	30	40	34		16	59	56	63	Screwworm fly (*Haematobia irritans*) (29)
17	36	35	28	33	32	31	29	31	29	30	31	30	33	41	41	16		62	60	66	Moth (*Samia cynthia*) (34)
18	63	62	64	64	64	64	62	66	61	59	61	62	65	61	72	58	59		56	62	Neurospora (*crassa*) (35)
19	56	57	61	60	59	59	59	58	62	62	62	61	64	61	66	63	60	57		41	Saccharomyces (*oviformis*) iso-1 (36)
20	66	65	66	68	67	67	67	68	66	66	66	65	67	69	69	65	61	61	41		Candida (*krusei*) (37)

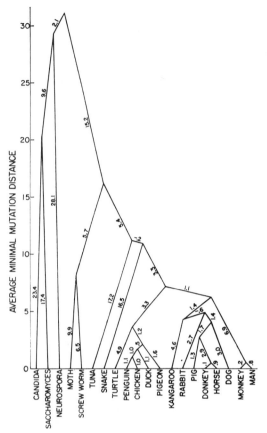

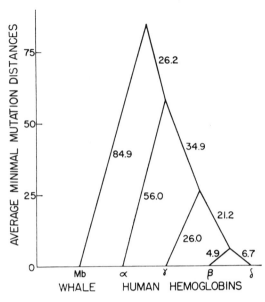

Fig. 2 (left). Phylogeny as reconstructed from observable mutations in the cytochrome c gene. Each number on the figure is the corrected mutation distance (see text) along the line of descent as determined from the best computer fit so far found. Each apex is placed at an ordinate value representing the average of the sums of all mutations in the lines of descent from that apex.

Fig. 3 (right above). A gene phylogeny as reconstructed from observable mutations in several heme-containing globins. See Fig. 2 for details. The percent "standard deviation" (7) for this tree is 1.33.

Table 4. Descent of the mammalian cytochromes. Changes in amino acids are shown in large capitals, with subscripts to indicate the number of mutations that had to occur to produce the indicated change. In general, unchanging amino acids are not repeated, but occasionally it has been necessary to relist an unchanged amino acid because a mutation appearing in one line of descent did not apply to other lines listed further down the page. Such unchanged amino acids are shown in small capitals. The lines of descent are shown on either side of the table. The last two columns give the sum of the mutations indicated in that row and the corresponding value from Fig. 2. The following rules were used in formulating each amino acid position of the ancestral sequences: Choose the amino acid so that the changes in the codon during descent require (i) the smallest overall number of mutations; (ii) the fewest segments containing multiple mutations (that is, two lines with one mutation each are preferred to one line with two mutations); (iii) the fewest sequential mutations (that is, one mutation in each of two lines following a branch point is preferred to one mutation before and one after the branch point); (iv) the fewest back mutations; (v) the fewest kinds of amino acids. Rule (i), where applicable, took priority over all others and rule (ii) took priority over the remainder. It was not found necessary to choose among the last three rules. The ancestral mammalian cytochrome c sequence shown was derived from the amino acid sequences of all 20 cytochromes c.

Amino acid No.	17	18	21	39	41	50	52	53	56	64	66	68	89	94	95	98	109	Listed in this table	From leg lengths in Fig. 2
Ancestral mammal	V	Q	L	H	U	P	O	S	A	E	Y	A	L	I	G	L	N		
Ancestral primate	W₁	M₂	S₁	.	.	.	.	.	L₁	.	.	.	V₁	.	.	.	.	6	6.9
Monkey	.	.	.	.	.	.	.	.	.	.	.	.	.	.	.	.	.	0	0.2
Man	.	.	.	.	.	.	.	.	W₁	.	.	.	.	.	.	.	.	1	.8
Kangaroo	V	Q	L	.	.	F₁	.	A	E	.	.	L	.	Y₁	.	.	.	2	1.4
	.	.	N₁	W₁	.	.	E₁	.	W₁	.	.	.	.	.	.	.	.	4	4.6
	.	.	H	U	.	.	S	.	E	.	.	.	.	.	.	.	.	0	−.6
Rabbit	.	.	.	.	V₃	.	.	.	.	.	.	.	.	A₁	.	.	.	3	2.7
	.	.	.	.	P	.	.	.	.	.	G₁	.	.	Y	.	.	.	1	1.4
Dog	.	.	.	.	.	.	.	.	.	.	.	E₁	.	.	I₄	.	.	2	3.0
Ancestral ungulate	.	.	.	.	.	.	.	.	.	.	.	.	I	.	Q₁	N	.	2	1.7
Pig	.	.	.	.	.	.	.	.	.	.	.	.	.	.	.	.	.	0	1.3
Ancestral perissodactyl	.	.	.	.	.	.	.	.	.	I₁	.	.	.	E₂	.	.	.	4	2.9
Donkey	.	.	.	.	.	.	.	.	.	.	.	.	.	.	.	.	.	0	0.1
Horse	.	.	.	.	.	E₁	.	.	.	.	.	.	.	.	.	.	.	1	.9

284

shown in Fig. 3. The overall result is as Ingram had previously indicated (15). A cautionary note may be derived from this. A wildly incorrect result could easily be obtained if the presence of multiple, homologous genes were not recognized and a phylogeny were constructed from sequences which were coded for, say, half by genes for alpha hemoglobin chains and half by genes for beta hemoglobin chains. This results from the speciation having occurred more recently than the gene duplication which permitted the separate evolution of the alpha and beta genes.

The method described can also be used to develop treelike relationships by employing data which are very different in character from mutation distances. For example, the physical characteristics of human beings have been used to construct a tree relating several ethnic groups (Fig. 4; 6).

Although we are examining the product of but a single gene, and a rather small one at that, the phylogenetic scheme in Fig. 2 is remarkably like that constructed in accord with classical zoological comparisons (5). There are only three noticeable deviations, discussed below, and these may well be changed as more species are added to the list. Of even greater value would be sequences from other genes, since special environmental effects may easily cause the convergence of one or several genes in phylogenetically disparate organisms. Hemoglobin amino acid sequences may soon be available in great enough numbers to prove useful in this respect.

Almost all the alternative phylogenetic schemes tested involved rearrangements within the groups birds (16, 17) and nonprimate mammals (14, 18, 19). With respect to the birds, it will be noticed that the penguin is closely associated with the chicken, whereas one might have expected that all the "birds of flight" (Neognathae) would be more closely related to each other than to the penguin (Impennae). This discrepancy is probably related to the very small numbers of mutations involved. In this regard, it is interesting to note that on the basis of a micro-complement-fixation technique using antisera to several purified enzymes, Wilson et al. (2) found that the duck is more closely related to the chicken than is the pigeon. This agrees with our findings.

In the second group, the kangaroo is shown closely associated with the nonprimate mammals, whereas most zoolo-

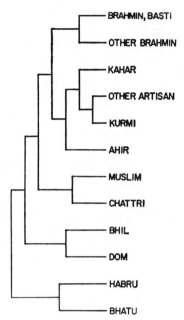

Fig. 4. Relations among various tribes and castes of India. The data used to construct this scheme are the D^2 values given by Rao (6). This figure is in principle like Fig. 2 except that, to prevent misinterpretation of the physical significance of the numbers one obtains, branching is shown as a rather uniform step function which preserves the relationships but obliterates the quantitative distances of the ordinate.

gists would maintain that the placental mammals, including the primates, are more closely related to each other than to the marsupials.

A third anomaly is that the turtle appears more closely associated with the birds than to its fellow reptile the rattlesnake. Although it is true that the snake is involved in seven of the nine instances where the reconstructed values differ from the input mutation distances by more than 4 mutations, this cannot account for the anomaly, which in fact results from the close similarity of the turtle's cytochrome c amino acid sequence to those of the birds.

Thus the phylogenetic tree in Fig. 2 is imperfect. Nevertheless, considering that only one gene product was analyzed and that no choices were made other than those dictated by the statistical analysis, the results are very promising, and a phylogeny based upon a quantitative determination of those very events which permit speciation, namely mutations, must ultimately be capable of providing the most accurate phylogenetic trees.

Elapsed Time and Evolutionary Change

It should be pointed out that the ordinate of Fig. 2 represents the minimum number of mutations observable. Since multiple mutations in a single codon are not likely to produce mutation values as large as the actual number of mutations sustained, Fig. 2 is greatly foreshortened with respect to the actual number of mutations (20). The possibility of obtaining an ordinate scale denoted as actual mutations by applying a correction factor, using the relative frequencies of codons observed to have sustained one, two, and three nucleotide changes, must await reliable statistical information on the relative probabilities that given amino acid substitutions will permit the progeny to compete successfully in their environment. Any meaningful correction of this sort is precluded at present by the lack of such statistical information, but its importance may be emphasized by noting that such a correction would yield an ordinate in Fig. 2 in which equal numbers of mutations would correspond to equal intervals of time, as long as the rate at which mutations are fixed, averaged for many lines of descent over very long periods of evolutionary history, does not vary appreciably (20).

It should be noted that the method does not assume any particular value for the rate at which mutations have accumulated during evolution. Indeed, from any phylogenetic ancestor, today's descendants are equidistant with respect to time but not, as computations show, equidistant genetically. Thus the method indicates those lines in which the gene has undergone the more rapid changes. For example, from the point at which the primates separate from the other mammals, there are, on the average, 7.5 mutations in the descent of the former and 5.8 in that of the latter, indicating that the change in the cytochrome c gene has been much more rapid in the descent of the primates than in that of the other mammals.

The method allows negative mutation distances, and a few were observed in some of the discarded phylogenetic schemes. Their absence from the best-fitting scheme would indicate that there were no significant evolutionary reversals in this gene.

One highly desirable goal is the reconstruction of the ancestral cytochrome c amino acid sequences. The procedure, though not difficult, is dependent upon the phylogenetic tree on which these sequence data are arranged. Given the present scheme (see Fig. 2) one can reconstruct the ancestral proteins. A reconstruction of the ancestral amino acid sequences for the mammalian portion of tree is shown in Table 4. One can then ask such a question as "What are the mutations required to account for the difference between the cytochromes c of the ancestral primate and of the ancestral mammal?" The data in Table 4 clearly identify the mutations as occurring in positions 17, 18, 21, 56, and 89. In a similar manner, the monkey and human lines are distinguished by a single mutation in the human line which resulted in the substitution of isoleucine for threonine at position 64.

There is presently no detectable relationship between the primary structures of cytochrome c and those of hemoglobins (12). Nevertheless, the reconstruction and comparison of the ancestral amino acid sequences may reveal a homology that cannot be detected in present-day proteins. The employment of such ancestral sequences may be generally useful for detecting common ancestry not otherwise observable.

Note added in proof. Since this article was accepted our attention has been called to several earlier papers which present some of the important concepts discussed here. Sokal and his collaborators (38) have for several years been studying various ways of producing treelike relationships from quantitative taxonomic information. In an interesting application of this type of technique, using the amino acid sequences of fibrinopeptides from several ungulates, R. F. Doolittle and B. Blombäck (39) constructed such a tree and specifically indicated how knowledge of the genetic code would be useful for more precise constructions. Jukes (40, fig. 3) has presented the Ingram scheme of the hemoglobin gene duplications and placed upon the

various legs estimates of the numbers of nucleotide substitutions. His figure is not essentially different from Fig. 3 of this article.

References and Notes

1. B. J. McCarthy and E. T. Bolton. *Proc. Natl. Acad. Sci. U.S.* **50**, 156 (1963).
2. A. C. Wilson, N. O. Kaplan, L. Levine, A. Pesce, M. Reichlin, W. S. Allison, *Fed. Proc.* **23**, 1258 (1964); C. A. Williams, Jr., in *Peptides of Biological Fluids*, H. Peeters, Ed. (Elsevier, New York, 1965), p. 62; M. Goodman, *ibid.*, p. 70; A. S. Hafleigh and C. A. Williams, Jr., *Science* **151**, 1530 (1966).
3. R. L. Hill, J. Buettner-Janusch, V. Buettner-Janusch, *Proc. Natl. Acad. Sci. U.S.* **50**, 885 (1963); R. L. Hill and J. Buettner-Janusch, *Fed. Proc.* **23**, 1236 (1964).
4. E. Margoliash, *Proc. Natl. Acad. Sci. U.S.* **50**, 672 (1963); E. L. Smith and E. Margoliash, *Fed. Proc.* **23**, 1243 (1964); E. Margoliash and E. L. Smith, in *Evolving Genes and Proteins*, V. Bryson and H. Vogel, Eds. (Academic Press, New York, 1965), p. 221.
5. A. S. Romer, *Vertebrate Paleontology* (Univ. of Chicago Press, Chicago, ed. 2, 1945).
6. Our procedure may be compared with the "cluster analysis" approach as formulated by A. W. F. Edwards and L. L. Cavalli-Sforza [*Biometrics* **21**, 362 (1965)]; their approach is, in one sense, the reverse of that we have used, since cluster analysis starts with all the elements as members of the same subset and proceeds to subdivide that subset into successively smaller but more numerous subsets until each element is the sole member of its own subset. In terms of Fig. 2, Edwards and Cavalli-Sforza constructed their tree from the top down, whereas we built ours from the bottom up. Edwards and Cavalli-Sforza report testing their method on C. R. Rao's data [*Advanced Statistical Methods in Biometric Research* (Wiley, New York, 1952)] on physical characteristics of 12 Indian castes and tribes. Rao had used these data to postulate relationships among the castes and tribes. Although the nature of these data is quite different from that of ours, the formal mathematical problems are very much alike, and we have used the D^2 values of Rao, as did Edwards and Cavalli-Sforza, to find the best tree. Edwards and Cavalli-Sforza's tree has a percent "standard deviation" (7) of 32.6. Our result, shown in Fig. 4, has a percent "standard deviation" of 29.2 and, except that it possesses greater detail, conforms to the conclusions drawn by Rao.
7. The quotation marks are placed around "standard deviation" because the data used in its formulation here are not statistically independent as is generally required. This is evident in that only 20 amino acid sequences determine the 190 mutation distances utilized.
8. The homology may be found by aligning the cysteine residues which bind the heme. Excellent examples of this may be seen in Fig. 10 of E. Margoliash and A. Schejter, *Advan. Protein Chem.* **20**, 114 (1965).
9. W. M. Fitch, *J. Mol. Biol.* **16**, 1 (1966).
10. M. G. Weigert and A. Garen, *Nature* **206**, 992 (1965).
11. S. Brenner, A. O. W. Stretton, S. Kaplan, *ibid.*, p. 994.
12. W. M. Fitch, *J. Mol. Biol.* **16**, 9 (1966).
13. It will be recognized that once the first tree is calculated, the number of computations required for alternatives becomes greatly reduced. For example, if instead of the tree shown in Fig. 2 one wishes to test a tree which differs only in the order in which the chicken, duck, and penguin are joined, the only legs in need of recalculation are those five descending to these birds from the avian apex.
14. The cow (18) and sheep (19) cytochromes c are identical with that of the pig (27).
15. V. M. Ingram, *The Hemoglobins in Genetics and Evolution* (Columbia Univ. Press, New York, 1963); A. B. Edmundson, *Nature* **205**, 883 (1965).
16. The cytochrome c of the turkey (29) is identical with that of the chicken (16).
17. S. K. Chan and E. Margoliash, *J. Biol. Chem.* **241**, 507 (1966).
18. T. Yasunobu, T. Nakashima, H. Higo, H. Matsubara, A. Benson, *Biochim. Biophys. Acta* **78**, 791 (1963).
19. S. K. Chan, S. B. Needleman, J. W. Stewart, E. Margoliash, unpublished results.
20. This is analogous to the relationship between numbers of amino acid replacements and the evolutionary time scale discussed by E. Margoliash and E. L. Smith in *Evolving Genes and Proteins*, V. Bryson and H. Vogel, Eds. (Academic Press, New York, 1965), p. 221.
21. B. Keil, Z. Prusik, F. Sorm, *Biochim. Biophys. Acta* **78**, (1963).
22. H. Matsubara and E. L. Smith, *J. Biol. Chem.* **238**, 2732 (1963).
23. J. A. Rothfus and E. L. Smith, *ibid.* **240**, 4277 (1965).
24. M. A. McDowall and E. L. Smith, *ibid.* p. 4635.
25. E. Margoliash, E. L. Smith, G. Kreil, H. Tuppy, *Nature* **192**, 1125 (1961).
26. O. F. Walasek and E. Margoliash, unpublished results.
27. J. W. Stewart and E. Margoliash, *Can. J. Biochem.* **43**, 1187 (1966).
28. C. Nolan and E. Margoliash, *J. Biol. Chem.* **241**, 1049 (1966).
29. S. K. Chan, I. Tulloss, E. Margoliash, unpublished results.
30. S. B. Needleman and E. Margoliash, *J. Biol. Chem.* **241**, 853 (1966).
31. S. K. Chan, I. Tulloss, E. Margoliash, *Biochemistry* **5**, 2586 (1966).
32. O. P. Bahl and E. L. Smith, *J. Biol. Chem.* **240**, 3585 (1965).
33. G. Kreil, *Z. Physiol. Chem.* **334**, 154 (1963).
34. S. K. Chan and E. Margoliash, *J. Biol. Chem.* **241**, 335 (1966).
35. J. Heller and E. L. Smith, *Proc. Natl. Acad. Sci. U.S.* **54**, 1621 (1965).
36. Y. Yaoi, K. Titani, K. Narita, *J. Biochem. Tokyo* **59**, 247 (1966).
37. K. Narita and K. Titani, *Proc. Japan Acad.* **41**, 831 (1965).
38. R. R. Sokal, *Syst. Zool.* **10**, 70 (1961); F. J. Rohlf and R. R. Sokal, *Univ. Kansas Sci. Bull.* **45**, 3 (1965); J. H. Camin and R. R. Sokal, *Evolution* **19**, 311 (1965).
39. R. F. Doolittle and B. Blombäck, *Nature* **202**, 147 (1964).
40. T. H. Jukes, *Advan. Biol. Med. Phys.* **9**, 1 (1963).
41. This project received support from grants from NIH (NB-04565) and NSF (GB-4017) to W.M.F. We thank Peter Guetter and Daniel Brick for valuable technical assistance.

19

Reprinted from *Syst. Zool.* **28**:49–62 (1979)

ALTERNATIVE METHODS OF PHYLOGENETIC INFERENCE AND THEIR INTERRELATIONSHIP[1]

JOSEPH FELSENSTEIN

The purpose of this paper is to examine the logical interrelationships between various methods of phylogenetic inference. The approach taken is to construct a probabilistic model of the evolution and interpretation of a character, a model which contains a variety of evolutionary events as well as an event corresponding to misinterpretation of the character by the taxonomist. Using this model, it is possible to use statistical inference methods to reconstruct the phylogeny of a group, given data on a number of characters. This is a difficult computational task. However, when particular parameters of the model are taken to be extreme, the maximum likelihood estimates of the phylogeny assume simpler forms. Three of these turn out to be equivalent to

[1] This report was prepared as an account of work sponsored by the United States Government. Neither the United States nor the United States Department of Energy, nor any of their employees, makes any warranty, express or implied, or assumes any legal liability or responsibility for the accuracy, completeness or usefulness of any information, apparatus, product or process disclosed, or represents that its use would not infringe privately-owned rights.

known methods of phylogenetic inference: two parsimony methods and one compatibility method. A new parsimony method, the polymorphism method, also emerges as one of the limiting forms of the method of maximum likelihood.

All of these methods of phylogenetic inference are justifiable within the same logical framework, and correspond to different assumptions about the relative probabilities of events. However, this justification is valid only if we are willing to assume extremely high or low probabilities for certain events. Alternative possibilities for providing a logical foundation for these methods of phylogenetic inference will be considered later in this paper.

Two of the existing methods which will emerge from the present model are parsimony methods. These are the method of Camin and Sokal (1965) and the Dollo method of Farris (1977). The third method is the compatibility method of Estabrook, Johnson, and McMorris (1976). All of these methods assume that we know which character state is the ancestral state in each character. For simplicity

TABLE 1. PROBABILITIES OF TRANSITION BE-
TWEEN POPULATION STATES DURING A SHORT IN-
TERVAL OF TIME dt.

From State:	To State:		
	0	01	1
0	$1 - a\,dt$	$a\,dt$	0
01	$b/2\,dt$	$1 - b\,dt$	$b/2\,dt$
1	0	$c\,dt$	$1 - c\,dt$

this paper will treat primarily characters
with two phenotypically distinguishable
states. This restriction can be relaxed relatively easily, as can the assumption that
the ancestral states are known. Other assumptions, such as the independence of
the characters and the special nature of
taxonomic errors of interpretaion, are not
so easily relaxed.

THE MODEL

The present model resembles that of
Farris (1977), but differs from it in several
ways: In Farris's model there were two
states of each character, 0 and 1, which
we may call *ancestral* and *derived* (or, if
one prefers, *plesiomorphic* and *apomorphic*). Farris assumed that transitions
of character state $0 \to 1$ and $1 \to 0$ had
different probabilities, and that these two
probabilities were the same in each segment of the phylogeny.

In contrast, in the present model let us
assume that the probabilities of evolutionary events are constant *per unit time*
rather than per segment of the phylogeny. In addition we add a third state, 01,
to the process. This is a state of the population in which both 0 and 1 are present,
so that there is a polymorphism. We shall
maintain a distinction between the character state, which is either 0 or 1, and the
state of the population, which may be 0,
01, or 1. Throckmorton (1968) has called
particular attention to the possibility that
apparent multiple origins of the same
character state may have resulted neither
from independent occurrence of the
state, nor from multiple reversions to the
character state, but instead from multiple
occurrences of loss of one or the other

character state from an ancestral population polymorphic for both states.

We thus have three population states,
0, 01, and 1 in the present model. We
assume that there is a constant probability $a\,dt$ that, in an infinitesimal interval
of time of length dt, state 0 will change
to state 01. This corresponds to the introduction of alleles coding for character
state 1 into the population, resulting in
polymorphism. Likewise, we assume a
constant probability $c\,dt$ that a population
having state 1 will change to state 01.
When a population is polymorphic, we
assume that loss of one or the other of its
two character states occurs with probability $b\,dt$, the chances of the two types
of loss being equal. In our model the tree
topology and segment lengths are regarded as given (rather than being generated
by some probabilistic model of speciation and extinction). All characters start
in population state 0 at the root of the
tree, and each character changes independently thereafter according to the
above probabilistic model. When a fork
in the tree is encountered, both descendant populations start with identical
states in all characters and thereafter
evolve independently. The transition
probabilities of this process are shown in
Table 1.

We now add to these evolutionary processes a process of human error in interpreting the characters. We assume that
after evolution has occurred, there is a
probability E that the character is misinterpreted by the taxonomist. Our model
of the process of misinterpretation is a
rather special and limited one. We assume that each character has the same independent probability E of misinterpretation. Furthermore we assume that the
misinterpretation, when it occurs, is so
serious as to render the character completely informationless, as if the character states coded by the taxonomist bore
no relation to the states actually present.
This is not a very good model of misinterpretation: it is chosen not for its real-
ism but because it is the one which most
readily allows us to obtain the compati-

bility method as a limit of maximum likelihood estimation of the phylogeny. Figure 1 gives a diagrammatic representation of our model.

LIKELIHOOD ANALYSIS

Our model allows us to explain a given set of data on a given phylogeny as resulting from a mixture of originations of character state 1, reversions to character state 0, losses of polymorphism, and presumed misinterpretations of characters. In principle, we could use this model to estimate the phylogeny by the method of maximum likelihood. We would need the quantities $P_{ij}(t)$, the probabilities of change from population state i to population state j during time t, which is the length of a segment of the evolutionary tree. The states i and j can each take the values, 0, 01, or 1. The Appendix gives a method for the numerical calculation of the $P_{ij}(t)$. In terms of these, the likelihood of the evolutionary tree (the probability of obtaining the observed data given the tree) can be written as

$$L = \prod_{i=1}^{\substack{\text{char-} \\ \text{acters}}} [EK_i + (1 - E) \sum_{j=1}^{\substack{\text{assign-} \\ \text{ments}}} \prod_{k=1}^{\substack{\text{tree} \\ \text{segments}}} P_{s_k s_{k'}}(t_k)], \quad (1)$$

where the summation (j) is over all possible ways in which states could be assigned to interior nodes (forks) of the tree. The quantity K_i is the probability that we would obtain the observed data if character i were misinterpreted. It is assumed to be a constant which may depend on the data but not on the tree. The indices s_k and $s_{k'}$ are the states assigned to the beginning and end of the kth branch of the tree, a branch whose length (in time) is t_k.

Equation (1) can be used for maximum likelihood estimation of the phylogeny, once we specify the quantities K_i. These depend on the details of our model of misinterpretation of characters. One particular choice for the K_i would be to have all configurations of character states in tip species be equiprobable if the character is misinterpreted. This corresponds to

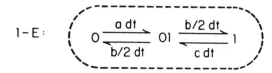

FIG. 1.—A diagrammatic representation of the mode of character state evolution of this paper. The probability per unit time of character state 1 arising is a, the probability per unit time that state 1 reverts to a polymorphic state is c, and the probability per unit time of loss of polymorphism is b. The probability of fundamental error in the interpretation of the character is taken to be E.

$K_i = (\frac{1}{2})^n$, where n is the number of tip species on the tree. Although (1) compactly summarizes the likelihood calculations, computing this expression is a daunting task. Each term in (1) is computed by summing the probabilities of obtaining the observed data for all possible interpretations of the events in the evolution of the character, including the event of misinterpretation of the character. For all but the smallest data sets, the number of possible assignments of states to interior nodes of the tree is very large. I have given elsewhere (Felsenstein, 1973) an algorithm which reduces greatly the amount of computation necessary. But the task is still difficult. To obtain a maximum likelihood estimate of the phylogeny we must make many evaluations of the likelihood, as we progressively alter the tree, trying to find the tree with the highest likelihood.

Another restriction on the use of the likelihood method is the need for us to know the parameters a, b, d, and E of the probabilistic model of evolution and interpretation. It would be convenient if we could find ranges of these parameters within which the maximum likelihood estimate of the phylogeny is not critically dependent on the values of the parameters.

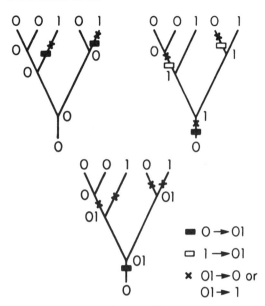

FIG. 2.—Three of the possible interpretations of the evolutionary events leading to a character not being unique and unreversed on an evolutionary tree. As in the text, 0 is the ancestral state, 1 the derived state, and 01 the polymorphic state.

IMPROBABLE EVENTS

In the likelihood function (1) the likelihood for one character is summed over all possible combinations of events which could have occurred in the evolution and interpretation of the character. If the probabilities of occurrence of the four types of event (origination, reversion, loss of polymorphism, misinterpretation) are very different, then we intuitively expect that of the many possible reconstructions of events, most of the likelihood will be contributed by one reconstruction. Figure 2 will serve as an example. It shows three reconstructions of evolutionary events in a character on a five-species phylogeny. There are many other possibilities. Now suppose that originations ($0 \rightarrow 01$) are rare but far more probable than reversions ($1 \rightarrow 01$) and that loss of polymorphism ($01 \rightarrow 1$, $\rightarrow 0$) is the most probable event of all, so that once polymorphism occurs it is virtually certain to be lost before the next node in the tree. Then when we see the data giv-

en in this case, we would intuitively prefer to explain it by assuming no reversions, assuming that whenever polymorphism occurs it is immediately lost, and assuming as few originations of character state 1 as possible. This reconstruction is shown in the upper left-hand part of the figure.

Alternatively, we might believe that origination was an improbable event, reversion also improbable but much less so, and loss of polymorphism very probable. In that case we would intuitively prefer to assume only one origination, followed by immediate loss of character state 0 from the resulting polymorphism, and as few reversions as possible, each followed by immediate loss of character state 1 from the polymorphism. This reconstruction of events is shown in the upper right-hand tree in the figure.

If on the other hand we believed that reversion was extremely improbable, origination less so but still improbable, and that retention of polymorphism was also improbable but not so much as the other two events, then we would prefer to make the reconstruction shown in the bottom tree in Fig. 2. In that tree there is only one origination and no losses, and the two occurrences of state 1 are explained by the persistence of polymorphism after the origination of character state 1, followed by loss of character state 0 in two lines and of character state 1 in two others.

Each of these is a reconstruction of events which, while invoking some improbable events, is able to explain the observed data while stretching our credulity the least. A fourth possibility is not shown in Fig. 2. This is the reconstruction of this character which explains the two occurrences of state 1 among the tips by asserting that the character has been misinterpreted. This is the reconstruction which we would make if we assumed that misinterpretation was a far more probable event a priori than having even one reversion, or one stretch of retained polymorphism, or one extra origination of state 1.

TABLE 2. ASYMPTOTIC VALUES OF TRANSITION PROBABILITIES FROM ONE POPULATION STATE TO ANOTHER, AS a, c, AND e^{-b} BECOME SMALL.

From:	To: 0	01	1
0	1	a/b	$at/2$
01	1/2	e^{-bt}	1/2
1	$ct/2$	c/b	1

A MORE PRECISE ACCOUNT

These four reconstructions of events are a few out of many possible ones, most of which involve combinations of events or superfluous origins or losses. For each reconstruction there will be a term in (1). We are seeking situations where only one reconstruction contributes the bulk of likelihood, in an attempt to verify the above intuitive arguments and to find out which phylogenetic methods may be maximum likelihood methods. First, consider the case where E, a and c are all very small, and where b, the rate at which polymorphism is lost is large. In this case the individual terms $P_{ij}(t)$ in (1) will each have a reasonably simple form. Table 2 shows what the asymptotic values of the $P_{ij}(t)$ will be as a, c, and e^{-b} all approach zero. Note that although the transition probabilities do not sum to one as we go across a row of Table 2, the discrepancy is infinitesimal since (for example) a/b and $at/2$ become infinitely small quantities as a approaches zero. The derivation of these quantities is straightforward: for example $P_{0,01}(t)$ is the probability of an origination $0 \rightarrow 01$ followed by retention of the polymorphic condition 01 for the remainder of an interval of total length t. This works out to be a/b as we make a, c, and e^{-b} small. Other more elaborate sequences of events contribute less and less to the probability $P_{0,01}(t)$ as we make these quantities small.

It is an extremely difficult task to consider all possible reconstructions of the character and their contributions to the likelihood expression (1). However, the logic of the present argument should be clear from consideration of four types of reconstructions. The first is that which corresponds to the upper left-hand tree

in Fig. 2. There the apparent parallelism in the character is explained by assuming transitions of population state $0 \rightarrow 1$ in a number of segments of the tree. Provided that no tips show polymorphism (which we assume), we can always make such a reconstruction. The term in (1) corresponding to it can be seen from Table 2 to be a product of terms, each of which is either 1 or $at_i/2$, where t_i is the length of a tree segment in which a change $0 \rightarrow 1$ occurs. For each such reconstruction, the contribution to (1) will be

$$L^* = (1 - E) \prod_j (at_j/2), \qquad (2)$$

where the product is over all segments j, of length t_j, which have a $0 \rightarrow 1$ change in character i. This can be rewritten as

$$L^* = a^{n_i}(1 - E) \prod_j (t_j/2). \qquad (3)$$

The quantity $1 - E$ is nearly equal to 1 by assumption. There will be many such quantities L^*, one for each possible placement of transitions $0 \rightarrow 1$ which leads to the observed data in character i. But since we are taking a to be very small, asymptotically only that reconstruction with the smallest value of n_i (the one requiring the fewest $0 \rightarrow 1$ changes) will contribute significantly to the term for character i in equation (1).

We can make a similar analysis of other types of reconstructions. For reconstructions like the upper right-hand one in Fig. 2, where there is only one change $0 \rightarrow 1$ followed by many changes $1 \rightarrow 0$, the likelihood contribution for character i is

$$L^* = (1 - E)(at_{i^*}/2) \prod_j (ct_j/2), \qquad (4)$$

where i^* is the index of the tree segment in which the transition $0 \rightarrow 1$ occurs and j indexes all the segments which have a transition $1 \rightarrow 0$ in character i. The contribution L^* of this reconstruction to the likelihood is proportional to ac^{m_i}, where m_i is the number of transitions $1 \rightarrow 0$ in the reconstruction of character i. As in the previous case, since a and c are small, only that reconstruction with the smallest

m_l contributes significantly to the likelihood term for character i.

For the reconstructions like that in the lower part of Fig. 2, with one origination $0 \rightarrow 01$ and multiple losses $01 \rightarrow 0$ or $01 \rightarrow 1$, the likelihood contribution for character i is

$$L^* = (1 - E)(at_{i*}/2) \prod_j e^{-bt_j} \prod_k (1/2), \quad (5)$$

where as before t_i^* is the length of the segment in which origination of state 1 occurs, where the index j runs over all tree segments in which retention of the polymorphism is assumed, and the index k runs over all tree segments in which loss of polymorphism occurs. Once again, only one of the reconstructions contributes significantly to the term for character i in (1). It is the one with the shortest extent of polymorphism, the smallest length of time in which this character must be assumed polymorphic. The likelihood contribution is proportional to $a \exp(-b \Sigma t_j)$, summation running over all tree segments in which polymorphism is assumed.

Finally, there is the reconstruction which specifies that an error has been made in interpreting the character. The contribution of character i to the likelihood expression (1) is simply EK_i, if i has been misinterpreted.

Even if our attention were restricted to these four reconstructions of events in character i, we would still have 4^C possible reconstructions of the events in all C characters. We now further restrict the parameters a, b, c, and E to reduce the number of reconstructions contributing significantly to the likelihood (1) to one. We examine four subcases of the case in which a, e^{-b}, c, and E are all small.

I. c, e^{-b}, E $\ll$ a. In this case there is far less contribution to the likelihood from reconstructions involving even one case of reversion, misinterpretation, or retention of polymorphism than there is from reconstructions involving only $0 \rightarrow 1$ changes. However, since a is small (though larger than the other three quantities), the likelihood for a given tree will

be contributed almost entirely by those reconstructions in which there are the fewest $0 \rightarrow 1$ changes. Also since a is small, the likelihood of the tree will be smaller, the more such changes it requires. The maximum likelihood estimate of the phylogeny is then easily seen to be that tree which requires the fewest $0 \rightarrow 1$ transitions to explain the observed data (with no misinterpretation, reversions, or retention of polymorphisms). This is precisely the Camin-Sokal parsimony criterion.

II. e^{-b}, E $\ll$ a $\ll$ c. When c is by far the largest of these four quantities, the only reconstructions which contribute significantly to the likelihood term for a particular character are those with no misinterpretations assumed, no retention of polymorphism, at most one $0 \rightarrow 1$ change per character, and within those constraints as few $1 \rightarrow 0$ transitions as possible. The likelihood of a tree will be smaller the more $1 \rightarrow 0$ changes are required on that tree to explain the observed character states in the tip species. The tree which will be the maximum likelihood estimate will be the one which requires the fewest $1 \rightarrow 0$ changes, given that there are no misinterpretations or instances of retention of polymorphism, and at most one $0 \rightarrow 1$ change per character. This is precisely the tree which will be found by the Dollo parsimony method presented by Farris (1977).

III. c, E $\ll$ a $\ll$ e^{-b}. In this case the bulk of the likelihood term for a given character will be contributed by reconstructions having no reversion or misinterpretations postulated, at most one origination of character state 1, and as short a stretch of the evolutionary tree as possible containing polymorphism for the character. In the maximum likelihood method which these assumptions give rise to, the quantity which is to be minimized in searching for the estimate is the extent of polymorphism. The likelihood will be proportional to $\exp(-bT)$, where T is the total length of all tree segments in which polymorphism needs to be retained, summed over all characters. This is a new parsimony method, which may be dubbed the polymorphism method.

IV. c, $e^{-b} \ll a^2 \ll E \ll a$. In this case, the reconstruction contributing the bulk of the likelihood for each character will be one having a single $0 \rightarrow 1$ transition or, if this is not possible, an assumed misinterpretation of the character. The likelihood will be smaller the more characters must be interpreted as having a misinterpretation. Thus the maximum likelihood estimates of the phylogeny will be the ones on which as many characters as possible require no more than one $0 \rightarrow 1$ transition. These will simply be the trees suggested by the largest set of mutually compatible characters. This objective was suggested by Le Quesne (1969), although the case treated here corresponds to the compatibility approach of Estabrook, Johnson, and McMorris (1976).

Thus three existing phylogenetic inference methods emerge as maximum likelihood estimation methods for extreme choices of parameters in our model, as well as a new method, the polymorphism parsimony method.

SOME COMPUTATIONAL VERIFICATION

The arguments given above are heuristic rather than rigorous. A full presentation of the details of the derivation is beyond the scope of this journal. It may be useful to present here some numerical computations verifying one of the steps of the argument, in particular some of the statements as to which reconstructions will contribute the bulk of the likelihood when the values of a, b, and c are extreme. Our example will be the tree shown in Fig. 3. Clearly state 1 cannot be unique and unreversed on this tree. Its evolution can be explained within the framework of our model only by invoking misinterpretation, reversion, or polymorphism. For simplicity we assume the lengths of all tree segments to be 1 (or alternatively, we assume the probabilities of change in the character to be constant per segment of the tree). The population states i and j of the character in

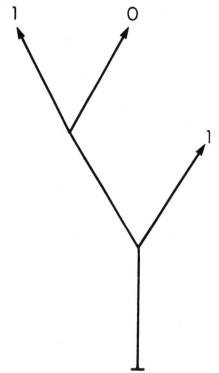

FIG. 3.—Evolutionary tree and character states used in computation of Table 4. All segments of the tree are of length 1.

the two interior nodes of the tree have not been observed. Let i be the state in the lower interior node of the tree. The likelihood of the tree with these data is given by summing over all reconstructions:

$$L = \tfrac{1}{8}E + (1 - E)$$
$$\cdot \sum_i \sum_j P_{0i}(1)P_{11}(1)P_{ij}(1)P_{j0}(1)P_{j1}(1), \quad (6)$$

where the summations in i and j each run through the three possible population states 0, 01, and 1. The $P_{ij}(t)$ may be evaluated by the method given in the Appendix. A computer program has been written to do this, given the values of a, b, c, and E. It is of particular interest to see what proportion of the likelihood is contributed by the reconstructions corresponding to the four phylogenetic methods mentioned. The contribution for the reconstruction involving misinterpretation is given by the term $\tfrac{1}{8}E$, derived under the particular assumption that

TABLE 3. FRACTION OF TOTAL LIKELIHOOD ACCOUNTED FOR BY THREE MAJOR METHODS OF TREE RE-
CONSTRUCTION FOR THE CASE SHOWN IN FIG. 3.

			Proportion of likelihood			
a	b	c	Camin-Sokal	Polymorphism	Dollo	All Other
0.0001	0.000001	0.00001	0.89	0.02	0.09	2×10^{-6}
0.0001	0.00001	0.00001	0.77	0.15	0.08	2×10^{-5}
0.0001	0.00005	0.00001	0.48	0.48	0.048	5×10^{-5}
0.0001	0.0001	0.00001	0.32	0.65	0.032	7×10^{-5}
0.0001	0.0005	0.00001	0.09	0.90	0.009	1×10^{-4}
0.0001	0.001	0.00001	0.05	0.95	0.005	1×10^{-4}
0.0001	0.01	0.00001	0.005	0.994	0.0005	1×10^{-4}
0.0001	0.1	0.00001	0.00063	0.999	0.00006	1×10^{-4}
0.0001	1.0	0.00001	0.0005	0.999	0.00005	3×10^{-4}
0.0001	2.0	0.00001	0.002	0.997	0.0002	1×10^{-3}
0.0001	4.0	0.00001	0.026	0.963	0.003	8×10^{-3}
0.0001	6.0	0.00001	0.234	0.691	0.023	0.05
0.0001	8.0	0.00001	0.65	0.18	0.065	0.10
0.0001	10.0	0.00001	0.80	0.025	0.08	0.09

when there is misinterpretation a random configuration of states in the tip species will be observed, with equal probability of 0 and 1 in each species. Three of the nine terms of the double summation correspond to the reconstructions made by the other methods: the Camin-Sokal parsimony method ($i = 0$, $j = 0$), the Dollo parsimony method ($i = 1$, $j = 1$), and the polymorphism method ($i = 01$, $j = 01$). The other six terms correspond to more complicated reconstructions.

The results agree closely with expectation. Table 3 shows the fractions of the likelihood contributed by these reconstructions, in a case in which b is varied while a and c are small and E is zero. When a is the largest of the four quantities, the Camin-Sokal reconstruction contributes most of the likelihood. As b is increased, the polymorphism reconstruction contributes more and more of the likelihood, the proportion declining only when e^{-b} becomes smaller than a. In this case, other reconstructions never contribute more than a small fraction of the likelihood. Note that although the polymorphism reconstruction contributes the bulk of the likelihood when b is intermediate in value, this does not mean that the polymorphism method is well justified in such cases. When parameters like b are intermediate in value, we have no guarantee that maximization of the

likelihood will correspond to minimization of the extent of polymorphism. The maximum likelihood method would presumably in such cases favor trees on which there was an intermediate degree of retention of polymorphism, in line with expectation.

Using the same computer program for a variety of values of a, b, c, and E, the following patterns emerge:

1. When a, e^{-b}, c, and E are all small, the four reconstructions taken together account for almost all of the likelihood of a tree.

2. When $a \gg c$, the Camin-Sokal reconstruction accounted for much more of the likelihood than the Dollo reconstruction, and vice versa when $c \gg a$.

3. When $e^{-b} \gg a, c$, the polymorphism reconstruction contributes the bulk of the likelihood.

4. When $E \gg a^2$, e^{-b}, c, the term corresponding to misinterpretation accounts for the bulk of the likelihood.

These patterns are what we would expect based on the discussion in the previous section of this paper.

RELATIONSHIP TO FARRIS'S DERIVATION

Farris (1977), in his derivation of the Dollo parsimony method, used a probabilistic model with two states, 0 and 1, allowing $0 \rightarrow 1$ and $1 \rightarrow 0$ changes to occur with different probabilities. He obtained the Dollo method as a maximum likelihood method under the assumption that $0 \rightarrow 1$ changes are improbable, but argued that the Dollo method emerges as the maximum likelihood method whatever the probability of $1 \rightarrow 0$ reversions. This seems in direct contradiction to the present argument, in which the Dollo method is obtained only when c is small.

It is important to realize that there is no contradiction. Farris's derivation differs from the present one in a fundamental way. He is not making a maximum likelihood estimate of the phylogeny, as in this paper, but is estimating an "evolutionary hypothesis" which consists of the phylogeny plus the states of all characters at all the interior nodes of the evolutionary tree. I have discussed the distinction between these two types of estimates in some detail elsewhere (1973) and have more recently shown (1978) that methods making a maximum likelihood estimate of this "evolutionary hypothesis" can fail to make a statistically consistent estimate of the phylogeny. It is both intriguing and alarming that in making a maximum likelihood estimate of the "evolutionary hypothesis," and then ignoring all aspects of the resulting estimate except for the phylogeny, one does not obtain a maximum likelihood estimate of the phylogeny.

VARIANTS OF THE MODEL

The present model has assumed that probabilities of evolutionary change in characters are constant per unit time. This assumption is easily relaxed. Farris (1977), in his presentation of the Dollo parsimony method, assumed constancy of probabilities of evolutionary change per segment of the evolutionary tree, rather than per unit time. This will frequently be a biologically reasonable assumption, since it assumes a higher rate of evolutionary change per unit time in those lineages which are undergoing speciation. This assumption can be incorporated into our framework simply by setting all of the lengths t_j equal. When this is done, the

same conditions as before yield the same phylogenetic methods, with one exception. The polymorphism method can now be interpreted differently. The extent (in time) of polymorphism in a character will now be directly proportional to the number of tree segments which show polymorphism.

A more serious problem with the present model is that it assumes that it is known, for each character, which is the ancestral state. If we assume instead that there are nonzero prior probabilities p, q, and r, that the initial state of the population is 0, 01, or 1, then the present approach gives different results. When we take $a = c \gg e^{-b}$, E with all these quantities small it turns out that the maximum likelihood estimate of the phylogeny is the tree found by Farris's unrooted Wagner tree parsimony method (1970). Even though we are estimating a rooted evolutionary tree, it is the form of the tree when the root is removed, not the position of the root, which has by far the greatest effect on its likelihood.

When $e^{-b} \gg a = c \gg E$, we obtain a variant polymorphism parsimony method. This involves finding the initial character state which requires the smallest extent of polymorphism on each given phylogeny, then choosing that phylogeny which achieves the minimum total extent of polymorphism needed to explain the observed data. When the probabilities of evolutionary change a, b, and c are taken to be constant per tree segment rather than per unit time, we also find that the estimate of the phylogeny is effectively unrooted, as in the case of Farris's unrooted Wagner tree method.

Finally, when we take $a = c \gg E \gg a^2 = c^2 \gg e^{-b}$, it turns out that the maximum likelihood method is an unrooted two-state compatibility method, the method discussed by McMorris (1977).

IMPLEMENTING THE POLYMORPHISM METHOD

The polymorphism parsimony method consists of evaluating each proposed phylogeny by reconstructing the state of each

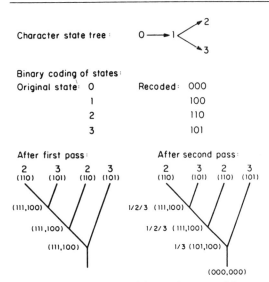

FIG. 4.—An example of the application of the reconstruction algorithm for multiple-state characters in the Polymorphism Method.

character at each interior node of the evolutionary tree, then taking as the measure of lack of merit of the phylogeny the total length (summed over all characters) of the segments of the tree which show polymorphism. A segment shows polymorphism for a character if the nodes at both ends of the segment are reconstructed as polymorphic. That phylogeny is chosen which has the smallest total extent of polymorphism.

A necessary component of this method is a rapid algorithm for assigning states to the populations at the nodes of an evolutionary tree. This can be done separately for each character, so it is only necessary to specify the procedure for reconstructing a single character, as follows. We start with states already assigned to the tips of the tree. We now make two passes through the tree, one down it (from tips to root) and one up it (from root to tips). (In computer science terminology, these are respectively a postorder tree traverse and a preorder tree traverse.) In the first pass, we want to assign to each node two states, each either 0 or 1. Call these M_i and m_i for node i. For the tips, let M_i and m_i be the

same and be equal to the state at the tip. Now we assign values to the interior nodes, working down the tree (in effect traversing it in postorder) such that if i is a node whose immediate descendants are j and k,

$$M_i = \max(M_j, M_k) \qquad (7a)$$

and

$$m_i = \min(m_j, m_k). \qquad (7b)$$

Now we assign to the root of the tree the states $M_r = m_r = 0$. We then proceed up the tree (in effect traversing it in preorder), altering the value of M_i as follows: if interior node i has immediate descendants j and k and ancestor A, we set $M_i = 0$ if $M_j + M_k + M_A \leq 1$, otherwise we leave it unaltered. The result is an assignment of states to nodes: if $M_i = m_i$ then both give the state assigned to the node. If $M_i = 1$ and $m_i = 0$, the node is polymorphic. It is essential that in the first pass down the tree that each node be encountered before its ancestor is, and that in the second pass which goes up the tree each node be encountered before its descendants are.

Preorder and postorder tree traverses are easily implemented in computer languages which allow recursive procedures. Some of these languages, such as PASCAL, SIMULA, and ALGOL 68, also have features which allow trees to be constructed and manipulated in a simple and natural fashion.

The same procedure also works with multi-state characters, as these can be reduced to a series of two-state characters by the binary recoding scheme introduced by Kluge and Farris (1969). The above procedure is simply applied to each of the resulting two-state characters. Figure 4 shows a character with four states as it is reconstructed on a four-species evolutionary tree by this method. Note that although there appears to be a new configuration (111) in two of the M_i arrays in the resulting tree, the number of states "segregating" in the resulting polymorphism can be correctly determined by taking one more than the number of differences between the M_i and the m_i arrays at each node i.

Once the reconstruction is made using this algorithm, it is a simple matter to compute the total length of polymorphic segments of the tree. This is summed over all characters to arrive at the measure of the lack of parsimony. The remainder of a polymorphism parsimony algorithm would consist of the strategy of searching among possible evolutionary trees to find the most parsimonious one. Since this problem of searching among trees to minimize an objective function is not unique to the polymorphism parsimony method, it will not be discussed here.

A PARADOX IN THE DOLLO METHOD

When a similar attempt is made to find an algorithm for reconstructing a multiple-state character according to the Dollo parsimony method, an interesting paradox is encountered which illuminates the logical interdependence of these various phylogenetic inference methods. Whereas in the polymorphism parsimony method we can always find a reconstruction of a character which assumes only a single origination of each state, this will not always be the case with the Dollo method. Consider the evolutionary tree, character-state tree and data shown in Fig. 4. At the uppermost fork, there is no state which can be assigned. If the state of the population there were 3, then state 2 could arise only by reversion to state 1 followed by a $1 \rightarrow 2$ transition. But there must have already been one such transition elsewhere on the tree to account for the other appearance of state 2. A precisely analogous argument rules out state 3 as having been present at this fork. If either state 0 or state 1 had been present, both states 2 and 3 would require two originations on the tree. There is thus no way to effect a reconstruction with only a single origin of each derived state, even if multiple reversions are allowed. The paradox cannot be resolved by decreeing that such an evolutionary tree would never be the estimate produced by the Dollo

parsimony method, for it is possible to arrange other characters so that, no matter what evolutionary tree is taken, the paradox arises in one or another character.

The resolution of the paradox is the realization that we must allow on such a tree certain events which are otherwise only reconstructed by the Camin-Sokal, polymorphism, or compatibility methods. We must allow either two originations of a state, a small stretch of retained polymorphism in this character, or assume a misinterpretation. This raises the question of how much penalty is to be assessed against this evolutionary tree for these extra events. The problem resolves itself naturally as long as we have obtained the Dollo method as a maximum likelihood estimate, for then the presence of these extra improbable events automatically penalizes the evolutionary tree by reducing its likelihood. The maximum likelihood framework is thus not only sufficient for deriving the Dollo parsimony method, but is also in a sense necessary for its use.

EXAMPLES OF USE OF THE POLYMORPHISM METHOD

The polymorphism parsimony method has already been applied once in the systematic literature. Inger (1967: fig. 6) presented a phylogeny of frog families which was generated using the polymorphism parsimony method. Figure 5 shows data from Inger's Table 1 coded as binary characters, and the phylogeny obtained when the polymorphism parsimony method is applied to them. The phylogeny may be clearer if the figure is rotated counterclockwise 45°. The variant of the polymorphism method used assumed that all tree segments were of equal length. The results differ somewhat from the figure presented by Inger, as the data used in that application were slightly different from those in Fig. 5. Inger's interpretation of the characters has been criticized by Kluge and Farris (1969), but this does not detract from the usefulness of his data as an example. The descrip-

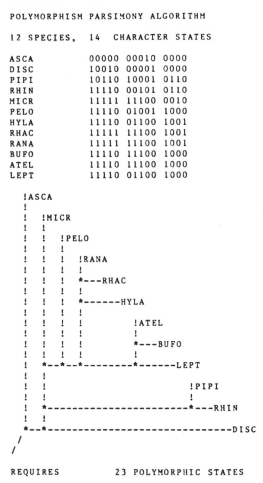

FIG. 5.—Data taken from Table 1 of Inger (1967) and the phylogeny which results when the polymorphism parsimony method is applied. Three-state characters have each been recoded as two two-state characters according to the method of Kluge and Farris (1969).

tion of the polymorphism method given by Inger is an attempt to convey some of its properties. Inger (1967, p. 80) wrote that the method "attempts to minimize the number of convergences and, where they cannot be avoided, keeps convergent taxa as close together as possible." Where the Camin-Sokal method would require two parallel changes, the polymorphism method will reconstruct a stretch of polymorphism between these two points on the evolutionary tree. In trying to reduce the total extent of polymorphism, this method will in effect try to move the points of occurrence of the parallel changes close together.

Another set of examples of the use of the polymorphism method is the estimation of phylogenies in the genus *Drosophila* by use of salivary chromosome banding information. In these studies, it is conventional to assume that each gene arrangement arose only once in evolution, and to arrange the phylogeny so as to minimize the extent of inversion polymorphism which must be assumed. A good example of this type of analysis is the paper of Stalker (1966). Frequently there is no information as to which of two sequences is ancestral and which the derived inversion, and in all cases one assumes that a = c, so that *Drosophila* taxonomists have usually been carrying out the unrooted character-state tree version of the polymorphism method.

THE LIMITATIONS OF LIKELIHOOD

We have seen that a common model can be developed, which yields both existing and new phylogenetic inference methods when we allow the parameters of the model to take extreme values. However this cannot be an entirely satisfactory justification for the use of these methods. For if we are willing to assume that (for example) origination of character state 1 is a rare event over the time span of our evolutionary tree, we expect little or no parallelism or convergence to appear in our data. This is rarely the case. It is not unusual to find that only a minority of characters could have all their changes of state be unique and unreversed. Le Quesne (1972) presented such an example.

If these methods are only maximum likelihood methods under conditions which our data render implausible, can we justify their use? We cannot simply forget these methods and use a general maximum likelihood method to estimate phylogenies. That would require knowledge of the values of the parameters of the model, which in our case are a, b, c, and E. It is unlikely that our knowledge of the details of the evolutionary process will be sufficiently precise for these to be known in the near future. Alternatively, one could estimate a, b, c, and E (or their equivalents) by maximum likelihood at the same time that we estimate the phylogeny. This involves enormous computational difficulties, but it is not an irrelevant approach to phylogenetic inference.

There is an alternative approach which seems more profitable. One can take the various parsimony and compatibility approaches, which are maximum likelihood methods under various extreme assumptions, as candidate methods for use when these conditions do not hold. While they would not then be maximum likelihood methods, they might nevertheless share with the likelihood method certain desirable statistical properties (notably consistency, sufficiency, and efficiency). I have elsewhere (Felsenstein, 1978) made a small start in this direction, by examining the statistical consistency of certain parsimony and compatibility methods in simple cases. They seem to have passed the few tests applied so far. However much more work along these lines remains to be done if we are to have a sound logical basis for our methods of estimating phylogenies.

ACKNOWLEDGMENTS

I am indebted to Lynn H. Throckmorton for discussions of his views of the evolution of characters, which led me directly to the polymorphism parsimony method. I am also indebted to Harvey Motulsky for writing programs to carry out the Camin-Sokal, polymorphism, and Dollo methods during the summer of 1971. This work was supported by Department of Energy contract No. EY-76 EY-76-S-06-2225 5 with the University of Washington.

REFERENCES

BELLMAN, R. 1970. Introduction to matrix analysis. Second Edition. McGraw-Hill, New York. 403 pp.
CAMIN, J. H., AND R. R. SOKAL. 1965. A method for deducing branching sequences in phylogeny. Evolution 19:311–326.
ESTABROOK, G. F., C. S. JOHNSON, JR., AND F. R. MCMORRIS. 1976. An algebraic analysis of cladistic characters. Discrete Math. 16:141–147.
FARRIS, J. S. 1970. Methods for computing Wagner trees. Syst. Zool. 19:83–92.
FARRIS, J. S. 1977. Phylogenetic analysis under Dollo's Law. Syst. Zool. 26:77–88.
FELSENSTEIN, J. 1973. Maximum likelihood and minimum-steps methods for estimating evolutionary trees from data on discrete characters. Syst. Zool. 22:240–249.
FELSENSTEIN, J. 1978. Cases in which parsimony or compatibility methods will be positively misleading. Syst. Zool. 27:401–410.
INGER, R. F. 1967. The development of a phylogeny of frogs. Evolution 21:369–384.
KLUGE, A. G., AND J. S. FARRIS. 1969. Quantitative phyletics and the evolution of anurans. Syst. Zool. 18:1–32.
LE QUESNE, W. J. 1969. A method of selection of characters in numerical taxonomy. Syst. Zool. 18:201–205.
LE QUESNE, W. J. 1972. Further studies based on the uniquely derived character concept. Syst. Zool. 21:281–288.
MCMORRIS, F. R. 1977. On the compatibility of binary qualitative taxonomic characters. Bull. Math. Biol. 39:133–138.
STALKER, H. D. 1966. The phylogenetic relationships of the species in the *Drosophila melanica* group. Genetics 53:327–342.
THROCKMORTON, L. H. 1968. Similarity *versus* relationship in *Drosophila*. Syst. Zool. 14:221–236.

Manuscript received June 1977
Revised August 1978

APPENDIX

Let $P_{ij}(t)$ be the probability that, given that it starts in state i at time 0, a population is in state j at time t. If A is the matrix

$$\begin{bmatrix} -a & b/2 & 0 \\ a & -b & c \\ 0 & b/2 & -c \end{bmatrix},$$

then it is well known (cf. Bellman, 1970, pp. 273, 169–170) that the matrix P of transition probabilities $P_{ij}(t)$ satisfies

$$P' = e^{At}. \tag{A1}$$

The spectral decomposition of A will be

$$A = U \wedge U^{-1}, \tag{A2}$$

where the eigenvalues λ_i are the solutions of the characteristic equation

$$\lambda^3 - \lambda^2(a + b + c) - \lambda(ac + ab/2 + bc/2) = 0 \tag{A3}$$

which has the three roots

$$\lambda = 0, \quad -\tfrac{1}{2}[(a + b + c) \pm ([a - c]^2 + b^2)^{1/2}]. \tag{A4}$$

If the three roots are denoted λ_1, λ_2, and λ_3, we find that the i-th column of U is the vector

$$[b(\lambda_i + c), \quad 2(\lambda_i + a)(\lambda_i + c), \quad b(\lambda_i + a)] \quad (A5)$$

Thus we can compute the λ_i and the elements of U. U^{-1} can be obtained numerically by a standard matrix inversion algorithm, and P can then be computed as the matrix product:

$$P' = U \left\{ \begin{array}{ccc} e^{\lambda_1 t} & 0 & 0 \\ 0 & e^{\lambda_2 t} & 0 \\ 0 & 0 & e^{\lambda_3 t} \end{array} \right\} U^{-1}. \quad (A6)$$

Note added in proof: Since this paper was submitted, Farris has presented (*SZ* 27:275–284) a somewhat different derivation of the polymorphism parsimony method. As with the Dollo parsimony method, he obtains it as the maximum likelihood estimate of a "phylogenetic hypothesis" consisting of the form of the evolutionary tree plus the character states at each fork of the tree. The comments made above (p. 57) apply in most particulars to his derivation of the polymorphism parsimony method.

Part VII

CRITIQUES OF CLADISTIC THEORY
AND METHODOLOGY

Editors' Comments
on Papers 20, 21, and 22

20   MAYR
Cladistic Analysis or Cladistic Classification?

21   ASHLOCK
The Uses of Cladistics

22   HULL
The Limits of Cladism

Any new approach to taxonomy invariably brings forth defenses of the traditional approaches and criticisms of the suggested innovations. The more that supporters claim a new method is likely to replace or supplant the old, the more the traditionalists tend to respond with criticism. A very few but very vocal and combative advocates aggressively promoted the presumptive virtues of cladistics, especially in the "Points of View" column of the journal *Systematic Zoology*. More traditional evolutionary systematists responded with critiques of cladistics, particularly as advocated by the more dogmatic practitioners.

One of the best critiques is that of Mayr (Paper 20), who also produced an insightful critique of numerical phenetics some years earlier (Mayr, 1965). Mayr makes the important distinction between (1) using cladistic analysis to help reconstruct phylogeny and (2) using only cladistic information in the construction of a classification. He clearly points out the desirability of the former and the unsuitability of the latter. In the same year Ashlock (Paper 21) followed similar lines of reasoning and also pointed to the useful applications of cladistics to biogeography. He gives a helpful list of common terms and their definitions.

Hull (Paper 22) provides a good philosophical overview of some of the issues of testability and phylogeny reconstruction as they relate to classification. He emphasizes that all types of knowledge are fallible, including cladistic knowledge, and he also stresses the need for an explicit approach to classification beyond just branching pattern information to include other dimensions of phylogeny (e.g., anagenetic divergence). These three papers raise most of the objections to advocating cladistics as the best or only approach to biological classification, and they will counterbalance the positive claims made by other workers (cf. Papers 1, 3, and 11).

REFERENCE

Mayr, E., 1965, Numerical Phenetics and Taxonomic Theory, *Syst. Zool.* **14:**73-97.

20

Reprinted from Z. zool. Syst. Evolut.-forsch. **12**:94–128 (1974)

Cladistic analysis or cladistic classification?

By Ernst Mayr

> Ein besonderes Anliegen ist es mir, zu betonen, daß die
> kritische Auseinandersetzung mit einem Autor dessen
> Verdienste nicht herabsetzen möchte. Das Gegenteil ist
> richtig. An unwesentlichen Arbeiten lohnt es sich nicht,
> Kritik zu üben. HENNIG, 1969

The choice of method in a scientific discipline depends to a large extent on the objectives of that discipline. If one wants to determine which of several methods of classifying animals and plants is most productive, one must first clarify one's concept of systematics. Quite rightly, therefore, HENNIG begins his *Grundzüge einer Theorie der phylogenetischen Systematik* (1950) with a discussion of the concept of systematics (pp. 1–12). Systematics, he says, is the ordering of the diversity of nature through constructing a classification which can serve as a general reference system. "Creating such a general reference system, and investigating the relations that extend from it to all other possible and necessary systems in biology, is the task of systematics" (HENNIG 1966, 7 [id. 1950, 10]).

The task of the creator of classifications, thus, is to find the best possible "general reference system". However, one can and should be more specific: a classification, in contradistinction to an identification scheme, functions as a biological theory (with all the explanatory, predictive, and heuristic properties of a theory) (MAYR 1969, 79–80); it must provide a sound foundation for all comparative studies in biology, and it must be able to serve as an efficient information storage and retrieval system (MAYR 1969, 229–244).

Generalizations in large parts of biology are derived from comparisons. Comparisons of groups, however, are meaningful in evolutionary studies only when such groups are correctly formed, that is, consist of "related" elements. The construction of efficient classifications, is, thus, as stated by HENNIG, a prerequisite for sound work in large parts of biology. WARBURTON (1967) has attempted to specify the criteria on which some classification can be judged to be superior to others in fulfilling the demand to serve as "general reference systems", as sound biological theories, and as efficient information storage and retrieval systems.

The 1930's and 1940's were dominated by the so-called "new systematics". Taxonomists concentrated their attention on the level of species and populations (*microtaxonomy*), which was also the area of principal concern of the newly emerging field of population genetics. The problems relating to the classification of higher taxa (*macrotaxonomy*) were largely neglected. There was, however, a significant minority of workers, particularly among the paleontologists and comparative anatomists, who felt that the seemingly so simple Darwinian credo that classifications should reflect "relationship" or "common descent" raised many unanswered questions. This is evident from several contributions to the volumes edited by HUXLEY (1940), HEBERER (1943), JEPSEN, MAYR and SIMPSON (1949), and more specifically from the writings of SIMPSON (1945) and RENSCH (1947). The intellectual ferment of this period led to the formulation of three competing theories of classification during the 1950's and 1960's, each of them claiming to be more objective and a better general reference system than the other two. These three theories, referred to by GÜNTHER (1971, 76) and described in more detail by MAYR (1969, 68–77) will now be characterized. (Unfortunately a considerable number of taxonomists have hardly any theory at all and deal with species and higher taxa purely descriptively, considering classification simply as identification systems.)

The three current theories of classification

a. Phenetic systematics (Phenetics): Organisms are classified, according to this theory, on the basis of "overall similarity". Similarity is calculated from the presence or absence of numerous unweighted characters or character states (SOKAL and SNEATH 1963). This method does not establish groups by inspection, but orders the lowest taxonomic units (usually species) into groups with the help of standardized procedures.

The methods and principles of phenetics have been critically analyzed elsewhere (MAYR 1965, 1969; JOHNSON 1970; HULL 1970).

b. Cladistic systematics (Cladistics)[1]*:* Organisms are classified and ranked, according to this theory, exclusively on the basis of "recency of common descent". Membership of species in taxa is recognized by the joint possession of derived ("apomorphous") characters. Grouping and ranking are given simultaneously by the branching points. (See below for a statement of the reasons why the designation "phylogenetic systematics" for this theory is misleading).

c. Evolutionary systematics: Organisms are classified and ranked, according to this theory, on the basis of two sets of factors, 1. phylogenetic branching ("recency of common descent", retrospectively defined), and 2. amount and nature of evolutionary change between branching points. The latter factor, in turn, depends on the evolutionary history of a respective branch, e. g., whether or not it has entered a new adaptive zone and to what extent it has experienced a major radiation. The evolutionary taxonomist attempts to maximize simultaneously in his classification the information content of both types of variables (1 and 2 above).

The synthetic or evolutionary method of classification thus combines components of cladistics and of phenetics, but in a rather different manner. It agrees with cladistics in the postulate that as complete as possible a reconstruction of phylogeny must precede the construction of a classification since groups that are not composed of descendants of a common ancestor are artifical and of low predictive value. More generally it agrees also with the cladists in the careful weighting of characters. It rejects, however, the "divisional" process of classification ("downward" classification), which is most evident in the cladists' definition of "monophyletic". Evolutionary classification rejects most of the conceptual axioms of phenetics, but agrees with it in the actual procedure of grouping by a largely phenetic approach. However, in contrast to the unweighted approach of the pheneticists, it bases its conclusions on the careful weighting of characters.

The method in which cladistic and phenetic components are combined was originated by DARWIN (see below).

---

[1] The nouns cladism and cladistics have been used interchangeably. Since the ending "-ics" corresponds to that of phenetics, systematics, and genetics I now prefer to use cladistics.

Is cladistics the best theory of classification?

The cladists are sincerely convinced that their theory produces the best classifications. HENNIG, for instance, states "that the claim of phylogenetic systematics for primacy among all possible forms of biological systematics has never been refuted even in the slightest" (1971, 9). GÜNTHER (1971, 38) likewise states: "W. HENNIG has elaborated and substantiated his theory of phylogenetic systematics to such an extent that it can be considered as irrefutable". GÜNTHER (l. c., p. 76) furthermore claims that, among the three prevailing conceptualizations of biological systematics, it is only the consistently phylogenetic (genealogical) concept which permits drawing phylogenetically unequivocal conclusions. Similar statements can be found in the writings of BRUNDIN, CROWSON, NELSON, SCHLEE and other cladists. GRIFFITHS, for instance, states that HENNIG's method "provides the only theoretically sound basis for achieving an objective equivalence between the taxa assigned to particular categories in a phylogenetic system" (1972, 9).

Considering this conviction of the superiority of their method cladists are genuinely puzzled "why there are nevertheless so many systematists who have not (or only with reservations) committed themselves to phylogenetic systematics" (HENNIG 1971, 9). HENNIG answers his own question by implying that it is simply insufficient familiarity with the objectives and methods of cladistics which has been in the way of a more general adoption. GÜNTHER, on the other hand, believes that it is the unavailability or neglect of three sets of facts which have prevented the more general application of cladistics: (1) the lack of sufficient available distinguishing characters, (2) the uncertainty as to which characters are ancestral and which derived, and (3) the difficulty of a clear recognition of convergences (1971, 77). In other words, both of these authors feel that empirical rather than conceptual reasons are responsible for the delay in a more rapid adoption of cladistics.

Is this conclusion really justified? Does a purely genealogical arrangement answer the demands of a "best classification"? Indeed, how do we determine which of several alternate classifications is the best?

There has long been agreement among the theoreticians of classification that in most cases those classifications are "best" which allow the greatest number of conclusions and predictions. MILL (1874, 466–467) expressed this, one hundred years ago, in the statement:

"The ends of scientific classification are best answered when objects are formed into groups respecting which a greater number of general propositions can be made, and those propositions (being) more important, than could be made respecting any other groups into which the same things could be distributed."

The opponents of cladistics claim that cladistic classifications do not satisfy MILL's criterion of a "best classification". The number of evolutionary statements and predictions that can be made for many holophyletic[2] groups (like birds and crocodilians) is often quite minimal, consisting ultimately only of a list of the synapomorphies. Indeed, the cladistic theory of classification would seem to suffer from several fundamental conceptual flaws.

The argument cannot be settled without a searching analysis of the theory (including all the underlying assumptions) on which cladistics is based.

There are many indications that most cladists have never given serious consideration to alternative theories of classification, particularly to the theory of evolutionary taxonomy. For how else could HENNIG (1971, 7) have classified the evolutionary

---

[2] A holophyletic group contains all the descendants of a stem species.

taxonomists in one group with those taxonomists who work without any theory at all? (See also BRUNDIN 1972, 111). Other cladists, in their arguments, proceed as if phenetics (classification simply based on similarity) is the only alternative to cladistics. Even GRIFFITHS (1972, 18) who clearly distinguishes between the three methods of classifying argues in his actual defense of cladistics only against "morphological-phenetic classifications." Objections to the cladistic theory are being brushed aside as being due to inconsistencies or as having a purely psychological basis (GÜNTHER 1971, 38).

There can be no hope for a meeting of the minds until the cladists face up to criticisms of their opponents and attempt to refute them, point by point. GRIFFITHS (1972) is the only cladist who has even attempted such a refutation.

In contrast to the flood of defenses of cladistics published in recent years (by BIGELOW, BRUNDIN, CRACRAFT, CROWSON, GRIFFITHS, GÜNTHER, HENNIG, KIRIAKOFF, NELSON, ROSEN, SCHLEE and others) there has been only a limited amount of critical analysis of their theory. SIMPSON (1961) and myself (1969) have dealt with it *en passant* in major textbooks. There have been several short critical book reviews, and certain specific points (like the definition of "monophyly") were criticized by ASHLOCK, COLLESS, FARRIS, GUTMANN, JOHNSON, MICHENER, PETERS, and others. But DARLINGTON's papers were really the only serious attempt of a broad criticism of cladistics. Even his criticism deals more with the application of cladistics to biogeography than with the underlying basic assumptions. Indeed DARLINGTON himself states that his criticism "is not a general consideration of cladism" (1970, 1). It is the objective of the present critical analysis to fill a serious gap in the taxonomic literature.

The components of cladistics

It is most important for the understanding of cladistics to realize that it actually consists of two quite different sets of operations:
1. the reconstruction of the branching pattern of phylogeny through *cladistic analysis,*
2. the construction of a *cladistic classification* based on this branching pattern.
The first of these two operations is important and largely unobjectionable. It is the second one which has encountered widespread criticism and will be carefully analyzed in the ensuing pages, with particular attention to the claim of the cladist that a classification should be a mirror image of the branching pattern of the phylogeny.

Reconstruction of the branching pattern of phylogeny (cladistic analysis)

The cladistic analysis starts from the basic assumption that a sound classification cannot be constructed without a thorough understanding of the phylogeny of the given group. The evolutionary taxonomist agrees, on the whole, with this assumption. All phylogeny, except in cases of reticulate evolution, is strictly genealogical. HENNIG is quite right when he states: "Phylogenetic research as biological science is possible only if it adopts the discovery of the genealogical relation of species as its first objective" (HENNIG 1969, 33).

But how is one to proceed if one wants to reconstruct the phylogeny of a group? As HENNIG has stated (1969, 19), this method rests, in the last analysis, on the simple realization "that all differences and agreements between various species originated in the course of phylogeny. During the splitting of a species its characters are transmitted to the daughter species either changed or unchanged". HENNIG is fully aware that all that can be inferred by this method is the sequence of splits but not at all their absolute chronology.

HENNIG has emphasized, and quite rightly so, that a phylogeny does not need to be based on fossils but can be inferred from a careful comparative analysis of morphological characters. This thesis is well substantiated by the classification of the Recent mammals. Our ideas of their relationships, based on a study of their comparative anatomy, have in no case been refuted by subsequent discoveries in the fossil record. On the other hand, the fossil record is entirely indispensable for the determination of absolute chronologies.

The most important step in the cladistic analysis is the attempt to separate characters into ancestral (plesiomorphous) and derived (apomorphous) characters. Only the latter are considered legitimate evidence for relationship, and taxa are therefore based on the joint possession of derived characters (synapomorphies). (For a consideration of the value of symplesiomorphies in the process of ranking see p. 118). Neither HENNIG nor any of his followers has claimed that this important principle was new when proclaimed by HENNIG: "The observation that taxa should only be characterized by apomorphous (derivative) conditions in their ground plan is, of course, by no means new and to many people seems self-evident" (GRIFFITHS 1972, 21). To mention only one example, TILLYARD's classification of the Perlaria (1921, 35–43) was based on this principle. In fact, one can say that most of the better taxonomists of former eras had applied this principle, as is quite evident from a study of their classifications.

Nevertheless, HENNIG deserves great credit for having fully developed the principles of cladistic analysis. The clear recognition of the importance of synapomorphies for the reconstruction of branching sequences is HENNIG's major contribution. The cladograms which can be constructed with the help of this method are as important for the evolutionary taxonomist as they are for the cladist. I have previously (MAYR 1969, 212–217) called attention to the extreme value of this method for the delimitation of taxa. The relative time sequence of the various branching points which the cladogram provides is of great value in many studies, particularly in zoogeographic ones, as HENNIG himself has demonstrated for the diptera of New Zealand (HENNIG 1960).

Cladistic analysis and cladistic classification

There is little argument between cladists and evolutionary taxonomists about the cladogram which results from the cladistic analysis. The argument arises over the relationship of such a cladogram to the classification that is to be based on it. Cladists assume that a one-to-one relationship exists between cladogram (phyletic diagram) and classification. The cladogram, once it is constructed, provides so to speak automatically also the classification. A cladogram and a classification are for the cladist merely two sides of the same coin. The evolutionary taxonomist, on the contrary, believes that a mere branching pattern cannot convey nearly as much interesting information as an evolutionary classification which takes additional processes of evolution into consideration (see below).

Traditionally, the first step in the classification of animals taken by the practicing taxonomist has always been the delimitation by inspection of seemingly "natural" groups. At first these are frankly based on the apparent "similarity" of the included species, that is on phenetic criteria. When HENNIG first proposed the cladistic method (1950) virtually all major higher taxa of animals were already known. He, therefore, automatically adopted the traditional method of taxonomy, of ranking and regrouping animal taxa which other authors had previously delimited. The validity of these provisional groups is subsequently tested in traditional taxonomy against a whole series of additional criteria, such as the homology of characters (similar and dissimilar ones), the presence of synapomorphies, the chronological relation to similar groups, an absence of conflict with the fossil record (when available), an absence of convergence (= spurious similarities), a meaningful geographic distribution, etc. The more experienced a taxonomist is, the more quickly and thoroughly he can undertake these tests. (The classifying procedure of the pheneticist is drastically different). Once established, a classification is thus constantly improved by the process designated by HENNIG (1950) as "reciprocal illumination", which, as HULL (1967) has shown, does not involve circular reasoning. Indeed, the method is nothing more than another application of the hypothetico-deductive approach (POPPER 1959, 1963), so commonly used in all branches of science and particularly (since DARWIN) in biology. As soon as a new (or improved) classification is proposed, it will generate new information which, in turn, will lead to a reanalysis and possibly an improvement of the classification. This traditional approach to classification was followed without serious criticism during much of the 19th and 20th centuries.

Unfortunately this approach by trial and error is sometimes rather inefficient and has led to frequent changes in classifications. Again and again the hope was therefore expressed for a more reliable approach. The better taxonomists agreed on two minimal conditions, the crucial importance of the right choice of characters (a point DARWIN already had emphasized), and the necessity of basing taxa on numerous characters. But this still left much uncertainty.

It was HENNIG's novel proposal simply to translate the cladogram into a hierarchical classification and thus do away with all the previous uncertainty. The proposal to construct a classification directly from the cladogram, however, does not convey the entire theory of cladistic classification. For this reason it would be highly desirable to present a detailed exposition of the entire theory in all of its ramifications, but this is rather difficult. Not only is HENNIG's original work (1950) written in a rather turbid style, but some of his earlier theses seem to be no longer maintained in more recent publications. Furthermore, some of his followers, like BRUNDIN, SCHLEE, and GRIFFITHS, seem to have added postulates which, although perhaps implicit in the original theory, were not explicitly made by HENNIG himself. Nevertheless, the major theses and postulates of cladistics have been stated sufficiently often to permit their enumeration. I shall try to list the more important ones, preferably by direct quotes from the works of cladists. I suspect that my listing is neither complete nor that each of the postulates is adopted by every cladist. However, I hope that this list can serve as a convenient basis for the ensuing analysis.

These are the more important postulates of cladistics (page reference in parenthesis refers to the more detailed discussion below):

1. that all taxa should be "monophyletic", with this term redefined in a novel way, in conflict with the traditional definition of this term (p. 103);
2. that the term phylogenetic be restricted to the branching (cladistic) component of phylogeny (p. 100);
3. that relationship be measured in terms of "recency of common descent", i. e., narrowly genealogically (p. 101–103);
4. that "there is only one dimension in phylogeny and that is the time dimension" (BRUNDIN 1966). Consequently, the splitting of phyletic lines (as reconstructed from the joint acquisition of derived characters) is admitted as the only legitimate evidence in the construction of classifications. To consider also similarities or the relative amount of ancestral (plesiomorphus) characters would lead to a "syncretistic system" which "robs the combination of any scientific value" (HENNIG

1966, 77). He quotes with approval BIGELOW's (1956) statement: "Classification must be based on one or the other (on "overall resemblance" *or* "recency of common ancestry"...), not on both, if philosophical confusion is to be avoided" (HENNIG, l. c.) (p. 102, p. 123);

5. that the categorical rank of a taxon is automatically given by the absolute geological age of the stem species, or (in a less rigorous formulation) by the "relative age" of the stem species (p. 114). (See also CROWSON, p. 251 and disclaimer by GRIFFITHS [1972, 10, 16]);

6. that species can be delimited in time by two successive events of speciation (p. 109);

7. that the splitting of lines is always a dichotomy, resulting in the production of two sister groups (p. 109);

8. that "homology ... is usually defined in terms of common origin in time" (GRIFFITHS 1972, 17). (This is simply not true. Except for the ancestor-descendant relationship, the concept of homology is completely independent of the time dimension. No other cladist has made such a claim. HENNIG himself adopts REMANE's (1952) concept of homology);

9. that "basically" all classifications should be horizontal classifications, valid only for a given time period (HENNIG 1950, 259) and that therefore the same taxon might be given different categorical rank in different geological periods (p. 115).

Objections to the theory of cladistic classification

The basic postulate of the cladistic theory, a complete congruence of cladogram and classification, can be satisfied only by making numerous assumptions and redefinitions and by ignoring numerous facts of evolution and of phylogeny (broadly defined). This results in major theoretical and practical shortcomings which will now be analyzed, point by point, in three major and a number of subordinate sections.

1. Arbitrary decisions

In order for their method of classification to work, cladists have to make a number of arbitrary decisions, involving a redefinition of well known terms, a re-interpretation of adaptive evolution, and the proposal of a new species definition. When these arbitrary decisions are rejected, very little support for cladistic classification remains.

a. Redefinition of well known terms

A large part of the controversy between cladistics and its opponents can be ascribed to the fact that the cladists have given an entirely new meaning to a number of widely used evolutionary terms which in a rather different sense had been in essentially consistent usage for about 100 years. The transfer of well known and universally understood terms to entirely new concepts cannot fail to produce confusion. This is particularly true for three terms: *phylogeny*, *relationship*, and *monophyletic*.

Phylogeny

Since the days of DARWIN and HAECKEL the term phylogeny has been applied to all aspects of descent. Any article or book in the last 100 years that has used the term phylogeny (and its adjective phylogenetic) has used it comprehensively for all phenomena revealed by Stammbaumforschung. The Collegiate Dictionary of Zoology, for instance, defines it as "1. Evolutionary relationships and lines of descent in any taxon. 2. The origin and evolution of higher taxonomic categories" (PENNAK 1971, 395). But HENNIG now attempts to restrict this term to a single aspect of phylogeny, that of branching. He states, "We will call 'phylogenetic relationship' the ... (genealogical) relations between different sections (in the diagram), each bounded by two cleavage processes in the sequence of individuals that are connected by tokogenetic relations" (1966, 20). Or, "We have defined the phylogenetic relationships ... as those segments of the stream of genealogical relationships that lie between two processes of speciation" (l. c. 29). SCHLEE (1971) defines phylogeny as "the origin of taxa, that is that part of evolution which is designated as its cladistic component."

HENNIG's definition and use of the term phylogenetic is clearly in conflict with the previously universal use of the term phylogenetic. His diagrams are cladograms and not at all phylogenetic trees which by the lengths and angles of their branches convey far more information than a cladogram. It would only aggravate the confusion if HENNIG's specialized theory of classification would continue to be designated as "phylogenetic" classification.

GISIN (1964) has proposed to designate HENNIG's method by the term "genealogical" classification, since genealogical (kinship) relationships are an important aspect of the HENNIG method. Unfortunately, the designation "genealogical" does not differentiate cladistics from certain other types of classification, because any classification in which each taxon consists of species that are derived from a common ancestor is a genealogical classification. DARWIN was, therefore, quite right in stating that evolutionary classifications are genealogies. In any phylogeny in which hybridization does not occur (and this is the case in nearly all animal phylogenies) there can be only one genealogy, because each speciational event is unique. This unequivocality of the branching component of phylogeny is rightly emphasized by the cladists. However, even an unequivocal genealogy can usually be converted into a number of different classifications. And this is where evolutionists and cladists have a parting of their ways. Those who allow for a different weighting of different adaptational processes and events (e. g., by giving greater weight to the occupation of a major adaptive zone) may arrive at a very different classification from someone who uses branching as his only basis (as do the cladists), even though the genealogies on which both base their classifications are identical. Since the term "genealogical" does not discriminate between evolutionary and cladistic classifications, I prefer the term "cladistic" for HENNIG's method because it applies to it unequivocally and not to any other system of classification. Furthermore, it conforms to the terminology proposed by RENSCH (1947) and CAIN and HARRISON (1960). It is the only term which accurately conveys the emphasis of the HENNIG method on branching and on branching alone.

When employed in phylogeny the term "genealogical" is obviously used in a somewhat generalized sense, with "taxa" corresponding to the "individuals" of a conventional genealogy. The only alternative would be to classify "generations" of individuals. This, indeed, would be logically impeccable, but useless for purposes of biological classification.

Relationship

The term relationship has been used in the systematic literature in many different senses. Recourse to a dictionary is of no help, since it lists extremely diverging definitions. The term relationship (or affinity) was widely used in the 18th century taxonomic literature, long before the evolutionary theory was adopted. For most authors, at that period, it simply meant similarity. Yet, even today, the term relationship is being defined in many different ways.

Pheneticists, in the beginning, were operating on the basis of the assumption that the phenotype accurately reflects the genotype and that an unweighted determination of "overall similarity" allows a correct determination of relationship. They are no longer as dogmatic about this as they were in 1963, but relationship still means similarity to them. The shortcomings of this interpretation have been pointed out by MAYR (1969) and many others.

The cladists go to the other extreme and restrict the term relationship to designate kinship in a strictly genealogical sense. According to HENNIG "the measure of phylogenetic relationship is the relative recency of common ancestry" (1966, 74). Relationship between two species is measured by the number of branching steps which separate them from the common ancestor (HENNIG 1950, 129). When CRACRAFT (1972, 381) claims that the cladistic method "is the best one available for determining relationships in a relatively unambiguous fashion", he has fallen victim to a circular argument because he uses the highly specialized definition of relationship of the cladists. To repeat, relationship for the cladist is genealogical kinship. But cladistic kinship alone, for an evolutionist, is a completely one-sided way of documenting relationship, because it ignores the fate of phyletic lines subsequent to splitting.

Let me explain. Since a person receives half of his chromosomes from his father, and his child again receives half of his chromosomes from him, it is correct to say that a person is genetically as closely related to his father as to his child. The percentage of shared genes (genetic relationship), however, becomes quite unpredictable, owing to the vagaries of crossing over and of the random distribution of homologous chromosomes during meiosis, when it comes to collateral relatives (siblings, cousins) and to more distant descendants (grandparents and grandchildren, etc.). Two first cousins (even two brothers, for that matter) could have one hundred times as many genes in common with each other than they share with a third first cousin (or brother) (among the loci variable in that population). The more generations are involved, the greater becomes the discrepancy between genealogical kinship and similarity of genotype, even though all these relatives still derive their genes from the gene pool of a single species.

In phylogeny, where thousands and millions of generations are involved, that is thousands and millions of occasions for a change in gene frequencies owing to stochastic processes, recombination, selection, and genetic revolutions, it becomes quite meaningless to express relationship only in terms of the genealogical kinship.

In addition to the cytogenetic processes there are also numerous aspects of selection which can result in a highly unequal degree of genetic change in different lines of descent. One of several phyletic sister lines may enter a new adaptive zone and there become exposed to severe novel selection pressures. As a result it will diverge dramatically from its cladistically nearest relatives and may become genetically so different that it would be biologically misleading to continue calling the sister groups near relatives. Yet being the joint descendants of a stem species they must be designated sister groups. And being sister groups they must be coordinate in rank, i. e., according to cladistic theory, they must have the same absolute categorical rank in the hierarchy (HENNIG 1966, 139). This decision ignores the fact that one is still very much like the stem species while the other has evolved in the meantime into a drastically different type of organism.

This situation is best illustrated by a diagram (Fig. 1). There will be a maximal genetic difference of 25 % between the genomes of B and C, but of 60 to 70 % between C and D. The cladist will say that C is more nearly related to D than to B, the evolutionist and the pheneticist that C is much closer to B than to D.

This independence of adaptive shifts from phyletic splitting is the reason why the evolutionary taxonomist adopts a very different definition of relationship. To him

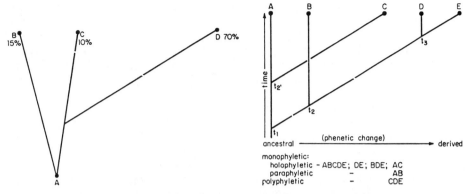

Fig. 1 *(left)*. Inferred percentual difference from ultimate ancestor (A). Taxon C is more closely related to B than to D, even though it shares a more recent common ancestor with D. – Fig. 2 *(right)*. Taxon C, through parallel evolution, has become more similar to D and E than to A with which it shares a nearer common ancestor. The converse is true for B (After Ashlock 1969)

relationship means the inferred amount of shared genotype, it means gene content rather than purely formalistic kinship. For what is of primary interest in a taxon, its evolutionary role, its system of adaptations, and all the correlations in its structure and characters, is ultimately encoded in its genotype.

Since distant relatives cannot be analyzed genetically and since a purely additive analysis of a complex system of adaptations would be quite meaningless anyhow, it is necessary to infer degree of genetic relationship on the basis of indirect evidence. In this approach use is made of every available clue, but primarily of the combined evidence from phyletic branching and from a carefully weighted phyletic analysis. The evolutionary taxonomist believes that an approach which superimposes a carefully weighted phenetic analysis on a preceding cladistic analysis is better able to establish degree of relationship than either a purely cladistic or an unweighted phenetic approach. And a classification based on such a multiple-based determination of relationship will be more reliable and more predictive than one based on one-sided criteria. The criteria which have to be employed to make such weighting meaningful have been discussed elsewhere (Mayr 1969, 217–228). These criteria permit measuring something that is more than mere "overall similarity."

Convergence, parallelism, and mosaic evolution, all these evolutionary phenomena underline the importance in evolution of the underlying invisible genotype. A diagram (Fig. 2) will illustrate the difficulties created by the concealed potential of the genotype. If species C owing to its concealed genotype acquires by parallel evolution very much the same characters as D and E, even though it branched off from the line leading to A, it would seem legitimate to classify it with D + E. The cladist would presumably call a taxon composed of C + D + E polyphyletic as if the similarity were due to convergence. Phenotypically, indeed, C + D + E form a polyphyletic group. Many evolutionary phenomena indicate the existence of a concealed genotype which cannot be read out completely and directly from the visible phenotype. Most of our difficulties with apparent polyphyly are due to the manifestations of such a concealed genotype.

Monophyletic

The traditional procedure of taxonomy is to recognize a higher taxon "intuitively", that is on the basis of shared characteristics, evolutionary role, etc. As Hennig stated it (1966, 146): "Taxonomy can begin its grouping task with the assumption that the degree of similarity between species corresponds with the degree of their phylogenetic kinship," an assumption to be thoroughly tested subsequently. There is thus agreement between HENNIG and the traditional taxonomist that the characterization and delimitation of groups has primacy in the classificatory operation (See above, w. 98). This primacy is confirmed by GRIFFITHS (1972, 7) who proposes "that the phylogenetic system should be expressed by revision of the traditional Linnaean system rather than by proposal of a separate classification."

Not until a taxon is established provisionally does the taxonomist ask whether it is truly a "natural" taxon, composed of species that are each other's closest relatives. One of the ways to satisfy this condition is to ask whether all members of the taxon have descended from a common ancestor, that is whether the taxon is monophyletic. The term monophyletic is a qualifying adjective for a noun, the noun *group* (or *taxon*). Among possible groups (and taxa) there are some which we can identify as being monophyletic. It is a distinctly "retrospective" term (MAYR 1969, 75). For this reason the term monophyletic, ever since HAECKEL, has been applied to groups which satisfied two conditions: 1. the component species, owing to their characteristics, are believed to be each other's nearest relatives, and 2. they are all inferred to have descended from the same common ancestor. The second qualification is needed to exclude unnatural groupings due to convergence.

When this traditional definition of monophyletic is applied to taxa, difficulties are rarely encountered: birds are monophyletic, crocodilians are monophyletic, and reptiles are monophyletic. The concept as such is entirely unambiguous, even though its application encounters occasional difficulties, as in the therapsid-mammalian transition (SIMPSON 1961, 124–125; but see CROMPTON and JENKINS 1973).

HENNIG has created enormous confusion by adding to the traditional definition of a monophyletic taxon the following qualification: ". . . and which includes all species descended from this stem species." This definition is the inevitable consequence of the elimination of all consideration of adaptive evolution from HENNIG's concept of phylogeny. Since birds and crocodilians (excluding all other living reptiles) are derived from a common ancestor, his method forces the cladist to recognize a taxon for birds and crocodilians together, even though this is a useless assemblage. HENNIG has transferred the qualification "monophyletic" from the taxon to the mode of descent. From a retrospective principle he has made monophyly a prospective criterion. This ignores, indeed it quite deliberately conceals, the most interesting aspects of evolution and phylogeny, those of adaptive radiation and the invasion of new adaptive zones (further discussed below under Grades).

His definition of monophyletic forces HENNIG to add another term to the phyletic terminology: *paraphyletic*[3]. A taxonomic group is paraphyletic if it has given rise to specialized side-lines which are not considered part of the group. For instance, the Reptilia are – for HENNIG – a paraphyletic group, because certain Reptilia were the stem mothers both of other reptilia and of the birds (and the same is true for the stem species that gave rise to the mammalian branch). To designate groups as paraphyletic strikes me as a purely formalistic approach. It is of no relevance whatsoever for the relationship between crocodilians and other reptiles to know that the branch leading to the crocodilians (the archosaurian lineage) produced an offshoot which eventually became the class Aves. The animal taxonomist does not classify a logician's schemata or diagrams, but actually concrete groups of organisms. It is of no relevance for our

---

[3] HENNIG designates a group as *paraphyletic* if the similarity of the composing taxa is based on symplesiomorphy. For instance, the Reptilia are a paraphyletic group, in contrast to the archosaurians (crocodilians and birds) and the therapsids (mammal-like reptiles and mammals), which are holophyletic.

judgment on the biological classification of the crocodilians whether or not a side branch gave rise to a drastically modified daughter group.

To take the traditional term monophyletic and transfer it to a new concept for which it had never been used before is contrary to sound language practices and to principles of scientific terminology. It strikes me as ludicrous when HENNIG's adherents criticize their opponents for an "illogical usage of the word monophyletic." If one wants to have a term for "the aggregate of all groups descended from a common ancestor," one must coin a new term. ASHLOCK (1971, 65; 1972) has recognized this quite clearly and has proposed for it the term "holophyletic." This corresponds to "monophyletic" as used by HENNIG in contrast to the traditional usage.

No cladist seems to have noticed some of the consequences of the redefinition of the term monophyletic. It forces him to abandon upward classification as traditionally practiced by the empirical taxonomists ever since DARWIN and even earlier, and replace it by "downward" classification. Although starting from entirely different premises, the cladist has methodologically returned to the "divisional" method of classification that was dominant from Cesalpino to Linnaeus. His criterion of division is of course very different from that of the adherents of ARISTOTLE's logical division, but the principle of classifying of both schools (cladists and logicians) is very much the same.

It may sound like a platitude to say that when classifying one ought to deal with entities which one has in front of himself. The pheneticist and evolutionist classify species and genera in this manner. Not so the cladist, who deals with the unknown quantities produced by phylogenetic splits. It is implicit in his principles that he is forced to make the prediction that sister lines derived from a stem species will have sufficiently similar evolutionary fates so that the resulting sister groups can be ranked at the same categorical level (= are coordinate). The case of the birds and crocodilians is a particularly convincing illustration of the thousands of occasions where this prediction does not come true. It is the abandonment of the principle of upward classification, dominant since DARWIN, and its replacement by ARISTOTLE's downward classification which is the fatal flaw in the philosophy of cladistic classification.

Those who would adopt the three terms *phylogeny, relationship,* and *monophyletic* in their new aberrant Hennigian definitions are forced to adopt drastic changes in the whole theory and practice of phylogeny and classification as compared to the traditional. It is quite true that one can operate in an entirely logical manner within the framework set by these new definitions. However, as GHISELIN has pointed out so often, one can operate in an entirely logical manner on the basis of totally wrong premises. Many, if not most, of the claims of the cladists go back to the consequences of their new definitions of these three terms.

b. The neglect of the dual nature of evolutionary change

DARWIN saw clearly that speciation involves two independent processes. One is the acquisition of reproductive isolation, a prerequisite of prevention of the hybridization between the two incipient species. The other one is the acquisition of niche differences resulting in "divergence of character" in order to overcome the effects of competition (DARWIN 1859, 111).

What is true at the level of the species, is equally true in macroevolution. We can distinguish two processes of evolution, that of the splitting of phyletic lines and that of the invasion of new adaptive niches and major adaptive zones by phyletic lines. Any theory of classification which pays no attention to the tremendous range of difference between shifts of phyletic lines into minor niches and into entirely new adaptive zones, is bound to produce classifications that are unbalanced and meaningless. But such a neglect of different kinds of phyletic evolution is precisely what the cladistic method demands.

The cladist proceeds in the construction of his classifications as if the splitting of lineages were the only phylogenetic process and as if all such splits were equivalent. All splits have equal weight for the cladist, just as all characters have equal weight for the pheneticist. His exclusive preoccupation with splitting has been confirmed by HENNIG on several occasions, but will be documented here only by the following quotation: "Decisive is the fact that processes of species cleavage are the characteristic feature of evolution; they are the only positively demonstrable historical processes that take place in supra-individual organism groups in nature." (HENNIG 1966, 235). This leads HENNIG to the claim that his method is the only one that gives historically correct answers.

The cladist states openly that branching is the only aspect of phylogeny of interest to him. That some of the resulting lineages may enter entirely new adaptive zones and then become extraordinarily different from other, more conservative lineages, is considered by him as irrelevant; if removed from the common ancestor by the same number of speciational steps they all will have to be given the same taxonomic rank. Cladistics treats any apomorphous character like any other. Those taxa are combined which have the greatest number of joint apomorphies (as being derived from the same stem species). So far as I can judge from the cladistic literature, no weighting of apomorphies is undertaken, and derived characters which signify entrance into an entirely new adaptive zone (as in the case of the birds) are given no more weight than the joint apomorphies of birds and crocodiles (which distinguish these living archosaurians from other living reptiles). The acquisition of minor specializations are given the same weight as major adaptive innovations. That some events in adaptive evolution are far more important than others is completely ignored. This is, perhaps, the most significant difference between cladistic and evolutionary classification.

It is evident that the cladist reveals great ambivalence in the treatment of divergence. He pays lip service to the fact that there are differences in the rate of evolution in various communities of descent, but does not draw any of the obvious conclusions from this observation. Neither HENNIG nor BRUNDIN nor any of their younger followers (i. e. SCHLEE, NELSON, GRIFFITHS) pays even the slightest attention to these differences in rates mentioned by HENNIG when they construct classifications. They likewise entirely ignore various other important evolutionary phenomena such as the existence of "grades" and of highly specialized side lines, all the phenomena of mosaic evolution, and all causal factors in evolution. How rapidly a new branch diverges, how it changes in relation to the "sister group", how many additional characters it acquires, which new adaptive zone it has invaded, etc., all these are questions which the cladist hardly ever mentions. By considering only genealogical distance the cladist acts as if he assumed that all lines diverge in an equivalent manner and that genealogical distance corresponds to genetic distance. By claiming that branching is the only historical process of consequence, he denies that other aspects of evolutionary change such as rate of evolution, adaptive radiation, the occupation of new adaptive zones, mosaic evolution, and many other macroevolutionary phenomena are eligible for the term "historical process."

Both components of phylogeny are potentially of equal importance for the evolutionary taxonomist, and both must be judiciously considered in the construction of classifications. Splitting as well as phyletic change occur simultaneously in evolution, but in most groups either one or the other process predominates at a certain geological period. Whenever there is massive splitting, such as, for instance, during the speciation of the 50,000 or 100,000 species of weevils (Curculionidae), or of the several thousand species of *Drosophila*, phyletic divergence is relatively insignificant. Among the invertebrates, and more specifically among the arthropods, there are numerous higher taxa in which abundant speciation has occurred without any impact on the basic morphology and without any shift into a new adaptive zone. All the species of these assemblages are repeated variations on a theme. This is in strong contrast to such memorable episodes in the history of the world as was the origin of the vascular plants, the angiosperms, the chordates, the vertebrates, the terrestrial tetrapods, the reptiles, or the birds.

RENSCH (1947), HUXLEY (1942, 1958), and SIMPSON (1959b, 1961), particularly, have emphasized the importance of these levels of adaptation, designated by HUXLEY as *grades*. All members of a grade are characterized by a well integrated adaptive complex. The successful evolution of a phyletic lineage toward and into a new adaptive zone is characterized by the stepwise acquisition (mosaic evolution) of a series of novelties to adapt it (and its descendants) to its new position in the ecosystem. Subsequently the basic new type of this phyletic line may undergo little evolutionary change but experience instead abundant adaptive radiation as a result of bountiful speciation and various modifications in the basic adaptive theme of the grade. In the history of the vertebrates we know many such cases of the formation of successful new grades such as the sharks, the bony fishes, the amphibians, reptiles, birds, and mammals. Each of these is characterized by a certain type of adaptation to the environment (BOCK 1965), regardless of the amount of cladistic break-up within the grade. It results in a great deal of loss of information to ignore the adaptive component of evolution expressed by the concept of grade and to limit one's attention only to the splitting of lines. But this is precisely what the cladists are doing.

Actually, the existence of minor and major grades is one of the most interesting phylogenetic phenomena, even though it is a phenomenon which we are still unable to understand adequately. Why is there so often such a uniformity of type within a higher taxon? There is a rich diversity of species of parrots but all of them from the smallest pygmy parrot or lorikeet to the largest cockatoo or macaw are characteristically parrots. And so it is in many, if not most, higher taxa. The reptiles represent a well defined grade between the amphibian level and that of the two derivatives of the reptiles, the birds and the mammals.

SIMPSON, in his various publications (1953, 1959b, 1961) has repeatedly discussed the contrast between clades and grades. Crocodiles have a more recent common ancestor with birds than with lizards. They belong, thus, to the same clade as birds but they do not belong to the avian grade, but rather to that of the reptiles. To which of these two aspects of evolution shall we give primacy? There are literally thousands of similar dilemmas in the evolution of animals and plants. For instance, the African apes (*Pan*) have a more recent common ancestor with man (*Homo*) than with the orang (*Pongo*). However *Pan* belongs to the same grade as *Pongo*, very different from that typified by man. The better the fossil record becomes known, the more often one encounters such dilemmas. To the evolutionary taxonomist the existence of grades seems often more significant and more meaningful biologically than the mere splitting of phyletic lines. How little some of the cladists appreciate the biological significance of grades is illustrated in a comment by BRUNDIN (1972, 111) who designates groups such as the reptiles as "timeless abstractions."

One senses two reasons for the deliberate neglect of evolutionary divergence by cladists. One is that this factor cannot be measured precisely and unequivocally. Indeed, rates and degrees of evolutionary divergence can usually be inferred only by extrapolation or other indirect approaches. Yet, by appropriate weighting (MAYR 1969, 220–228) one can draw meaningful probabilistic inferences, which, although not entirely precise, are far more valuable than the advice to ignore evolutionary divergence altogether. A second reason is that cladists seem to think that they have to make a choice, in the delimitation of taxa, between basing them *either* on branching points *or* on degrees of evolutionary divergence. They fail to appreciate the added amount of information by utilizing *both* sources of evidence.

Cladists, when criticized for the neglect of evolutionary divergence, try to defend themselves by referring to HENNIG's *deviation rule* "When a species splits, one of the two daughter species tends to deviate more strongly than the other from the common stem species" (1966, 207). The establishment of this rule, say the cladists, proves that they do not ignore phyletic evolution. Several aspects of this rule are remarkable. First of all, it is in flat contradiction to HENNIG's assertion that splitting is the only historical process in phylogeny. Unequal deviation is a historical process which – as such – is independent of splitting. Secondly, although the deviation rule was pronounced by HENNIG already in 1950 (p. 111) and confirmed in 1966 and 1969 (p. 43), and although it is in principle adopted by most cladists (e. g. BRUNDIN 1972, 108), it seems to play no role whatsoever in the construction of any of their classifications. The dendrogram illustrating the deviation rule (1950, Fig. 25) is one of HENNIG's very few diagrams in which the angles of all clades are not the same. One almost gets the impression that it is the whole purpose of the deviation rule to permit cladists to defend themselves against the accusation of having ignored evolutionary divergence altogether. For the consequences of the deviation rule are completely neglected. They would become obvious, if the process of unequal deviation were to occur after each step of speciation. Then it would become evident how important phyletic divergence is. For instance, if one of the newly arisen "sister groups" is very much like the parental group or even identical with it (as implied by HENNIG 1966, 59) while the other deviates strongly, the terminology "sister groups" is no longer applicable since one now deals with a continuing parental group from which a daughter group has split off (HENNIG 1966, Figs. 14, 15). But beyond this formal objection there is the much more serious one that a greatly accelerated rate of divergence in one branch (one sister group) while subsequent branches of the other sister group diverge only slightly, would lead in time to such an unbalance of the system as to destroy completely the usefulness of the dichotomous cladogram. Such asymmetric branching happens in evolution very frequently and is easily accommodated in the classifications of the evolutionary taxonomists but fits only very awkwardly (if at all) into the cladistic classifications.

Some of HENNIG's followers have recognized these contradictions. SCHLEE (1971, 5, 30, 37) sees rather clearly that the deviation rule is dispensable: "It neither forms an argument for the justification of HENNIG's method nor a prerequisite for the work with HENNIG's principle." SCHLEE adds a mysterious interpretation by HENNIG himself (l. c. 6): "that the 'deviation rule' must be understood in a special genealogical sense, but not in a morphological-biological sense." Actually, the opposite is true: in a strict genealogical sense, there can be no deviation. If there is an unequal deviation, it must be in "a morphological-biological sense." Such morphological deviation is the normal situation in phylogeny. Whenever there is a splitting of phyletic lines, almost invariably one line will diverge more rapidly and more widely than the other, in fact, one of them may not change at all. The complete neglect of the frequent occurrence of this process is one of the fatal errors in the translation of the cladistic analysis into a classification.

Since the consideration of grades in classification is sometimes referred to in the cladistic literature as a typological approach, I would like to call attention to a somewhat different usage of the term typological (typologisch) in the American and the German literature. Typological in American usage is a straight synonym of essentialistic, referring to the abstraction of an underlying *eidos* (essence) and the neglect of the existence and importance of variation. In the German phylogenetic literature

Typus or Bauplan is also an abstraction, representing either the inferred ancestral "type" or the "ideal" Bauplan of a major taxon. To recognize the reptiles as a legitimate taxon, means recognizing a generalized reptilian Bauplan, and is then referred to as a typological approach. The emphasis in the German usage is on the typological philosophy of idealistic morphology. Typological in the German usage often also implies "phenetic" (broadly defined). The confusion between the two concepts of "typological" is well illustrated in a discussion by SCHINDEWOLF (1967).

c. A purely formalistic species definition

HENNIG's species definition in his original treatise (1950) is concerned only with the delimitation of species in the time dimension. A species is simply the distance between two branching points of the phylogenetic tree (p. 111). This concept is retained in his 1966 book (pp. 56–65). For instance: "The limits of the species in a longitudinal section through time would consequently be determined by two processes of speciation: the one through which it arose as an independent reproductive community, and the other through which the descendants of this original population ceased to exist as a homogeneous reproductive community" (p. 58). Although the last words of this definition have a slight flavor of the biological species definition, the diagrams in which it is illustrated (particularly his Fig. 15, p. 60) show how purely formalistic HENNIG's species concept is. For instance species B and D_1 differ in no way from the stem species A, but must be called different species, because C and E had in the meantime branched off this stem. In contrast species D_2 is different from species D_1 ("morphologically" says HENNIG, but his argument would be the same if they were "reproductively" isolated), but must be called the same species, because no branch had budded off from this stem in the meantime.

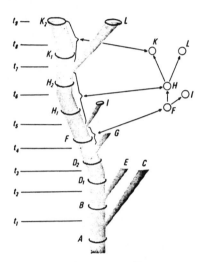

Fig. 3. The relation, according to HENNIG, between branching and speciation (From HENNIG 1966, Fig. 15)

I am calling attention to HENNIG's species concept, not because it is the decisive capstone in the HENNIG theory (it is *not*, HENNIG's own claims notwithstanding), but merely to give another example of the arbitrariness and purely formalistic nature of the major components of HENNIG's cladistic theory. HENNIG's species concept is so obviously unbiological and unrealistic that it has been rejected by numerous recent critics. PETERS, for instance, refutes HENNIG's formulations very effectively (1970, 28–30), showing that among all the possible properties of species their duration in the time dimension is the least meaningful. CAIN (1967, 412) also exposes the biological meaninglessness of HENNIG's formalism.

HENNIG's concept of the process of speciation is strongly affected by his species concept. The various possibilities implied in HENNIG's discussions can be formulated as three alternatives:

A. Parental species eliminated by
1. the splitting of the mother species into two daughter species (dichotomy), or
2. the simultaneous splitting of the mother species into more than two daughter species, or

B. Parental species continuing
3. after the origin of one of several side branches from the essentially unchanged phyletic mainline.

In his 1950 book HENNIG allows only for alternative 1. and among cladists this is still the favored alternative because it fits most easily into the cladistic scheme. Cladists are not unaware of alternative 3. but find a formal solution which does away with this inconvenient phenomenon. If a species *a* throws off a species *c*, species *a* will have to be called species *b* from the branching point on, in order to satisfy the cladistic postulates, even when *b* is biologically quite indistinguishable from *a*. In kinship studies, writes HENNIG (in SCHLEE 1971, 28): "the question of the biological identity of different species from different chronological horizons becomes totally irrelevant." This illustrates well to what extremes the dogmatic formalism is carried by HENNIG. Because, no matter what he says, if *a* and *b* are biologically identical, then they simply are not different species but the same species from which species *c* has branched off at some time. And this is of critical importance for the discrimination between sister and daughter groups. HENNIG's solution is thoroughly misleading. The production of a side branch, a new phyletic lineage, does not change the parental species.

Dichotomy or not. Modern speciation studies permit us to determine at which relative frequency alternatives 1. and 2. occur. All the indications are that a simple dichotomy into two daughter species is not the rule. Polytypic species almost invariably have more than two subspecies. Far better evidence is provided by superspecies which consist of groups of allospecies. In North American at least 48 (= 40%) of 126 superspecies of birds have more than two allospecies (MAYR and SHORT 1970). Among 94 Northern Melanesian superspecies 61 (= 65%) are non-dichotomous, containing three to 13 allospecies (MAYR and DIAMOND in press). The dichotomous "standard" is also refuted by all species-rich genera and by the frequency of clusters of sibling species. What is usually found in speciation studies is that the maternal species undergoes relatively little evolutionary change while numerous daughter species bud off at the periphery. HENNIG himself is not unaware of this situation as shown by his presentation of geographic variation in the snake *Dendrophis pictus* in which the large central population is surrounded by six peripherally isolated populations (HENNIG 1966, Fig. 16, p. 61). It is increasingly realized by biologists (MAYR 1942, 1963) that peripheral budding is the most frequent process of speciation, even though most of these daughter species are extremely short-lived. Under these circumstances I fail to comprehend the logic of NELSON's (1971, 374) assertion: "The use of dichotomous speciation as a methodological principle is required before an hypothesis of multiple speciation is even tentatively acceptable".

The difference between splitting and budding might seem a purely semantic one, if it were not for the fact that the cladists base such far-reaching conclusions on the postulate of consistent dichotomy. DARLINGTON (1970, 2–4) has presented an incisive critique. Even some cladists are beginning to abandon the principle of obligatory dichotomy (e. g. SCHLEE 1971, 27), but they have not yet faced up to the consequences this poses for their theory of sister groups. If a non-dichotomous split has produced three or four independent phyletic lines, it would mean that each of the resulting groups has two or three different sister groups rather than a single one. Many of the discussions of sister groups found in the cladistic literature would become meaningless under these circumstances.

In his *Stammesgeschichte der Insekten* (1969) HENNIG recognizes a number of higher taxa which consist of more than two sister groups. However, he emphasizes that this is a purely provisional arrangement, acceptable only "as long as the exact relationships are still uncertain."

The role of *extinction* is greatly underestimated in the cladistic constructions. Pairs of related taxa (HENNIG's sister groups) are indeed frequently encountered. Their existence, however, is usually due to the extinction of numerous intermediate phyletic lines rather than to a peculiar process of speciation, i. e., the splitting of a mother species into two daughter species. There is little evidence that this occurs frequently, and none whatsoever that this is the universal process of phylogeny.

d. The mode of origin of higher taxa

His phylogenetic theory forces the cladist to propose an unrealistic mode of origin for higher taxa. Since he recognizes branching as the only phylogenetic process, he has to give his branching points two properties: they are the origin of new species and also of new higher taxa. This arbitrary assumption in no way corresponds to the facts, as correctly pointed out by DARLINGTON (1970, 2). Speciation, that is the acquisition of reproductive isolating mechanisms between populations, and the acquisition of phylogenetically significant new apomorphous characters are two largely independent processes. The study of groups of sibling species and of most species-rich genera shows that the acquisition of reproductive isolation is often (if not usually) without effect on the morphological criteria that a taxonomist or evolutionary biologist would associate with the origin of new higher taxa. It is the exception rather than the rule that one of the daughter species acquires during speciation a character which is of potential significance for the characterization of a new higher taxon. The appearance of new apomorphous characters is correlated with the invasion of new niches and adaptive zones rather than with speciation (SIMPSON 1956b; BOCK 1965). The occurrence of phylogenetic dichotomy, thus, becomes more plausible when it is divorced from speciation, because the probability that several daughter lines shift simultaneously into the same new adaptive zone is small. It is, however, not nil, since a change of climate and vegetation or the arrival of a new predator may, indeed, cause a simultaneous identical shift in several related lines.

Probably more frequent and potentially more troublesome is the occurrence of the simultaneous acquisition of the same apomorphous characters in different lines in a cluster of sister groups, let us say the descendants of a highly polytypic superspecies. Extinction often provides a practical solution to this dilemma. The evolution of a new higher taxon from the line initiated by the new species is a later event, representing the second phylogenetic process, evolutionary divergence, and has nothing to do with splitting as such. It is a misleading formulation to say that higher taxa split. The evidence for this assertion is this: A higher taxon is a collective assemblage, which comprises numerous species and lesser evolutionary lines. Very few of them ever diverge to the extent that they form a separate higher taxon. But every once in a while one of these lesser branches diverges to such an extent that it must be removed eventually as a separate higher taxon.

Two aspects are important: The "Adam" of the new phyletic line almost invariably belongs to the ancestral taxon. The first member of the phyletic line that eventually led to the birds (long before *Archaeopteryx*) was presumably an otherwise rather conventional dinosaur, but with feathers or feather-like scales. Is a single derived character enough to throw such an "Adam" into the new taxon which will eventually emerge from his lineage? Paleontologists have long been concerned about this problem, which has been discussed in great detail by SIMPSON (1961 and elsewhere).

ASHLOCK (1971, 67) makes the sensible suggestion that the delimitation of the ancestral taxon against the derived taxon should be arrived at as follows: "The unique innovations found in the living members should be traced through the available fossils and the break placed somewhere between the first appearance of one of these characters and the first appearance of all of them . . . The specific assignment of the boundary

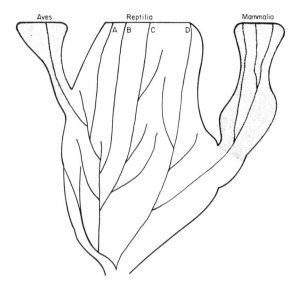

Fig. 4. The independence of the emergence of the avian and mammalian grades from the branching pattern of the reptilian grade. A (= crocodilians) belongs cladistically with the Aves, but is still a characteristic member of the reptilian grade. The origin of birds and mammals does not affect the categorical status of the reptilian branches from which they arose.

should be phenetically determined, weighting characters if appropriate, so as to establish attainment of the grade of the derived group." He admits that failure of preservation of the soft parts and an absent fossil record may make this task difficult, if not impossible, but in principle this is certainly the appropriate procedure.

The other important conclusion is that the origin of a side branch is of no evolutionary consequence for the main branch (except for possible competition). For instance, the Class Reptilia, a well characterized grade of tetrapods, has existed since the Carboniferous and survives in four living orders. Some time in the Triassic one of the numerous reptilian side branches (the cynodonts among the therapsids) evolved into the mammals and a little later another one gave rise to the birds (Fig. 4). A rigid application of their dogma forces the cladists to break up the reptilian grade into many separate "classes" and to designate particular reptilian lineages as the "sister groups" of birds

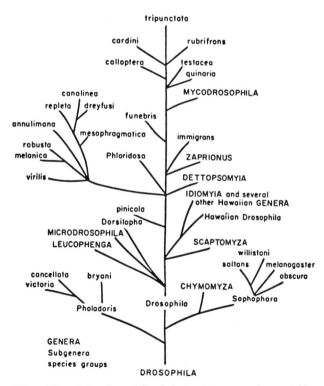

Fig. 5. The origin of specialized derivative genera from within the Drosophila phyletic tree (From THROCKMORTON 1965)

322

and mammals. The fact that no one would place the crocodilians outside the reptiles, if birds did not exist, reveals how artificial and arbitrary this procedure is. The essential unity of the reptiles is best illustrated by the continuing argument among paleontologists as to which particular orders of reptiles are most closly related to which others.

One of the major sources of the cladistic difficulty is the assumption that the origin of any new higher taxon requires the disappearance of the parental taxon. Even though there is massive evidence that many, if not most, higher taxa are lateral derivatives (side branches) of other taxa that had existed long previously and have continued to exist long after the split, his particular formalism forces the cladist to make such unrealistic claims as: "The stem species belongs neither to one nor to the other (daughter) group (to which it has given rise): it cannot be assigned to either of the two" (HENNIG 1969, 33).

How unsuitable this cladistic approach is can be demonstrated best by the study of a richly diversified group which is still in existence and actively speciating. The modern family of Drosophilidae represents such a case (Fig. 5). In this higher taxon one has such rich information from morphology, chromosomes, behavior and other characteristics that one can reconstruct the probable phylogeny with a considerable degree of reliability. The resulting dendrogram shows that several specialized genera originated from various locations within the *Drosophila* dendrogram. The origin of a specialized side line, to which one accredits separate generic status, affects in no way the taxonomic status of the main line of *Drosophila* which continues as the genus *Drosophila*. This is not a unique case. The origin of multiple side lines, each specializing in a somewhat different way while the major stem group continues essentially unchanged, is a very common occurrence in phylogeny.

2. A misleading conceptualization of ranking

The process of classification consists essentially of two steps, 1. the *grouping* of lower taxa (usually species) into higher taxa, and 2. the assignment of these taxa to the proper categories in the taxonomic hierarchy (*ranking*). The considerations which govern these two processes are quite different from each other (MAYR 1969, cap. 10). The cladists, following the erroneous assumption that phylogeny is a unitary process (consisting only of branching) assert that classifying likewise is a single-step procedure and that the grouping of taxa simultaneously also supplies their rank. Traditionally the rank of a taxon is determined by such criteria as degree of difference, uniqueness of the occupied adaptive zone, or amount of adaptive radiation within (MAYR l. c. 233). According to cladistics rank is given automatically by time of origin, and the same rank must be given to sister groups.

This erroneous conclusion is reinforced by HENNIG's frequent confusion of the terms taxon and category. In 1950 such confusion was excusable because up to that time no terminological distinction had been made. By 1966 when the difference had been made abundantly clear by numerous writers, the confusion was no longer defensible. Yet, even in 1966 (pp. 77–83) HENNIG speaks of the "reality and individuality" of categories, when he means taxa, and this confusion thoroughly obfuscates his discussion of the ranking procedures. The whole process of classification is based on the clear discrimination of *taxon* and *category:* zoological groups (taxa) have "reality" in nature because taxa names like birds, butterflies, or bats, are unambiguously names for clearly distinguishable groups. It is equally evident that the ranking of these taxa in the Linnaean hierarchy of taxonomic categories is rather arbitrary and often highly controversial. What one author considers a tribe, a second may call a subfamily, and a third a family or even superfamily. (See below, p. 122).

HENNIG, unfortunately, believes (or at least originally believed) that a knowledge of branching points would permit him to determine the categorical rank of a taxon. He utilizes two criteria: the geological age of the branching point (leading to the nearest sister group) and the number of subsequent branching points. He has stated this without reservations at numerous places in his writings. For instance, the taxa in the hierarchy "are subordinated to one another according to the temporal distance between their origins and the present; the sequence of subordination corresponds to the 'recency of common ancestry'" (1950, 83). "...in the phylogenetic system... the absolute rank order cannot be independent of the age of the group since... the coordination and subordination of groups is by definition set by their relative age of origin" (1966, 160).

Both of HENNIG's criteria have been frequently criticized. PETERS (1970), for instance, recognizes that among two evolutionary lines, which evolve at the same rate, one might give off a large number of side branches, while the other might maintain a monolithic singularity. It would make no sense to grant a higher rank to the frequently branching line than to the non-branching line. After all, branching and evolutionary divergence are two independent processes. HENNIG and some of his followers now speak occasionally of "relative chronological age" instead of "absolute geological age" but in 1966 HENNIG still maintained that the location of the branching point on the geological time scale determines categorical rank (Fig. 6).

Accordingly, taxa that originated from a split in the pre-Cambrian are ranked as phyla; between Cambrian and Devonian as classes; between Mississippian and Permian as orders; between Triassic and Lower Cretaceous as families; between Upper Cretaceous and Oligocene as tribes; and in the Miocene as genera. HENNIG continues "then the mammals would have to be called an order... the Marsupialia and Placentalia would have to be downgraded to families, and the 'orders' of the Placentalia would be tribes" (p. 187). Similar statements can be found in the writings of other cladists. For instance: "Species are to be ranked 1. according to their relative time of origin or 2. such that sister groups are given equal rank" (NELSON 1972, 366). "Relative age" is a term frequently referred to in the cladistic literature but no one has yet proposed an operational method of determining relative age.

If one would really rank taxa on the basis of their (geological) time of origin, one would be forced to adopt an entirely unbalanced system. Since the genus *Lingula* originated earlier than either the class Aves or the class Mammalia, one would have to place *Lingula* (and other contemporary, still surviving genera) into higher categories than birds and mammals. Even a cladist would presumably rather be inconsistent than go to

Geological time periods		Categorical rank
VI	Miocene	Genus
V	Oligocene	Tribe
	Upper Cretaceous	
IV	Lower Cretaceous	Family
	Triassic	
III	Permian	Order
	Mississippian	
II	Devonian	Class
	Cambrian	
I	Precambrian	Phylum

Fig. 6. The assignment by Hennig of categorical rank on the basis of the absolute geological age of the stem species (From HENNIG 1966, Fig. 58 and text p. 187)

such extremes. The discussions of GRIFFITHS (1972, 10) and CROWSON (1970, 250–254) reveal their difficulties.

Curiously, HENNIG has cited the classification of parasites as support for the cladistic method (1950, 261–269). He confirms that "host group and group of parasites must be given the same systematic rank ... when a group of parasites, which is restricted to a certain host group, originated simultaneously with it, perhaps in the manner that the stem species of the group of parasites lived as parasites in the host group" (l. c. 265). And yet, as OSCHE (1961) has pointed out, the chronology of parasites and their hosts actually refute the cladistic method. It is highly probable that many vertebrate parasites originated at the same geological period as their host taxa. This would necessitate, if one were to follow HENNIG strictly, that a genus of cestodes be raised to the rank of a family or order because it parasitizes a family or order of vertebrates or reciprocally that the host taxon be reduced to the rank of a genus. It would also mean that the superfamily Ascaroidea of the nematodes be given the same categorical rank as the class Cestoda because the stem species of both taxa are believed to have invaded the vertebrates at the same time.

The more carefully phylogenies are studied the more difficulties for the HENNIG principle of ranking are discovered. Various auxiliary devices proposed by HENNIG do not improve the situation. By assuming branching to be a regular process, with an approximately even rate, HENNIG believes that the number of species contained in a taxon provides an approximate measure for its age: "Monophyletic groups with large numbers of species cannot be very young" (1966, 182). Recent evolutionary researches have clearly demonstrated that this conclusion does not hold. The correlation between age of a taxon and number of contained species is extremely loose. The Hawaiian drosophilids, a clearly monophyletic group, is not only very recent (probably less than 4 million years old) but also extremely rich in species (at least 600 to 800). Most of the species flocks of African cichlid fishes are likewise very young, a product of the last couple of million years. Most families of rodents, rich in species, are geologically much younger than most of those families of mammals that contain only two or three genera. The monotremes with four to five species are as old or older than the eutherians with more than 3,000 species (PETERS 1970, 31). Well known types of marine invertebrates go back 300 to 400 million years without ever having produced rich species flocks.

HENNIG and his supporters claim that the great merit of the cladistic method is that it provides non-arbitrary definitions of the higher categories. GRIFFITHS (1972, 16), for instance, praises HENNIG for having proposed a logically unobjectionable definition "in terms of the age of origin of the stem species." I have shown above how unrealistic this proposal is and how utterly it fails to provide a sound procedure of ranking. Invertebrate paleontologists, who can demonstrate extremely different evolutionary fates for different phyletic lines (derived from the same stem species) have frequently pointed out to what unbalanced classifications the cladistic method of ranking would lead. SIMPSON (1961, 142–144) likewise has described what the application of this principle would do to vertebrate classification. CROWSON (1970, 260) admits "If phylogenetic classification proceeds usually by dichotomous divisions, and very unequal ones at that, it will necessitate the usage of many more categories [and taxa names] than were needed for older, 'formal', systems." SIMPSON is fully justified in calling such classifications "completely impractical." HENNIG himself (1969, 10) now realizes that this method leads to an absurd scale of ranking and has abandoned in his book on the phylogeny of insects any attempt to provide categorical ranks for the higher taxa. But what method of classification is this which cannot rank higher taxa?

3. Operational neglect of evident facts

Many aspects of phylogeny create difficulties for cladistics. HENNIG and his followers are not unaware of these facts and occasionally discuss them freely. However, they ignore these difficulties in the construction of their classifications or at least they offer no operational instructions on how to cope with them. I have referred to this already in the discussion of the "deviation rule" (see above) in which the unequal divergence of phyletic lines is acknowledged, but without drawing the obvious consequences for classification. Other facts that are neglected are the following.

a. The difficulty of determining the direction of evolutionary sequences

Taxa are constructed by HENNIG on the basis of synapomorphies (derived characters). The direction of the evolutionary change must therefore be determined for each character sequence. Sometimes it is easy to make such a decision, in many other cases it is quite difficult. HENNIG is well aware of this problem and has formulated four rules (1950, 172 ff.) which help to determine which of several alternate characters are more ancestral and which derived. Most of these rules, reformulated by PETERS and GUTMANN (1971, 242), are based on practices that go back to DARWIN or even earlier periods. PETERS and GUTMANN (l. c. 256) emphasize correctly that any formalism in phylogenetic research (including cladistics) tends to lead to superficial and often unreliable conclusions. It must be replaced by a far more biological attitude toward the morphological evidence. The phylogenetic reconstruction must be based on an analysis of function and adaptive significance, such as is practised in evolutionary morphology (BOCK 1965, 1969). I refer to the essay of PETERS and GUTMANN (l. c.) for a very perceptive statement of this weakness of cladistics. Their considerations make it obvious that the magnitude of an adaptive innovation is of the utmost importance in classification, a fact consistently ignored by the cladists.

b. The discrimination between parallelism and convergence

The common possession of the same character in two different taxa may have one of four possible causes (MAYR 1969, 202):
1. *Plesiomorphous similarity.* The sharing of characters with an ancestor (see the cladistic literature for more precise specifications).
2. *Synapomorphous similarity.* The unique sharing of characters derived from a stem species, e. g., all birds, but no other organisms, have feathers.
3. *Similarity due to parallelism.* Characteristics produced by a shared genotype inherited from a common ancestor.
4. *Similarity due to convergence.* These can be either convergent acquisitions (wing in pterodactyls and bats) or convergent losses (leglessness in snakes and worm lizards).
To discriminate between these four possibilities is not nearly as easy as it may sound, because all that is accessible to the student of phylogeny are phenotypes, while phylogeny actually consists of a change in genotypes.

The traditional recipe for the discrimination between parallelism and convergence is the analysis of homology. Cases where a homology between characters can be established, that is a derivation from the equivalent character in the common ancestor, are the result of parallel evolution. Similarities in non-homologous characters are the result of convergence (BOCK 1963, 1967). To follow this recipe in practice, is, however, very difficult, not only because the establishment of homology is often fraught with difficulties (HULL 1967), but also because in distant relatives the line between parallelism and convergence is often not sharp (For a discussion of parallelism see also

SIMPSON 1961, 103–106). OSCHE (1965) has given a perceptive analysis of the difficulties caused by the potentialities of the hidden genotype.

Cladists pay little attention to the various possibilities: "In deciding whether corresponding characters of several species are to be regarded as synapomorphies, convergences, homologies, or parallelisms we must determine whether the same character was already present in a stem species that is common only to the bearers of the identical characters" (HENNIG 1966, 120). Contrary to HENNIG's claim this does not permit a discrimination between the four possibilities but only between synapomorphy on one hand and the other three on the other hand. Here and elsewhere (for instance l. c. 121) HENNIG is quite outspoken in his lack of interest in making a distinction between parallelism and convergence (see also PETERS 1972, 168). And yet such a distinction is of crucial importance in deciding degree of relationship between taxa.

Taxa are classified in cladistics on the basis of the presence or absence of derived characters. Indeed HENNIG has emphasized repeatedly that nothing counts in classification but such characters. Unfortunately, only part of the genotype is expressed in the visible phenotype and yet the hidden part of the genotype is often as important for the future evolution of a phyletic line than that which is revealed in the visible phenotype. HENNIG (1950, 176) himself has shown that a potential for stalked eyes is widespread among the acalyptrate dipterans but is realized only in scattered species and genera. A secondary jaw articulation originated at least 14 times independently in the class Aves (BOCK 1969). In the genus *Drosophila* the repeated manifestation of concealed potentialities is extremely frequent (THROCKMORTON 1962, 1965). There is hardly a higher taxon known in which such derived characters do not occur scattered through the system. In any particular case the presence or absence of the character is not in the least determined by the presence or absence of the character in the common ancestor. To be sure the genotype of the common ancestor has the potential for the development of this character but its realization is unpredictable.

When such an independent realization of incipient tendencies occurs with several characters, there is no way to determine in what sequence these derived characters were acquired in different lines and in what sequence the various related lines branched from each other. This can be shown in a diagram (Fig. 7). The only way the cladist

Fig. 7. Contradictory information provided by different sets of characters owing to mosaic evolution. According to character set 1 the lineages B and C form a sister group to A; according to character set 2 the lineages A and B form a sister group to C

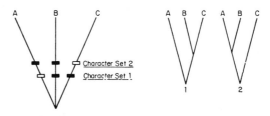

can cope with this problem is to say that it will disappear if one takes enough characters. Unfortunately this is not necessarily true. Usually there are not enough such characters available and, when only few are available, the decision will have to be made on the basis of careful, weighting, often hardly distinguishable from being arbitrary. To imply that there really is not much difference between convergence and parallelism (HENNIG 1966, 121) reveals a peculiar reluctance to come to grips with one of the major weaknesses of the cladistic system. These are not purely theoretical problems. The recent effort by INGER (1967) and of KLUGE and FARRIS (1969) to arrive at a satisfactory phylogeny of the anurans depends to a considerable extent on decisions concerning the relative primitiveness of various characters.

GRIFFITHS (1972, 24–26) has a lengthy discussion on the question of whether or not convergence poses a serious problem for cladistics. He is surely on safe grounds in his belief that highly complex characters or character combinations are not apt to be acquired independently in unrelated phyletic lines. (For a critique of the complexity criterion see PETERS 1972, 168–170). However characters (even rather complex ones) can be lost independently in different lines and rather simple characters can be acquired convergently. Unfortunately GRIFFITHS misinterprets DARWIN in the following sentence: "Doubts have been raised about the validity of DARWIN's distinction between 'adaptive' and 'non-adaptive' characters for purposes of evolutionary evaluation, and I therefore do not employ such a criterion" (p. 25). Actually DARWIN makes no such distinction. He speaks only of *ad hoc* specializations which indeed have low phyletic weight and can be acquired convergently. This is what CRACRAFT (1972) has overlooked in his recent discussion of ratite evolution. If several groups of running birds lose their power of flight independently, acquire large size and specialize entirely in running, one would expect them to acquire the ratite complex of characters even if the stem species of this assemblage did not have these characters. This consideration is quite independent of the question whether or not the families of birds which lost their power of flight were a closely-knit group of related genera or only distantly related to each other. Potentially the same objection applies to acquisition of diving adaptations by *Hesperornis*, grebes and loons. The arguments of the cladistic school do not weaken the validity of DARWIN's warning against relying too much on *ad hoc* specializations (see also MAYR 1969, 220, 223).

HENNIG believes that one can distinguish convergence (and parallelism) from synapomorphy by "taking into account as many characters as possible" (1966, 121). This is true in principle, but often impossible to implement. SCHLEE (1971, 23) lists some additional criteria which, indeed, are quite helpful. However, when well established phylogenies are carefully analyzed, it becomes obvious that an unequivocal decision can only rarely be reached. Owing to a very scanty fossil record of insects, as compared, for instance, to that of mammals, the insect taxonomists are not nearly as aware, as they ought to be, of the frequency of parallelism. For instance "within the most advanced group (of therapsid reptiles), the cynodonts, several mammalian features (e. g., dentary-squamosal jaw articulation, loss of alternate tooth replacement, complex occlusion, and double-rooted cheek teeth) are known to have evolved independently in several phyletic lines" (CROMPTON and JENKINS 1973, 138). Most workers also accept that the incorporation of the quadrate and articular into the middle ear must have occured independently in therian and non-therian mammals. The frequency of the independent acquisition of identical adaptations in birds, for instance a secondary jaw articulation (BOCK 1959) or various specializations in different lines of woodpeckers (BOCK and MILLER 1959), or of independent phyletic advances in *Drosophila* (THROCKMORTON 1965) highlight the difficulties (see BOCK 1967 and PETERS 1972, 168).

More disturbing is the fact that cladists, when actually constructing classifications, never seem to come to grips with the difficulties caused by parallelism and convergence. They simply ignore them.

c. The information content of ancestral characters

Cladists rightfully criticize the adoption of "mere similarity" as the only criterion of taxon delimitation, but so have evolutionary taxonomists for many years. GHISELIN (1969b), SIMPSON (1961) and myself (1969) have consistently pointed out the pitfalls of basing a classification entirely on similarity, particularly (as proposed by the pheneticists) on unweighted similarity. However, this does not justify going to the other extreme by rejecting all consideration of similarity in the construction of a classification. To do so would result in the loss of a great deal of taxonomically important information. After all similarity, when properly evaluated, is an important index of the amount of shared genotype and is the basis for the determination of homology. Cladists completely ignore this and as a result overlook the fact that the retention of a large number of ancestral characters is just as important an indicator of "relationship" (traditionally defined) as the joint acquisition of a few "derived" characters.

Two extremes have been proposed with respect to the relative importance of conservative (ancestral) and advanced (derived) characters. The cladists consider only the latter, while it is sometimes said that it is the study of conservative characters which is most apt to reveal relationship. The evolutionary taxonomist is convinced (and has acted on this basis for the last 100 years) that one must evaluate information from both types of characters in order to be able to construct a sound classification. Nothing could be further from the truth than the claim that "primitive similarities contain no phylogenetic information" (CRACRAFT 1972, 383).

Let me illustrate this with two examples: There are three major families of living gallinaceous birds. Among these the Megapodiidae have the greatest number of primitive characters while the Phasianidae have the greatest number of derived characters. The South American Cracidae are intermediate. They share a few derived characters with the advanced Phasianidae but a far greater number of primitive characters with the Megapodiidae. Traditionally, because they seemingly share to such a large extent the same genotype, the Cracidae have been said to be more closely related to the megapodes than to the Phasianidae. As a cladist, however, CRACRAFT (l. c. 283) insists that the Cracidae are "more closely related" to the Phasianidae because they share with them a few derived characters. I consider this to be a misleading statement. In the aggregate of their characters Cracidae are obviously much closer to the Megapodiidae than to the Phasianidae. To know that the Phasianidae branch off the same branch that leads to the Cracidae is important, but only part of the evidence that leads to a classification.

Let me cite another example which HENNIG has recently used in order to illustrate the superiority of the cladistic approach (1971, 12–14) (See Fig. 8). In the Canadian amber of the Upper Cretaceous two fossil diptera were found which show relationship to two modern families, to the Phoridae, a cosmopolitan family with some 2,500 species, and to the Sciadoceridae with two Recent species, both in the southern continents. In which of the two Recent families should one place these fossils? The original

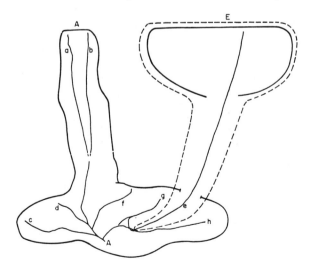

Fig. 8. Two different interpretations of the taxonomic assignment of fossil flies. According to the traditional interpretation (solid outline) the fossils (e) belong to the Sciadoceridae (A); according to the cladistic interpretation (broken outline) they belong to the Phoridae (E). a–h = sublines of the Sciadoceridae with varying evolutionary potentials, most of which (c, d, f, g, h) have become extinct.

describers of the fossils (MCALPINE and MARTIN 1966) acknowledged distant relationship with the Phoridae but placed the fossils in the Sciadoceridae because they are far
more similar to species of that family than to any recent Phoridae.

The two Canadian fossils (*Sciadophora* and *Prioriphora*) differ from the living
Sciadoceridae consistently only in a single apomorph character, the absence of a discal
cell. In two other apomorph characters (absence of anal cell, dorsal arista) the recent
Sciadoceridae are variable. One of the two Canadian fossils (*Prioriphora*) acquired
some additional phorid-like apomorph characters (e. g., loss of proscutellum, apically
enlarged palpus, insertion of R_{4+5} far from the wing tip, m_1 not joining m_2) but in all
these characters the other Canadian fossil (*Sciadophora*) retains the plesiomorph
(Sciadoceridae-like) condition. Although the Canadian fossils (as a group) show one
or two derived characters, they agree in the majority of their characters far better with
the living Sciadoceridae than with the Phoridae. In particular, they lack the additional
fusion and reduction in the length of wing vein r_3 and the loss of the second basal cell.
Also in typical phorids the coxae and femora are stouter and the whole habitus is
stockier.

In this case, and in several others I have studied, the entire cladistic re-classification
rests on one or two characters. This is implicitly nothing but a return to a "single-
character classification." In spite of their exhortations to base classifications on the
holomorph (= totality of characters), in practice virtually all dichotomies in classifications are based on exceedingly few characters, often a single character pair. This has
been rightly criticized by DARLINGTON (1970, 17).

d. Mosaic evolution

As far as ancestral versus derived character states are involved, cladistics assumes that
once "evolution" has "decided" to give one phyletic line a primitive character state
and the "sister group" the derived character state, this difference will be perpetuated
forever. In many instances, this is, indeed, the case. In many others, and this is completely ignored by the cladists, the potential for the derived character exists in all
daughter lines of the original ancestor, as I have just pointed out, but is realized in
the various daughter lines irregularly and at different rates, leading to parallel
evolution. The claim "if derived character states can be identified, then monophyletic
lineages can be constructed" (CRACRAFT 1972, 381), is clearly not justified.

Cladists have not been able to overcome the difficulties caused by mosaic evolution.
Yet it is not true (as has been claimed) that they believe that dichotomies create
primitive and advanced *groups*. They realize, like all good taxonomists, that most
groups possess a mixture of primitive and derived characters. Even the most primitive
group of living mammals, the monotremes, have a number of derived characters.

However, a difficulty is created by the fact that newly acquired characters are
sometimes lost again in subsequent evolution. Such a double apomorphy (= secondary
primitiveness) would be masquerading as a plesiomorphy. Although some cladists are
aware of this possibility, I do not recall that it was ever taken into consideration in
the construction of cladistic classifications.

Darwin and classification

Evolutionary taxonomists have long been convinced that they strictly follow Darwinian principles of classification by giving equal consideration to branching *and* to
phyletic change. As SIMPSON (1961, 52) has said: "Evolutionary taxonomy stems

explicitly and almost exclusively from DARWIN" (See SIMPSON 1959a, for a more detailed discussion of DARWIN's theory of classification). In recent years cladists have tried to claim DARWIN for their side and NELSON has gone so far as to state: "If, indeed, there is a cladistic school DARWIN is its founder and chief exponent" (1971, 375). He has referred to cladistics as "the DARWIN–HENNIG classification" (1972, 370). Is there any justification in this claim?

DARWIN's *Origin* (1859) was the first major publication to propose evolution through common descent, at that time an entirely novel concept (quite different from LAMARCK's concept of evolution). Genealogical language quite naturally played an important role in this volume. Like the modern evolutionary taxonomist (see above, p. 99) DARWIN started out on a strictly genealogical basis: "The arrangement . . . must be strictly genealogical in order to be natural." Since any group of animals can have only one genealogy, as we have seen above, DARWIN's postulate is an axiom for any evolutionary taxonomist. However this was only the first step for DARWIN, because he continues: ". . . but that the amount of difference in the several branches or groups, though allied in the same degree in blood to their common progenitor, may differ greatly, being due to the different degrees of modification which they have undergone; and this is expressed by the forms being ranked under different genera, families, sections, or orders" (1859, 420) (See also GHISELIN 1969a, 84). It is significant that DARWIN at this point refers back to his famous diagram in the fourth chapter (opposite page 116). Here he shows that three congeneric Silurian species (A, F, and I) evolved into 15 modern genera. These represent three "sister groups" derived from the three Silurian genera. The descendants of A and of I now constitute distinct families or even orders (p. 125). "But the existing genus F 14 may be supposed to have been but slightly modified; and it will then rank with the parent genus F; just as some few still living organic beings belong to Silurian genera" (p. 421). No more explicit statement could be wished for to refute the claim that DARWIN was an exponent of cladistic classification.

GHISELIN and JAFFE (1973) have shown how frequently DARWIN in his classification of the Cirripedia deviated from a cladistic classification: He places *Alcippe* in the Thoracica even though it is on the Abdominalia stem; *Pachylasma,* on the branch leading to the Balaninae, is included with the Chthamalinae; *Pollicipes* gives rise to the Lepadidae (with which it is included) and biphyletically to the stalkless cirripedes (Verrucidae and Balanidae). He could have adopted none of these classifications if he had followed the cladistic definition of monophyly. In each case DARWIN established what the cladists would call a paraphyletic group.

In contrast to NELSON's is BRUNDIN's claim (1972, 107) that the HENNIG method had revealed the "weaknesses of current neo-Darwinistic theory." No other cladist makes such claims. There are several indications in BRUNDIN's writings that he fails to comprehend the Darwinian theory. Does he perhaps believe in some sort of orthogenesis, as implied by the statement: "The evolutionary process is far more orderly than admitted by the neo-Darwinists of today" (l. c. 119)?

Cladistic objections to the methods of evolutionary (combined) classification

GRIFFITHS (1972, 16) has stated that the "combined" (= evolutionary) grouping should be rejected because it "raises serious logical difficulties." HENNIG, furthermore, has claimed repeatedly that a "synthetic" (1971, 11) or "syncretistic" (1966, 77) systematics is unable to elaborate a consistent system ("robs the combination of any scientific value") and leads to serious error. GRIFFITHS (1972, 15–17), however, is the only author who has tried seriously to enumerate and classify what he considers to be the shortcomings of the evolutionary method. In order to minimize future misunderstandings, let me attempt to refute some of his objections.

GRIFFITHS states that I reject the endeavor to achieve "an unequivocal correspondence between phylogeny and classification above the species level" (l. c. 16). On the contrary, I object to HENNIG's narrow definition of phylogeny, which forces the cladist to neglect at least half of the information which phylogeny provides. It is my contention that the traditional method of classification reflects phylogeny to a far greater extent than HENNIG's.

GRIFFITHS further states that it is impossible to measure rates of evolutionary change accurately. Correct! Perhaps it is inappropriate to say a classification should reflect "rates of evolution." However, a classifier should not and can not ignore the *results* of highly uneven rates of evolution. Even if it should be impossible to compare numerically the (slow) rate of evolution between the stem species and the modern crocodilians with the (rapid) rate between the same stem species and the modern birds, every beginner can see how much more drastically the birds differ from the common ancestor than the crocodilians. To ignore this altogether, because it can not (yet) be measured accurately, would seem a poor escape from a difficulty.

A third objection is that the evolutionists fail to provide an objective definition of the categories above the species level. Their definition compares very unfavorably, claims GRIFFITHS (1972, 7–8) with "HENNIG's proposal to define categories above the species level in terms of the origin of the stem species of the member taxa." GRIFFITHS forgets that HENNIG has traded biological meaning for a hoped-for logical consistency. As DARWIN showed (see above), the modern descendants of two Silurian sister species may be, biologically speaking, respectively an order and a genus. To give them the same categorical rank to satisfy HENNIG's formalistic definition may be logically impeccable, but is simply wrong biologically. HENNIG himself, in the mean, time, has abandoned the claim to be able to give a non-arbitrary definition of the higher categories and refrains from placing higher taxa of insects into categories (see above, p. 115). GRIFFITHS' objections to the evolutionary criteria of ranking are based on the special cladistic definitions of monophyly and phylogeny. They become invalid and irrelevant when these definitions are rejected.

It is sometimes stated that the evolutionary taxonomists fail to provide definitions for categories. This is not correct. All I have emphasized is that the species category is the only category for which a non-arbitrary definition is possible. Both SIMPSON (1961) and I (1969) have provided formal definitions for the higher categories, even though for reasons we have stated such definitions have limits to their usefulness.

These are GRIFFITHS' specified criticisms. More generally the cladist criticizes the evolutionary taxonomist for failing to provide simple criteria for making decisions in classification. It would seem to me that the nature of the material precludes a simplistic approach. The number of variables that must be considered in the construction of a classification is so large that simple methods will not work. This is the reason why the evolutionary taxonomist carefully weights the evidence and uses his judgment in arriving at conclusions. He asks what role a higher taxon plays in the economy of nature. He considers the nature of the adaptive break-through which gave rise to the taxon (PETERS 1972). In short, he insists on approaching his material as a biologist and evolutionist, rather than looking for automatic answers. In the short run this may create difficulties and uncertainties, but who would want to question that a classification, which utilizes all potentially available information, is more informative, more predictive, and indeed more truly reflecting past evolution than a classification which arbitrarily restricts itself entirely to the information provided by the branching pattern?

Synopsis

A sound classification of a group of organisms can not be devised without a well considered reconstruction of its phylogeny.

One component of such a reconstruction is an establishment of the branching pattern of the various phyletic lines, and the design of a cladogram. HENNIG has demonstrated that this can be done in a relatively unequivocal manner by classifying characters into apomorphous (derived) and plesiomorphous (ancestral) characters.

The high information content of derived (apomorph) characters was appreciated by many taxonomists long before HENNIG, but never sufficiently stressed, and, in fact, entirely ignored by some authors. The emphasis on the proper weighting of synapomorphies, under the impact of HENNIG's cladistic theory, has been a healthy development in systematics.

It would require an impartial analysis to determine how many recent improvements in the classification of fishes, insects, and other groups were the result of a rigorous application of cladistic analysis. CRACRAFT (1972) has recently asserted that application of the cladistic method would revolutionize avian taxonomy. However, after 12 pages of discussion, he failed to produce even a single case in which he could demonstrate that the currently accepted classification was wrong. All he was able to show was that a classification based only on branching points is sometimes different from a classification in which phylogenetic divergence is given primacy. The case of the fossil Sciadoceridae (see pages 119–120 above) is another illustration of a change in classification, but not necessarily an improvement. The reasons why the cladistic method cannot show more achievements should be obvious from the previous discussion.

Cladistic grouping encounters many difficulties even if we exclude unacceptable decisions in ranking. For instance, very often there are not enough apomorphous characters available, or else there is doubt as to which of two alternative character states is ancestral and which is derived, and, finally, owing to mosaic or parallel evolution, there may be conflict between the information provided by different apomorphous characters.

As valuable as the cladistic analysis is, it does not automatically provide a classification. By making use of only one of the two available sets of phylogenetic data, cladistics produces classifications which are less able to serve as general reference systems than evolutionary classifications, because they have a poorer information content. Evolutionary taxonomists, by weighting the information from both sources, arrive at classifications which may be criticized as being more "subjective", but which reflect evolutionary history more accurately and are therefore more meaningful biologically. The cladist believes that the simple adding up of synapomorphies will provide the correct classification, so to speak, automatically. The evolutionary taxonomist, in contrast, feels that only a careful weighting of all the evidence will reveal meaningful degrees of relationship, in the sense of inferred genetic relationship. He feels, furthermore, that a classification must pay attention to major adaptive events in evolutionary history, like becoming terrestrial or airborne, since these are of greater importance for the ranking of taxa than the mere splitting of phyletic lines. JOHNSON (1970) and MICHENER (1970, 20–22) are other systematists who have recently stated the case in favour of evolutionary (synthetic) systematics.

GRIFFITHS (1972, 17) has proposed that if one wants to express the effects of different degrees of evolutionary divergence in different phyletic lines, one should for this purpose construct an entirely separate classification, in other words that one should have several (or at least two) sets of classifications. This proposal strikes me as altogether impractical. PETERS and GUTMANN (1971, 256) likewise reject the purely formalistic approach of the HENNIG school and demand that it be replaced by a biological attitude toward the morphological evidence. This includes giving due consideration to the size of adaptive breakthroughs.

An eclectic classification which considers with equal care the branching points in phylogeny and all aspects of phylogenetic divergence would seem the best way to generate biologically meaningful classifications, permitting the greatest number of broad generalizations. DARWIN's advice to use both these sources of information was adopted by the most successful classifiers of the last 100 years.

In conclusion, it is evident that, no matter how useful cladistic analysis is, it cannot be automatically translated into a classification.

Acknowledgements

The present version of my analysis is the result of a great deal of revision, made possible by the most generous cooperation of numerous friends and correspondents. Earlier drafts were read and extensively criticized by PETER D. ASHLOCK, WALTER BOCK, ARTHUR CAIN, STEPHEN GOULD, OTTO KRAUS, GÜNTHER OSCHE, D. S. PETERS and DONN ROSEN. Their comments have led to an extensive rewriting and, hopefully, to clarification. GARETH NELSON, although he thoroughly disagrees with most of my conclusions, has been most helpful in supplying me with references and reprints that are difficult to obtain. I deeply appreciate his generosity.

In a field as difficult and controversial as the theory of classification no one can expect to be always right. I would be completely gratified if my analysis were to contribute to a clarification of the issues.

Summary

Three theories of classification compete with each other at the present time, each claiming to be best suited for meaningful and reliable classification. One of these, HENNIG's theory of cladistics, uses the "recency of common descent" as the primary criterion for the delimitation and categorical ranking of taxa.

The opponents of cladistics, however, raise the objection that one must make a difference between cladistic analysis and cladistic classification. Admittedly, since every natural group (taxon) must consist of closest relatives, it is indispensable in the delimitation of such a group to make sure that all members are derived from a common ancestor, that is that the group is monophyletic. HENNIG's method of classifying characters in plesiomorphous and apomorphous ones is excellently suited to lend precision to an unequivocal determination of the common ancestor. HENNIG's cladistic analysis is thus an important contribution to systematic methodology.

The possibility of converting the results of such analysis directly into a classification, as is demanded by the cladists, is however questioned by many systematists. Categorical rank ("family," "order," etc.) of a taxon is determined according to cladistic theory by the branching point of the dendrogram, and so-called sister groups must be given the same categorical ranking. This means that the branching point fixes irrevocably the categorical rank of subsequently evolving taxa without any consideration of evolutionary events that happen later (except branching).

Such a "downward" classification is unable to take into consideration the sometimes rather drastically different fates of several phyletic lines that are derived from the same common ancestor. A group which had invaded an entirely new adaptive zone (as for instance, the birds) is given the same rank as a sister group (like the crocodilians) which had remained in the ancestral adaptive zone. Cladistics classifies exclusively on the basis of branching points instead of an evaluation of the characteristics and adaptive complexes of taxa. The characters of taxa are taken into consideration only as far as this is necessary for the determination of the branching points.

The redefinition of three widely-used technical terms, phylogenetic, relationship, and monophyletic, by the cladists has caused a great deal of confusion in the literature. There is no reason to abandon the traditional definitions.

Multiplication of species is not a dichotomy in every case and in periods of intensive speciation it occurs sometimes that three or more sister groups originate simultaneously.

The new diagnostic characteristics of a phyletic line emerge often only long after the original branching-off from the ancestor. A downward classification is misleading in such cases. It is unrealistic to separate the older taxa of a new phyletic line from the parental group as long as they are still joined to it by the common possession of critical characters. It is equally unrealistic to act during classification as if the occurrence of a sidebranch resulted in the extinction of the main line.

It leads to striking contradictions if one determines the categorical rank of a taxon on the basis of the absolute geological time that has passed since the original branching. Only a slight improvement is achieved by accepting "relative" age as criterion.

The occurrence of parallel evolution, of convergence, and of mosaic evolution is on the whole not considered in the classifications of the cladists. The information content of plesiomorph characters is likewise ignored. As a result the information content of cladistic classifications is greatly reduced.

The traditional method of classification, according to which both processes of phylogeny, branching, and subsequent divergence are considered equally, is the method which was proposed by DARWIN in 1859 and was used by him in his own systematic works, for instance in his monograph of the Cirripedia.

The objections which have been raised by the opponents of cladistics lead to the conclusion that it is not possible to translate the results of a cladistic analysis, that is the cladogram, directly into a classification.

Zusammenfassung

Kladistische Analyse oder kladistische Klassifikation

Zur Zeit werden drei Theorien der Klassifikation unterschieden, von denen jede behauptet in der Lage zu sein, das beste „allgemeine Ordnungssystem" schaffen zu können. Eine dieser Theorien ist die kladistische Theorie, von W. HENNIG begründet, die den Abstand von der nächsten gemeinsamen Stammart (Vorfahren) als Maßstab für die Abgrenzung und kategorische Eingliederung von Taxa annimmt.

Die Gegner der kladistischen Klassifikation erheben dagegen den Einwurf, daß man zwischen kladistischer Analyse und kladistischer Klassifikation unterscheiden muß. Da jede natürliche Gruppe (Taxon) aus engsten Verwandten bestehen muß, ist es für die Abgrenzung einer solchen Gruppe unerläßlich nachzuweisen, daß alle Mitglieder von einem gemeinsamen Ahnen abstammen, d. h. daß die Gruppe monophyletisch ist. HENNIGS Methode, Merkmale in plesiomorphe und apomorphe zu trennen, ist hervorragend geeignet, die eindeutige Bestimmung des gemeinsamen Vorfahrens zu präzisieren und die anzuerkennenden Gruppen einheitlicher zu gestalten. HENNIGS kladistische Analyse ist somit ein wichtiger Beitrag zur systematischen Methode.

Die Möglichkeit, die Ergebnisse dieser Analyse unmittelbar in eine Klassifikation zu übersetzen, wie es von den Kladisten vorgeschlagen wird, wird jedoch von vielen Systematikern bezweifelt. Der kategorische Rang ("Familie", „Ordnung" usw.) eines Taxons wird gemäß der kladistischen Theorie durch den Zweigpunkt des Stammbaumes bestimmt, und sogenannte Schwestergruppen müssen in dieselbe Kategorie eingereiht werden. Das bedeutet, daß der Verzweigungspunkt unwiderruflich den kategorischen Rang der später entstehenden Taxa festlegt, ohne Rücksicht auf später erfolgende Evolutionsgeschehnisse.

Solch eine „Abwärts-Klassifikation" kann die unterschiedlichen Geschicke der mehreren Deszendenzlinien, die von dem gemeinsamen Stammart abzweigten, nicht berücksichtigen. Eine Gruppe, die in eine völlig neue Adaptationsebene eingewandert ist (wie z. B. die Vögel), wird genauso behandelt, wie eine Schwestergruppe (wie z. B. die Krokodile), die in der ancestralen Adaptationszone verblieben ist. Statt Taxa aufgrund ihrer Merkmale und Adaptationskomplexe zu klassifizieren, klassifiziert die Kladistik ausschließlich aufgrund der Verzweigungspunkte. Sie berücksichtigt die Merkmale der Taxa nur so weit, als das für die Erfassung der Verzweigungspunkte nötig ist.

Die Umdefinierung der drei wohlbekannten Fachausdrücke phylogenetisch, Verwandtschaft und monophyletisch durch die Kladisten hat viel Verwirrung in der Literatur verursacht. Es besteht kein Grund, die traditionellen Definitionen aufzugeben.

Artbildung ist nicht notwendigerweise eine Dichotomie, und in einer Periode von intensiver Artbildung können manchmal drei oder noch mehrere Schwestergruppen entstehen.

Die charakteristischen neuen Eigenschaften einer Stammlinie treten oft erst lange nach der ursprünglichen Abzweigung in Erscheinung. Eine Abwärts-Klassifikation ist in diesen Fällen irreführend. Es ist unrealistisch, die älteren Taxa einer neuen phyletischen Linie von der Elterngruppe zu trennen, solange sie mit ihr noch durch gemeinsamen Merkmalbesitz verbunden sind. Ebenso unrealistisch ist die Annahme, daß das Abzweigen einer phyletischen Linie die Stammlinie sozusagen auslöscht.

Den kategorischen Rang eines Taxons aufgrund der absoluten geologischen Zeit seit der ursprünglichen Abzweigung zu bestimmen, führt zu großen Widersprüchen. Das „relative" Alter als Maßstab anzunehmen ermöglicht nur eine geringe Verbesserung.

Das Vorkommen von paralleler Evolution, von Konvergenz und von Mosaikevolution wird in den Klassifikationen der Kladisten im allgemeinen nicht berücksichtigt. Der Informationsgehalt plesiomorpher Merkmale wird gleichfalls ignoriert. All das mindert den Informationsgehalt kladistischer Klassifikationen.

Die traditionelle Methode der Klassifikation, die die beiden Vorgänge der Evolution, Verzweigung und nachfolgende Divergenz gleichmäßig berücksichtigt, ist diejenige, die DARWIN 1859 vorschlug und in seinen eigenen systematischen Arbeiten (z. B. in der Monographie der Cirripedia) benutzte.

Die Einwürfe, die von den Gegnern der Kladistik erhoben werden, deuten darauf hin, daß man eine Klassifikation nicht einfach aus dem Ergebnis der kladistischen Analyse, d. h. dem Kladogram ablesen kann.

Literature

ASHLOCK, P., 1971: Monophyly and associated terms. Syst. Zool. 20, 63–69.
— 1972: Monophyly again. Syst. Zool. 21, 430–438.
BIGELOW, R., 1956: Monophyletic classification and evolution. Syst. Zool. 5, 145–146.
BOCK, W., 1959: Preadaptation and multiple evolutionary pathways. Evolution 13, 194–211.
— 1963: Evolution and phylogeny in morphologically uniform groups. Am. Nat. 97, 265–285.
— 1965: The role of adaptive mechanisms in the origin of higher levels of organization. Syst. Zool. 14, 272–287.
— 1967: The use of adaptive characters in avian classification. Proc. 14th Internat. Ornith. Congr. Oxford: Blackwell, p. 61–74.
— 1968: Review of HENNIG 'Phylogenetic systematics'. Evolution 22, 646–648.
— 1969: Comparative morphology in systematics. p. 411–448. In: Systematic Biology. Washington: National Academy of Sciences.
BOCK, W.; MILLER, W., 1959: The scansorial foot of the woodpeckers, with comments on the evolution of perching and climbing feet in birds. Amer. Mus. Novit. 1931.
BRUNDIN, L., 1966: Transantarctic relationships and their significance, as evidenced by chironomid midges. With a monograph of the subfamilies Podonominae and Aphroteniinae and the austral Heptagyiae. K. svenska Vetensk. Akad. Handl. 4, 11.
— 1972: Evolution, causal biology, and classification. Zool. Scripta. 1, 107–120.
CAIN, A. J., 1967: One phylogenetic system, a review of 'Phylogenetic Systematics' by W. HENNIG. Nature 16, 412–413.
CAIN, A. J.; HARRISON, G. A., 1960: Phyletic weighting. Proc. Zool. Soc. London 135, 1–31.
CRACRAFT, J., 1972: The relationships of the higher taxa of birds: Problems in phylogenetic reasoning. Condor 74, 379–392.
CROMPTON, A. W.; JENKINS, F. A. Jr., 1973: Mammals from reptiles: a review of mammalian origins. 131–155. In: Annual Review of Earth and Planetary Sciences.
CROWSON, R. A., 1970: Classification and biology. London: Heinemann Educ. Books.
DARLINGTON, P. J., 1970: A practical criticism of Hennig-Brundin 'Phylogenetic systematics' and Antarctic biogeography. Syst. Zool. 19, 1–18.
DARWIN, C., 1859: On the origin of species by means of natural selection, or the preservation of favoured races in the struggle for life. London: John Murray.
FARRIS, J., 1966: Estimation of conservatism of characters by consistency within biological populations. Evolution 20, 587–591.
— 1967: The meaning of relationship and taxonomic procedure. Syst. Zool. 16, 44–51.
GHISELIN, M. T., 1969a: The triumph of the Darwinian method. Berkeley: Univ. of California Press.
— 1969b: The principles and concepts of systematic biology. p. 45–55. In: Systematic Biology. Washington: National Academy of Sciences.
GHISELIN, M. T.; JAFFE, L., 1973: Phylogenetic classification in DARWIN's Monograph on the subclass Cirripedia. Syst. Zool. 22, 132–140.
GISIN, J., 1964: Synthetische Theorie der Systematik. Z. zool. Syst. Evolut.-forsch. 2, 1–17.
GREENWOOD, P. H.; ROSEN, D. E., 1971: Notes on the structure and relationships of the Alepocephaloid fishes. Amer. Mus. Novit. 2473, 1–41.
GRIFFITHS, G. D. C., 1972: The phylogenetic classification of Diptera Cyclorrhapha with special reference to the structure of the male postabdomen. The Hague: Dr. W. Junk N. V., Publishers.
GÜNTHER, K., 1956: Systematik und Stammesgeschichte der Tiere. Fortschr. Zool. 10, 37–55.
— 1971: Abschließende Zusammenfassung der Vorträge und Diskussionen. p. 76–87. (see also page 38). In: SIEWING (ed.).

HEBERER, G. (ed.), 1943: Die Evolution der Organismen: Ergebnisse und Probleme der Abstammungslehre. Jena: Gustav Fischer.

HENNIG, W., 1950: Grundzüge einer Theorie der Phylogenetischen Systematik. Berlin: Deutscher Zentralverlag.

— 1960: Die Dipteran-Fauna von Neuseeland als systematisches und tiergeographisches Problem. Beitr. z. Entomologie 10, 15–329.

— 1966: Phylogenetic Systematics. Urbana: University of Illinois Press.

— 1969: Die Stammesgeschichte der Insekten. Senckenberg-Buch 49. Frankfurt/M.

— 1971: Zur Situation der biologischen Systematik. p. 7–15. In: SIEWING (ed.).

HULL, D., 1967: Certainty and circularity in evolutionary taxonomy. Evolution 21, 174–189.

— 1970: Contemporary systematic philosophies. Ann. Rev. Ecol. Syst. 1, 19–54.

HUXLEY, J. (ed.), 1940: The new systematics. Oxford: Clarendon Press.

— 1942: Evolution: The modern synthesis. London: Allen and Unwin.

— 1958: Evolutionary processes and taxonomy, with special reference to grades. Uppsala Univ. Arsskr. 6, 21–39.

ILLIES, J., 1961: Phylogenie und Verbreitungsgeschichte der Ordnung *Plecoptera*. Verh. D. Zool. Ges. Bonn Zool. Anz. Suppl. 25, 384–394.

INGER, R., 1967: The development of a phylogeny of frogs. Evolution 21, 369–384.

JEPSEN, G.; MAYR, E.; SIMPSON, G. G., 1949: Genetics, paleontology, and evolution. Princeton: Princeton University Press.

JOHNSON, L. A. S., 1970: Rainbow's end: the quest for an optimal taxonomy. Syst. Zool. 19, 203–239.

KIRIAKOFF, S. G., 1959: Phylogenetic systematics versus typology. Syst. Zool. 8, 117–118.

KLUGE, A.; FARRIS, J., 1969: Quantitative phyletics and the evolution of Anurans. Syst. Zool. 18, 1–32.

MAYR, E., 1942: Systematics and the origin of species. New York: Columbia University Press.

— 1963: Animal species and evolution. Cambridge, Mass.: The Belknap Press, Harvard University.

— 1965: Numerical phenetics and taxonomic theory. Syst. Zool. 14, 73–97.

— 1969: Principles of Systematic Zoology. New York: McGraw-Hill.

MAYR, E.; SHORT, L., 1970: Species taxa of North American birds. Cambridge, Mass.: Nuttall Ornithological Club.

MCALPINE, J.; MARTIN, J., 1966: Systematics of Sciadoceridae and relatives with descriptions of two new genera and species from Canadian amber and erection of family Ironomyiidae (Diptera: Phoroidae). Canad. Ent. 98, 527–544.

MICHENER, C. D., 1970: Diverse approaches to systematics. Evol. Biol. 4, 1–38.

MILL, J. S., 1874: A system of logic, ratiocinative and inductive, being a connected view of the principles of evidence and the methods of scientific investigation. 8th ed. London: Longsman, Green and Co.

NELSON, G., 1971: Cladism as a philosophy of classification. Syst. Zool. 20, 373–376.

— 1972: Comments on HENNIG's 'Phylogenetic Systematics' and its influence on ichthyology. Syst. Zool. 21, 364–374.

OSCHE, G., 1961: Aufgaben und Probleme der Systematik am Beispiel der Nematoden. Verhandl. Deutsch. Zool. Gesell. Bonn 1960, 329–384.

— 1965: Über latente Potenzen und ihre Rolle im Evolutionsgeschehen. Zool. Anz. 174, 411–440.

— 1971: Discussion comments. p. 85. In: SIEWING (ed.).

PENNAK, R., 1964: The collegiate dictionary of zoology. New York: Ronald Press.

PETERS, D. S., 1970: Über den Zusammenhang von biologischem Artbegriff und phylogenetischer Systematik. Frankfurt: Aufsätze u. Red. Senckenberg. naturforsch. Ges.

— 1972: Das Problem konvergent entstandener Strukturen in der anagenetischen und genealogischen Systematik. Z. f. zool. Syst. Evolut.-forsch. 10, 161–173.

PETERS, D. S.; GUTMANN, W. F., 1971: Über die Lesrichtung von Merkmals- und Konstruktions-Reihen. Z. f. zool. Syst. Evolut.-forsch. 9, 237–263.

POPPER, K. R., 1959: The logic of scientific discovery. London: Hutchinson.

— 1963: Conjectures and refutations. London: Routledge and Kegan Paul.

REMANE, A., 1952: Die Grundlagen des natürlichen Systems, der vergleichenden Anatomie und der Phylogenetik. Leipzig: Geest und Portig.

RENSCH, B., 1947: Neuere Probleme der Abstammungslehre. Die transspezifische Evolution. Stuttgart: Ferdinand Enke.

SCHINDEWOLF, O. H., 1967: Über den ‚Typus‘ in morphologischer und phylogenetischer Biologie. Abh. Akad. Wiss. Lit. Math. Naturw. Klasse, Mainz, N. R. 4, 57–131.

SCHLEE, D., 1971: Die Rekonstruktion der Phylogenese mit HENNIGs Prinzip. Frankfurt: Aufsätze u. Reden. Senckenberg. naturforsch. Ges.

SIEWING, R. (ed.), 1971: Methoden der Phylogenetik. Erlanger Forschungen Reihe B, 4.

128   E. Mayr

SIMPSON, G. G., 1945: The principles of classification and a classification of mammals. Bull. Amer. Mus. Nat. Hist., **85**, 1–350.
— 1953: The major features of evolution. New York: Columbia Univ. Press.
— 1959a: Anatomy and morphology: classification and evolution, 1859 and 1959. Proc. Amer. Phil. Soc. **103**, 286–306.
— 1959b: The nature and origin of supraspecific taxa. Cold Spring Harbor Symp. Quant. Biol. **24**, 255–271.
— 1961: Principles of animal taxonomy. New York: Columbia Univ. Press.
SOKAL, R.; SNEATH, P., 1963: Principles of numerical taxonomy. San Francisco: Freeman.
THROCKMORTON, L., 1962: The problem of phylogeny in the genus *Drosophila*. Studies in genetics II, 207–343. University of Texas Publication 6205.
— 1965: Similarity versus relationship in *Drosophila*. Syst. Zool. **14**, 221–236.
TILLYARD, R., 1921: A new classification of the order Perlaria. Canad. Ent. **53**, 35–43.
WARBURTON, F. E., 1967: The purposes of classification. Syst. Zool. **16**, 241–245.

Author's address: Professor Dr. ERNST MAYR, Museum of Comparative Zoology, Harvard University, Cambridge, Massachusetts 02138

21

Reprinted from *Annu. Rev. Ecol. Syst.* **5**:81–99 (1974)

THE USES OF CLADISTICS[1]

Peter D. Ashlock

Department of Entomology, University of Kansas, Lawrence, Kansas 66045

The usefulness of cladistics derives from the fact that cladogenesis, the branching component of phylogeny, is a part of the theory of evolution. I am an evolutionary systematist, a member of the Simpson-Mayr school of systematics, which has profound objections, principally in the area of classification, to the cladistic or so-called phylogenetic school of Hennig (30). Nonetheless, I think cladists are quite right when they complain that their very real and important contributions to biogeography and to chronistics and coevolution have been ignored or seriously misunderstood. It is the purpose of this discussion to review and enlarge on these areas. Accepting the tenets of the cladistic school on biological classification is neither necessary nor desirable, but cladistic analysis is a prerequisite for an evolutionary classification.

TERMINOLOGY

Simpson (46) characterized Hennig's terminology as idiosyncratic. The years since have demonstrated that this terminology is a mixture of valuable concepts and terms occasionally misapplied, badly defined, or not defined at all. The list below is meant to correct this situation and to provide a vocabulary for evolutionary systematics. Deviations from Hennig are identified and equivalent terms provided. Ashlock (1, 2) has discussed the terms related to monophyly, and Tuomikoski (47) has provided helpful discussion of some concepts.

Cladistic: Pertaining to the branching sequence in evolution.
Anagenetic: Pertaining to the accumulation of changes in ancestor-to-descendent lineages.
Cladistic analysis: Analysis of the characters of organisms to infer the evolutionary branching sequence of a group's phylogeny (*phylogenetic analysis* of Hennig).
Cladistic classification: Classification in which only holophyletic (q.v.) taxa are permitted and categorical rank is determined by the group's age.

[1]Preparation of this review was supported by University Research Grants, University of Kansas. Contribution No. 1537 from the Department of Entomology, University of Kansas.

Monophyly: A monophyletic group is one whose most recent common ancestor is a cladistic member (q.v.) of that group.

Holophyly: A holophyletic group is a monophyletic group that contains all descendents of the most recent common ancestor of that group (*monophyly* of Hennig).

Paraphyly: A paraphyletic group is a monophyletic group that does not contain all descendents of the most recent common ancestor of that group.

Polyphyly: A polyphyletic group is one whose most recent common ancestor is not a cladistic member (q.v.) of that group.

Cladistic member: A cladistic member of a group is any recent member of a holophyletic group, as demonstrated by one or more synapomorphous characters, any fossil that shares these characters, and all inferred ancestors within the group.

Apomorphous: The relatively derived state of a sequence of homologous characters.

Plesiomorphous: The relatively primitive state of a sequence of homologous characters.

Synapomorphous: Uniquely derived apomorphous character that is found in two or more taxa under consideration. Such characters serve to demonstrate the holophyly of groups of taxa that possess them.

Autapomorphous: Apomorphous characters found in a single taxon not being considered for further subdivision.

Sister-group: In a dichotomous cladogram, the two holophyletic groups that are descendent from any inferred ancestor.

Phylogeny: The evolutionary history of organisms, to include both cladistic and anagenetic information (in Hennig's usage, the cladistic aspects of evolution).

CLADISTIC ANALYSIS

A cladogram is a hypothesis, the best explanation of the distribution of characters, be they morphological, behavioral, or other, in the organisms under study, using all of the facts available. It cannot be proved, although it may be supported by external evidence.

Hennig and his followers employ for cladistic analysis what has come to be known as "Hennig's principle": Only synapomorphous characters delimit monophyletic taxa. For evolutionary systematists, the principle as stated presents some problems which are solved if it is restated: Only synapomorphous characters delimit holophyletic groups. The word monophyletic as used by Hennig (all the descendents of the most recent common ancestor) indicates a concept of very great theoretical importance and utility, but the concept is far from the traditional meaning of monophyletic and is unsuitable for evolutionary systematics. The term holophyletic was coined by Ashlock (1, 2) for Hennig's concept. The word *groups* is used instead of *taxa* since, unlike cladists who may proceed directly from cladistic analysis to classification, evolutionary systematists require additional anagenetic analysis before formal taxa can be delimited.

Hennig has never really defined *synapomorphous,* at least in English, nor has its complementary term *plesiomorphous* been defined. In reading the works of cladists, one quickly realizes that apomorphous characters must be derived characters, while plesiomorphous characters are primitive. It is not so immediately apparent that not all derived characters are synapomorphous. While loss of wings in holometabolous insects is derived, and fewer than five toes in tetrapods is derived, use of such characters in cladistic analysis will result in preposterous groupings. (Generally one should be suspicious of "loss" characters.) Reliable synapomorphous characters (sometimes referred to as true synapomorphies) are unique and complex. For example, feathers and a horny bill on birds are excellent for making an inference that the common ancestor of modern birds had these characters as well. Other good examples are the tube feet of the Echinodermata, the halteres of Diptera, and retractible claws of cats.

It is also obvious that the systematist must arrange the characters of the group under study into primitive-to-derived sequences (transformation series of cladists; primitive-to-derived character states of numerical taxonomists), and that the parts of these sequences must be evolutionary homologues.

Estabrook (22) has reviewed computer approaches to cladistic analysis. Methods are based on two principles: parsimony and compatibility. Parsimony (the cladogram with the fewest steps is the most probable one) makes the assumption that evolution usually takes the shortest route. Rogers, Fleming & Estabrook (43) have criticized the principle on mathematical grounds. I would add that while no law of evolutionary theory requires evolutionary parsimony, the principle is not without use. Compatibility (the largest collection of compatible characters is best evidence for the true cladogram) more accurately reflects evolutionary theory. The compatibility matrix of Camin & Sokal (14) and the successive approximations approach of Farris (23) are two such methods. Though both include parsimony in their calculations, the trees produced are not always the most parsimonious; rather, they reflect a high degree of compatibility.

Computer cladistic analysis can be helpful in producing cladograms when no cladistic hypotheses are clearly evident from the data, or in producing alternate hypotheses. Computer-produced cladograms, however, should always be checked for synapomorphy [using, for example, Wilson's (48) consistency test]. Other ways to check the validity of a cladogram are suggested elsewhere in this review. Strict operational methodologies are notably inappropriate (31) to cladistic analysis since cladogenesis is a theoretical, not an empirical, concept.

BIOGEOGRAPHY

Direct and Indirect Biotic Connections

To my mind the most important paper to appear in recent years on the subject of historical—as opposed to ecological—biogeography is Hennig (27), translated into English by Wygodzinsky under the title, "The Diptera fauna of New Zealand as a problem in systematics and zoogeography" (29). The paper is at once brilliant and badly conceived. Its title and organization are such that few but dipterists would be attracted, yet the principles covered are of use to all systematists and biogeographers.

In his introduction, Hennig writes (29)

> Many a taxonomist who writes a monograph of his group leaves questions unanswered
> which he, and only he, could have answered and only because he had been unaware of
> the existence of these questions. Such questions consequently remain unanswered, because
> it is too difficult to obtain again for study the material (types, rare species, etc.) which
> the monographer had at his disposal.

The general questions that interest Hennig are: From which other areas has a given, relatively isolated land surface (large island or continent) received its faunal elements, and with which other areas has it therefore been connected? When, how long, and in what sequence have these connections existed?

To attack these questions, Hennig says that it is necessary to ask about each endemic species in the area concerned (here New Zealand): which is its sister-species and where does it occur. It is likely that the first sister-group, which may or may not be a single species, will also be found in New Zealand; but continuing cladistic analysis will yield a sister-group on another land mass and eventually in all biotic areas of the world. Hennig refers to this method of analysis as "search for the sister-group."

One would expect highly vagile organisms to be distributed in both New Zealand and Australia. The relative proximity of the land masses forms an adequate explanation for the distribution. A more exciting possibility is that the sister-group of a New Zealand or Australian group occurs in South America. Hennig refers to such groups as A-S (Australian-South American) groups.

In listing A-S genera of Diptera, Hennig notes that such highly derived and recent large groups as the Schizophora are poorly represented, while more primitive flies are well represented. He feels that this is no accident. Hennig suggests three possible routes between southern land masses of the eastern Old World (A) and the New World (S): (a) across the southern Pacific; (b) across the Antarctic continent; and (c) across northern hemisphere land masses. Routes a and b Hennig terms direct routes, while c is designated as indirect.

Before discussing how one establishes whether or not a given group followed a direct route of distribution, Hennig dismisses two superficially attractive indications that are, in fact, insufficient to distinguish between direct and indirect faunal connections. These are: (a) members of an A-S group that are more similar to one another than either is to any other group in the world; and (b) members of an A-S group that form a holophyletic group.

Neither case provides proof of direct distribution, since the A-S group might have had its origin in the northern hemisphere, migrated to different parts of the southern hemisphere, and later been replaced by more advanced groups in the northern hemisphere.

Hennig proposed three criteria that are adequate to show a direct connection between disjunct parts of A-S groups: the progression rule, the phylogenetic intermediate rule, and the multiple sister-group rule.

THE PROGRESSION RULE   Hennig's well-named progression rule refers to a geographic sequence of taxa whose direction of progression is indicated by a series of

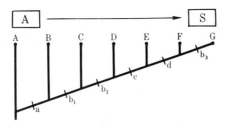

Figure 1   Hennig's progression rule.

increasingly derived synapomorphous characters. Figure 1 illustrates Hennig's hypothetical example, where taxon A is found in New Guinea, B in Queensland, C in Victoria, D in Tasmania, E in New Zealand, F in Tierra del Fuego, and G in Chile. If it is established that group A through G is holophyletic, and further, if cladistic analysis convincingly demonstrates a sequence of holophyletic groups BCDEFG, CDEFG, DEFG, EFG, and FG as established by a series of synapomorphous characters (either unrelated characters such as a, c, and d in Figure 1 or a sequence of progressively more derived homologous characters such as b_1, b_2, and b_3), then one would have to agree that the group A through G did indeed progress from New Guinea through Australia, Tasmania, New Zealand, and on directly to South America. To hypothesize an alternate route through the northern hemisphere, one would have to suppose a migration of the derived taxa (or their ancestors) through an area of more primitive species, without having left traces.

The progression rule can be used only when the distribution of the organisms is not seriously disturbed. However, that a group of organisms fits the progression rule is evidence that it has not been seriously disturbed. It is highly unlikely that a group of organisms would be disturbed in such a way as to give a false logical progression of synapomorphous characters.

Examples   No published account has come to hand that employs the progression rule in establishing a direct A-S connection. However, such an account is currently in preparation for the mutillid wasps by D. J. Brothers. The progression rule may also be used to arrive at biogeographic conclusions not involving A-S routes. A routine revision of the ischnorhynchine lygaeid bug genus *Neocrompus* (3) made no mention of the progression rule, but it clearly applies. *Neocrompus* was originally described for a single species from Samoa distinguished partially by the widely flaring rear portion of the seventh abdominal segment, a character unique in the Lygaeidae. The 1966 revision added four species to the genus. One, from New Guinea, has the seventh segment unflared and similar to all others in the family. Two of the species, from Fiji and the Austral Islands, are flared like the Samoan species, and the last, from Tahiti and nearby islands, has the flared seventh segment with an additional lobe shaped somewhat like a protruding thumb. Clearly, the New Guinea species is the most primitive of the five, and the Tahiti species the most derived, indicating an eastward progression of the genus across the South Pacific.

THE PHYLOGENETIC INTERMEDIATE RULE   Hennig did not provide a name for this rule, but he headed the discussion "Demonstration of phylogenetically intermediate forms on islands situated on the direct line of connection between South America and Australia-New Zealand."

Evidence of a former direct connection between two continents (such as Australia and South America) would be provided if forms more primitive than Old World and New World representatives were found on islands on or near the shortest direct route between the continental areas concerned (Figure 2). The phylogenetic intermediate rule is really a special case of the progression rule in which members of a group proceed in two directions from an origin that is no longer inhabitable (Antarctica). The same strictures apply to acceptance of the phylogenetic intermediate rule hypothesis as to acceptance of a progressive rule hypothesis.

No examples of use of the phylogenetic intermediate rule have come to hand.

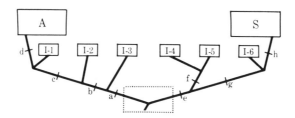

Figure 2   Hennig's phylogenetic intermediate rule.

THE MULTIPLE SISTER-GROUP RULE   Again, Hennig provided no short name for this model. The rule may be stated: If a monophyletic (i.e. either holophyletic or paraphyletic) group can be demonstrated to have multiple sister-group relationships between the areas under discussion, then a direct connection between these areas has been established (Figure 3). With an A-S multiple sister-group relationship, one must give serious consideration to Antarctica as the home of the group's common ancestor, although either Australia or South America remain as possibilities.

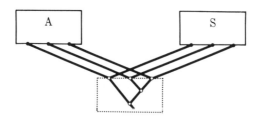

Figure 3   Hennig's multiple sister-group rule.

344

Figure 3 demonstrates three or six transoceanic crossings of the monophyletic group, depending upon possible involvement of Antarctica. Hennig feels, as was pointed out earlier, that a single sister-group relationship between taxa in Australia and South America is not sufficient evidence for a direct connection. However, as the number of A-S sister-group relationships within a single monophyletic group increases, the credibility of a northern origin decreases. If their origin were northern, the distribution of the organisms in Figure 3 would require six southern migrations and subsequent loss of all trace of the origin.

Examples   It is with applications of the multiple sister-group rule that the most successful uses of cladistics have come. Notable is Brundin's work with chironomid midges, for it was the first cladistic analysis that dealt with the concepts of continental drift. The major work was published in 1966 (10) but the earliest discussion was in 1963 (8), and he provided summaries in 1965 (9) and 1967 (11). The major work and the two summaries carry Brundin's cladogram that, if accurate, demonstrates twenty transantarctic relationships. Such massive demonstration of the multiple sister-group rule clearly establishes direct southern faunal connections. These were not merely A-S relationships, but included Africa as well. Especially striking in Brundin's cladogram is the sequence of sister-group relationships. The sister-group of an Australian organism is found in South America, not New Zealand as one might expect from the proximity of the land masses. New Zealand midges are the sister-group of Australian and South American midges together. Brundin (9) does not hesitate to conclude that he is on the track of the real nature of transantarctic relationships. He states:

> The conclusion is inescapable that the transantarctic relationships developed during a period when the southern lands were directly connected with each other. There is no reason to speculate on island stepping-stones or chance dispersal over wide stretches of ocean. We have to accept as a fact that the transantarctic relationships and the distribution patterns of the chironomid midges are orderly. And they show very clearly that the connections between the southern lands were broken according to a certain sequence which started with the separation of southern Africa. The next event was the break of the links between Antarctica and New Zealand, which obviously never had any direct connections with Australia. Later separation between Antarctica and Australia occurred. The last connections between Antarctica and Patagonia were cut still later, probably not very long ago. We have indeed the right to be confident, since this sequence of events, as indicated by the chironomid midges, is in good accordance with modern opinions among the geologists concerning the disruption of Gondwanaland.

Another such study was done by Edmunds (21) in the mayflies, and he in a similar manner summarized his results:

> From the total evidence from Ephemeroptera plus minimal geological evidence as noted, the suggested sequence of the breakup of Gondwanaland is as follows. 1. India drifted to the north (evidence largely geological). 2. South Africa plus Madagascar drifted to the north with the continent pivoting so that the break with South America widened most rapidly at the south leaving Africa and Brazil attached or close together in the

tropics. The presence of many African groups in Madagascar suggests that the split of Madagascar from Africa was one of the last breaks. 3. New Zealand plus New Caledonia drifted to the north. 4. New Caledonia separated from New Zealand. 5. Australia drifted north and Antarctica drifted south. (The direction and sequence of 5 in relation to 6 are based on geological evidence.) 6. South America drifted northwest in relation to Antarctica.

Edmunds' study confirms the results of Brundin's study. Both are in the best tradition of the scientific method: both use one kind of evidence (synapomorphous characters) to arrive at an hypothesis (their respective cladograms) and then find their hypotheses to be confirmed not only by each other but also by external evidence (the sequence of the breakup of Gondwanaland as postulated from geological evidence).

Hennig's (29) discussion of the multiple sister-group rule does not include continental drift, but continental drift provides the best theoretical explanation for the multiple sister-group rule. In Figure 3, the land mass with the dotted outline may be considered to be a predrift supercontinent upon which a holophyletic group has evolved three widely distributed lineages. If the supercontinent split into two daughter continents (A and S in Figure 3), both would carry the three lineages. The continental splitting may continue. If the three lineages on daughter continent A were each to develop into two lineages, and continent A split in two, each fragment (call them A-1 and A-2) would carry all six lineages. The sister-group of a lineage found on fragment A-1 would be found on fragment A-2. The sister-group of a lineage on continent S, however, would be found on both A-1 and A-2, not on one fragment only. The sequence of the splitting of continents from Gondwanaland explains the orderly splitting that Brundin and Edmunds found.

THE DRIFT SEQUENCE RULE   The Brundin and Edmunds studies have introduced what amounts to a fourth biogeographic rule which can be applied to the problem of direct versus indirect connections between various land masses. The rule may be stated: If the sister-group relations of a monophyletic group conform to the sequence of continental drift of at least three continents, then direct biotic connections between the continents have been established. At least three continents are specified because sister-group relations between two only are, as Hennig has pointed out, inconclusive. As stated, the multiple sister-group rule need not be a part of the evidence for direct connections. For example, if three holophyletic groups were each confined to New Zealand, Australia, and South America, and the sister-group of the New Zealand group was found to be in Australia and South America together, conforming to the last few splits from Gondwanaland, then requirements of the drift sequence rule would have been met. Requirements of the multiple sister-group rule, however, would not. The Brundin and Edmunds studies are, of course, all the stronger for having met both of these rules.

It is possible that the drift sequence demonstrated from geological evidence may not be matched by the cladistic branching of a particular group of organisms, even though they were on a supercontinent before breakup. Such discrepancies may be caused by uneven distribution of the organisms over the supercontinent, by such barriers as mountains, epicontinental seas, or ice, or by extinction on one or more of the fragments of the supercontinent. It is hoped that someone will prepare an atlas of continental drift to show not only the positions of the land masses through geological time but also what is known of climatic and other possible barriers.

Several authors have discussed the sequence and timing of the breakup of Pangaea. The broad movements have been clearly established, but changes in the interpretation of details must be expected for several years to come. Literature on the subject should be watched carefully as it appears, but the following papers will serve as a starting point (4, 5, 16, 20, 24, 42).

An obvious criticism of all of these biogeographical models is that the models themselves (such as the Gondwanaland breakup sequence) were used to "discover" the characters used in the study and thereby compromise the validity of the conclusion. There is no denying that the best demonstration of a biogeographic model would come from a study in which the investigator assembled a convincing array of strong synapomorphous characters without reference to the distribution of the organisms under study. Such an ideal is too much to ask, however. Good synapomorphous characters are very difficult to find, and a systematist welcomes anything that aids him in looking for them. The validity of such studies must be judged by assessing the quality of the characters establishing the groups and the possibility of alternate hypotheses based on the same or additional data.

Other Biogeographic Uses

PLACE OF ORIGIN   Darlington (19) and Brundin (12) have exchanged opinions on various points, including the validity of Hennig's biogeographic rules and of Brundin's application of them. Part of the exchange dealt with whether the more primitive or more derived members are likely to be found at the center of origin of groups. Brundin, in the more perceptive of the discussions, noted that Darlington's arguments are based on single, arbitrarily selected characters, when such questions as the center of origin of a group should be answered by means of sister-group or cladistic analysis of the group. Nelson's (38) discussion of the subject provides some specifics of the methodology.

The first place to look for the probable origin of a group, i.e. the original distribution of its common ancestor, is within the area of distribution of the monophyletic group to which the common ancestor belongs (including geographically holophyletic and paraphyletic groups). There are limitations, however. Too many of the species may have a wide distribution. Extinction can cause very serious errors. If members of a group migrate more than once from area A to area B, and the group then becomes extinct in area A, then the distribution of living forms will be misleading. A suitable fossil found in area A must be taken into account. If, on the other hand, a single species from a group that had its origin in area A migrated to area B, speciated, and became the founder of a new large group of organisms, while all members of the group remaining in area A became extinct, then, in terms of living species, area B is the place of origin of the group. Where a monophyletic group has been found, as by Brundin and Edmunds, to occupy several southern continents, the area of origin is the sum of these continents, or Gondwanaland. Confirmation that such a southern distribution is actually a Gondwanaland distribution would have to come from at least one of the four rules previously discussed.

NUMBER OF INTRODUCTIONS   The determination of how many times a particular taxonomic group has crossed geographic barriers to account for its present distribution is a problem whose solution requires application of some obvious principles, and some that are not so obvious as well.

One example can be found in the Hawaiian birds of the endemic family Drepaniidea (the honeycreepers), well known for their bizarre beak modifications. Each species of Hawaiian honeycreeper has as its closest relative another Hawaiian honeycreeper, and only collectively are they related, more distantly, to a mainland group. The consensus of those who have discussed the group is that the honeycreepers resulted from a single introduction to the Islands. The various species, or some intermediate ancestors, certainly could have evolved elsewhere, migrated to the Islands, and left no trace of their origin, but there is no evidence of it. Thus the simplest inference, that there was one introduction to the Islands, stands.

The Drepaniidae is probably a monophyletic (and holophyletic) group. If so, the example can be stated cladistically: A holophyletic group that is distributed only in a limited geographic area probably had its place of origin within that area. As there is no evidence that the most recent common ancestor lived elsewhere, the group is not only holophyletic but also geographically holophyletic, and represents a single introduction.

Geographic holophyly is illustrated in Figure 4a where the taxa labeled O are found at the place of origin and those labeled X represent the geographically holophyletic group. The migration from the place of origin must have taken place during the time period indicated by internode 2–4 on the cladogram, as shown by the arrow.

If another related group of Hawaiian land birds such as the honeyeaters (*Moho* and *Chaetoptila,* another holophyletic group with mainland relatives thought to represent a single introduction to the Islands) is added to the Drepaniidae, then the assemblage becomes geographically polyphyletic: that is, the most recent common ancestor could not have lived in the Islands. Thus, in a geographically polyphyletic

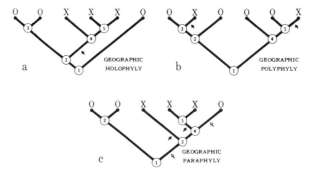

Figure 4   Cladograms illustrating number-of-introduction principles.

group, the minimum number of introductions is equal to the number of geographically holophyletic subgroups that make up the larger geographically polyphyletic group. Geographic polyphyly is illustrated in Figure 4*b*. Arrows again indicate the internodes where migrations must have taken place.

An additional possibility, geographic paraphyly, requiring involvement of at least three groups, is illustrated in Figure 4*c*. This group had its origin and development in one area and has a descendent group in a second area. If a third group descendent from the second is found in a third area, one has an example of the progression rule. If this third group is found in the same region as the original ancestral group, conditions for geographic paraphyly have been met.

Geographic paraphyly involves an element of ambiguity. If in Figure 4*c*, the O's are found in Siberia, the X's in North America, and the ancestor of 1 is firmly established as having an Old World origin (descendent of a complex of Old World forms), it is obvious that the group migrated from the Old World to the New across the Bering Straits and that two migrations have taken place. The question is, in which direction? The group may have migrated from the Old World to the New twice, as indicated by the black arrows. It is equally possible that the group migrated to the New World and later evolved a group that migrated back to the Old World, as indicated by the white arrows.

Additional information is needed to remove the ambiguity of geographic paraphyly. For example, if the group represented in Figure 4*c* had the taxa labeled O in the Americas and those labeled X in the Hawaiian Islands, then one would postulate that two migrations had taken place from the Americas to the Islands, based on the assumption that an insular form would have a difficult time competing with continental forms.

The minimum number of migrations a group has made to island archipelagos is very important to MacArthur & Wilson's *Theory of Island Biogeography* (34) and to Leston's "Spread potential and the colonization of islands" (33).

CHRONISTICS

Chronistics, the study of the age of biological groups, is most often left to paleontologists. As Hennig (30) has said, the paleontological method has the reputation of being the most reliable, if not the only, method of determining the age of groups. But, like most methods, it has limitations. Spottiness in the fossil record limits the method to giving only minimal ages of groups, not absolute ages. Moreover, fossils show fewer characters than recent organisms, making them more difficult to use for cladistic analysis.

However, properly dated fossils can be more significantly assessed with cladistic analysis than without. Hennig (30) demonstrated this with three groups of Diptera, and a generalized example is illustrated in Figure 5. The figure shows three groups of recent organisms, X, Y, and Z, whose cladistic analysis was based upon characters (or groups of characters) a, b, and c, demonstrating that X, Y, and Z are progressively more derived.

If a properly dated fossil relative of group X-Y-Z is discovered showing only the characters symbolized by a, little can be said about the age of the group. The fossil

Figure 5   Cladogram illustrating chronistic principles.

might have come from the internode ancestral to node 1, the internode between 1 and X, somewhere along internode 1–2 before characters b evolved, or from a side branch leading to an extinct taxon from any of the three mentioned internodes. At best, all this fossil can do is to establish a date after which the characters a could not have evolved.

If a dated fossil bearing characters b is found, however, then the investigator has a date after which the inferred ancestor at node 1 could not have originated. Such a fossil, then, would help determine the age of the primitive branch to X. Similarly, a fossil bearing characters b and c must have evolved subsequent to node 2, and provides a date after which ancestors at both nodes 1 and 2 could not have originated. Favorable fossils, then, can help to date various parts of a cladogram, even lineages for which no fossils are available.

Barriers to migration can be used to date parts of cladograms, just as fossils are used. For example, both Brundin on chironomid flies and Edmunds on mayflies have much to say about the ages of their organisms. Both of these groups, by the multiple sister-group rule and the drift sequence rule, have been established as present on Gondwanaland at the latest before Africa drifted away. The date when Africa separated from Gondwanaland, then, is the latest possible time for the origin of the groups. Similarly, the New Zealand break establishes a date of origin for the New Zealand elements of the groups and for their sister-groups.

Thus, if a group of organisms can be established to have crossed what is now a barrier to their migration, and a date is known for the initiation of that barrier, then the latest possible date of origin, not only for the group that crossed the barrier but for its sister-group as well, is the date of the barrier. Barriers may be of many sorts: epicontinental seas, deserts, mountains, forests, gaps in strings of islands preventing island-hopping; effectiveness of all barriers depends on the vagility of the organisms involved.

CO-EVOLUTION

Hennig (30) discussed what he called the parasitological method, dependent upon Fahrenholz's Rule: "In the case of permanent parasites, the relationship of the host can usually be inferred directly from the systematics of the parasites." The assumption is that the evolution of the hosts is directly paralleled by that of the parasites (Figure 6a). Hennig points out a number of reasons why this assumption can lead to false conclusions. The parasite group may have joined its host group well after

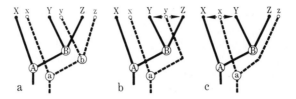

Figure 6 Cladograms illustrating problems possibly encountered in the study of host-parasite or symbiont lineages.

the host group began to evolve. Even if the two evolved together, some host lineages may have escaped from their parasites, or a parasite may have transferred from one host lineage to another. Unequal rates of evolution in various lineages of host and parasite may also cause problems if similarity rather than cladistics is used as a criterion for relationship.

Even without these problems, and with a proper cladistic analysis of the parasites, some less obvious difficulties can be created by incomplete parallelism of speciation in the host and parasite lines. In Figure 6*b*, a speciation of the host at B is not paralleled by a speciation of the parasite, and the same parasite y is found on both hosts Y and Z. One would here, on the basis of the parasite, correctly conclude that the hosts Y and Z form a holophyletic group. In Figure 6*c*, however, the parasite again has not speciated in a parallel manner with the host at A, but the presence of parasite x on both hosts X and Y here leads to an incorrect assumption that X and Y form a holophyletic group.

Perhaps a better way to look at Fahrenholz's Rule is to treat it as a question rather than a rule. If a cladistic analysis is made on both the host and parasite (or other symbiotic system) groups, then one could ask: Did the parasite evolve with the host? If not, was there a partial co-evolution of the host and parasite? The degree of co-evolution would be shown by the degree that the cladogram of the host matches the cladogram of the parasite. A study by W. Ramírez (in preparation) on *Ficus* (to subgeneric level) and the fig-wasp pollinator (Agaonidae) demonstrates a nearly complete congruence between cladograms of the figs and of the wasp pollinators, indicating a nearly complete parallelism of the evolution of the two groups. Ramírez can even demonstrate to some degree how the figs and wasps adapted to one another by comparing the characters that appear on equivalent internodes of the congruent cladograms.

Another possible co-evolutionary use of cladistics is in the study of hard-to-associate morphotypes within the species of a single group: immatures versus adults in groups with a pronounced metamorphosis, and males versus females where these differ grossly. If sets of morphotypes are complete, cladistic analysis in each set within the group should yield identical cladograms. It would be especially interesting to apply cladistic analysis to those cases where the original association was made on biological grounds, for then differing rates of evolution, if any, could be studied in the morphotypes.

CLASSIFICATION

Probably the most important and controversial use of cladistics is in classification. The phenetic school of taxonomy does not concern itself with cladistic analysis, feeling that phylogeny is inherently unknowable. There are, of course, numerical taxonomists interested in computer methods of cladistic analysis, and such persons should not be confused with pure pheneticists. Another nameless group, whose best spokesman is Blackwelder (6, 7), and which is not concerned with cladistic analysis, feels that taxonomy is something one does, not something one thinks about. Cladistic analysis, then, is of concern to the cladistic and evolutionary schools of systematics.

The cladistic school of Hennig will allow only holophyletic groups of organisms in its classifications and rejects both polyphyletic and paraphyletic groups. Sistergroups are to be given the same categorical rank, and rank is determined by the age, i.e. date of origin, of the group. The age criterion is apparently less stringent than others, for Hennig (30) has suggested that vertebrate zoologists need not be held to the age-rank system he suggests for insects, and Nelson (40) suggests some measure of relative difference be incorporated while maintaining the other precepts. Resulting classifications are most often grossly different from those produced by other schools of systematics.

Evolutionary systematics, in contrast, will allow both holophyletic groups and paraphyletic groups, both of which have traditionally been considered to be monophyletic units. Polyphyletic groups, or those based upon convergence, are excluded, as they are not genealogical groups. Formal taxa are recognized as genealogical groups of relative internal homogeneity that are separated from phylogenetically related groups by decided gaps. Categorical rank is based upon tradition, with changes made conservatively. In general, the precise definitions of categorical ranks must remain unsolved problems. Classifications in evolutionary systematics will be similar to those of the pheneticists and to traditional classifications, but will be based upon an estimate of genetic similarity rather than phenetic similarity.

The most important work supporting the cladist's view on classification, of course, is Hennig (30). Some other discussions that, together with Hennig (30), give most of the arguments in favor of cladistic classification and against evolutionary classification are (10, 13, 15, 17, 18, 26, 28, 39–41, 44, 45). Evolutionary systematists have been less vocal than cladists. The principal works of the Simpson-Mayr school are Simpson (46) and Mayr (37). Other discussions that support evolutionary systematics and argue against cladistic classification are (1, 2, 19, 32, 35, 36, 47). Darlington's (19) discussion is unfortunate, since he did not have an adequate understanding of the cladist's viewpoint, but I can agree with some of his statements. Hull (32) agrees with the philosophical position of evolutionary systematics, but he criticizes its weak methodologies. Pheneticists have also entered the discussion, giving arguments against both the cladistic and evolutionary schools, often confusing them.

Cladists have developed a number of arguments defending their system against evolutionary and other schools of systematics. For example, speciation is the only decisive process in evolution, and as cladistic branching is a direct result of past speciation events, classifications that depend only on cladistic analysis of evolution are the most precise possible and will reflect nature's own hierarchy. Only with cladistic classifications can the phylogeny of higher taxa be read directly from the formal classification, and only with a cladistic classification can the mysteries of historical biogeography be understood and studied. Cladistic classifications, furthermore, have historical precedence, because Darwin was the first cladist.

The cladists go on to say that classification systems dependent upon anagenesis are just as typological as those of the phenetic school. Mayr's "genetic similarity" is indistinguishable from "taxonomic distance" of the numerical taxonomists, and similarity itself is a composite based upon plesiomorphous, apomorphous, convergent, parallel, and reversed characters. The measurement of similarity depends upon atomization or unit treatment of characters, which is unrealistic. Any of the proven benefits (e.g. biogeographic) of phylogenetic systematics (cladistics) are impossible under other systems, and any compromise system (phenetic plus cladistic) is bound to lead to confusion. Higher taxon names of the noncladist do not give you the characters of the group or seem to serve any other purpose.

Finally, evolutionary systematics involves a great deal of "art" in its methodology, and evolutionary classifications can only be inferred. Since there is no real method involved, the Simpson-Mayr school deserves no further consideration.

It should be emphasized that these arguments are a composite of those advanced by cladists. No one cladist would use all of them, and many could probably think of more.

As a theoretical and practical science, systematics has many tasks. Some are discussed in this review, and it is to the credit of the cladists and especially of Hennig's admirable logic that new depths have been added to the science of systematics. Other tasks are the traditional ones of describing and explaining groups of similar organisms, and of providing an information retrieval system for all concerned with these organisms. All science is, as Hull (32) has so well demonstrated, most productive when it is based in well-formulated theory. Systematics in the fulfillment of its tasks is most productive when it uses evolutionary theory to its fullest, and it is with evolutionary theory that systematics makes its ultimate explanations, with classification providing the framework for these explanations. Omissions of major parts of evolutionary theory from the systematic process can lead only to loss of information in resultant classifications.

All schools (evolutionary, cladistic, and phenetic) are complex bodies of theory and methodology. All, within their own contexts, can stand improvement. It has often been repeated that evolutionary systematics contains too much art, and Hull (32) has called for a reduction of this art. Evolutionary systematists are indebted to both pheneticists and cladists for forcing a reexamination of evolutionary systematics and pointing to better ways of approaching problems.

For example, it has been the practice of evolutionary systematists to treat the phylogenetic dendrogram as merely a summary of an already completed classification. The phenetic and cladistic schools, on the other hand, treat dendrograms or other graphic displays as necessary steps in the process of formulating a classification. In fact, the dendrogram has become an integral part of their classification. The lesson for evolutionary systematics, if it is to achieve its stated goal of maximal use of evolutionary theory, is that in forming a classification, one must first approach the problem of classification with cladistic analysis to establish the branching patterns of phylogeny and then establish relative amounts of evolution on each internode of the cladogram. The cladogram is thus preserved in its entirety, and the major anagenetic gaps on the tree are established, providing a means to delimit the holophyletic and paraphyletic groups that become formal taxa. The benefits of such a methodology, I believe, will be very great.

Because the new evolutionary dendrogram will have a cladogram at its core, it can be put to all the uses of the cladogram discussed in this review as well as any that might be thought of in the future. When cladists claim that only cladistic classifications allow understanding of biogeography, and that such studies are forever lost to the Simpson-Mayr school of evolutionary systematics, they are only partially correct. Evolutionary systematics without cladistic analysis cannot do these things. On the other hand, it is not the list of formal taxonomic names that is the result of cladistic classification which permits cladistic conclusions, it is the cladogram that does so.

Since in any dichotomous dendrogram there is one less branching point than there are terminal points, the number of names needed for a complete description of the cladogram is one less than the number of species contained in the group, in addition to the names needed for the species themselves. Any time a cladistic classification contains a group with three or more immediately subordinate taxa, the classification is ambiguous in its ability to reflect the cladogram. A trifurcation can represent a true evolutionary trifurcation or, more probably, any of three possible unresolved dichotomously branched systems. Four subordinate taxa could imply any of 17 possible interpretations. I don't believe that the cladists want to burden systematics with the number of names needed to make formal cladistic classifications completely describe cladograms. Thus formal cladistic classifications, like evolutionary ones, are not without ambiguity in their ability to describe cladograms. Both systems need to refer to cladograms for the benefits of cladistic analysis, which both have available. It is the cladistic analysis, not the cladistic classification, that provides the potential uses of cladistics.

The inclusion of relative anagenetic information in the dendrogram leads to a more complete understanding of taxa. In a cladistic analysis, only those characters with a high cladistic information content (synapomorphies) are needed or wanted. Schlee (44) reports that he used only 33 of some 500 characters studied in a classification of the sternorrhynchous Homoptera. Such a drastic paring of characters is right and proper for cladistic analysis, but also represents a huge information loss in the final dendritic or formal classification. As the cladogram was established on the basis of those relatively few characters, anagenetic analysis would attempt to place as many more characters on the dendrogram as possible. The basic cladogram would be available to help with decisions regarding the ancestral or derived nature of cladistically weaker characters. This methodology will require unitization of characters, which should not cause serious difficulty. It must be assumed that one's ability to discover characters is proportional to the actual amounts of evolution that have taken place throughout the history of the group. Such methods, which will require appropriate character weighting, will not only establish major gaps in the phylogenetic dendrogram useful in delimiting formal taxa, but also will demonstrate most fully the evolutionary history of taxa and their attributes.

Mayr (37) has emphasized that evolutionary systematics classifies organisms on the basis of genetic rather than phenetic similarity. The cladists are quite right when they say that the concept of similarity is a composite one, but it is only with cladistic plus anagenetic analysis that the various kinds of similarity can be sorted out. Such analysis will count convergent, parallel, and reversed similarities as the genetic differences they really represent. Phenetic studies treat these genetic differences as similarities. Apomorphous and plesiomorphous similarities are the source of most genetic and phenetic similarities. This fact explains why an evolutionary classification as outlined will resemble phenetic and traditional classifications to a large degree. Since cladistic classifications are concerned only with apomorphous similarity, they are often drastically unlike those produced by any other method.

For the dubious advantage of being able to read off "phylogeny" from a formal listing of taxa, cladists are willing to pay, as Hull (32) has said, a price too high for many biologists. Species that split off in the Precambrian but gave rise to no other species would have to be classed as phyla. Such classifications would be highly monotypic and highly asymmetrical. In Nelson's (39) classification of the vertebrates, jawless fishes are treated at the same hierarchical rank as all other vertebrates put together, and birds are included in a group with crocodilians, well below their customary ranking in the taxonomic hierarchy. Such classifications, which group highly dissimilar taxa and separate similar ones, seriously restrict generalizations that may be made about the members of formal taxa and greatly interfere with nonspecialists' recognition of taxonomic groups. Cladistic classifications are inherently unstable; the discovery of a single character can establish that a group formerly thought to be holophyletic is paraphyletic, making it invalid in spite of a high degree of homogeneity.

I fail to see that a classification employing both cladistic and anagenetic information is confusing. Far from being a compromise between cladistic and phenetic systems, it incorporates the best of both. Because the methods are based upon theoretical rather than empirical considerations, it can do a far better job of explaining similarity than a pure phenetic approach can do.

Ghiselin & Jaffe (25) have shown that Darwin in his classification of barnacles knowingly accepted paraphyletic groups, and so the claim that Darwin was the first cladist is incorrect. Such arguments are trivial at best. Presumably, present workers know more about the subjects of phylogeny and classification than Darwin did. His major contributions to biology are widely recognized while his mistakes (such as his ideas about inheritance) are unimportant.

To say that the higher category names of noncladists are worthless because they do not give you the phylogeny or characters or anything else is really an attack on human language. The word *chair* meant nothing, either, until there was human consensus about what it would mean. The valid criticism that no good methodology exists for evolutionary systematics will become invalid as soon as such a method is published. I would suggest that it is inevitable that such methods will appear within the next few years.

Literature Cited

1. Ashlock, P. D. 1971. Monophyly and associated terms. *Syst. Zool.* 16:63–69
2. Ashlock, P. D. 1972. Monophyly again. *Syst. Zool.* 21:430–38
3. Ashlock, P. D., Scudder, G. G. E. 1966. A revision of the genus *Neocrompus* China. *Pac. Insects* 8:686–94
4. Axelrod, D. I. 1972. Ocean-floor spreading in relation to ecosystematic problems. *Occas. Pap. Univ. Arkansas Mus.* 4:15–68
5. Axelrod, D. I., Raven, P. H. 1972. Evolutionary biogeography viewed from plate tectonic theory. *Challenging Biological Problems,* ed. J. H. Behnke, 218–36. New York: Oxford Univ. Press. 502 pp.
6. Blackwelder, R. E. 1967. *Taxonomy.* New York: Wiley. 698 pp.
7. Blackwelder, R. E. 1969. The nature of taxonomic data. *Pan-Pac. Entomol.* 45:293–303
8. Brundin, L. 1963. Limnic Diptera in their bearings on the problem of transantarctic faunal connections. *Pacific Basin Biogeography: A Symposium,* ed. J. L. Gressitt, 425–34. Honolulu: Bishop Museum Press. 563 pp.
9. Brundin, L. 1965. On the real nature of transantarctic relationships. *Evolution* 19:496–505
10. Brundin, L. 1966. Transantarctic relationships and their significance, as evidenced by the chironomid midges, with a monograph of the subfamily Podonominae, Aphrotaeninae and the austral Heptagiae. *Kgl. Sv. Vetenskapsakad. Handl.* (4)11:1–472
11. Brundin, L. 1967. Insects and the problem of austral disjunctive distribution. *Ann. Rev. Entomol.* 12:149–68
12. Brundin, L. 1972. Phylogenetics and biogeography. *Syst. Zool.* 21:69–79
13. Brundin, L. 1972. Evolution, causal biology, and classification. *Zool. Scr.* 1:107–20
14. Camin, J. H., Sokal, R. R. 1965. A method for deducing branching sequences in phylogeny. *Evolution* 19:311–26
15. Cracraft, J. 1972. The relationships of the higher taxa of birds: Problems in phylogenetic reasoning. *Condor* 74:379–92
16. Cracraft, J. 1973. Continental drift, paleoclimatology, and the evolution of birds. *J. Zool. London* 169:455–545
17. Cracraft, J. 1974. Phylogenetic models and classification. *Syst. Zool.* 23:71–90
18. Crowson, R. ·A. *Classification and Biology.* New York: Atherton. 350 pp.
19. Darlington, P. J. Jr. 1970. A practical criticism of Hennig-Brundin "phylogenetic systematics" and Antarctic biogeography. *Syst. Zool.* 19:1–18
20. Dietz, R. S., Holden, J. C. 1970. The breakup of Pangaea. *Sci. Am.* 223:30–41
21. Edmunds, G. F. Jr. 1972. Biogeography and evolution of Ephemeroptera. *Ann. Rev. Entomol.* 17:21–42
22. Estabrook, G. F. 1972. Cladistic methodology: A discussion of the theoretical basis for the induction of evolutionary history. *Ann. Rev. Ecol. Syst.* 3:427–56
23. Farris, J. S. 1969. A successive approximations approach to character weighting. *Syst. Zool.* 18:374–85
24. Fooden, J. 1972. Breakup of Pangaea and isolation of relict mammals in Australia, South America, and Madagascar. *Science* 175:894–98
25. Ghiselin, M. T., Jaffe, L. 1973. Phylogenetic classification in Darwin's monograph on the sub-class Cirripedia. *Syst. Zool.* 22:132–40
26. Griffiths, G. C. D. 1972. *The phylogenetic classification of Diptera Cyclorrhapha with special reference to the structure of the male post abdomen.* The Hague: Dr. W. Junk N.V. 340 pp.
27. Hennig, W. 1960. Die Dipteren-Fauna von Neuseeland als systematisches und tiergeographisches Problem. *Beit. Entomol.* 10:221–329
28. Hennig, W. 1965. Phylogenetic systematics. *Ann. Rev. Entomol.* 10:97–116
29. Hennig, W. 1966. The Diptera fauna of New Zealand as a problem in systematics and zoogeography. *Pac. Insects Monogr.* 9:1–81. Transl. P. Wygodzinsky
30. Hennig, W. 1966. *Phylogenetic Systematics.* Urbana: Univ. Illinois Press. 263 pp.
31. Hull, D. L. 1968. The operational imperative, sense and nonsense in operationalism. *Syst. Zool.* 17:438–57
32. Hull, D. L. 1970. Contemporary systematic philosophies. *Ann. Rev. Ecol. Syst.* 1:19–54
33. Leston, D. 1957. Spread potential and the colonization of islands. *Syst. Zool.* 6:41–46
34. MacArthur, R. H., Wilson, E. O. 1967. *The Theory of Island Biogeography.* Princeton: Princeton Univ. Press. 203 pp.

35. Mayr, E. 1965. Numerical phenetics and taxonomic theory. *Syst. Zool.* 14:73–97
36. Mayr, E. 1965. Classification and phylogeny. *Am. Zool.* 5:165–74
37. Mayr, E. 1969. Principles of Systematic Zoology. New York: McGraw. 428 pp.
38. Nelson, G. J. 1969. The problem of historical biogeography. *Syst. Zool.* 18:243–46
39. Nelson, G. J. 1969. Gill arches and the phylogeny of fishes, with notes on the classification of vertebrates. *Bull. Am. Mus. Natur. Hist.* 141:475–552
40. Nelson, G. J. 1971. "Cladism" as a philosophy of classification. *Syst. Zool.* 20:373–76
41. Nelson, G. J. 1972. Comments on Hennig's phylogenetic systematics and its influence on ichthyology. *Syst. Zool.* 21:364–74
42. Raven, P. H., Axelrod, D. I. 1972. Plate techtonics and Australasian paleobiogeography. *Science* 176:1379–86.
43. Rogers, D. J., Fleming, H. S., Estabrook, G. 1967. Use of computers in studies of taxonomy and evolution. *Evolutionary Biology,* ed. T. Dobzhansky, M. K. Hecht, W. C. Steere, 1:169–96. New York: Appleton-Century-Crofts. 444 pp.
44. Schlee, D. 1969. Hennig's principle of phylogenetic systematics, an "intuitive, statistico-phenetic taxonomy"? *Syst. Zool.* 18:127–34
45. Schlee, D. 1971. *Die Rekonstruktion der Phylogenese mit Hennig's Prinzip.* Frankfurt: Kramer. 62 pp.
46. Simpson, G. G. 1961. *Principles of Animal Taxonomy.* New York: Columbia Univ. Press. 247 pp.
47. Tuomikoski, R. 1967. Notes on some principles of phylogenetic systematics. *Ann. Entomol. Fenn.* 33:137–47
48. Wilson, E. O. 1965. A consistency test of phylogenies based on contemporaneous species. *Syst. Zool.* 14:214–20

22

THE LIMITS OF CLADISM

David L. Hull

"Our classifications will come to be, as far as they can be so made, genealogies; and will then truly give what may be called the plan of creation. The rules for classifying will no doubt become simpler when we have a definite object in view" (Darwin, 1859:486).

Charles Naudin's "simile of tree and classification is like mine (and others), but he cannot, I think, have reflected much on the subject, otherwise he would see that genealogy by itself does not give classification" (Darwin, 1899, 2:42).

One possible goal for biological classification is to "represent" phylogeny, to construct classifications so that certain features of phylogeny can be read off unequivocally. However, Darwin was quite correct when he noted that genealogy by itself does not give classification. Different taxonomists may select different aspects of phylogeny to represent. One might choose order of branching, another degree of divergence, another amount of diversity, and so on. Whether or not the rules for classifying become simpler, as Darwin hoped they would, depends on the features of phylogeny which the taxonomist chooses to represent and the particular set of principles which he formulates to represent them. A classification which attempts to represent only one aspect of phylogeny is likely to be simpler than one which attempts to represent two simultaneously. Given any particular goal, the simplicity of the rules capable of accomplishing this goal depends both on the inherent limitations of the mode of representation and the ingenuity of the taxonomist. For example, the traditional Linnaean hierarchy might lend itself to representing certain features of phylogeny more readily than others, but nothing precludes a taxonomist from introducing additional representational devices to remove these limitations. The end result is, of course, more complex rules for classifying. Systematists are thus forced to strike a balance between how much they wish to represent and how complicated they are willing to make their principles of classification and resulting classifications.

In recent years a school of taxonomy has arisen whose members have seriously taken up the challenge of attempting to represent explicitly and unambiguously *something* about phylogeny. The school is cladism, the aspect of phylogeny which the cladists have chosen to represent is order of branching. The cladists have given two reasons for choosing this particular feature of evolutionary development to represent. First, the Linnaean hierarchy, as a list of indented names, lends itself more naturally to expressing discontinuous than continuous phenomena. Whether or not evolution is gradual or saltative, phylogeny is largely a matter of splitting and divergence. A hierachy of discrete taxa names is well-calculated to represent successive splitting. It is not as well-calculated to represent varying degrees of divergence. Second, the cladists argue that order of branching can be ascertained with sufficient certainty to warrant the inclusion of such information in a classification while other aspects of phylogeny cannot be. The method which cladists have devised to discover order of branching is cladistic analysis.

Cladists argue that, by and large, all a systematist has to go on is the traits of the specimens before him, whether these specimens represent extant or extinct forms. By studying his specimens, he can discover nested sets of characters. Unique derived characters distinguish a monophyletic group from its nearest relatives; shared derived characters combine these monophyletic groups into more inclusive monophyletic groups. In this way the systematist can ascertain nested sets of sister-groups. This method, so cladists argue, cannot be used to discern a variety of other relations—speciation without the appearance of at least one unique derived character, reticulate evolution, multiple speciations, and the ancestor-descendant relation—nor can any other method.

Several questions arise at this juncture. If the Linnaean hierarchy cannot represent certain features of evolutionary development very well, is not that a limitation of the Linnaean hierarchy? Instead of refusing to represent what a particular system of representation has difficulty in representing, a better alternative might be to abandon or improve that system of representation. This is one avenue which the cladists themselves have taken (Griffiths, 1976; Hennig, 1966; Løvtrup, 1973; Nelson, 1972a, 1973c; Patterson and Ro-

sen, 1977; Wiley, 1979). Do the evolutionary phenomena which cladistic analysis cannot discern actually exist? As I understand evolution and evolutionary theory, two of these phenomena clearly occur (ancestors giving rise to descendants and reticulate evolution), one is quite likely (multiple speciation), and the fourth is doubtful (speciation without deviation, although the unique derived character may be all but undetectable). Is not the inability of cladistic analysis to discern the preceding phenomena, assuming they take place, a limitation of the methods of cladistic analysis? Instead of declining to deal with such phenomena because the methods of cladistic analysis cannot discern them, perhaps a better strategy might be to attempt to improve upon the methods of cladistic analysis or to supplement them with other methods. The crucial question with respect to cladism, as I see it, is what is cladistic analysis? What are its limits? Is cladistic analysis *one* way of ascertaining phylogeny, the *only* way of ascertaining phylogeny, the only way of obtaining knowledge of *any* phenomena?

One fact about cladism which complicates attempts to answer the preceding questions is that cladism, like all scientific movements, is neither immutable nor monolithic. At any one time, cladists can be found disagreeing with each other about particular principles and conclusions. In addition, if one traces the development of cladism through time, from its inception in the works of Hennig (1950), through its introduction to English-speaking systematists (Hennig, 1965, 1966; Brundin, 1966; Nelson, 1971a, 1971b, 1972a, 1972b, 1973a, 1973b, 1973c, 1974) to the latest publications of present-day cladists, significant changes can be discerned. The most significant change which has taken place in cladism is a transition from *species* being primary to *characters* being primary. For Hennig (1966, 1975) and Brundin (1972a), the basic units of cladistic analysis are species, characterized by at least one unique derived character. Now, for a growing number of cladists, any monophyletic group which can be characterized by appropriate traits can function in cladistic analysis, regardless of whether that group is more inclusive or less inclusive than traditionally-defined species. Cladistic classifications do not represent the order of branching of sister-groups, but the order of emergence of unique derived characters, whether or not the development of these characters happens to coincide with speciation events. It is not the emergence of new species which is primary but the emergence of new traits (Tattersall and Eldredge, 1977:207; Rosen, 1978:176). In general, cladists seem to be moving toward the position that the particulars of evolutionary development are not relevant to cladism. It does not matter whether speciation is sympatric or allopatric, saltative or gradual, Darwinian or Lamarckian, just so long as it occurs and is predominantly divergent (Cracraft, 1974; Bonde, 1975; Rosen, 1978; and forthcoming works by Nelson and Platnick, Eldredge and Cracraft, and others).

The writings of most cladists have been concerned with developing methods of ascertaining the cladistic relations between biological taxa, but Platnick and Cameron (1977) have shown that these same techniques can be used to discern the cladistic relations exhibited by languages and texts (see also Kruskal, Dyen, and Black, 1971; Haigh, 1971; and Nita, 1971). In general cladistic analysis can be used to discover the cladistic relations between any entities which change by means of modification through descent (Platnick, 1979). As general as this notion of cladistic analysis is, it still retains a necessary temporal dimension. Transformation series must be established for characters, not just an abstract atemporal transformation series like the cardinal numbers or the periodic table, but a series of actual transformations in time. For example, the vast majority of matter in the universe happens to be hydrogen. In isolated pockets of the universe, more complex molecules have developed, but

they have not developed by progressing up the periodic table, from hydrogen to helium, lithium, beryllium, etc.

From the beginning, Gareth Nelson (1973c) seems to have been developing two notions of cladistic analysis simultaneously, one limited to historically developed patterns (cladism with a small 'c'), the other a more general notion applicable to all patterns (Cladism with a big 'C'). His method of component analysis is a general calculus for discerning and representing patterns of all sorts. For example, in a recent discussion, Nelson (1979:28) states that a "cladogram is an atemporal concept . . . a synapomorphy scheme." It remains to be seen whether all of the terms of cladistic analysis from "sister group " to "synapomorphy" can be defined so that none of them necessarily presupposes a temporal dimension. Although I think that the principles of cladistic analysis *can* be extended to any system which genuinely evolves (i.e., changes through time by means of modification through descent), I will limit myself in this paper to phylogeny and biological evolution. I also will not address myself to Nelson's more general notion. Attempting to discuss the principles of cladistic analysis as they apply to phylogeny is a sufficiently difficult task without attempting to present interpretations of these principles which are also consistent with Nelson's more general program.

METHODOLOGICAL PRINCIPLES AND OBJECTIONS

In spite of certain observations which cladists have made periodically about the evolutionary process, the principles of cladistic analysis do not concern evolution but the goals of biological classification and the means by which they can be realized. These principles are primarily methodological. As I see it, the goal of cladism is to represent cladistic relations in both cladograms and classifications as explicitly and unambiguously as possible. Thus, cladistic methodology has two parts: rules for discerning cladistic relations and rules for representing them in cladograms and classifications. For example, Nelson raises two objections to the ancestor-descendant relation: first, "ancestral species cannot be identified as such in the fossil record" (Nelson, 1973b:311), and second, "they are also inexpressible in classifications" (Nelson, 1972a:227).

One source of the confusion which has accompanied the controversy over cladism is the interpretation of their methodological principles as empirical beliefs about evolution. The claim that dichotomous speciation never occurs is an empirical claim to be tested by empirical means. The claim that we can never distinguish between dichotomous speciation events and more complex sorts of speciation is a methodological claim about the limits of our methods of scientific investigation. Finally, the claim that multiple speciation, if it occurs and if it can be discerned, cannot be represented unambiguously in cladograms and/or classifications is a comment about the limitations of these modes of representation. Once the principles of cladism are recognized for what they are, *methodological principles*, the logic of the cladistic position on a variety of issues becomes much clearer.

Scientists are interested in truth, not Absolute Truth, but truth nevertheless. Although philosophers have yet to produce a totally unproblematic analysis of science as the pursuit of truth (see, for example, Laudan, 1977), I think that scientists are correct in the focus of their attention. Scientists also present arguments to buttress their empirical claims and try to make their arguments as good as possible. Unfortunately, scientists are rarely able to present their arguments in the unproblematic forms thus far analyzed successfully by mathematicians and logicians. The reasoning which goes into the formulation and testing of scientific theories is a good deal more complex and informal than any explicit rational reconstruction has yet been able to capture. Scientists are interested in both

truth and cogency of argument, but they are a good deal more interested in truth than in cogency of argument. If a line of reasoning which led to a particular conclusion turns out to be somewhat less than perfect, it really does not matter all that much, just so long as the conclusion depicts reality with greater fidelity than previous attempts.

Scientists have limited patience when it comes to discussing arguments. That patience is even more limited when the arguments are methodological. As strange as it may sound coming from a philosopher, I am highly sceptical of methodological principles and prescriptions, especially when they are presented in the midst of scientific disputes. They tend to be suspiciously self-serving, designed to put one's opponents at a disadvantage while shoring up one's own position. Invariably the advocates of a particular methodology can know what they need to know while their opponents can never hope to know what they need to know. For example, the pheneticists claim that we can analyze traits into unit characters, but we can never hope to establish genuine, evolutionary transformation series among characters. None too surprisingly, the pheneticists need to know the former but not the latter. Only the cladists and evolutionists need to know the unknowable. Similarly, the cladists claim that we can establish transformation series with sufficient certainty to warrant the role which they play in cladistic systematics but that ancestor-descendant relations are unknowable. As luck would have it, cladists need to know the former but not the latter. Only the evolutionists need to know the unknowable.

As cynical as the preceding remarks may sound, I think they have some validity. It is no accident that all four of the phenomena which the cladists claim cannot be known, cannot be discerned by the traditional means of cladistic analysis and cause problems for the explicit and unambiguous representation of sister-group relations in both cladograms and classifications. It is also no accident that one of the phenomena which cladists claim cannot be known (the ancestor-descendant relation) plays a central role in the research program of their chief rivals—the evolutionists. I do not mean to imply that the preceding remarks apply uniquely to the cladists. They apply to scientists in general. If I were investigating the pheneticists or the evolutionists, comparable observations would apply as readily to them. These are the sorts of games which scientists (not to mention philosophers) play with each other.

In general, I think it is very bad strategy for proponents of a particular scientific research program to stake their future on epistemological considerations, especially on our inability to know something. Phenomena which scientists in one age claim can never be known often become common knowledge at a later date. The history of science is littered with the bodies of scientists who staked the success of their movements on what we can never know. I agree with Einstein (1949:684), who, in response to philosophical criticisms of his work, stated:

> Science without epistemology is—insofar as it is thinkable at all—primitive and muddled. However, no sooner has the epistemologist, who is seeking a clear system, fought his way through to such a system, than he is inclined to interpret the thought-content of science in the sense of his system and to reject whatever does not fit into his system. The scientist, however, cannot afford to carry his striving for epistemological systematic that far.

I hardly want to argue against methodological rigor in science, but I also do not want to see scientific progress sacrificed to it. Invariably methodological rigor is a retrospective exercise, carried on long after all the Nobel prizes have been won. From the point of view of current methodologies, scientists will, as Einstein (1949:684) noted, appear to the "systematic epistemologist as a type of unscrupulous opportunist." Instead of cladists insisting that certain aspects of phylogeny can never be known and could not be represented cladistically if

they could be known, a wiser strategy would be to attempt to devise methods of analysis capable of discerning these features of evolutionary development and representational techniques sufficient to represent them. In this paper, I have attempted to produce an internal criticism of cladism; that is, I have accepted the goals of cladism and have set myself the task of deciding which methodological principles are central to the undertaking, which peripheral, and which could be modified or abandoned without loss and possibly with some gain.

CLADOGRAMS, PHYLOGENETIC TREES, AND EVOLUTIONARY SCENARIOS

A curious feature of scientific development is the frequency with which a new movement is named by its opponents. Social Darwinists no more wanted to be called Social Darwinists than cladists have wanted to be called cladists. Twenty years ago, Julian Huxley (1958) coined the terms "clade" and "grade" to distinguish between groups of organisms with a common genetic origin and groups distinguished by different levels of organization. Later Mayr (1965:81) and Camin and Sokal (1965:312) introduced the term "cladogram" to refer to a diagram "depicting the branching of the phylogenetic tree without respect to rates of divergence" (Mayr, 1978:85). Finally, Mayr (1969:70) invented the term "cladism" as a substitute for "phylogenetic systematics," the name preferred by Hennig and his followers. Gradually the cladists themselves have come to use the term to refer to themselves, albeit grudgingly.

Whether the Hennigian school of systematics is called "phylogenetic" or "cladistic" is not very important, but the precise nature of cladograms as they function in cladistic analysis is. In fact, uncertainty over what it is that cladograms are supposed to depict and how they are supposed to depict it has been the chief source of confusion in the controversy over cladism. Like all terms, the meaning of "cladogram" has changed through the years. It no longer means to cladists what Mayr, Camin and Sokal proposed. The meanings of "cladism" and "cladistic analysis" have changed accordingly. In an attempt to reduce the confusion over the meaning of "cladogram," cladists have introduced the distinction between cladograms, phylogenetic trees, and evolutionary scenarios.[1] Once again, the particular terms used to mark these distinctions are not important; the distinctions themselves are. It is also true that the cladists have devised these distinctions for their own purposes. Others might want to draw other distinctions or to draw these distinctions differently. In any case, all three appellations refer to representations, not to the phenomena being represented. "Cladogram" and "tree" refer to two sorts of diagrams, while "scenario" refers to a historical narrative couched in ordinary biological language.

Phylogenetic trees are designed to depict phylogenetic development, indicating which taxa are extinct, which extant, which gave rise to which, degrees of divergence, and so on. As diagrams they do not include discussions of the methods used to construct them, the natural processes which produced the phenomena they depict, and a variety of other considerations of equal importance. A tree is a diagram, not an entire taxonomic monograph. All sorts of conventions have been devised to represent phylogeny in a diagrammatic form; for example, lines usually represent lineages, forks represent speciation events, the slant and length of a line reflect the rapidity with which divergence took place, and the termination of a line indicates extinction. Other techniques of representation have also been used on occasion; for example, dots indicating questionable connections, lines of varying thicknesses reflecting relative

numbers of organisms, and so on. However, as a diagram in two-dimensional space, a tree can include only so much information before a point of diminishing returns sets in. Eventually attempts to include additional sorts of information result in the loss of information. If trees are supposed to be systems of information storage and retrieval, the information must be retrievable.

Scenarios, as cladists use the term, are not diagrams. They are historical narratives which attempt to describe not only which groups gave rise to which (the sort of information contained in trees) but also the ecological changes and evolutionary forces which actually produced the adaptations which characterize the organisms discussed. Romer's (1955:57) story of the role which the drying up of ponds and streams played in the transition from the crossopterygians to early amphibians is by now a classic example of an evolutionary scenario. Because scenarios are presented in ordinary language—supplemented with a host of technical biological terms—the phenomena which scenarios can describe are limited only by the limitations of language.

Hennig's early cladograms (Hennig, 1950:103; 1966:59, 71) give every appearance of being highly stylized trees. The circles arrayed along the top of the diagram apparently represent extant species, while those at the nodes represent extinct, common ancestors (see Fig. 1a). At one stage in the development of the cladogram, Nelson (1972b, 1973b) argued that the circles at the nodes did not represent real common ancestors but "hypothetical ancestors." The term is problematic. All taxa that ever existed are equally real. Only our ability to discern them varies. We can usually discern extant species with greater certainty than extinct forms, but the postulation of any taxon, whether extant or extinct, involves highly complex inferences and requires evidence which is frequently quite difficult to obtain. (Recall the objections which pheneticists raised to the biologi-

---

[1] The distinction between cladograms, trees, and scenarios discussed in Tattersall and Eldredge (1977) was taken from a manuscript by Gareth Nelson. See Mayr (1978) for early definitions of such terms as "cladogram," "phenogram," and "phylogram."

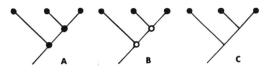

FIG. 1.—The evolution of the cladogram. (a) A cladogram in which the darkened circles at the termini represent extant species, while those at the nodes represent common ancestors. (b) A cladogram in which the darkened circles at the termini represent species, both extant and extinct, while the circles at the nodes represent "hypothetical common ancestors," i.e., morphotypes. (c) A cladogram in which the darkened circles at the termini represent species, both extant and extinct, while the nodes represent speciation events and/or the emergence of unique derived characters.

cal species concept even when it is applied to extant forms.) If extinct forms are "hypothetical" because of their inferential basis, then so are extant forms. The difference between extant and extinct species is not between observation and inference, but between inferences of varying degrees of certainty.

Another interpretation of "hypothetical ancestor" is that the circles which appear at the nodes of cladograms are not supposed to be taxa at all but "morphotypes," rational constructs characterized by the defining traits of the taxa listed at the termini of the cladogram and no others. As Platnick (1977c:356) observes, the "nodes of cladograms represent only inferred species (or, more accurately, only minimum sets of synapomorphic characters)." In this sense, hypothetical ancestors are "hypothetical," but they are not "ancestors." They are not even taxa (see Fig. 1b). In present-day cladograms, all taxa, whether extinct or extant, appear along the top of the cladogram. Cladograms would be much less misleading if they included *nothing* at the nodes (see Fig. 1c).

If the nodes in a cladogram do not represent real taxa (ancestral or otherwise), what do the lines and forks in cladograms represent? Two answers have been given to this question, which may or may not be reducible to the same answer. A fork in a cladogram can represent either a spe-

ciation event (the splitting of one species into two), or the emergence of an evolutionary novelty (a unique derived character), or both. If the emergence of a unique derived character occurs always and only at speciation, then the two answers always coincide. If not, not. Similar observations hold for the term "cladistic relation." It can refer to order of splitting of taxa, or to order of appearance of evolutionary novelties, or both. The decision hinges on how "operational" one wishes to make cladistic analysis. Since speciation events are inferred by means of the appearance of evolutionary novelties, one inferential step can be eliminated by talking only about the emergence of evolutionary novelties and not speciation events.

CERTAINTY AND LEGITIMATE DOUBT

One reason for the cladists' introducing the distinction between cladograms, trees, and scenarios was to clarify the technical sense in which they were using the term "cladogram." Too many people were misinterpreting cladograms as bizarre, highly stylized trees. A second reason was to establish the logical and epistemological priority of cladograms over trees and of trees over scenarios. As cladists define these terms, an inclusion relation exists between them. Scenarios include all the information contained in trees, and more besides. Trees include all the information contained cladograms, and more besides. Thus, any knowledge required for a cladogram is required for a tree, and any knowledge required for a tree is required for a scenario, but not vice versa. Thus, cladograms are more certain than trees, and trees more certain than scenarios. As harmless as the distinction between cladograms, trees, and scenarios may seem on the surface, it results in cladistic analysis being in some sense basic to all evolutionary studies.

The relations set out above *do* obtain between cladograms, trees, and scenarios—as the cladists define these terms. That is one reason why non-cladists might prefer other definitions. But even if one were to accept the cladists' definitions, certain conclusions which cladists have claimed follow from the preceding line of reasoning do not. For example, in claiming that "trees should always be based on cladograms and that scenarios should follow from trees," Tattersall and Eldredge (1977:205) are confusing logical and epistemological order with temporal order. The conclusion to an argument follows logically from its premises. That does not mean that people must think of the premises first and only then think of the conclusion. Sense data are epistemologically prior to all knowledge, but that does not mean that scientists should begin every scientific study with an extensive investigation of their own sense data. If scientists had actually proceeded in this fashion, we would still be awaiting Ptolemaic astronomy, not to mention relativity theory. The inferences which take place in the actual course of scientific investigations are very intricate and frequently "feed back upon each other," as Tattersall and Eldredge (1977:205) note. In retrospect, there is considerable point to unraveling these inferences and setting them out in some logically coherent fashion, but the insistence that scientists must always begin at the beginning and proceed step by step according to some sort of logical or epistemological order would stop science dead in its tracks.[2]

Cladists also make much of the differences in certainty between cladograms, trees, and scenarios. In actual practice, scientists wander from one level of analysis to another in the course of their investigations. As a result knowledge ac-

quisition is a process of both "reciprocal illumination" (Hennig, 1950; Ross, 1974) and "reciprocal blundering." Getting one element right helps improve our understanding of the other elements, but errors feed through the system just as readily. Although cladists might well admit that knowledge acquisition in general is a process of reciprocal illumination, they also tend to be so sceptical of trees and scenarios that they see all the illumination going in a single direction—from cladograms to trees and scenarios. At times cladists seem to argue that not only should evolutionary studies *begin* with cladograms but also they should *end* there as well. Cladistic relations are knowable; everything else about phylogeny is unknowable.

For example, Platnick (1977d:439) argues that the *only* way that phylogenetic trees can be tested is by the same means used to test cladograms, "by evaluation in the light of newly discovered characters." Any three taxa can be ordered into 22 different trees. Some of these trees can be rejected by discovering appropriate characters, i.e., the appropriate autapomorphies and synapomorphies. Other trees can be rejected only by claiming that no autapomorphies exist for the relevant taxa. Since the most that a systematist can legitimately claim is that he has yet to detect such autapomorphies, he is not justified in rejecting these trees. Platnick (1977d:440) concludes that "phylogenetic trees are not testable by character distributions and thus that scientific phylogeny reconstruction is *not* possible at the level of phylogenetic trees and must be restricted to the level of cladograms." But the same sort of argument can be presented against the cladists. They do not need to know ancestor-descendant relations, but they do need to establish transformation series and to determine the polarity of these series. They do need to individuate characters and decide which are autapomorphic, synapomorphic, etc. For instance, synapomorphies can be used to test cladograms *if* they are synapomorphies, but no one has yet to sug-

---

[2] The pheneticists presented similar arguments for the epistemological priority of phenetic classifications. According to the pheneticists, the only place at which systematists can legitimately begin is phenetic characters and the construction of purely phenetic classifications. Once such a phenetic classification is erected, then a systematist could put various "interpretations" on his classifications and possibly produce one or more "special purpose" classifications; see e.g., Sneath and Sokal (1973).

gest any infallible means for deciding whether or not a trait is actually a synapomorphy. The same can be said for such relations as the polarity of transformation series. In fact, Tattersall and Eldredge (1977:206) state, "In practice it is hard, even impossible, to marshal a strong, logical argument for a given polarity for many characters in a given group." The sort of argument from negative evidence which the cladists use against others at the level of taxa can be used against them at the level of traits. Of course, their opponents also must individuate and categorize traits. Thus, the cladists' position is refuted at only one level of analysis, while the position of their opponents is refuted at two. Put more directly, such arguments refute no positions whatsoever.

The cladists have weakened the force of their arguments by presenting them in a needlessly dichotomous fashion. The only thing that they need to know to produce cladograms are cladistic relations, and they are knowable. Others who wish to produce trees need to know much more; e.g., ancestor-descendant relations, the existence of multiple speciation and reticulate evolution, etc. These phenomena are totally unknowable. The cladists need not present such an extreme (and suspiciously self-serving) position. Estimations of cladistic relations are inferences and as such carry with them the possibility of error. However, it is certainly true that everything which the cladists need to know the evolutionists also need to know, while the evolutionists need to know more besides. These additional phenomena also carry with them a certain degree of uncertainty. Hence, evolutionary classifications are bound to be more uncertain than cladistic classifications. The question remains whether all evolutionary phenomena other than cladistic relations are so uncertain that no attempt should be made to include information about them in a classification. In the succeeding sections of this paper, I will investigate various phenomena which cladists argue cannot

be known (or at least cannot be known with sufficient certainty) to see why they are so difficult to discern. Is it that they cannot be known, or that they cannot be known by means of cladistic analysis? Is there no way to acquire knowledge of the world other than by cladistic analysis? The point of this section is, however, that the difference in our ability to ascertain cladistic relations and other evolutionary phenomena is one of degree, not kind. Differences in degree of certainty are neither very neat nor aesthetically pleasing, but they are the most that cladists can justifiably claim. What they lose in elegance and simplicity, they gain in plausibility.

Certain cladists, in more recent publications, seem to be moving in precisely this direction. For example, Eldredge (1979:169) states that, contrary to his earlier opinions:

> I no longer oppose the construction of phylogenetic trees outright, or for that matter, scenarios (which are, after all, the most fun), but merely point out that, in moving through the more complex levels, we inevitably become further removed from the original data base in adding assumptions and *ad hoc* (and largely untestable) hypotheses. As long as we understand precisely what we are doing at each step in the analysis, which includes having an adequate grasp of the probability that we are wrong and of what the assumptions are what we have added along the way, there no longer seems to me any reason for anyone to tell anyone else what *not* to do.

Although those workers engaged in the production of trees and scenarios might differ with Eldredge on the situation being as extreme as he makes it out to be, they too are aware of the difficulties. For example, Lucchese (1978:716) concludes a paper on the evolution of sex chromosomes with the following remarks:

> . . . for an individual to profess an insight into the biological changes that have occurred through time remains an act of faith of considerable magnitude. [Yet, within] the limitations just set forth, the purpose of this article has been to spin a relatively reasonable evolutionary tale in the hope of expanding the current perception of a highly significant example of genetic regulation in eukaryotes.

THE PRINCIPLE OF DICHOTOMY

Few principles attributed to cladistic taxonomists have caused more consternation and confusion than the claim that all cladograms and classifications must be strictly dichotomous. When cladism was first introduced to English-speaking systematists, cladists and anti-cladists alike agreed that the principle of dichotomy was "essential to the philosophy of Hennig and Brundin " (Nelson, 1971a:373; see also Darlington, 1970:3; Mayr, 1974:100; Bonde, 1975:302; Platnick, 1977d:438). Although Cracraft (1974:79) agrees that "phylogenetic classification *sensu* Hennig and Brundin would appear to require the assumption of dichotomous branching," cladistic classification in a more general sense "does not necessitate dichotomous branching, and the exact pattern of branching is determined by the manner in which shared derived character-states cluster taxonomic units." The question remains why the principle of dichotomy has seemed so central to most cladists and continues to seem so to some.

Although no cladist has ever maintained that the principle of dichotomy is an empirical claim about the speciation process, few cladists have been able to resist the temptation to justify it by reference to empirical considerations. For example, Hennig (1966:210, 211) begins by stating that dichotomy is "primarily no more than a methodological principle" and then goes on to add, "A priori it is very improbable that a stem species actually disintegrates into several daughter species at once." Cracraft (1974:79) interrupts the discussion quoted above with the remark that the principle of dichotomy "can be justified on theoretical grounds not associated with classification." Bonde (1975:302) agrees. Although dichotomy is a methodological principle, in nature "speciations are probably nearly always dichotomous."

Whether multiple speciation seems possible or impossible depends on the unit of time one selects to define "simultaneous." As Hennig (1966:211) notes, the issue of dichotomous speciation can be trivialized by defining "simultaneous" too narrowly. If it is defined in terms of split seconds, then multiple speciation is impossible. Conversely, if "simultaneous" is defined too broadly, everything can be made to occur "at the same time." The question of dichotomous speciation can be made significant only by the specification of a unit of time which makes sense in the context of the evolutionary process. In the absence of such a unit, the notion of simultaneity can be expanded or contracted at will and the issue decided by fiat (see Cracraft, 1974:74).

Whether multiple speciation as an empirical phenomenon seems plausible or implausible depends on the model (or models) of speciation which one holds. If speciation is viewed as the gradual splitting and divergence of large segments of a species, as Hennig (1966:210) maintains, then multiple speciation looks less likely than if it is viewed as a process in which small, peripheral populations become isolated and develop into separate species, as Wiley (1978:22) supposes. But, as the cladists have reminded us often enough, dichotomy is a methodological principle. If speciation were always dichotomous, this fact about evolution would certainly lend support to dichotomy as a methodological principle, but good reasons might exist for producing exclusively dichotomous cladograms and classifications even if speciation is not always dichotomous.

Cladists have presented two sorts of justification for their preference for the principle of dichotomy: our inability to *distinguish* dichotomous speciation from more complex sorts of speciation and our inability to *represent* unequivocally more complex sorts of speciation in cladograms and classifications. If one limits oneself just to traditionally-defined traits (e.g., presence of various sorts of appendages, dentitions, etc.) and if one assumes that these traits have been appropriately individuated and identified, then the argument which the cladists put forth is reasonably straightforward. If a cladist starts with any two species chosen at random and compares a third species to them, two of these species will be cladistically more closely related to each other than either is to the remaining species. If the cladist continues to add species to his study one at a time, he will produce a consistently dichotomous cladogram until one of two things occurs: either he happens upon a genuine instance of multiple speciation or else he is unable to resolve this complex situation into the appropriate sequence of successive dichotomies. Because dichotomy is the simplest hypothesis, it should always be preferred, evidence permitting. But as the cladists see it, given the sort of data available to the systematist and the resolving power of his methods of analysis, he will always end up where he begins—with doubt or with dichotomy. Never is a systematist justified in concluding that a speciation event is genuinely trichotomous because he is never justified in dismissing the possibility that an apparent trichotomy is really a pair of unresolved dichotomies.

If the preceding is a fair characterization of the cladists' argument, it has two weaknesses. First, the cladists are selective in acknowledging possible sources of error. After all, a systematist cannot guarantee that a trait which he considers to be unique and derived actually is. There is always the possibility that he is dividing a single group into two or lumping two groups into one. If a trichotomy represents either a genuine trichotomy or two unresolved dichotomies, then a dichotomy could just as well represent either a genuine dichotomy, a lumped trichotomy, or a single lineage divided mistakenly into two. When a systematist claims that a particular speciation event is trichotomous, he may be mistaken, but he may also be mistaken in claiming that it is dichotomous. During the early stages of investigation, doubts about the actual cladistic relations which characterize the

groups under study are legitimate. Many trichotomies are very likely to be resolvable into dichotomies after further study, many of the groups treated as single may have to be split into two or more groups, and vice versa. However, as the groups are studied more exhaustively, the legitimacy of continued doubt decreases. If after exhaustive study, three species continue to appear to be related trichotomously, the claim that they might actually be related by a pair of dichotomies begins to ring rather hollow. Descartes has shown where that sort of mindless scepticism leads. If a systematist can become reasonably certain that he has correctly discerned a genuine dichotomy, I see no reason why he cannot also attain that same level of certainty in the identification of trichotomies—even using just the traditional methods of cladistic analysis. But systematists are not limited just to the study of ordinary taxonomic traits. There is no reason for not using the methods of cladistic analysis on other sorts of evidence, e.g., the data of historical biogeography. As Platnick and Nelson (1978:10) remark:

> Trichotomous cladograms are of no significance as tests (or as initial hypotheses) unless the cladograms for all available test groups are trichotomous, in which case we may suspect that an event disconnected an area into three smaller areas simultaneously. Testing this hypothesis by biotic relationships seems impossible (because we have no way of distinguishing those cladograms reflecting actual trichotomies from those reflecting only our failure to find the relevant synapomorphies that would resolve a dichotomous cladogram), but it should be subject to independent testing by data from historical geology, which can either be in accordance with such a synchronous tripartite disconnection of the total area or not.

If multiple speciation can occur in nature and if in certain circumstances it can be discerned, then cladists would be wise to devise methods of representing multiple speciation in their cladograms and classifications. One theme which the cladists repeatedly voice is that anything represented in a cladogram and classification should be represented explicitly and unambiguously. If cladists were willing to argue that trichotomous speciation never took place, then trichotomies in cladograms would not be ambiguous. They would always represent a pair of unresolved dichotomies. As things stand now, they are ambiguous, representing either a genuine trichotomy or else a pair of unresolved dichotomies. Whether or not cladists think that instances of multiple speciation can be discerned, the adoption of such an ambiguous mode of representation seems counter-productive. A better alternative would be to reserve trichotomies for representing trichotomous speciation events the way that dichotomies are used to represent dichotomous speciation events, epistemological doubts to one side, and to devise another way of indicating incomplete knowledge. For example, if a systematist is reasonably sure that two species are each other's closest relatives, he can indicate this by a dichotomy. If he thinks they might be but is not sure, he can connect them by dots. Similar remarks apply to trichotomies (see Fig. 2). No system of representation can represent everything, but if doubt is important enough to represent in a cladogram, it is important enough to represent unequivocally.

Similar remarks are equally applicable to classifications. For example, Nelson (1973c) distinguishes between the two sorts of representation commonly used in the Linnaean hierarchy—subordination and sequencing—and shows how combinations of these two conventions can represent a variety of evolutionary phenomena including trichotomous speciation. More recently, Wiley (1979) has suggested using *sedis mutabilis* whenever the order of listing taxa in a classification means nothing. In addition to the difficulties which the recognition of multiple speciation raises for cladograms and classifications, it also poses problems for the proper definition of monophyly and related terms (Wiley, 1977; Platnick, 1977c).

FIG. 2.—The ambiguity between genuine trichotomies and unresolved dichotomies. (a) A cladogram which represents unambiguously a pair of successive dichotomies; epistemological doubts, if any, are not indicated. (b) A cladogram which is commonly used to represent a pair of unresolved dichotomies, a genuine trichotomy, common ancestry and reticulate evolution. (c) A cladogram which indicates doubt about the actual relations between the species indicated. Such cladograms might in time resolve to (a) a pair of dichotomies or (b) a genuine trichotomy. Common ancestry and reticulate evolution must be represented in some other but equally unambiguous fashion.

ANCESTOR-DESCENDANT RELATIONS

Although some disagreement exists over the occurrence of multiple speciation, everyone agrees that certain species which once existed no longer exist and that some of these species gave rise to Recent species. The cladists' complaints about the ancestor-descendant relation cannot possibly be interpreted as empirical. They are clearly methodological. Once again the distinction must be made between difficulties in discerning ancestor-descendant relations and difficulties in representing them. Farris (1976:272) acknowledges this distinction when he presents two reasons for not treating fossil species as ancestral to later species:

> First, there is no obvious way of using a classification to represent ancestor-descendant relationships. A taxonomic hierarchy with its well-nested taxa is naturally suited only to the representation of sister-group relationships. Second while sister-group relationships may more or less readily be established through the detection of apomorphous similarities, ancestor-descendant relationships may not be so established.

A second distinction of equal importance is between the recognition of extinct species as species and deciding which species gave rise to which. Those cladists who are willing to recognize extinct species as species while maintaining that the ancestor-descendant relation is unknowable are put in the position of explaining how we can know the former but not the latter (Engelmann and Wiley, 1977). Those cladists who argue that neither is knowable are spared this dilemma but are faced with the problem of what to do with fossils. Løvtrup (1973) argues that extinct species are so poorly known that they should be totally excluded from classifications. Crowson (1970) sees the differences in our ability to recognize extinct and extant forms to be sufficiently great that they should be included in separate classifications. Nelson (1972a:230) reasons that fossil groups should be fitted into classifications which also contain recent groups but distinguished by some convention. Finally, Rosen (1978:176) agrees that extant and extinct forms should be distinguished but that they should be treated as the same sorts of entities—groups distinguished by unique derived characters. In this section I will deal first with difficulties in discerning species and ancestor-descendant relations and then with problems of representation.

Hennig (1966, 1975) and Brundin (1972b) maintain that the biological species concept is central to phylogenetic systematics. From the beginning, critics of the biological species concept have argued that it is not sufficiently "operational" even for extant species. Perhaps it is possible to ascertain which organisms actually have mated with one another to produce fertile offspring and which organisms cannot mate with each other and/or produce fertile offspring, but it is impossible to discover which could have done so but did not. When the species under investigation are extinct, so the critics argue, nothing can be discovered directly about their breeding relations. It is certainly true that the biological species concept is not very operational, but as I have argued elsewhere, no theoretically significant concept in science is (Hull, 1968, 1970). The interesting question is whether biological species can be discerned often

enough and with sufficient certainty to justify systematists' attempting to represent them in their classifications. An even more fundamental question is whether biological species are sufficiently important to the evolutionary process to warrant the attention they are given. But in any case, the existence and extent of biological species are almost always inferred indirectly from character distributions. Hence, mistakes will be made. When these species are extinct, mistakes are even more likely and our ability to correct them even more limited. No one could possibly claim that all pterodactyls belonged to a single species, but finer distinctions are highly problematic.

Thus, those cladists who continue to maintain that biological species are the ultimate units in their investigations are forced to admit that their cladograms and classifications might be mistaken. They might have lumped several species into one or divided a single species into two or more species. For extinct forms, these mistakes are difficult, if not impossible, to remedy. Cladists who have abandoned the biological species concept are spared this problem. They are not classifying species but are grouping organisms according to the possession of particular traits. The only doubt associated with such an exercise is whether the organisms actually have the trait attributed to them, whether or not it is actually *a* trait and not several, and whether it has the distribution attributed to it. The question remains whether it is any easier to individuate traits than it is to individuate theoretically significant taxa. The two seem both conceptually and inferentially very closely connected.

If extinct species cannot be recognized as genuine groups of some sort, they cannot be recognized as sister groups. But assuming that extinct forms can be recognized as extinct species or as extinct groups delimited by a particular trait, can ancestor-descendant relations between extinct and extant forms also be discerned? Cladists are all but unanimous in claiming that they cannot. As Nelson

(1973b:311) states, "Indeed, I have assumed that ancestral species cannot be identified as such in the fossil record, and I have pointed out that this assumption is fundamental to Hennig's phylogenetic systematics." Two considerations seem central to the cladists' claim that ancestor-descendant relations are unknowable. First, any distribution of characters which would imply that one taxon was ancestral to another would equally imply that one was the sister group of the other, and no other way exists for deciding ancestor relations. Second, given the vast number of species which must have existed from the beginning of time and the relatively few which have left records, it seems very unlikely that very many actual common ancestors have been discovered.

I find the second argument more persuasive than the first. It is always *possible* that what appears to be a genuine ancestor is really a sister group and that the two share an unknown common ancestor. Of course, it is always *possible* that a trait which has been recognized as a single trait is actually two, that a trait which is considered to be derived really is not, that a derived character which is thought to be unique really is not, etc., etc. The *possibility* of error pervades all of science. What is needed is some reason to believe that not only is it possible for the two groups to be sister groups instead of one being ancestral to the other, but that one hypothesis is more probable than the other. Reference to the vast number of species which have existed and the few which have left any traces of their existence does just that. If paleontologists claim that they have identified an extinct species which is in a direct line of descent to some earlier or later species, then such claims are on the face of it extremely unlikely.

For example, when Romer claims that amphibians arose from ancient crossopterygians, he surely could not have meant that one of the fossils already collected represented the actual stem species from which all amphibians arose and that he knew precisely which fossils these were. Perhaps paleontologists have been overly enthusiastic in identifying missing links and common ancestors, but I find it hard to believe that they have been this foolhardy. As Harper (1976) and others have pointed out, paleontologists rarely specify a particular species as a common ancestor. Almost always a genus or higher taxon is specified. When paleontologists claim that one genus or other higher taxon arose from another, they do not mean that higher taxa "evolve," although at one time macroevolution was more popular than it is now. All they mean is that the species which is ancestral to this higher taxon, if it were ever discovered, would be placed in the taxon which has been specified as being ancestral. For example, whether or not specimens of the actual species which gave rise to amphibians have been discovered, whether or not they could be recognized as such if they were, Romer is committed to the view that they would be placed among the crossopterygians. Nor are paleontologists necessarily committed to Simpson's definition of monophyly. Even though they may be willing to specify only a genus or other higher taxon as a common ancestor, they may still have the goal of making all higher taxa monophyletic at the species level and would redraw the boundaries of their higher taxa if they became convinced that a higher taxon was descended from two or more species, even though these species were all included in a higher taxon of equal or lower rank. But it must also be admitted that claiming a higher taxon as a common ancestor is not as empirical as claiming that a particular species is. Higher taxa, as they are usually constructed, are largely a function of the principles of classification adopted (Engelmann and Wiley, 1977).

Harper (1976) also presents eight principles which he thinks can help distinguish sister-group relations from ancestor-descendant relations. All of Harper's principles cannot be discussed here, but one is sufficiently important to warrant mentioning—the role of fossils. Does not the existence of fossils in certain strata and their absence in others, especially when a fairly extensive fossil record is available, imply something about possible ancestor-descendant relations? Hennig thinks it does. For example, he (Hennig, 1966:169) argues that the "sequence in which the characters in question evolved" is "sometimes clarified by fossils." He also thinks that fossils can help determine the minimum age of the monophyletic groups to which they belong (Hennig, 1975:112). Nelson (1969:245) is a good deal more sceptical: "If a given fossil could be demonstrated to have been a representative of a population ancestral to a Recent species," it might be of some significance, but such an "ancestor-descendant relationship strictly speaking cannot be demonstrated." Even though Nelson (1972b:367–370) admits that data concerning "relative stratigraphic position" might narrow the "range of possible relationships held by the taxa in question," he recommends divorcing "problems of relationships from data concerning stratigraphic distribution of fossils." He concludes that ancestor-descendant relations are both unknown and unknowable in an "empirical sense," that is, "by way of inference from observation of study material" (Nelson, 1973b:311).

Following Popper (1959, 1972), several cladists have argued that hypotheses about ancestor-descendant relations are unscientific because they are unfalsifiable (Wiley, 1975; Engelmann and Wiley, 1977; Platnick and Gaffney, 1977, 1978; Cracraft, 1978; Nelson, 1978a; Patterson, 1978). Because these hypotheses are unscientific, they should play no role in science. At one time Bonde (1975) agreed with this line of reasoning, but in the meantime he (Bonde 1977:772) has noticed something else about ancestor-descendant hypotheses: they are bolder than sister-group hypotheses. Because cladograms claim less than trees and trees claim less than scenarios, scenarios are bolder than trees and trees are bolder

than cladograms. And, according to Popper, bolder hypotheses are preferable to less bold hypotheses. Thus, two of Popper's desiderata come into conflict. If all the relevant data were available, bolder hypotheses would be easier to falsify because they imply more. The problem is that in this case, the evidence necessary to test the bolder hypotheses is much harder to obtain than the evidence necessary to test the less bold hypotheses. And what is worse, mistakes in ascertaining ancestor-descendant relations introduce much more serious errors into a study than mistakes in ascertaining sister groups (Engelmann and Wiley, 1977).

As I mentioned earlier, I am highly suspicious of scientists claiming that something cannot be known, especially when they do not need to know it and their opponents do. In this case, I think that the cladists have exaggerated the difficulties in making reasonable inferences about probable ancestors for a second reason as well—difficulties in representation which common ancestors pose for both cladograms and classifications. If common ancestors cannot be recognized with reasonable certainty, then the evolutionists are in trouble. If they can be, then cladists are in trouble. As Tuomikoski (1967:144) remarks, "Hennig and Brundin are aware of the fact that such a linear classification of monophyletic groups is capable of expressing only horizontal sister relationships, not vertical mother-daughter relationships." If ancestor-descendant relations can never be known with sufficient certainty to warrant postulating them, then the inability of cladograms and classifications to represent them is no great drawback, but Wiley (1977) has suggested that a better strategy would be to devise representational devices capable of representing them just in case. So far, several systems of representation have been suggested for distinguishing extinct from extant groups, and the difficulties which attempts to distinguish between sister-group relations and ancestor-descendant relations in cladograms and classifications have been pursued at great length (Nelson, 1973c; Cracraft, 1974; Patterson, 1976; Patterson and Rosen, 1977; Wiley, 1979). For example trichotomies in cladograms are already ambiguous, representing either trichotomous speciation events or a pair of unresolved dichotomies. Attempting to represent common ancestry in this same way would introduce yet another dimension of ambiguity. But I see no reason why some system of representation cannot be developed to distinguish these three possible situations in both cladograms and classifications.

FIG. 3.—Three possible patterns of evolution. (a) A phylogenetic tree in which a single lineage changes in time without splitting. (b) A phylogenetic tree in which a single ancestral lineage splits into two descendant lineages and both descendant lineages diverge from the ancestral lineage. (c) A phylogenetic tree in which splitting occurs but only one of the descendant lineages diverges from the ancestral lineage.

PHYLETIC EVOLUTION AND SPECIATION

Cladists are frequently interpreted as claiming that phyletic evolution does not occur. Once again, the actual thrust of their discussion of phyletic evolution is methodological, not empirical. Figure 3 represents three possible patterns of evolution: gradual change in a lineage without the lineage splitting, splitting followed by both descendant lineages diverging, and splitting followed by only one of the descendant lineages diverging. Simpson (1961) has argued that sometimes lineages change sufficiently through the course of their development so that later stages should be considered distinct species even though no splitting has taken place. The inevitable operationalist objections have been raised against Simpson's suggestion: that no non-arbitrary way exists for dividing a continuously evolving lineage into distinct species and that lineages cannot be established in the first place because there is no way of knowing which organisms are ancestral to which. I have already discussed these topics at sufficient length both in this paper and elsewhere (Hull, 1965, 1967, 1970). If the continuity of phyletic evolution were sufficient to preclude "objective" subdivision, then all our measurements of physical space would be equally suspect because space is even more continuous. Perhaps the establishment of lineages is a risky business, but it is also an extremely important undertaking, not for the purposes of clas-

sification but for the purposes of formulating and testing hypotheses about the evolutionary process. As important as scientific classifications are, scientific theories are even more important.

Eldredge and Gould (1972) have argued that phyletic evolution rarely, if ever, occurs and that the evidence for it is largely illusory. However, regardless of whether or not phyletic evolution occurs, the cladists maintain that a single lineage should not be divided into species no matter how much it might change. New species are to be recognized only when splitting occurs. According to Hennig (1966, 1975), species are not classes of similar organisms but well-integrated gene pools. In phyletic evolution, the lineage changes through time but retains its integration. Hence, it should be considered a single species. As Platnick (1977a:97) argues:

> To attempt to divide a species between speciation events would indeed be arbitrary: we would not call an individual person by one name at age 10 and a different name at age 30. Dividing species at their branching points, however, becomes not only non-arbitrary but necessary: we would not call a son by the same name as his father.

In this instance, I think Hennig is right, not because of any epistemological or methodological reasons, but because his decision is consistent with the most theoretically appropriate definition of "species." Species are integrated lineages developing continuously through time (Simpson, 1961; Mayr, 1963; Ghiselin, 1974; Hull, 1976, 1978; Wiley, 1978). Cohesiveness at any one time and continuity through time are what matter, not phenotypic or even genotypic similarity. Some lineages may diverge extensively through time without splitting; some not. It does not matter. A continuously evolving lineage should no more be divided into distinct species than an organism undergoing ontogenetic development should be divided into distinct organisms. These same considerations apply to the other situations shown in Fig. 3.

Cladists argue that new species should be recognized *only* when splitting occurs. In non-saltative speciation, temporal continuity is maintained, but the cohesiveness of the lineage is disrupted. They also argue that *whenever* splitting takes place, the ancestral species must be considered extinct. Bonde (1975:295) characterizes Hennig's position as follows:

> By the process of speciation the ancestral species is split into *two new species* called *sisterspecies* (or daughterspecies) .... That the two sisterspecies are *new* and the ancestral species becomes extinct at the speciation is most practical for a consistent terminology, and it also follows from Hennig's species concept, and from the definition of phylogenetic relationship below (cf. Brundin, 1972a:118).

According to Hennig, the two sorts of splitting depicted in Figs. 3b and 3c are to be treated in exactly the same way; i.e., as a mother species becoming extinct as it divides into two daughter species. It does not matter that in one instance both daughter species diverge from the mother species, while in the second only one does. As much trouble as it may cause the systematist attempting to distinguish ancestors from descendants, it simply does not matter that the mother species is indistinguishable from one of its daughter species. I agree with Hennig's reasoning but not his conclusion. If integration is what matters, then the differences indicated in Figs. 3b and 3c are irrelevant for the individuation of species.

If divergence is irrelevant *without* splitting, it should be just as irrelevant *with* splitting. On Hennig's view, what should really matter is the integration of the gene pool. At speciation, so Hennig claims, the ancestral gene pool totally disintegrates, resulting in the extinction of the ancestral species. According to Bonde (1977:754), Hennig's species concept is the "only logical extension in time of the concept of the integrated gene pool. At speciation this gene pool is disintegrated and two (or more) new sister species originate, while the original species becomes extinct." Wiley (1978: 21–22) agrees that "in most cases the methodological necessity of postulating extinction of ancestral species in phylogeny reconstruction as advocated by Hennig (1966) is biologically (as well as methodologically) sound," but not always. According to his own evolutionary species concept, an ancestral species can survive a split, if it can "lose one or more constituent populations without losing its historical identity or tendencies."

If speciation can occur only by the massive disintegration of the parental gene pool, then Hennig's decision always to treat ancestral species as extinct is well-founded biologically. However, if Mayr (1963), Carson (1970), and Eldredge and Gould (1972) are right and speciation always (or usually) takes place by the isolation of small, peripheral populations, then it seems very unlikely that such an event will totally disrupt the organization of the parental gene pool, and Hennig's convention of treating ancestral species as always going extinct upon speciation loses its empirical support.[3] As interesting as these empirical considerations are, as I have mentioned several times previously, cladists are currently disassociating themselves from any particular

---

[3] The arguments presented by Eldredge and Gould (1972) support Hennig's contention that new species should be recognized only when splitting occurs. These same arguments, however, count against Hennig's contention that ancestral species always disintegrate upon speciating.

views about the evolutionary process. For example, Bonde (1977:793) states that "contrary to my earlier beliefs, the details of the process of speciation are not important for the phylogenetic systematic theory because the patterns resulting from allo-, para- and sympatric speciations can be analyzed in the same way in terms of degrees of phylogenetic relationship." If so, then facts about the evolutionary process cannot be used to cast doubt on the cladists' research program. Such considerations cannot thereby be used to support it either, and Hennig's arguments must be abandoned.

The only remaining consideration is "consistency of terminology." Hennig seems to think that calling a species by the same name before and after speciation would cause all sorts of terminological confusion. However, comparable terminological difficulties are easily surmounted at the level of organisms. When a single *Paramecium* splits down the middle to form two new organisms, each is considered a distinct organism. If we were prone to name such entities, we would give each a separate name. However, *Hydra* can continue to exist while budding off other *Hydra*. Once again, if we were inclined to, we could give each of these organisms its own name. The parent *Hydra* would retain its name even though it budded off other descendant *Hydra*. Finally, in sexual reproduction, two parental organisms produce gametes which come together to form a third. Organisms need not cease to exist when they mate. Queen Victoria and Prince Albert remained the same organisms as they produced child after child. Occasionally, we do call a son by the same name as his father. Usually we have the good sense to add "Jr.," but whether we do or not, they remain distinct organisms. If such practices occasion so little confusion at the level of organisms, there seems no reason for them to introduce insurmountable terminological difficulties at the level of species.

In this section I have argued that if one accepts Hennig's conception of species

as integrated gene pools, then species changing gradually through time should be considered single species and not divided successively into distinct species. New species should be recognized *only* at splitting, but at splitting the ancestral species does not *always* go extinct. How many species are to be recognized at splitting depends on what happens to the integration of the parental gene pool. If it remains largely unaffected, then the ancestral species continues to exist as it buds off descendant species. If not, it becomes extinct. Whether or not the daughter species diverge after splitting is irrelevant for the individuation of species as integrated gene pools. Divergence may matter for a host of other reasons, but not for the individuation of species. Of course, none of the preceding is relevant to those cladists who think that the recognition of species as some sort of significant evolutionary unit plays no role in cladistic systematics. It is also true that in the absence of some sort of divergence, the systematist is unlikely to notice that speciation has occurred (see later discussion of the Rule of Deviation).

MONOPHYLY AND RETICULATE EVOLUTION

The term "monophyly" has had an extremely varied history. According to Mayr (1942:280) all taxa should be monophyletic. By this he means that all organisms included in a taxon should be "descendants of a single species." He does not say, however, whether *all* the descendants of a single species should be included in a single higher taxon, nor whether this stem species itself should be included in this higher taxon or excluded from it. In later discussions, Mayr (1969) states that he intended no such implications. He also states explicitly that, as far as animals are concerned, reticulate evolution may be disregarded. He justifies his position by reference to how rare genuine hybrid species are among animals. Botanists might choose to make another decision. Hennig (1966:73) defines "monophyly" somewhat differently: "A monophyletic group is a group of species descended from a single ("stem") species, and which includes all species descended from this stem species." On Hennig's definition, all and only those species (both extant and extinct) which are descended from a single species must be included in a single higher taxon (see also Bonde, 1975:293; 1977:757). Hennig (1966:207–208) also recognizes that the hybrid origin of species would produce "special complications" for the principle of monophyly and echoes Mayr in his response that luckily hybridization is rare in animals. Among plants he is willing to countenance polyphyletic species but not polyphyletic higher taxa (Hennig 1966:208). Hennig (1966:64, 70–72, 207) is not clear about the recognition and placement of "stem species" in his classifications.

As much as Mayr and Hennig differ with respect to the principle of monophyly, they agree that all species included in a single higher taxon must be descended from a single stem *species*. Simpson (1961:124) presents a much broader definition: "*Monophyly is the derivation of a taxon through one or more lineages* (temporal successions of ancestral-descendant populations) *from one immediately ancestral taxon of the same or lower rank.*" According to Simpson, a taxon is strictly monophyletic if it arises from a single immediately ancestral species; it is minimally monophyletic if it arises from a single immediately ancestral taxon of its own or lower rank. Thus, a genus is *strictly* monophyletic if it arises from a single immediately ancestral species. It is only *minimally* monophyletic if it arises from two or more species which are included in the same genus. It is polyphyletic if it arises from two or more species not included in the same genus. As broad as Simpson's definition is, it too is confronted by the difficulty of accommodating hybrid species (Hull, 1964).[4] His response is the

same as that of Mayr and Hennig: hybrid species may occur in plants, but they are rare in animals. Simpson also does not specify any uniform way in which stem species are to be treated.

Several points are at issue in the preceding discussion: (1) should all taxa be monophyletic at the level of species, (2) if so, how are hybrid species to be treated, (3) should all the descendants of a single stem species be included in the same higher taxon, and (4) if so, what should be done with the stem species itself? The cladists find themselves in agreement with respect to (1) and (3). All species in any higher taxon must be descended from a single ancestral species. Conversely, all the descendants of any one stem species must be included in a single higher taxon. Some disagreement exists over the proper treatment of (2) and (4), hybrid species and stem species. For those cladists who maintain that the only kind of event in the past which can be reconstructed with reasonable certainty is speciation (or at least the production of unique derived characters), common ancestry, whether single or multiple, cannot be ascertained. The distribution of characters resulting from hybridization is indistinguishable from that resulting from a trichotomy, which in turn is indistinguishable from a pair of unresolved dichotomies, which in turn is indistinguishable from common ancestry. Since all three phenomena can produce the same distribution of traits, all such distributions of traits should be interpreted as unresolved dichotomies—other things being equal.

The claim that we can never distinguish between a genuine common ancestor and a closely related sister group has some plausibility. The claim that we can-

monophyletic further weakens the already loose relationship between phylogeny and biological classifications which Simpson proposes, Bonde (1975, 1977) interprets me to be defending Simpson's definition of "monophyly." There is a point to making one's polemics as polite as possible, but when an attack is interpreted as a defense, that point has been passed.

---

[4] Although the main purpose of Hull (1964) was to argue that allowing taxa to be only minimally

not discern hybrid species does not. We know with as much certainty as we know any empirical phenomenon that certain species are of hybrid origin. We did not observe the hybridization event in the past, but we can infer its occurrence with a high degree of certainty. For example, if a species were discovered at the boundary between two species with the combined chromosomes of these two species, the likelihood is very high that this species is a hybrid of the other two. Perhaps the outcome of a cladistic analysis of these three species would be a pair of unresolved dichotomies, but that is a problem for cladistic analysis.

Another reason which cladists have for rejecting both stem species and hybrid species as unknowable is the difficulties which both relations pose for unequivocal representation in both cladograms and classifications. For example, Nelson (1979:8) observes:

> An instance of hybridization would be represented in fundamental cladograms by non-combinable components that exhibit non-random replication, and in the general cladogram by tri- or polytomies that present conflicting, but non-random, possibilities for dichotomous resolution (as exemplified in fundamental cladograms).

Instead of introducing yet another dimension of ambiguity into polytomous cladograms, a better alternative would be to devise distinct methods of representation for each of these distinct phenomena. Perhaps we will never know which species are actually stem species, perhaps we will never know which speciation events are trichotomous, perhaps we will never know which species are of hybrid origin, but just in case someday we can, why not devise methods of representing these phenomena in cladograms and classifications? Wiley (1978, 1979) proposes to do just this.

In the previous sections I have argued that certain methodological principles attributed to the cladists really do not contribute all that much to the long-term goal of cladism—the representation of cladistic relations explicitly and unambiguous-

ly. If speciation is on occasion trichotomous and such events can be discerned, why not represent them? If common ancestors and reticulate evolution can be discerned, why not represent them? However, the principle of monophyly *is* important to cladism. In recent years a large literature has grown up over the "proper" definition of "monophyly" and related terms (Ashlock, 1971, 1972; Nelson, 1971b, 1973b; Farris, 1974; Platnick, 1976, 1977b, 1977c; Wiley, 1977). Scientific terms change their meanings as science develops. "Gene" did not mean the same thing at the turn of the century as it does today. One important function of the history of science is to trace such semantic changes. However, the results of such inquiries cannot be used to designate certain meanings as "proper" and others as "improper." I'm not sure what Darwin meant by the term "monophyly." I'm not even sure he ever used the term. Changing the definitions of terms in science haphazardly introduces needless confusion. From the point of view of making assertions about classifications, the term "holophyly" will do as well as "monophyly." The desire on both sides to retain the term "monophyly" for their concept stems from the need to establish continuity in scientific development. Just as continuity matters in individuating lineages in biological evolution, it matters in individuating scientific lineages in conceptual evolution. The sort of fighting over words which is so common in science is not an idle exercise.

Terminological squabbles to one side, the dispute between Ashlock (1971, 1972) on the one hand and Nelson (1971b, 1973b) and Farris (1974) on the other hand has some substantive points. Cladists want to represent cladistic relations by nested sets of derived traits. To do so, they must exclude what they term "paraphyletic" taxa. As far as assertive content is concerned, which definition of "monophyly" wins out is irrelevant; the structure of the resulting classifications is not. Whether one claims that all taxa must be holophyletic (sensu Ashlock) or monophyletic (sensu the cladists) is important sociologically. What matters substantively is that the taxa themselves be grouped appropriately.

However much cladists and evolutionists disagree about the proper definition of "monophyly," they agree on one point: hybrid species pose a problem. Although I hate to destroy this rare instance of unanimity, I think they are mistaken. The goals of neither school of taxonomy require that all species arise from single ancestral species. All that is needed is that all species have a single origin. In most cases, species do have their origins in the speciation of single ancestral species. It follows that these species also have single origins. But sometimes a new species arises from two immediately ancestral species. They nevertheless have their origins in a single speciation event, and that is all that is necessary. Both schools need to require that all taxa be monophyletic, but only as monophyletic as nature allows.

THE DEVIATION RULE

According to Hennig (1966:207), "When a species splits, one of the two daughter species tends to deviate more strongly from the other from the common stem species" (see also Hennig, 1950:111). Brundin (1972b:71) claims that the deviation rule is "one of the most fundmental aspects of the principle of life" and essential to cladism. Bonde (1975:296) views deviation as a common (though not universal) pattern of evolution, which is important (though not necessary) to cladism. Schlee (1971:5) finds it "unessential" to the Hennigian approach, while Nelson (1971a:373) concludes that it is both unessential and a methodological principle. If the rule of deviation is supposed to be an empirical claim about evolution, it has some plausibility. As Bonde (1975:296) notes, if speciation takes place only when an "atypical" population becomes isolated in an "atypical" environment, one might reasonably expect this new daughter species to diverge more from the ancestral species than its sister does.

The rule of deviation also has a methodological form. Whether or not speciation is always accompanied by the production of at least one unique derived character in one of the groups,[5] if no such character is discernible, the groups will not be discernible by the methods of cladistic analysis. Of course, they will not be discernible by any other methods either. It is unlikely (probably impossible) for speciation to occur without at least one new character developing, but if "character" is used in its traditional sense to refer to such things as tooth structure, feather color, type of blood proteins, peculiarities in mating dance, etc., it may be the case that speciation can occur without the production of a distinguishing feature of the sort that a systematist is likely to notice. He can know that the two groups are *two* groups because they do not mate. He may even be able to tell which group is which because they inhabit different ranges. But he will not be able to decide which species a specimen belongs to merely by examining it. (Of course, one might redefine the term "character" to include spatial location or some such.)

For those cladists who see their task as the recognition of species and the grouping of these species into more and more inclusive taxa, the discovery of the appropriate unique derived characters and shared derived characters is crucial. It is very likely that the relevant characters exist. The task is to discover and categorize them appropriately. The cladists

---

[5] The number of characters necessary to distinguish a new taxon depends on a variety of decisions. For example, if one assumes that the ancestral species has already been delineated and can bud off new species without itself going extinct, a single unique derived character will do. If the ancestral species also must be delineated, a second unique derived character is necessary. If ancestral species are always considered to be extinct upon speciation and speciation is always dichotomous, still another character is required, and so on. Of course, in traditional cladistic classifications, all these "ancestral" species will be treated as sister groups.

share this task with all systematists. For those cladists who see their task as the recognition of sequences of unique derived characters and shared derived characters, regardless of the coincidence of the emergence of these characters with speciation events, the rule of deviation becomes a tautology. It reduces to the claim that when a unique derived character is recognized a new group is recognized; otherwise not. By terming the rule of deviation, interpreted in this way, a tautology, I do not mean to denigrate it.

CONCLUSION

In this paper I have examined some of the principles which cladists have suggested to facilitate the representation of sister-group relations in cladograms and classifications. Currently some disagreement exists among the cladists themselves over the precise makeup of these principles, but eventually a single, generally accepted, reasonably simple system of conventions is likely to materialize. The consistent application of these principles will result in classifications which allow anyone who wishes to read off from them the sister-group relations which went into their construction. I have also raised questions concerning some of the prescriptions which cladists have enunciated about what cannot be known about phylogeny or represented if it were known. All scientific knowledge is fallible, including the recognition of sister groups. Perhaps ancestors, hybrid species and multiple speciation events cannot be discerned with the same degree of confidence as sister-group relations, but the contrast is not between fact and fantasy. Cladists have selected sister-group relations because they are relatively easy to discern using the methods of cladistic analysis and just as easy to represent in cladograms and classifications. Opponents have complained that the price is too high. Being able to infer sister-group relations from biological classifications is not worth the increase in complexity and asymmetry of the resulting classifications. They argue

that the cladists' rules for classifying are simple enough, but that the resulting classifications are not.

Another avenue of attack, employed by some taxonomists, is to agree that genealogy is worth representing in classifications but object to the aspect selected by the cladists. Divergence is also an important feature of evolutionary development. Instead of constructing classifications solely on the basis of cladistic relations, varying degrees of divergence should also be reflected. One problem with this response is that no methods have been set out thus far which permit the inclusion of both sorts of information in a single classification in such a way that both are retrievable. It is one thing to let a variety of considerations influence how one constructs a classification. It is quite another to formulate a set of principles so that others can retrieve this information from the classifications. If classifications are to be systems of information storage and retrieval, information must actually be *retrievable*. So far, only Ashlock and Brothers (1979) have taken up the challenge. A second problem with this alternative is that it is likely to increase the complexity of both the rules of classification and the resulting classifications. If classifications which represent only sister-group relations are too complicated, classifications which represent sister-group relations, ancestor-descendant relations *and* degrees of divergence are likely to be even more complicated.

Another possible response is that cladists have taken Darwin's suggestion too literally. Classification should be in some vague sense "phylogenetic," but biological classifications cannot be made to reflect very much about phylogeny without frustrating other functions of scientific classification. The most that systematists can hope to do is to weigh a variety of conflicting goals and produce the best possible compromise. Clear decisions between these various alternative goals are not easy, but if the cladists have done nothing else, they have shown the sorts of rules which are necessary if biological classifications are to represent explicitly and unambiguously a particular feature of phylogenetic development. They have also posed clearly and forcefully a challenge to their fellow systematists: if the goal of biological classification is *not* to represent one or more aspects of phylogenetic development, what *is* the goal of biological classification?

ACKNOWLEDGMENTS

I would like to thank Walter Bock, Niels Bonde, Joel Cracraft, Steve Farris, Ernst Mayr, Gareth Nelson, Norman Platnick and Ed Wiley for long and heated battles of the most enjoyable sort over the exact nature of cladistic analysis and cladistic representation. Needless to say, our meeting of minds has yet to attain the isomorphism characteristic of cladograms and cladistic classifications. The research for this paper was supported in part by NSF grant SOCX75-03535 A01.

REFERENCES

ASHLOCK, P. D. 1971. Monophyly and associated terms. Syst. Zool. 20:63–69.

ASHLOCK, P. D. 1972. Monophyly again. Syst. Zool. 21:430–437.

ASHLOCK, P. D., AND D. J. BROTHERS. 1979. Systematization and higher classification in evolutionary systematics through cladistic and anagenetic analysis. Manuscript.

BONDE, N. 1975. Origin of "higher groups": viewpoints of phylogenetic systematics. Problemes actuels de paleontologie—evolution des vertebres. Coll. Internat. C.N.R.S., no. 218:293–324.

BONDE, N. 1977. Cladistic classification as applied to vertebrates. In Hecht, M. K., P. C. Goddy, and B. M. Hecht (eds.), Major patterns in vertebrate evolution. Plenum Publishing Corporation, New York, pp. 741–804.

BRUNDIN, L. 1966. Transantarctic relationships and their significance, as evidenced by chironomid midges. K. svenska Vetensk. Akad. Handl. (4)11:1–472.

BRUNDIN, L. 1972a. Evolution, causal biology, and classification. Zool. Scripta. 1:107–120.

BRUNDIN, L. 1972b. Phylogenetics and biogeography Syst. Zool. 21:69–79.

CAMIN, J. H., AND R. R. SOKAL. 1965. A method for deducing branching sequences in phylogeny. Evolution 19:311–326.

CARSON, H. L. 1970. Chromosome tracers of the origin of species: some Hawaiian *Drosophila* species have arisen from single founder individuals in less than a million years. Science 168:1414–1418.

CRACRAFT, J. 1974. Phylogenetic models and classification. Syst. Zool. 23:71–90.

CRACRAFT, J. 1978. Science, philosophy, and systematics. Syst. Zool. 27:213–215.

CROWSON, R. A. 1970. Classification and biology. Atherton Press, New York.

DARLINGTON, P. J., JR. 1970. A practical criticism of Hennig-Brundin "phylogenetic systematics" and antarctic biogeography. Syst. Zool. 19:1–18.

DARWIN, CHARLES. 1859. On the Origin of Species, a facsimile of the first edition (1859) with introduction by Ernst Mayr (1966). Harvard University Press, Cambridge, Mass.

DARWIN, F. 1899. The Life and Letters of Charles Darwin. D. Appleton and Company, New York.

EINSTEIN, A. 1949. Einstein: reply. In Schilpp, P. A., (ed.), Albert Einstein: Philosopher-Scientist. Open Court, New York.

ELDREDGE, N. 1979. Cladism and Common Sense. In Cracraft, J., and N. Eldredge (eds.), Phylogeny and Paleontology. Columbia University Press, New York.

ELDREDGE, N., AND S. J. GOULD. 1972. Punctuated equilibria: an alternative to phyletic gradualism. In Schopf, T. M. J., (ed.), Models in paleontology. Freeman, Cooper, and Co., San Francisco, pp. 82–115.

ENGELMANN, G. F., AND E. O. WILEY. 1977. The place of ancestor-descendant relationships in phylogeny reconstruction. Syst. Zool. 26:1–11.

FARRIS, J. S. 1974. Formal definitions of paraphyly and polyphyly. Syst. Zool. 23:548–554.

FARRIS, J. S. 1976. Phylogenetic classification of fossils with Recent species. Syst. Zool. 25:271–282.

GHISELIN, M. 1974. A radical solution to the species problem. Syst. Zool. 23:536–544.

GRIFFITHS, G. C. D. 1976. The future of Linnaean nomenclature. Syst. Zool. 25:168–173.

HAIGH, J. 1971. The manuscript linkage problem. In Hodson, F. R., D. G. Kendall, and P. Tăutu (eds.), Mathematics in the Archaeological and Historical Sciences. Univ. Press, Edinburgh, pp. 396–400.

HARPER, C. 1976. Phylogenetic inference in paleontology. J. Paleo. 50:180–193.

HENNIG, W. 1950. Grundzüge einer Theorie der phylogenetischen Systematik. Deutscher Zentralverlag, Berlin.

HENNIG, W. 1965. Phylogenetic systematics. Ann. Rev. Ent. 10:97–116.

HENNIG, W. 1966. Phylogenetic systematics. University of Illinois Press, Urbana.

HENNIG, W. 1975. "Cladistic analysis or cladistic classification?": a reply to Ernst Mayr. Syst. Zool. 24:244–256.

HULL, D. L. 1964. Consistency and monophyly. Syst. Zool. 13:1–11.

HULL, D. L. 1965. The effect of essentialism on taxonomy. Brit. J. Phil. Sci. 15:314–326; 16:1–18.

HULL, D. L. 1967. Certainty and circularity in evolutionary taxonomy. Evolution 21:174–189.

HULL, D. L. 1968. The operational imperative—sense and nonsense in operationalism. Syst. Zool. 17:438–457.

HULL, D. L. 1970. "Contemporary systematic philosophies." Ann. Rev. Ecol. Syst. 1:19–54.

HULL, D. L. 1976. Are species really individuals? Syst. Zool. 25:174–191.

HULL, D. L. 1978. A matter of individuality. Phil. Sci. 45:335–360.

HUXLEY, J. 1958. Evolutionary processes and taxonomy with special reference to grades. Uppsala Univ. Arssks., pp. 21–38.

KRUSKAL, J. B., I. DYER, AND P. BLACK. 1971. The vocabulary method of reconstructing language trees. In Hodson, F. R., D. G. Kendall, and P. Tǎutu (eds.), Mathematics in the Archaeological and Historical Sciences. Univ. Press, Edinburgh, pp. 361–380.

LAUDAN, L. 1977. Progress and its problems: toward a theory of scientific growth. University of California Press, Berkeley.

LØVTRUP, S. 1973. Classification, convention and logic. Zool. Scripta. 2:49–61.

LUCCHESI, J. C. 1978. Gene dosage compensation and the evolution of sex chromosomes. Science 202:711–716.

MAYR, E. 1942. Systematics and the origin of species. Columbia University Press, New York.

MAYR, E. 1963. Animal species and evolution. Harvard University Press, Cambridge, Mass.

MAYR, E. 1965. Classification and phylogeny. Amer. Zool. 5:165–174.

MAYR, E. 1969. Principles of systematic zoology. McGraw-Hill, New York.

MAYR, E. 1974. Cladistic analysis or cladistic classification. Z. f. zool. Systematik u. Evolutionsforschung. 12:94–128.

MAYR, E. 1978. Origin and history of some terms in systematic and evolutionary biology. Syst. Zool. 27:83–87.

NELSON, G. 1971a. "Cladism" as a philosophy of classification. Syst. Zool. 20:373–376.

NELSON, G. 1971b. Paraphyly and polyphyly: redefinition. Syst. Zool. 20:471–472.

NELSON, G. 1972a. Phylogenetic relationship and classification. Syst. Zool. 21:227–230.

NELSON, G. 1972b. Comments on Hennig's "phylogenetic systematics" and its influence on ichthyology. Syst. Zool. 21:364–374.

NELSON, G. 1973a. The higher-level phylogeny of vertebrates. Syst. Zool. 22:87–91.

NELSON, G. 1973b. "Monophyly again?": a reply to P. D. Ashlock. Syst. Zool. 22:310–312.

NELSON, G. 1973c. Classification as an expression of phylogenetic relationship. Syst. Zool. 22:344–359.

NELSON, G. 1974. Darwin-Hennig classification: a reply to Ernst Mayr. Syst. Zool. 23:452–458.

NELSON, G. 1978. Classification and prediction: a reply to Kitts. Syst. Zool. 27:216–217.

NELSON, G. 1979. Cladistic analysis and synthesis: principles and definitions, with a historical note on Adanson's Familles des Plantes (1763–1764). Syst. Zool. 28:1–21.

NITA, S. C. 1971. Establishing the linkage of different variants of a Romanian chronicle. In Hodson, F. R., D. G. Kendall, and P. Tǎutu (eds.), Mathematics in the Archaeological and Historical Sciences. Univ. Press, Edinburgh, pp. 401–409.

PATTERSON, C. 1976. The contribution of paleontology to teleostean phylogeny. In Hecht, M. K., P. C. Goody, and B. M. Hecht (eds.), Major Patterns in Vertebrate Evolution, Plenum Press, New York.

PATTERSON, C. 1978. Verifiability in systematics. Syst. Zool. 27:218–221.

PATTERSON, C., AND D. E. ROSEN. 1977. Review of ichthyodectiform and other Mesozoic teleost fishes and the theory and practice of classifying fossils. Bull. Amer. Mus. Nat. Hist. 158:81–172.

PLATNICK, N. I. 1976. Are monotypic genera possible? Syst. Zool. 25:198–199.

PLATNICK, N. I. 1977a. Review of concepts of species. Syst. Zool. 26:96–98.

PLATNICK, N. I. 1977b. Paraphyletic and polyphyletic groups. Syst. Zool. 26:195–200.

PLATNICK, N. I. 1977c. Monotypy and the origin of higher taxa: a reply to E. O. Wiley. Syst. Zool. 26:355–357.

PLATNICK, N. I. 1977d. Cladograms, phylogenetic trees, and hypothesis testing. Syst. Zool. 26:438–442.

PLATNICK, N. I. 1979. Philosophy and the transformation of cladistics, Syst. Zool. 28:537–546.

PLATNICK, N. I., AND H. D. CAMERON. 1977. Cladistic methods in textual, linguistic, and phylogenetic analysis. Syst. Zool. 26:380–385.

PLATNICK, N. I., AND E. GAFFNEY. 1977. Systematics: a Popperian perspective. Syst. Zool. 26:360–365.

PLATNICK, N. I., AND E. GAFFNEY. 1978. Evolutionary biology: a Popperian perspective. Syst. Zool. 27:137–141.

PLATNICK, N. I., AND G. NELSON. 1978. A method of analysis for historical biogeography. Syst. Zool. 27:1–16.

POPPER, K. R. 1959. The logic of scientific discovery. Basic Books, New York, 480 pp.

POPPER, K. R. 1972. Objective knowledge. Oxford University Press, Oxford.

POPPER, K. R. 1976. Unended quest: an intellectual autobiography. Open Court Press, La Salle, Illinois, 255 pp.

ROMER A. S. 1955. The vertebrate body. W. B. Saunders Company, Philadelphia and London.

ROSEN, D. 1978. Vicariant patterns and historical explanation in biogeography. Syst. Zool. 27:159–188.

ROSS, H. H. 1974. Biological systematics. Addison-Wesley Publishing Company, Reading, Massachusetts.

SCHLEE, D. 1971. Die Rekonstruktion der Phylogenese mit Hennig's Prinzip. Aufsätze u. Red Senckenberg. Naturforsch. Ges. 20:1–62.

SIMPSON, G. S. 1961. Principles of animal taxonomy. Columbia University Press, New York, 247 pp.

SNEATH, P. H. A., AND R. R. SOKAL. 1973. Numerical taxonomy. W. H. Freeman Co., San Francisco.

TATTERSALL, I., AND N. ELDREDGE. 1977. Fact, theory, and fantasy in human paleontology. American Scientist 65:204–211.

TUOMIKOSKI, R. 1967. Notes on some principles of phylogenetic systematics. Ann. Ent. Fenn. 33:137–147.

WILEY, E. O. 1975. Karl R. Popper, systematics, and classification: a reply to Walter Bock and other evolutionary taxonomists. Syst. Zool. 24:233–243.

WILEY, E. O. 1977. Are monotypic genera polyphyletic?: a response to Norman Platnick. Syst. Zool. 26:352–354.

WILEY, E. O. 1978. The evolutionary species concept reconsidered. Syst. Zool. 27:17–26.

WILEY, E. O. 1979. An annotated Linnaean hierarchy, with comments on natural taxa and competing systems. Syst. Zool. 28:308–337.

AUTHOR CITATION INDEX

SUBJECT INDEX

About the Editors

THOMAS DUNCAN is associate professor of botany and director of the University and Jepson Herbaria at the University of California, Berkeley. His research interests include the systematics and evolution of *Ranunculus* and related genera, and the application of quantitative methods in systematics and evolution.

TOD F. STUESSY is professor of botany and director of the Herbarium at Ohio State University. His research interests include monographic and evolutionary studies of various genera of the Compositae, principles and application of cladistic analysis, and the evolution of island floras.